DMU 0414954 01 4

AF598480

MYCOTOXINS
in
AGRICULTURE
and
FOOD
SAFETY

BOOKS IN SOILS, PLANTS, AND THE ENVIRONMENT

Soil Biochemistry, Volume 1, edited by A. D. McLaren and G. H. Peterson
Soil Biochemistry, Volume 2, edited by A. D. McLaren and J. Skujiņš
Soil Biochemistry, Volume 3, edited by E. A. Paul and A. D. McLaren
Soil Biochemistry, Volume 4, edited by E. A. Paul and A. D. McLaren
Soil Biochemistry, Volume 5, edited by E. A. Paul and J. N. Ladd
Soil Biochemistry, Volume 6, edited by Jean-Marc Bollag and G. Stotzky
Soil Biochemistry, Volume 7, edited by G. Stotzky and Jean-Marc Bollag
Soil Biochemistry, Volume 8, edited by Jean-Marc Bollag and G. Stotzky
Soil Biochemistry, Volume 9, edited by G. Stotzky and Jean-Marc Bollag

Organic Chemicals in the Soil Environment, Volumes 1 and 2, edited by C. A. I. Goring and J. W. Hamaker
Humic Substances in the Environment, M. Schnitzer and S. U. Khan
Microbial Life in the Soil: An Introduction, T. Hattori
Principles of Soil Chemistry, Kim H. Tan
Soil Analysis: Instrumental Techniques and Related Procedures, edited by Keith A. Smith
Soil Reclamation Processes: Microbiological Analyses and Applications, edited by Robert L. Tate III and Donald A. Klein
Symbiotic Nitrogen Fixation Technology, edited by Gerald H. Elkan
Soil–Water Interactions: Mechanisms and Applications, Shingo Iwata and Toshio Tabuchi with Benno P. Warkentin
Soil Analysis: Modern Instrumental Techniques, Second Edition, edited by Keith A. Smith
Soil Analysis: Physical Methods, edited by Keith A. Smith and Chris E. Mullins
Growth and Mineral Nutrition of Field Crops, N. K. Fageria, V. C. Baligar, and Charles Allan Jones
Semiarid Lands and Deserts: Soil Resource and Reclamation, edited by J. Skujiņš
Plant Roots: The Hidden Half, edited by Yoav Waisel, Amram Eshel, and Uzi Kafkafi
Plant Biochemical Regulators, edited by Harold W. Gausman
Maximizing Crop Yields, N. K. Fageria

Transgenic Plants: Fundamentals and Applications, edited by Andrew Hiatt
Soil Microbial Ecology: Applications in Agricultural and Environmental Management, edited by F. Blaine Metting, Jr.
Principles of Soil Chemistry: Second Edition, Kim H. Tan
Water Flow in Soils, edited by Tsuyoshi Miyazaki
Handbook of Plant and Crop Stress, edited by Mohammad Pessarakli
Genetic Improvement of Field Crops, edited by Gustavo A. Slafer
Agricultural Field Experiments: Design and Analysis, Roger G. Petersen
Environmental Soil Science, Kim H. Tan
Mechanisms of Plant Growth and Improved Productivity: Modern Approaches, edited by Amarjit S. Basra
Selenium in the Environment, edited by W. T. Frankenberger, Jr., and Sally Benson
Plant–Environment Interactions, edited by Robert E. Wilkinson
Handbook of Plant and Crop Physiology, edited by Mohammad Pessarakli
Handbook of Phytoalexin Metabolism and Action, edited by M. Daniel and R. P. Purkayastha
Soil–Water Interactions: Mechanisms and Applications, Second Edition, Revised and Expanded, Shingo Iwata, Toshio Tabuchi, and Benno P. Warkentin
Stored-Grain Ecosystems, edited by Digvir S. Jayas, Noel D. G. White, and William E. Muir
Agrochemicals from Natural Products, edited by C. R. A. Godfrey
Seed Development and Germination, edited by Jaime Kigel and Gad Galili
Nitrogen Fertilization in the Environment, edited by Peter Edward Bacon
Phytohormones in Soils: Microbial Production and Function, William T. Frankenberger, Jr., and Muhammad Arshad
Handbook of Weed Management Systems, edited by Albert E. Smith
Soil Sampling, Preparation, and Analysis, Kim H. Tan
Soil Erosion, Conservation, and Rehabilitation, edited by Menachem Agassi
Plant Roots: The Hidden Half, Second Edition, Revised and Expanded, edited by Yoav Waisel, Amram Eshel, and Uzi Kafkafi
Photoassimilate Distribution in Plants and Crops: Source–Sink Relationships, edited by Eli Zamski and Arthur A. Schaffer
Mass Spectrometry of Soils, edited by Thomas W. Boutton and Shinichi Yamasaki
Handbook of Photosynthesis, edited by Mohammad Pessarakli
Chemical and Isotopic Groundwater Hydrology: The Applied Approach, Second Edition, Revised and Expanded, Emanuel Mazor
Fauna in Soil Ecosystems: Recycling Processes, Nutrient Fluxes, and Agricultural Production, edited by Gero Benckiser
Soil and Plant Analysis in Sustainable Agriculture and Environment, edited by Teresa Hood and J. Benton Jones, Jr.
Seeds Handbook: Biology, Production, Processing, and Storage: B. B. Desai, P. M. Kotecha, and D. K. Salunkhe

Modern Soil Microbiology, edited by J. D. van Elsas, J. T. Trevors, and E. M. H. Wellington

Growth and Mineral Nutrition of Field Crops: Second Edition, N. K. Fageria, V. C. Baligar, and Charles Allan Jones

Fungal Pathogenesis in Plants and Crops: Molecular Biology and Host Defense Mechanisms, P. Vidhyasekaran

Plant Pathogen Detection and Disease Diagnosis, P. Narayanasamy

Agricultural Systems Modeling and Simulation, edited by Robert M. Peart and R. Bruce Curry

Agricultural Biotechnology, edited by Arie Altman

Plant–Microbe Interactions and Biological Control, edited by Greg J. Boland and L. David Kuykendall

Handbook of Soil Conditioners: Substances That Enhance the Physical Properties of Soil, edited by Arthur Wallace and Richard E. Terry

Environmental Chemistry of Selenium, edited by William T. Frankenberger, Jr., and Richard A. Engberg

Principles of Soil Chemistry: Third Edition, Revised and Expanded, Kim H. Tan

Sulfur in the Environment, edited by Douglas G. Maynard

Soil–Machine Interactions: A Finite Element Perspective, edited by Jie Shen and Radhey Lal Kushwaha

Mycotoxins in Agriculture and Food Safety, edited by Kaushal K. Sinha and Deepak Bhatnagar

Additional Volumes in Preparation

MYCOTOXINS in AGRICULTURE and FOOD SAFETY

edited by

Kaushal K. Sinha
T. M. Bhagalpur University
Bhagalpur, India

Deepak Bhatnagar
United States Department of Agriculture
Agricultural Research Service
New Orleans, Louisiana

MARCEL DEKKER, INC. NEW YORK • BASEL • HONG KONG

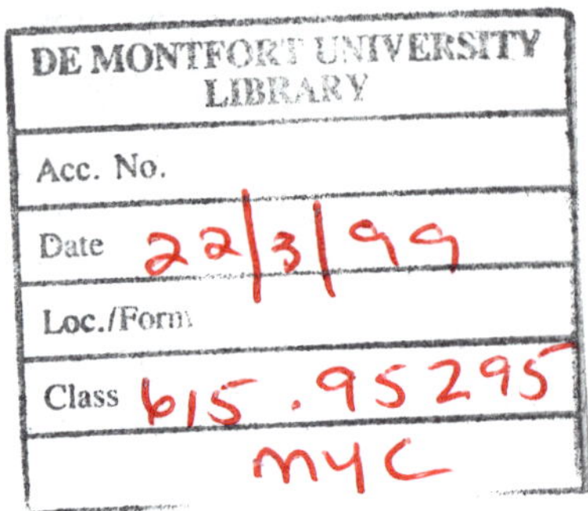

ISBN: 0-8247-0192-5

This book is printed on acid-free paper.

Headquarters
Marcel Dekker, Inc.
270 Madison Avenue, New York, NY 10016
tel: 212-696-9000; fax: 212-685-4540

Eastern Hemisphere Distribution
Marcel Dekker AG
Hutgasse 4, Postfach 812, CH-4001 Basel, Switzerland
tel: 44-61-261-8482; fax: 44-61-261-8896

World Wide Web
http://www.dekker.com

The publisher offers discounts on this book when ordered in bulk quantities. For more information, write to Special Sales/Professional Marketing at the headquarters address above.

Current printing (last digit):
10 9 8 7 6 5 4 3 2 1

PRINTED IN THE UNITED STATES OF AMERICA

In Memoriam

Krishna S. Bilgrami
(1933–1996)

Krishna S. Bilgrami was born on July 9, 1933, in Bilgram (Dist. Hardoi, U.P. India). He obtained his high school and college education at Allahabad in India, and also obtained D.Phil. and D.Sc. degrees in Botany from Allahabad University in 1956 and 1962, respectively. Professor Bilgrami worked at the Universities of Allahabad and Jodhpur as Assistant Professor and Reader, respectively. From 1970 to 1993, he was Professor and Head of the Department of Botany at Bhagalpur University, and twice he was the Dean of the Faculty of Science. He was also Pro-Vice Chancellor and Acting Vice-Chancellor of Bhagalpur University from 1978 to 1979. He was a CSIR Emeritus Scientist at Bhagalpur University at the time of his death. His research work in mycotoxins is internationally recognized. He was a Fellow of various Scientific Academies and Societies in India, including the Indian National Science Academy. He was the President of the Botany Section of the Indian Science Congress (ISCA, 1987) and the Indian Botanical Society (1988). He delivered a number of special and invited lectures.

Professor Bilgrami published more than two hundred original research papers in various areas of Botany in Indian and international journals. As Principal Investigator, he had completed technical reports of a large number of research projects funded by various national and international agencies on different aspects of fungal toxins and fungal physiology. More than thirty-five research students obtained their Ph.D. degrees under his supervision. He was a member of several scientific and research committees of the Indian government. He wrote six university textbooks for graduate-level studies and seven research monographs. His recently published pictorial narrative entitled *The Living Ganga* has received excellent reviews.

Dr. Bilgrami's significant contributions to the understanding of the preharvest mycotoxin contamination process are recognized by the mycotoxin community.

Preface

Fungal-induced food toxicity has existed since early civilization. But mycotoxins and mycotoxicoses were relatively obscure and neglected in the scientific literature until the discovery of aflatoxins as the causative agent of Turkey-X disease in 1960 in England. The death of 100,000 turkey poults within a short period of time intensified interest and research on these aspects of food safety. Periodic outbreaks of several human mycotoxicoses have also been reported from different parts of the world since that discovery. Because of health risks involved from mycotoxin contamination of crops and commodities, considerable attention is now being given to the prevention of contamination or detoxification of contaminated commodities to ensure a safe food and feed supply.

An overview of the information on natural contaminants of food and feed shows that mycotoxins are particularly important in agriculture and public health because their presence exerts a negative impact on relevant commerce, in addition to the health risks. Therefore, it is imperative to study the mycotoxin phenomenon to maintain a competitive edge in contemporary agriculture.

Mycotoxins are different from several other food contaminants with respect to their biological origin. Production of mycotoxins is influenced by several biotic and abiotic factors, and thus the intensity of contamination also varies with the specificity of host and locality. The explosion in research in mycotoxins, particularly in the last decade, has resulted in a vast increase in information about the mechanisms of actions of these toxins. It is not possible to cover every single aspect of this issue in one volume; however, an attempt has been made to cover as many significant topics as possible. In developing this volume we decided that the treatment of the subject should be comprehensive but concise. It was also our goal to provide the reader with the most contemporary status of mycotoxicology, with specific emphasis on the developments in the last five years or so. Toward this end, we have attempted to bring together a range of specific expertise to provide a detailed overview of mycotoxins in terms of theoretical, methodological, empirical, biosynthetic, and regulatory considerations.

Elucidation of the mycotoxin phenomenon clearly requires an integrated approach. *Mycotoxins in Agriculture and Food Safety* is a genuine testament to the creative and audacious nature of mycotoxicologists, with the unusual cross-section of scientific information presented on various aspects of mycotoxins, including recent advances in the methods of mycotoxin analysis, toxic effects of these toxins on human and animal health as well as on crop physiology, the factors involved in the transmission of fungal inocula, and mycotoxin development. Management of mycotoxin contamination by detoxification of contaminated commodities in the short run, and by enhancement of host resistance to fungal growth and toxin formation in the long run, have also been proposed. Food safety regulations, their implementation through various governmental and international agencies, and the economic significance of food safety have also been included as individual chapters in the book. These chapters have been prepared by active researchers in mycotoxicology and food safety, and researchers interested in mycotoxins and students of biology in general will be rewarded by careful perusal of the presentations and fascinated by their extraordinary extrapolations. This book will be useful to scientists, teachers, and students associated with the fields of agriculture, plant pathology, microbiology, biochemistry, biotechnology, veterinary and medical sciences, and food technology, as well as all who are interested in obtaining a safe food supply.

We are extremely grateful to all of the contributors for taking the time and effort to provide, in a timely manner, a comprehensive examination of their respective subject areas. In addition, the contributions of so many scientists who have in the past and are at present researching this field are greatly appreciated. The editors also want to thank numerous colleagues who provided critical comments and suggestions in the preparation of this text; their efforts were critical to the success of this project. We are grateful to the staff at Marcel Dekker, Inc., in particular Rod Learmonth and Jeanne McFadden, for careful review of the material. K.K.S. also acknowledges the support and encouragement of Professor S. K. Varma, Head of the University Department of Botany, Bhagalpur University, India.

Kaushal K. Sinha
Deepak Bhatnagar

Contents

TOXICITY OF MYCOTOXINS

FACTORS IN MYCOTOXIN FORMATION

MANAGEMENT OF MYCOTOXIN CONTAMINATION

FOOD SAFETY REGULATIONS PERTAINING TO MYCOTOXINS

Contributors

David Abramson, Ph.D. Research Scientist, Cereal Research Center, Agriculture and Agri-Food Canada, Winnipeg, Manitoba, Canada

Glenn A. Bennett National Center for Agricultural Utilization Research, United States Department of Agriculture, Agricultural Research Service, Peoria, Illinois

Deepak Bhatnagar, Ph.D. Research Geneticist, Food and Feed Safety Research Unit, Southern Regional Research Center, United States Department of Agriculture, Agricultural Research Service, New Orleans, Louisiana

Krishna S. Bilgrami, Ph.D.[†] Emeritus Scientist, CSIR, Department of Botany, T. M. Bhagalpur University, Bhagalpur, India

Robert L. Brown, Ph.D. Research Plant Pathologist, Food and Feed Safety Research Unit, Southern Regional Research Center, United States Department of Agriculture, Agricultural Research Service, New Orleans, Louisiana

Antonio Bottalico, Ph.D. Professor, Institute of Plant Pathology, University of Sassari, Sassari, Italy

Jeffrey W. Cary, Ph.D. Research Molecular Biologist, Food and Feed Safety Research Unit, Southern Regional Research Center, United States Department of Agriculture, Agricultural Research Service, New Orleans, Louisiana

[†] *Deceased.*

J. Chelkowski, Ph.D. Professor, Institute of Plant Genetics, Polish Academy of Sciences, Poznań, Poland

Ajoy K. Choudhary, Ph.D., F.P.S.I. Research Associate, Department of Botany, T. M. Bhagalpur University, Bhagalpur, India

Thomas E. Cleveland, Ph.D. Supervisory Research Microbiologist, Food and Feed Safety Research Unit, Southern Regional Research Center, United States Department of Agriculture, Agricultural Research Service, New Orleans, Louisiana

Raymond D. Coker, C.Chem., M.R.S.C., Ph.D. Research Leader, Food Security Department, Natural Resources Institute, University of Greenwich, Kent, England

Patrick F. Dowd Research Entomologist, Bioactive Agents Research Unit, National Center for Agricultural Utilization Research, United States Department of Agriculture, Agricultural Research Service, Peoria, Illinois

Antonio Logrieco Researcher, Institute of Toxins and Mycotoxins of Plant Parasites, Italian National Research Council, Bari, Italy

Gary A. Lombaert Natural Toxins Chemist, Health Protection Branch, Health Canada, Winnipeg, Manitoba, Canada

Rebeca López-García, Ph.D. Department of Food Science, Louisiana State University, Baton Rouge, Louisiana

Gerald G. Moy, Ph.D. Program of Food Safety and Food Aid, World Health Organization, Geneva, Switzerland

Douglas L. Park, Ph.D. Professor and Head, Department of Food Science, Louisiana State University, Baton Rouge, Louisiana

Gary A. Payne, Ph.D. Professor, Department of Plant Pathology, North Carolina State University, Raleigh, North Carolina

Ronald T. Riley, Ph.D. Research Pharmacologist, Toxicology and Mycotoxin Research Unit, United States Department of Agriculture, Agricultural Research Service, Athens, Georgia

Gabriele Sabbioni, Ph.D. Professor, Walther-Straub-Institut for Pharmacology and Toxicology, Ludwig-Maximilians-Universtät München, München, Germany

Ovnair Sepai, Ph.D. Lecturer, Department of Environmental and Occupational Medicine, The Medical School, University of Newcastle, Newcastle upon Tyne, England

Arun Sharma, Ph.D. Scientific Officer (G), Food Technology Division, Bhabha Atomic Research Centre, Mumbai, India

Kaushal K. Sinha, Ph.D. Reader, Department of Botany, T. M. Bhagalpur University, Bhagalpur, India

Eric W. Sydenham, Ph.D. Program on Mycotoxins and Experimental Carcinogens, Medical Research Council, Cape Town, South Africa

Mary W. Trucksess, Ph.D. Supervisory Chemist, Division of Natural Products, Center for Food Safety and Applied Nutrition, Food and Drug Administration, Washington, D.C.

David M. Wilson, Ph.D. Professor, Department of Plant Pathology, Coastal Plain Experiment Station, University of Georgia, Tifton, Georgia

Garnett E. Wood, Ph.D. Research Chemist, Division of Programs and Enforcement Policy, Center for Food Safety and Applied Nutrition, Food and Drug Administration, Washington, D.C.

MYCOTOXINS in AGRICULTURE and FOOD SAFETY

1

Mycotoxins in Preharvest Contamination of Agricultural Crops

Krishna S. Bilgrami and Ajoy K. Choudhary
T. M. Bhagalpur University, Bhagalpur, India

I. INTRODUCTION

Ecophysiological conditions are substantially different during pre- and postharvest stages of agricultural crops. The mold invasion, infestation, and mycotoxin elaboration to a great extent depend upon various environmental factors prevailing in the vicinity. Preharvest period begins with the emergence of seedling and continues up to maturity, finally ending with the harvesting of the crops. It is well established that a large number of fungi, during their growth on plant and plant parts, produce mycotoxins which get accumulated in different concentrations in the affected plants or their parts. Cole and Cox [1] listed around 120 metabolites of common hyphomycetous fungi that were toxic to higher animals. The data collected so far reflect that mycotoxins elaborated by species of *Aspergillus*, *Fusarium*, *Alternaria*, and *Penicillium* are widespread in various food items and some of them (e.g., aflatoxins) don't get destroyed even at very high temperatures (260°C or so). It is also established that a large number of storage fungi (responsible for postharvest mycotoxin production) get associated with food grains from the fields (i.e., preharvest stage). However, the fungi that are associated with standing crops may or may not be able to grow in fields because conditions required for successful mold growth are less favorable in fields than in warehouses and godowns. Moreover, only such molds are able to grow under fields that exhibit certain degree of parasitism for invading the receptive parts of the susceptible varieties of plants. There are more than a dozen "host-specific" plant pathogenic toxins, but they have not been classed as mycotoxins because they are not known to have deleterious effects on human or animal systems. Such restricted classification of fungal toxins appears to be rather artificial [2]. In this communication the term mycotoxin is used to designate fungal secondary metabolites which occur naturally as contaminants of agricultural and other consumable products and which show toxicity in animals via a natural route of administration.

Some preconditions are absolutely essential for preharvest production of mycotoxins. The foremost are the availability of toxigenic fungal strain, susceptible host, and favorable agroclimatic niche. In standing crops, host-fungus-environment interaction is critical in predisposition of mycotoxin contamination. Drought and/or temperature stress at critical times in the life-cycle of the susceptible crops is one of the dominant factors. Agronomic practices and insects play major role in exacerbation of mycotoxin contamination under field conditions [3,4]. Once the crop becomes infected under field conditions, the fungal growth continues usually with increasing vigor at postharvest stages and in storage.

Aflatoxin research was initially focused on postharvest mycology, pathology, and animal toxicology. However, in mid 1970s, aflatoxin was discovered before harvest in U.S. and Indian corn [5,6]. In the present decade aflatoxin has become a major problem because of its extensive preharvest contamination of maize [7], peanut [8,9], cottonseed [10], treenuts [11], mustard [12], linseed [13], and sorghum [14].

The role of fusarial toxins in various ailments of animal and human beings came to light due to widespread occurrence of alimentary toxic aleukia (ATA) in western Siberia during the second world war. This disease was attributed to trichothecenes group of mycotoxins elaborated by *Fusarium sporotrichioides*, *F. poae*, and some other fungi. Common fusarial toxins elaborated as contaminants of food and feed substances include zearalenone (ZEA), deoxynivalenol (DON), nivalenol, and T-2 toxins. DON caused heavy infestation of food grains in Kashmir Valley, which resulted in high incidence of fusariotoxicosis [15]. Other reports of fusarial toxins are in peanut and sorghum from India [16]. Though genus *Alternaria* is known to be responsible for various plant diseases such as early blight of potato and various leaf spots and fruit rot, since the dawn of the present century its role in animal and human disease was not suspected till the middle of the present century. It may be mentioned that among the various molds responsible for ATA in Siberia, some species of *Alternaria* were also obtained from the infested grains. About 70 secondary metabolites belonging to diverse chemical groups are elaborated by species of *Alternaria*. However, only five of the toxic metabolites occur as natural contaminants [17]. These include tenuazonic acid (TEA), altertoxin-I (ATX-I), alternariol (AOH), alternariol methyl ether (AME), and altenuene (ALT). Occasionally, other mycotoxins produced by either field fungi (i.e., ergot alkaloids) or storage fungi (ochratoxin A) occur in some regions. Moldy grains may contain complex mixtures of numerous and diverse mycotoxins. Some toxins have not been identified and their toxicology is not well understood.

We shall discuss the preharvest contamination of agricultural crops in relation to the nature of mycotoxins and the role of agroclimatic factors influencing mycotoxin production in that conditions.

II. PREHARVEST CONTAMINATION OF AFLATOXINS

Aflatoxins are basically difuranocoumarin compounds which include aflatoxin B_1, B_2, B_{2a}, B_3, G_1, GM_1, G_2, G_{2a}, M_1, M_2, M_{2a}, GM_2, P_1, Q_1, R_0, RB_1, RB_2, AFL, AFLH, AFLM, and methoxy, ethoxy, and aceto derivates. However, only a few of them, most importantly aflatoxin B_1, have been reported as naturally occurring compounds. Toxigenic strains of *A. flavus* rather than *A. parasiticus* and *A. nomius* [18,19] are found most frequently distributed worldwide in air, soil, and numerous organic substances [20]. The first report of *A. flavus* infection in preharvest crop was reported in corn in 1920 by Taubenhaus [21]. His conclusion that *A. flavus* can infect only damaged kernels was based on the field observation of the damaged kernels with visible spore production. However, Ilag [22] was able to demonstrate *A. flavus* infection without any apparent injury to the corn kernels. Recent studies have clearly indicated the association of *A. flavus* even with undamaged preharvested crops [7–9,23,24]. In this communication the aflatoxin problem in field crops will be discussed mainly in the cereals and oilseeds.

A. In Cereals

Aflatoxin contamination in cereals has caused important economic losses. Among the cereals, maize is one of the richest substrates for aflatoxin elaboration and cobs even in standing crops get high degrees of infestation. An understanding of how, when, and why aflatoxin contamination occurs is necessary in order to develop effective control measures. Invasion of corn by *A. flavus* occurs via silks. *A. flavus* has been isolated from the silks in the field [25,26]. In field studies colonization of *A. flavus* was observed mostly 1 week after silking. In a comparison of three silk stages (green unpollinated, yellow, and brown), Marsh and Payne [27] found yellow silk to be most susceptible because silk at this stage starts senescence and becomes succulent. Hesseltine and Bothast [28] found that as silks senesced, they become a suitable media for microbial growth and provide entry for fungi into the ear. *A flavus* grows down the silk very rapidly. In the phytotron at day/night temperature of 34/30° C, *A. flavus* was recovered from the tip of the ear 2 days after inoculation and from the base 4 days after inoculation [27]. Fungus from the infected silk may move down the silk internally. Colonization of external and internal silk follows the similar pattern. In a field study *A. flavus* was observed on external silk in 30% of uninoculated ears 18 days after silk emergence. However, after 27 days of silking, 72% of the ears had internally colonized silks and the fungus was present on the surface of the kernels in 47% of the cobs. Colonization of a particular kernel, however, did not require it to be adjacent to the infected silk, and the fungus was found spread across the glume and kernel surfaces. Superficial spreading of *A. flavus* mycelium among kernels, especially near the tip, was observed by Rambo et al. [29]. Colonization of silk shortly after pollination and the rapid growth of *A. flavus* down the silk suggested that the fungus infection to kernels may be following the same path as does the pollen tube, i.e., stylar canal. However, the timing of silk colonization precludes infection of kernel through stylar canal since by that time abscission layer is formed at the silk attachment site. On the basis of consistent association of infection with the tip of the kernel (pedicel), it was suggested that *A. flavus* may enter into the kernel through that region. SEM studies have also shown *A. flavus* mycelium in the tip cap (pedicel) [30]. They pointed out that colonization of corn kernels may take place from pedicel region into the germ since this route is common for many ear-infecting fungi. Recently Zummo and Scott [31] investigated how kernels in maize ears inoculated with *A. flavus* become infected and the role of the cob in that process. The relatively low levels of pedicel infection of kernel by *A. flavus* (7%) compared to combined infection of other kernel segments (45%) lead them to conclude that *A. flavus* penetrated maize kernels mainly through the pericarp. Further investigations in this direction are required to understand the precise infection process necessary to identify the source of resistance to *A. flavus* in field maize crops.

Maize seeds infected with aflatoxin-producing fungus or naturally contaminated with aflatoxins resulted in reduced germination [32]. Virescent leaf pattern was observed on maize seedlings grown from seeds infected with *A. flavus* [33,34]. On the basis of SEM studies Kelly [35] has suggested that *A. flavus* becomes systemic in young maize seedlings grown from contaminated seed. The isolate may also cause seedling death. In a young plant the organism may follow the meristem of the plant; however, the incidence of this organism was suppressed in mature plants perhaps due to the host tolerance.

Other cereal crops, viz. wheat, barley, oat, and sorghum, are not very susceptible to extensive preharvest aflatoxin contamination. Out of 3489 samples of these small grains analyzed by Stoloff [36] during 1968–1975, only 19 samples contained detectable levels of aflatoxins. Levels of aflatoxin were also very low in positive samples. Preharvest sorghum in 1980 and 1981 in Georgia's coastal plain and in Mississippi was analyzed for aflatoxin and zearalenone [37]. No toxin was detected in samples of Mississippi; however, incidence of aflatoxin in Georgia was rather high (56%) but the concentrations were low (0 to 55 ppp). Hagler et al. [38] examined North Carolina grain sorghum in crop years 1981–1995. Aflatoxin incidence was 44% but the concentration was still low, from 0 to 13 ppp. In India also the occurrence of all the four aflatoxins in the head molds of sorghum has been reported [14,39]. The magnitude of aflatoxin problem in these crops is less and, therefore, little investigated.

1. Agroclimatic Conditions Influencing Infection and Aflatoxin Production in Cereals

For infection of *A. flavus* and subsequently aflatoxin production one of the prerequisites is the availability of inoculum potential. On the basis of statistical analysis of data, Bilgrami and Choudhary [40] concluded that there was a significant positive correlation between toxigenic conidia of *A. flavus* in the aerosphere and incidence of aflatoxin in preharvested maize cobs. Incidence was much higher in flood-prone diara lands of Bihar state. Besides a higher percentage of toxigenic conidia in diara than in nondiara areas, the percentage of viable conidia of toxigenic *A. flavus* in wet months is comparatively greater than in dry or cold months. Percentage incidence of *A. flavus* was affected by season and location. Location alone, however, had lesser effect. Correlation between air borne inoculum of *A. flavus* and aflatoxin contamination in fields has been reported also [23,41]. Higher incidence of aflatoxin in standing maize crops during monsoons and summers than in winters is obviously due to high frequency of toxigenic conidia of *A. flavus* in aerosphere. The frequency of toxigenic *A. flavus* strains over crop canopy is sometimes as high as 50% during the monsoon season, which was lesser during summers and least in winters. In fact, it is the richness of toxigenic conidia in the aerosphere during the rainy season that seems to be one of the important factors for high-level contamination of aflatoxins in Kharif maize crops (August-September) than the Rabi crop (Table 1).

Jones et al. [42] correlated the levels of airborne conidia with environmental factors by daily collection of *A. flavus* conidia in irrigated and nonirrigated plots. Higher levels of airborne conidia in the nonirrigated plots correlated well with an increased number of infected kernels in those plots. In another investigation, Wicklow [43] observed that sporogenic conidia of *A. flavus* may be the major source of primary inoculum in corn fields in the American South, where aflatoxin is a recurrent problem. The fungus appears to form many sclerotia in insect-damaged kernels before harvest. These sclerotia are dispersed into the soil during the harvest. Conidiophores and conidia are produced in the spring from exposed sclerotial surface.

Moisture and temperature together play the most significant role for planning any control strategy for microbial development. In greenhouse experiment silk inoculation of corn grown between 32 and 38°C had 73% infected kernels [25], whereas corn grown between 21 and 26°C, only 2.5% kernels were infected. The critical time required for high temperature to favor infection was between 16 and 24 days after inoculation at silking stage [44], when kernel moisture was approximately 30%. The time of inoculation is also important. A comparatively lesser number of infected kernels was observed when the inoculations were made 1 and 5 weeks after silk emergence than when the inoculations were made 2 and 3 weeks after silk emergence [25,45]. Interregional experiments showed that the

Table 1 Seasonwise and Locationwise Percentage (Average) Incidence of *Aspergillus flavus* in the Aerosphere of Maize Fields

Seasons	Percent incidence of *A. flavus* in different seasons		
	Locality A (L_1)	Locality B (diara area L_2)	Mean
Winter	23.17	23.00	23.08
(January–March)	(28.53)	(28.14)	(28.34)
Summer	35.50	52.67	44.08
(April–June)	(32.26)	(46.53)	(41.40)
Monsoon	66.08	59.83	62.96
(July–September)	(54.63)	(50.76)	(52.70)
Mean	41.58	45.17	43.48
	(39.81)	(41.81)	(40.81)

Figures in parentheses indicate transformed values since the original values were in percentage and hence these were transformed into degrees and then analysis of variance was done (Angle = Arc sin $\sqrt{\text{Percentage}}$); 5% CD for season = 5.322; 5% CD for season × location = 7.526.

highest aflatoxin levels were associated with the highest 3 months average temperature. Higher incidence and severity of aflatoxin contamination were found in corn grown in the South than in the North cornbelt of USA [46].

High moisture and high relative humidity (RH) are essential for spore germination and fungus proliferation. Maize plants exposed to drought stress are likewise more susceptible to infection by *A. flavus* than plants not under stress [47]. Bilgrami [7], on the basis of field experiments on maize and mustard crops, suggested that the increased level of infection was due to high levels of inoculum and to reduced leaf area in the nonirrigated plots which made the reproductive parts of the crops more accessible to fungal species. In a seasonwise survey of the mycoflora associated with maize grains in standing maize crops, it was shown that the environmental conditions had a marked effect on the quality and density of molds [24]. In the winter, when the moisture ranged between 22% and 30%, species of *Furasium* were dominant. Incidence of *A. flavus* was comparatively lesser (17%) in colder months. However, with the rise in temperature during summers, when grain moisture was 18% to 22%, *A. flavus* dominated and had 72% incidence. Its incidence was highest (85%) in rainy season when grain moisture was up to 40% and also RH of the environment was very high. Moisture content in relation to mold invasion and aflatoxin contamination in developing maize crops were also determined [48]. Mold association was almost uniform in all three locations (tips, middles, and the basal portions) of the maize cobs. This was attributed to the morphological nature of the cob husks (loose and complete husks) and the warm and humid climate at the site. Occurrence of *A. flavus* begins from early dough stage. At harvest, moisture percentage of Kharif crop was higher and it varied from 25.2% to 28.2%. Moisture content (MC) and aflatoxin contamination were negatively correlated in developing maize kernels of Kharif and Rabi crops. Jones et al. [42] reported that maize kernels harvested at 18% moisture content had comparatively higher amounts of aflatoxin than those harvested at 28% MC. It has been suggested that in developing corn kernels at moisture content below 35%, *A. flavus* cannot compete sufficiently with other microorganisms. Based on these facts it has been recommended to harvest the crop at high moisture level and subsequently expose the crop for natural and quick drying up to safe moisture level to reduce the chances of aflatoxin production. However, in Bihar, Kharif crop is harvested during monsoon rains, which delay the natural drying of the kernels [49,50]. Risk of further accumulation of toxin, therefore, cannot be ruled out.

Although damage is not a prerequisite for aflatoxin formation, the incidence of *A. flavus* and aflatoxin contamination were high in damaged kernels [51–53]. Insect-damaged kernels provide an opportunity for the fungus to circumvent the natural protection of the integument and establish infection sites in vulnerable interior [23,54–56]. Wounding by insects may provide infection courts and allow kernels to dry down to level of moisture more favorable for growth of *A. flavus* and aflatoxin production than for other fungi. Lillehoj [57] pointed out that insects collected from corn at various geographical locations show a general internal presence of *A. flavus*. He suggested that *A. flavus* infection of developing corn kernels is linked to the development of the second generation European corn borer (*Ostrinia nubilalis*) and that this development period coincides with the time span during which corn kernels are susceptible to *A. flavus* infection. Based on these observations, he also contended that aflatoxin produced by *A. flavus* may function as chemical signals between species in a particular econiche and they could perform a function in establishing the species or cultivar within the niche. A survey of maize-growing areas of Bihar was conducted (1985–1990) to study the correlation between crop damage and levels of aflatoxin contamination in standing maize crop [7]. Altogether, 21 types of insects were collected of which seven insect species belonged each to the order Coleoptera and Hemiptera, four to Lepidoptera, two to Isoptera, and only one to Diptera. The frequency of occurrence of these insects varied considerably. The most destructive insect pests were the stem borer, *Chilo partellus* Swinh; rice weevil, *Sitophilus oryzae* Linn; and the termites *Odontotermus obesus* Rambur, *Microtermus obsi* Holm, and *Nezara viridula* Linn.

The role of other microorganisms in plant-fungus interaction has important bearing in specified econiche. Depending on the environmental conditions, aflatoxin may or may not be formed in the presence of other fungi. In the individual maize kernels of Kharif and Rabi maize crops association patterns of various fungi along with *A. flavus* was determined through chi square test (2 × 2 table) [58]. Significant (positive and negative) associations were noted. *Fusarium moniliforme* and *Trichoderma viride* had significant negative correlation in almost all the association with various coinhabiting fungi. *Rhizopus nigricans* had also negative correlations with different mycoflora except with *Mucor* species. Significant negative correlations here mean that whenever these fungi are present in a particular microhabitat, the chances of occurrence of other fungi (or *A. flavus*) are very limited or rare. Widstrom et al. [59], when they coinoculated *A. flavus* and *F. moniliforme* simultaneously through cob husks, observed that *F. moniliforme* reduced aflatoxin formation by *A. flavus* isolates approximately by two-thirds. *F. moniliforme* had no significant effect on naturally occurring aflatoxin contamination by *A. flavus*. This may be due to timing of infection by both fungi. They contended that *A. flavus* and *F. moniliforme* respond differently to the environment. Nontoxigenic strains of *F. moniliforme* with sufficient aggressiveness to compete with the toxigenic strains of *A. flavus* may be utilized for biological control. Similarly, Cotty and Bhatnager [19] utilized atoxigenic strain of *A. flavus* to prevent aflatoxin contamination through competitive exclusion of the toxigenic strains. However, in vitro activity did not predict the ability of a nontoxigenic strain to prevent contamination of developing crops since some of the atoxigenic strains had the ability to convert several precursors of aflatoxin into aflatoxin B_1.

The status of fungal colonists as interference competitors and the sequence in which these colonists become established within the kernels, may contribute to the variation in the aflatoxin contamination among field/storage samples. A correlation between the seasonal incidence of competing fungi with *A. flavus* and various environmental factors was also made [60]. The incidence of *A. flavus* on individual maize kernels was significantly correlated to the moisture content of the substrate as well as percent RH. Incidence of *A. flavus* in competition with *Penicillium* spp. and also with *A. niger* was high, and these had significant positive correlations with the highest temperature of the season. However, temperature was negatively correlated with *Fusarium* spp.

The coinhabiting mycoflora with *A. flavus* on individual kernels were evaluated for their influence on aflatoxin biosynthesis [61]. All 13 types of associations of fungal species with *A. flavus* inhibited aflatoxin B_1 and G_1 production up to varying degree (34.3% to 100%). Inhibition in radial growth of *A. flavus* by *Fusarium moniliforme* (59.8%), *Trichoderma viride* (72.5%), and *Rhizopus nigricans* (42%) could be directly correlated to the percent inhibition of aflatoxin production. It was suggested that the reduction in aflatoxin production by these fungi might be due to physical competition of nutrients (Fig. 1). With *A. niger*, *Cladosporium herbarum*, *F. oxysporum*, and *A. candidus*, inhibition in radial

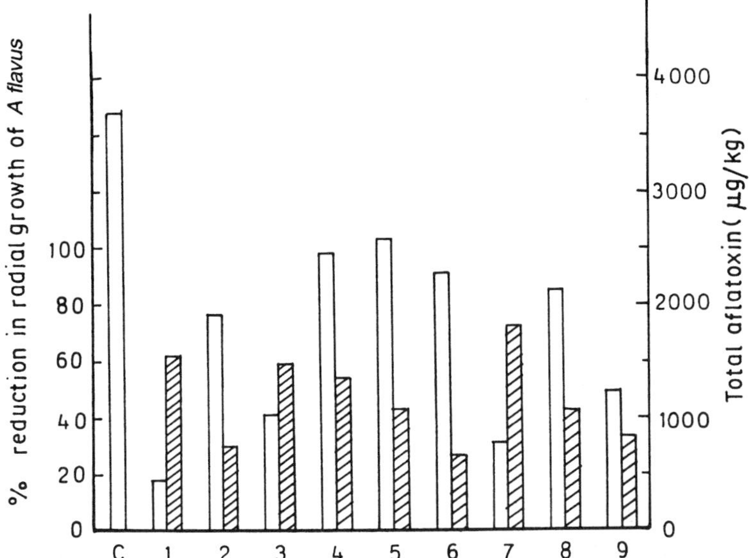

FIGURE 1 Histogram showing the influence of different coinhabiting mycoflora on growth and aflatoxin production by *Aspergillus flavus*. (1) *A. niger*; (2) *Fusarium oxysporum*; (3) *F. moniliforme*; (4) *Penicillium chrysogenum*; (5) *Alternaria tenuis*; (6) *A. candidus*; (7) *Trichoderma viride*; (8) *Rhizopus nigricans*; (9) *Cladosporium herbarum*. □ Aflatoxin; ▨ radial growth; C = control.

growth was 63%, 33.6%, 30%, and 26.5%, respectively, while percent reduction in aflatoxin was 88.6%, 68.6%, 52.3%, and 43.1%, respectively. Therefore, in this case, the reduction in aflatoxin production may not be attributed to competition only. The possibility of the competitor inhibiting aflatoxin production or degrading most of the aflatoxins cannot be ruled out. However, in case of *A. niger* it has been suggested that *A. niger* lowers substrate pH sufficiently to suppress aflatoxin synthesis [62]. The reason for reduction in aflatoxin level can, therefore, be attributed to one or a combination of the factors: (1) physical competition for space and nutrition; (2) test fungus may compete with *A. flavus* for a substrate required for toxin production; (3) presence of test fungus may be affecting a change in the biochemical environment influencing the metabolic pathway; (4) degradation or detoxification of aflatoxin.

Aflatoxin contamination in standing maize crops has led to the evaluation of agronomic practices in order to reduce the impact of the problem. The concept of a disease triangle involving host, pathogen, and environment is useful to apply when evaluating control strategies for standing crops. Selection of proper fields may sometimes be useful in minimizing aflatoxin contamination in maize crops.

Either due to economic reasons or for consumption purposes, growers get motivated to cultivate maize instead of other crops. In certain regions some soils are not conducive for maize cultivation, and aflatoxin elaboration is thus the consequence of unsuitability. In topical countries, particularly in the regions with prolonged warm and humid climate, mycotoxin contamination in food and feed is a serious problem. In Bihar state, where Kharif crop is harvested during monsoon rains, it is more prone to aflatoxin contamination [48].

Field studies have been carried out in the U.S., India, and other countries in order to screen maturity group and severity of aflatoxin production. Certain factors that can regionally be associated with reduced aflatoxin levels (husk cover, pericarp thickness, kernel hardness, etc.) may be useful resource of polygenic resistance to aflatoxin contamination. A statistically designed field experiment was carried out with three varieties of maize [63]. These three varieties, viz. suwan composite (SC), diara composite (DC), and M_9, had different maturity periods—i.e., 95, 80, and 105 days, respectively. These varieties were also different with regard to husk characteristics. The husk was complete but loose in DC variety, while variety SC had tight and incomplete husk. In the case of M_9 the husk was tight and complete. It was observed that SC was more susceptible to aflatoxin elaboration under natural conditions. Inoculated crops of M_9 were found to be more prone to aflatoxin production. The morphological characteristics of variety DC and M_9 which had complete husks seemed to help them to some extent in preventing *A. flavus* infestation. However, greater susceptibility of M_9 facilitated aflatoxin production under inoculated condition. Earlier, Widstrom et al. [64] found that aflatoxin levels were twice as high in early-maturing, loose-husked maize hybrids as in late-maturing, tight-husked. Similarly, in a field experiment, Barry et al. [65] evaluated the complex relationship among insects and aflatoxin contamination in preharvested maize as affected by difference in husk tightness. Results demonstrated that kernels of tight-husked hybrids contained significantly less aflatoxin contamination than did those of loose-husked.

Plant density on field crop is also one of the factors that could potentially have an effect on the epidemiology of *A. flavus* infection and aflatoxin production. Plant density could minimize the infection by *A. flavus* by reducing the exposure of maize ears and silks [25,27,66]. In a field experiment with three cultivation rates, single-line weeding (i.e., SLW with single spading, SLW with two spading, and SLW with three spading and three planting densities [56,000 plants ha^{-1}, 67,000 plants ha^{-1}, 83,000 plants ha^{-1}]) was carried out [67]. It was found that the inoculation had substantial impact upon planting density and varieties. Natural occurrence of aflatoxin was lesser in densely cultivated plants (67,000 plants ha^{-1}, 83,000 plants ha^{-1}). Plant population of 56,000 plants ha^{-1} resulted in high incidence of aflatoxin.

The influence of planting dates has been investigated at diverse locations. Widstrom et al. [68] reported a rise in the percent of aflatoxin-contaminated samples in maize planted on April 19 as against the maize planted on May 2. Late planting shifts the ear development phase of the crop to higher temperatures, high inoculum load, and increased insect activities. In some geographical regions, planting date may be selected to take advantage of periods of higher rainfall that occur most frequently at some particular time.

Fertility and drought stress have been found as contributing factors in preharvest aflatoxin contamination in corn. Positive correlation was observed between high aflatoxin levels in corn and plots receiving low levels of nitrogen [66]. Inheritance studies have also been carried out for genetic control of resistance to aflatoxin production in maize [69,70]. However, inability to repeat differences among hybrids over locations and years has been a major factor in not being able to recommend one hybrid over another for the control of aflatoxin contamination.

B. In Oilseeds

Among different oilseed crops, aflatoxin is a serious preharvest problem in groundnut [30,71], cottonseed [10,72], and mustard [12,73]. Besides providing a high-quality cooking oil, groundnut is also eaten raw, roasted, and made into confectionery and snack foods. It is also added to soup and other dishes. Mustard is consumed by human beings in the form of oil and condiments. It is fed to cattle as a feed supplement in the form of oil cake (known in India as Khalli). Cottonseed is mainly grown for oil and feed for cattle.

In groundnut, field contamination is considered to be a significant source of inoculum while storage contamination is of comparatively lesser magnitude [74]. Groundnut is widely grown in tropical and subtropical regions where the occurrence of drought in the later phases of the growing season of the crop is common. This plant is unusual because flowers are formed and fertilized above the soil and subsequent fruit development takes place in the soil. The subterranean pod is, therefore, in close contact with soil microorganisms for an extended period, and many species of soil fungi including *A. flavus* have been isolated from healthy and damaged seeds. Preharvest invasion of groundnut seeds by *A. flavus* was earlier thought to depend on physical and/or biological damage to pods by parasitic fungi and insects [75,76]. It is now established that groundnuts without obvious damage can also be invaded by *A. flavus* and *A. parasiticus* and contaminated with aflatoxin in field before harvest. Physical damage of the embryo and cotyledons by the invading fungal mycelium as well as biochemical changes of the stored carbohydrates, protein, and oil, normally reduce the value of these ingredients to the young seedling. In some cases, mycelium of the fungus remains viable when the seed is sown and may contribute to either seed rot or seedling disease (aflaroot). In aflaroot disease, the radicle does not develop in secondary root.

It is well established that in a standing peanut crop, *A. flavus* invasion can occur in soil during pod development and maturation. However, the exact mode of infection of peanut fruit has not been fully elucidated. Some researchers [77–79] have suggested that *A. flavus* may invade the flowers, penetrate down the pegs, and subsequently establish in the developing seeds. Peanut flowers inoculated with washed conidia of *A. flavus* were readily colonized by the fungus [80]. Freshly opened flowers were inoculated by pulling down the keels, exposing the stigma and stamens and gently dusting the exposed parts with conidia [79]. In 48 hours, the conidia had germinated and considerable hyphae had formed, but they were confined primarily to the stigma. Only a few conidia germinated on the style. Some hyphae entered the style through the stigma and ramified in styler tissue near the stigma. Further, the hyphae of *A. flavus* can grow down the style, eventually reaching the top of the ovary. Two days after inoculation, sporulation was also visible on anther and distal portion of the filament.

Some investigations have been carried out to assess the possibilities of infection via pegs. Aerial pegs of peanut plants, grown under gnotobiotic conditions after flower inoculation with conidia of *A. flavus*, were readily colonized by fungus without apparent damage to the developing embryo [80]. The incidence of aerial and subterranean infection of pegs by *A. flavus* was 75% and 80% at 128 and 144 days after planting, respectively, in drought treatment [81]. At harvest, peg infection was 40% and 45% in cool (24.8°C) dry soil and heated (34.5°C) irrigated soil, respectively, as compared to 2% to 3% infection in the cool (25.2°C), irrigated soil.

However, further studies in Australia [82] have failed to establish a definite link between flower and peg invasion and between peg and fruit invasion. More research is needed to answer the question "Can flower and aerial peg invasion lead to significant invasion of groundnut fruit?," and also "Can this occur under both normal and drought stress conditions?" In comparative studies of the invasion of flowers, aerial pegs and kernels by wild-type and mutant strains of *A. flavus* or *A. parasiticus*, Cole et al. [83] have suggested that preharvest *A. flavus* infection and subsequent aflatoxin contamination originate mainly from the soil. Direct invasion of the developing peanut fruit by *A. flavus* in soil in the geocarposphere has generally been assumed as likely route for eventual contamination of the kernel with aflatoxin after penetration of the pod wall and tastae and infection of the kernels by the fungus. However, *A. flavus* may be present in the developing ovary of peanuts at the tip of the peg even before it is pushed into the soil.

Cotton plants can become infected by *A. flavus* through the fresh cotyledonary leaf scar of the seedling at the six- to eight-leaf stage [84]. Cotton bolls naturally contaminated with *A. flavus* on the boll surface or lint were collected at maturity [85]; 78% of bolls with seeds contaminated by *A. flavus* also had the fungus in subtending stem and peduncles, whereas only 31% of the naturally contaminated bolls with no *A. flavus* in the seed showed the fungal infection in the stem or peduncle. On the other hand, bolls inoculated through carpel walls 30 days after anthesis had *A. flavus*-infected seed at maturity, but only 11% of their subtending stems and peduncles were infected. This difference explains that the fungus grows upward from supporting stem or peduncle to the bolls. The possibilities of downward movement (bolls to peduncle or supporting stem) has been ruled out. Upward movement of the fungus was further substantiated in the greenhouse experiments where cotton seedlings were inoculated at the cotyledonary leaf scars with *A. flavus*, the fungus was being subsequently isolated from these leaf scars and also from flower buds, developing bolls, and stem sections in the upper parts of the plants but never from roots or stem section below the cotyledonary node.

In cotton bolls infection occurs much earlier in unopened bolls prior to maturity but before opening of the bolls, and aflatoxin may not be formed as a result of infection through sutures. Aflatoxin levels and the ratio of toxic and nontoxic seeds were similar in naturally infected bolls as well as in bolls inoculated 30 days after anthesis. Locks of bolls with suture opening 1 to 3 mm were prone to infection, whereas those open 10 mm or more showed no BGYF on the lint. Lee et al. [86], following suture inoculation, also confirmed the presence of BGYF on lint; however, no aflatoxins were detected in the seed. *A. flavus* produced the most aflatoxin in seed from bolls of most of the 12 cultivars inoculated 30 days after anthesis with less toxin occurring in 20- and 40-day inoculated bolls [87]. Twenty days and 30 days inoculated bolls were harvested 10, 15, 20, 25, and 30 days after inoculation. Aflatoxin was found in unopened immature bolls 10 to 20 days after inoculation. The amount of toxin increased with subsequent time interval. The presence of aflatoxin in seed in unopened (closed) bolls explains that for *A. flavus* infection and aflatoxin contamination it is not the requisite for the bolls to be opened.

Cotton flowers and developing bolls were inoculated with spores on the nectaries at anthesis and 7, 18, 25, 32, and 35 days postanthesis, with the bolls being harvested at maturity [88]. Percent incidence of infected seeds was considerably higher when inoculation was up to 25 days postanthesis compared to the longer days of postanthesis inoculations. The sharp decline in susceptibility to *A. flavus* infection between 25 and 30 days after anthesis was attributed to the degeneration of the funiculus (which attaches to the developing seed to placentae) that occurs 30 days after anthesis or with changes in biochemical factors in developing bolls [89].

In mustard, preharvest infection of *A. flavus* has been investigated by Bilgrami et al. [12,90]. These experiments revealed that on artificial inoculation of *A. flavus* the level of aflatoxin production had considerably increased. In field experiments, the spraying of *A. flavus* was carried out when the entire inflorescence of the crop of a particular plot had matured to the fruiting stage. This period in Varuna and BR-40 varieties of mustard was attained after 70 and 60 days of planting, respectively [90]. In field inoculation experiment, percent incidence

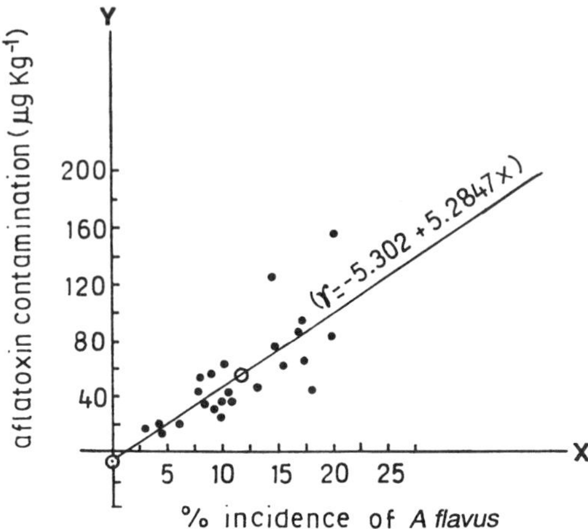

FIGURE 2 Line of regression showing significant positive correlation between percent incidence of *A. flavus* and aflatoxin contamination.

of *A. flavus* had significant positive correlation ($P > .01$) with aflatoxin contamination (Fig. 2).

1. Agroclimatic Conditions Influencing Infection and Aflatoxin Production in Oilseeds

In peanuts, late-season drought stress, particularly in the semiarid tropics, is a major factor associated with aflatoxin contamination [91,92]. The relationship between late-season drought stress and increased *A. flavus* invasion and aflatoxin contamination in groundnuts was observed in different countries—viz. South Africa [76], Nigeria [93], and U.S.A. [81,94,95]. Aflatoxin levels in kernels harvested from rain-fed plots 120 and 130 days after sowing was in a range of 694 to 10,240 μg/kg^{-1}; however, zero to trace amount of aflatoxins were detected in kernels from irrigated plots [96,97]. Drought stress has a vital role in accelerating *A. flavus* infection and aflatoxin production in sound mature kernels [98]. Contamination with aflatoxin in kernels from three growers field which had no, moderate, and severe drought stress averaged 6, 73, and 444 μ/kg^{-1}, respectively. On the contrary, some studies have revealed that drought stress alone was not responsible for aflatoxin contamination since drought-stressed groundnuts were not always contaminated with aflatoxins [53,99].

Further studies have provided additional information about the influence of soil temperature on the extent of *A. flavus* infection and aflatoxin contamination in groundnuts [74,100,101]. Kernels from undamaged pods grown under drought conditions at mean soil temperatures (5 cm depth) of 24.8°C or lower were not contaminated with aflatoxin. Under similar conditions, kernels grown at 25.7, 26.3, and 27.8°C had lower concentrations of aflatoxins. However, at 29.0, 29.6, and 30.5°C, kernels were heavily contaminated with aflatoxin. Kernels from pods grown at 31.3°C were free from aflatoxin, as were kernels from pods grown under irrigation and irrigated heated plots with a mean temperature of 34.5°C. It has been demonstrated that 45°C temperature prevents the growth of *A. flavus*, and 2 to 4 hours at 50°C suppressed growth for nearly 24 hours [102,103]. Optimum mean pod zone temperature has been determined in a range of 28 to 30°C for aflatoxin production in drought conditions during the last 30 to 50 days of the growing season. Cole and co-workers, however, did not record any aflatoxin contamination in kernels of undamaged pods in plots with adequate irrigation (irrespective of pod zone soil temperature) or from drought-stressed crops when the mean pod zone soil temperature during the last 30 to 50 days before harvest remained <25°C or >32°C [104,105]. Their results suggested that groundnuts subjected to drought stress may not be contaminated with aflatoxin unless drought is accompanied by mean pod zone soil temperature of 25 to 31°C during late stages of pod development. Even a small mean temperature change may significantly influence aflatoxin production in drought-stressed groundnuts.

More than 20 days but probably less than 30 days of drought stress at soil temperature optimum for aflatoxin development is required for preharvest aflatoxin contamination. Therefore, the threshold for aflatoxin contamination was between 20 and 30 days of drought before harvest at optimum temperature [104]. Neither drought nor elevated temperature alone resulted in aflatoxin formation by *A. flavus* in sound mature kernels [51]. Extended drought below optimum temperature could result in an amount of aflatoxin similar to that attained over a shorter period of time at more optimal temperature for aflatoxin formation [106].

No aflatoxin was detected in heavily *A. flavus*-infected kernels from irrigated plots (with mean geocarposphere temperature of 34.5°C) [53]. This was interpreted as adequate irrigation prevents aflatoxin production. However, further investigations suggested that a drought-stressed soil temperature mean of 31.3°C was apparently too high for aflatoxin production even though kernels were heavily invaded with *A. flavus* [105]. This may be due to the effect of high or low temperature on fungus metabolism rather than level of irrigation. Drought and temperature stress conditions in the pod zone and not in the root zone predispose groundnuts to contamination with aflatoxin [100]. Reduced metabolic activity associated with decreased pod moisture content under drought stress seems to increase the susceptibility of groundnuts to *A. flavus* infection. Another possible role of drought stress in preharvest fungal infection could involve suppression of microbial competitors of the aflatoxin-producing fungus by elevating the soil temperature in the pod zone. *A. niger*, a common competitor of *A. flavus* and also favored by high temperature, was more prevalent in irrigated plots (no aflatoxin) than drought plots, where kernels were colonized by high frequency of *A. flavus* and aflatoxins. Pod splitting is another factor contributing to aflatoxin contamination. Pod maturing under fluctuating soil moisture conditions during a season of inadequate or irregular rainfall is prone to pod splitting. Seeds in split pods are frequently invaded by *A. flavus* and subsequently become contaminated with aflatoxins [107].

Temperature also plays key role in preharvest cottonseed infection and aflatoxin contamination. In greenhouse experiments, Gilbert et al. [108] observed that with the diurnal cycles, infection of cottonseed increased as the duration of daily maximum temperature of 30°C or higher increased and as the number of diurnal maximum temperature cycles of 30°C increased. Low temperature and low rainfall have been correlated with the low levels of aflatoxin [72,109]. In Arizona, aflatoxin contamination was considerably high in cotton plants grown at low elevations compared to those at higher elevations (above 55 cm). At low elevations daily minimum temperature of 24°C in combination with precipitation exceeding 2 to 3 cm resulted in high levels of aflatoxins [30].

Moisture appears to be more important than temperature for boll opening in cotton plants. *A. flavus* infection did not occur in bolls opened more than 1 to 3 mm. This aspect has been attributed to moisture-related factor [72]. Most of the bolls of lower one-third of the plant contained considerably higher levels of *A. flavus* infection than those of middle and the top [110]. This pattern has also been suggested as moisture-related effect. Ashworth et al. [72] suggested that the lack of aflatoxin in cottonseed in San Joaquin valley was due to the prevailing low RH which resulted in rapid opening and quick drying of the bolls and thereby, bolls escape from infection. On the other hand, considerably high RH results in high pH (8.0 to 10.2) of aqueous extracts of cotton fibre that is less favorable for growth and aflatoxin production by *A. flavus* [111]. The threshold level of RH factor influencing *A. flavus* infection remains still to be explained. High RH around late opening bolls promoted by late-season irrigation could be more favorable for *A. flavus* infection and subsequent aflatoxin contamination [112]. Lee and Russell [113] investigated the influence of water stress on *A. flavus* infection. The highest incidence of seed infection was observed between −1.6 and −1.9 MPa (−16 and −19 bars) water potentials which are the levels associated with moderate to severe water stress (drought). Moderately low water potentials appear to influence the infection of *A. flavus*.

In India, mustard plant is grown in the rabi season (November to April). The climatic conditions during fruiting-maturing period (January to March) of pods were moderate cold and foggy night alternating with moderate hot day. High temperature (28 to 31°C) and congenial RH (50% to 94%) prevailing during the fruiting stage of the crop has been suggested favorable for *A. flavus* infection in field mustard crops [12,114].

Insect damage is not the prerequisite for *A. flavus* infection; however, many investigators have concluded that peanut pods with shells that were damaged while the crop was in the soil were most likely to contain toxic kernels than were pods with undamaged shells [75,115,116]. A number of soil-inhabiting pests including pod borers, millipeds, mites, white grubs, termites, and nematodes have been implicated in *A. flavus* infection of groundnuts before harvest. A lesser cornstalk borer (*Elasmopalus lignosellus* Zeller), a common pest of groundnut in USA, predisposes groundnut fruit to *A. flavus* infection [117,118]. Kernels from undamaged pods often contain very high levels of aflatoxins. Another serious pest of groundnut, southern corn rootworm (*Diabrotica undecimpunctata howardi* Barber), has been commonly associated with increased fungal infection. Pod damage in the groundnut by termites has also been reported [76,119]. Termites (*Microtermes* spp. and *Odontotermes* spp.) are associated with pod scarification and facilitating invasion by *A. flavus*. Potential involvement of mites and nematodes has been implicated in aflatoxin problem in field groundnut crops [120,121]. Mites penetrate groundnut pods, feed on kernels, and disseminate spores of *A. flavus*.

The relationship between pinkboll and other insects to *A. flavus* infestation in cotton bolls and aflatoxin contamination has been well investigated [110,122, 123]. The exit holes made by mature larvae predispose bolls to infection by *A. flavus* and other fungi. Increased bollworm infestation resulted in an increase in infection by *A. flavus* and aflatoxin contamination in cottonseed [112]. In bollworm-infested crops insecticidal application resulted in decrease in aflatoxin contamination [123]. Scavenger beetles (Nitidulidae) and birds may also introduce *A. flavus* propagules into the bolls via the exit holes created by bollworms [10].

In mustard crops, the significance of insect damage under field condition has been evaluated [67]. Insect (*Lipaphis erysime*) incidence varied from 9% to 22% and 67% to 100% in crops sown on first (15 October) and third (15 November) planting dates, respectively. Incidence of insect in preharvested crop had significant negative correlation ($r = -.9354$) with yields (Fig. 3). Delayed planting resulted in significantly low yields, where the incidence of aphids was significantly high. This aphid was found associated with the crop during flowering to the fruiting stages. It appears in mid-December and population starts increasing from January onward. This aphid is known to suck the sap from the plants and devitalize them [124]. The attacked flowers fail to produce pods and even if pods are formed, seed setting is low. Insect-damaged samples become an ideal substrate for *A. flavus* colonization and subsequent elaboration of aflatoxin. Besides, the insects also act as vectors.

Cultural control of aflatoxin contamination of groundnut, mustard, and cottonseed must take into consideration all varied environmental and agronomic factors that influence pod/boll and seed infection by the aflatoxin-producing fungi and aflatoxin production. These factors can vary considerably from one location

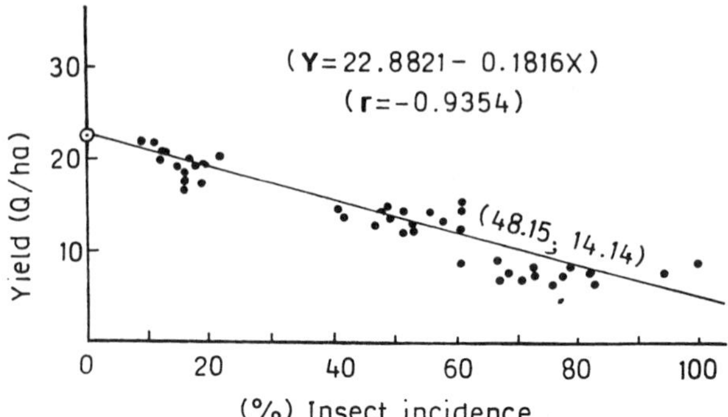

FIGURE 3 Line of regression showing relationship between insect incidence (%) versus yields (Q/hectare) of the mustard cultivars sown on three planting dates.

to another and between seasons in the same location. Some environment may be particularly favorable to fungal infection and subsequent aflatoxin contamination of crop, and this may even raise the question as to whether the crop should be grown in such places or not. However, in most of the situations it may be possible to devise cultural practices that can to some extent be able to reduce or eliminate aflatoxin contamination in these crops.

Groundnut cultivation continuously on the same land may lead to buildup of high populations of *A. flavus* and/or *A. parasiticus* in the soil, which in turn increases the probability of seed infection and aflatoxin contamination [125,126]. Limited research has been done on the effects of crop rotations on fungal infection and aflatoxin contamination in groundnuts [127]. Joffe and Lisker [125] observed the effect of crop sequence and soil types on the mycoflora of groundnut kernels. They observed that the incidence of *A. flavus* was not influenced by any crop sequence. Total kernels were consistently higher on medium and heavy soil than on light soil. In India, the effect of the previous season's crop (vegetables, rice, or groundnut) on population of *A. flavus* and other fungi in the soil, rhizosphere, and geocarposphere of groundnuts and of the shell and seeds at various stages of the development and aflatoxin contamination at harvest was investigated [126]. *A. flavus* populations in the rhizosphere and geocarposphere were high where groundnut was the previous crop and the infection of shells and seeds was also high. The effects of various crop sequences of maize, soybean, groundnut, green manuring with sorghum, sudangrass, fallow, and minimum tillage soybean cropping on *A. flavus* group population in soil were examined in field plots on a farm in Virginia from 1975 to 1979 [127]. Planting of maize in 1975 and groundnut in 1976 was associated with significant increases in the population of *A. flavus* group in soil in the following years, compared with fallow treatment. Populations increased primarily in the lower half of the plough layer following maize planting. Other crop sequences did not significantly affect the populations of *A. flavus*.

Some researchers have pointed out that the incidence of *A. flavus* infection and aflatoxin contamination is likely to be much higher in groundnuts planted on light sandy and red sandy loam soils than in groundnuts planted on vertisols [107]. This appears to be related mainly to the water-holding capacity (water potential) and aeration in the soils. Light sandy and red sandy loam soils have lower water-holding capacities, and groundnuts grown on these soils are more prone to drought stress than those grown on vertisols that have higher water-holding capacity. Light sandy and red sandy loam soils favor rapid proliferation of aflatoxigenic fungus, especially under conditions of low water potential at which the activity of other microorganisms is minimal.

The possible effect of calcium on preharvest aflatoxin contamination of groundnut has been established [105,128]. The groundnut obtains its calcium requirements from the surrounding soil, and this may be difficult under drought conditions. If a relationship existed, it could be complex one involving drought, available calcium, pod development, and *A. flavus* infection. In another experiment, Davidson et al. [98] reported that application of gypsum to a soil in Georgia reduced aflatoxin contamination. Cole et al. [105] and Wilson et al. [128], however, did not observe any such effect.

Cropping pattern has also been found to influence the aflatoxin contamination in field mustard crops [12]. Aflatoxin contamination in two varieties of mustard, Varuna and BR-40, sown on three planting dates (1, 15, and 30 November) under two different cropping patterns, viz., monocropping and mixed cropping (along with UP-262 variety of wheat), was examined in Rabi (winter) crops of 1987–88 and 1988–89 (Tables 2 and 3). Analysis of variance showed a significant effect of planting date and cropping pattern as well as combined effect of planting date and cropping pattern. The level of aflatoxin was low when mustard was cultivated along with wheat (mixed cropping). The maximum number of fungi was recorded in mixed cropping samples. The low incidence of *A. flavus* in mixed cropping samples due to high incidence of various fungi (which resulted in negative correlation) might be attributed to increased competition between toxigenic and nontoxigenic molds or due to change in agroecosystem niche suitable for *A. flavus*, which ultimately resulted in low levels of aflatoxins (Fig. 4). Microbial degradation or detoxification of aflatoxin can also not be ruled out [61,129].

A. flavus is frequently found associated with several other fungi in groundnut pods and seeds [130]. Premature death of plants, particularly during pod development and maturity, from root and stem infections by pathogens such as *Rhizoctonia solani*, *Sclerotium rolfsii*, and *Fusarium* spp. increases the chances of seed contamination with aflatoxin [116,131]. Lesions produced by these pathogenic fungi facilitate invasion of seeds by *A. flavus* [116,132]. Some viral diseases

TABLE 2 Main Effect of Aflatoxin Contamination (μg/kg) in Mustard Crops (1987–1988 and 1988–1989)

Treatment	1987–1988	1988–1989	Pooled
D_1	106.25 (8.777)[a]	35.42 (4.686)	70.84 (6.73)
D_2	110.58 (8.931)	56.08 (6.567)	83.34 (7.75)
D_3	272.08 (14.721)	279.25 (13.811)	275.67 (14.27)
CD at 1%	3.99	2.948	2.0986
C_1	354.72 (17.109)	155.00 (10.447)	254.86 (13.78)
C_2	158.94 (11.916)	154.94 (9.204)	156.94 (10.56)
C_4	82.78 (7.732)	111.44 (7.656)	97.11 (7.70)
C_5	55.44 (6.478)	72.94 (6.111)	64.20 (6.29)
CD at 1%	1.97	2.162	2.1519
I_0	100.31 (8.351)	57.28 (5.075)	78.79 (6.71)
I_1	225.74 (13.269)	189.89 (11.634)	207.76 (12.45)
CD at 1%	1.76	3.387	1.8539

[a]Figures in parentheses are transformed values.
D_1 = first planting date (1 November); D_2 = second planting date (15 November); D_3 = third planting date (30 November); C_1 = Varuna (monocropping); C_2 = BR-40 (monocropping); C_4 = Varuna, wheat UP-262 variety (mixed cropping); C_5 = BR-40, wheat UP-262 variety (mixed cropping); I_0 = Uninoculated; I_1 = inoculated. (Ds are main plots represented by D_1, D_2, and D_3; Cs are subplots for cropping pattern represented by C_1, C_2, C_4, and C_5; Is are plots within subplot for inoculation represented by I_0 and I_1.)

TABLE 3 Interaction of Planting Dates and Cropping Pattern of Pooled Mean of 1987–1988 and 1988–1989

Planting dates	Cropping pattern				Mean
	C_1	C_2	C_4	C_5	
D_1	116.42 (9.08)	70.25 (6.97)	48.42 (5.16)	40.25 (5.73)	70.84 (6.73)
D_2	152.42 (11.05)	96.00 (8.86)	51.75 (6.18)	33.17 (4.90)	83.34 (7.75)
D_3	495.75 (21.20)	304.58 (15.85)	191.17 (11.76)	111.17 (8.26)	275.67 (14.27)
Mean	254.86 (13.78)	156.94 (10.56)	97.11 (7.70)	64.20 (6.29)	

CD at 1%, C at same D = 2.4253; D at same C = 1.8587; D_1 = first planting date (1 November); D_2 = second planting date (15 November); D_3 = third planting date (30 November); C_1 = Varuna; C_2 = BR-40; C_4 = Varuna, wheat UP-262 variety; C_5 = BR-40, wheat UP-262 variety.

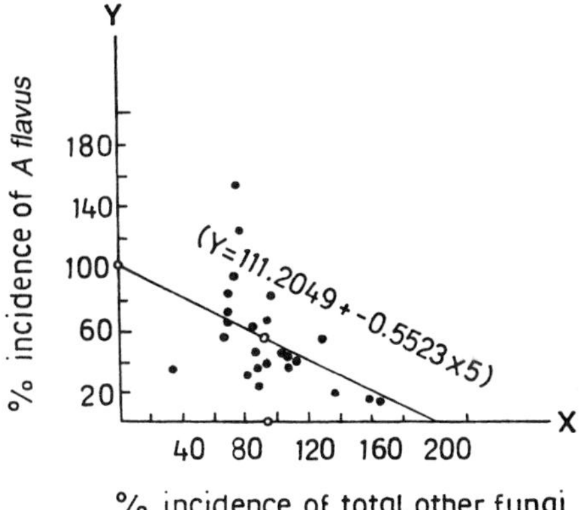

FIGURE 4 Line of regression showing negative correlation between percent incidence of total fungi and *A. flavus*.

such as groundnut rosette and bud necrosis may also predispose groundnuts to invasion by *A. flavus* [115]. Similarly, in cotton plants boll infection by *A. niger* and *Rhizopus nigricans* has been reported to cause carpel necrosis, resulting in premature separation of carpels exposing the boll fibres to *A. flavus* infection [10]. Contrary to this, it is also hypothesized that interactions between fungi as they compete for the substrate might under favorable environmental conditions restrict invasion of groundnuts by *A. flavus* and/or contamination with aflatoxin [133]. *A. niger* and *Rhizoctonia solani* appeared to limit the development of *A. flavus* and aflatoxin production in the substrate. In Israel, Joffe [134] observed that a large number of viable propagules of *A. niger* in the geocarposphere and moderate invasion of seeds by this species was associated with very limited invasion of seeds by *A. flavus*.

The weather, as well as crop age, may also have an effect on fungal infection and aflatoxin contamination. McDonald and Harkness [75] made harvesting trials during three consecutive years (1962, 1963, and 1964) at Mokwa and Kano Agricultural Research Station, North Nigeria, to investigate the occurrence of aflatoxin in groundnut crops at harvest and when lifted before, at, and after normal maturity. Crops harvested at or earlier than the normal time were free from aflatoxin, but late harvesting usually resulted in aflatoxin contamination. In the Kano trials there was an increase in the toxicity of the crop at harvest with increase in age. It appeared that the time of appearance of aflatoxin in the Kano crop depended on the occurrence of late-season drought stress [75]. At Mokwa, the long wet season apparently delayed the appearance of aflatoxin in the crop. At Mokwa, groundnut is normally harvested before the end of the rains, while in Kano it is harvested either at the end of the rains or later. In India, shells and kernels of groundnut were examined in plants from 60 days old to harvest in the rainy (Kharif) and postrainy (Rabi) seasons. Most of them were free from fungal infection at first, but nearly all shells were infected by harvest time. There was no aflatoxin in kernels from undamaged pods from Rabi season, but aflatoxins (118 to 128 $\mu g/kg^{-1}$) were found in samples from rain-fed plots in Kharif season. In kernels from damaged pods, higher levels of aflatoxin were detected in samples from rain-fed plots (780 to 6700 $\mu g/kg^{-1}$) than from irrigated plots (820 to 1840 $\mu g/kg^{-1}$) in Kharif season. Low levels of aflatoxins were found in damaged pods from Rabi season. Besides weather, age of crop has also been attributed to influence the infection. Much higher percentage of *A. flavus* invasion occurred in overmature seeds [135]. In India, Mehan et al. [96] showed that levels of *A. flavus* and aflatoxin B_1 were much higher in seeds from overmature pods of several groundnut genotypes than in seeds from immature and mature pods, especially under drought stress conditions. Seeds become susceptible to *A. flavus* invasion when seed moisture content is below 30% [75,136]. Drought stress, lowered seed moisture content, overmaturity, and decreased plant vigor in groundnuts are interrelated and moisture-related, and these factors contribute to increased susceptibility to *A. flavus* invasion and aflatoxin contamination [75,92,115].

Mechanical damage of peanut pods during field cultivation and at the time of harvest has been widely associated with the rapid invasion of pods by *A. flavus* [115,116]. Kernels from pods with mechanical damage and growth cracks had higher levels of aflatoxin than those with rot and insect injury [132]. Damage to shell or kernel provides an increased probability of rapid and direct invasion of kernel by *A. flavus*, which in turn increases the probability of aflatoxin formation. Damage also increases nutrient availability for rapid growth of *A. flavus*.

Considerable work has been done on the possibility of possible genetic resistance in groundnuts to seed infection and aflatoxin contamination in field [96,98,137]. For this purpose it is necessary to establish the relevance of in vitro "seed colonization" resistance to *A. flavus* invasion of developing pods in the field. A laboratory method to screen live groundnut for resistance to aflatoxin production was used at ICRISAT [138] to test 502 genotypes. None was totally resistant to aflatoxin, but highly significant differences in aflatoxin production were found [96]. Two Valencia-type genotypes, PI337394F and PI337409, had high levels of resistance to in vitro seed colonization by *A. flavus* and *A. parasiticus*. Six more breeding lines (GFA1, GFA2, AR1, AR2, AR3, and AR4) were later reported to be resistant [139]. Cultivar J11 has been found resistant to seed infection in multilocational field trials in India and USA [137,140]. PI337409 showed resistance in Senegal but was susceptible in USA [141]. Certain conditions should have to be met when screening groundnuts for resistance to infection by the aflatoxigenic fungi or aflatoxin production under field conditions. First, only undamaged pods should be analyzed, as any kind of damage is likely to override resistance. Second, a test site where late-season drought stress is of common occurrence would be most effective since temperature and moisture stress during pod maturation are important.

Screening of mustard varieties against aflatoxin contamination under field conditions was carried out twice during 1988 to 1990 [73]. During 1988–89, varieties Varuna, Kranti, and TR-46 were most susceptible ones to aflatoxin elaboration, whereas Tori and BR-23 were comparatively less susceptible. However, under field conditions, none of the varieties of mustard was found to be

TABLE 4 Main Effect of Aflatoxin Contamination (μg/kg) in Preharvested Mustard Varieties (1988–1989)

Treatment	D_1	D_2	D_3	Mean	I_0	I_1
Varuna	52.52	137.20	189.57	126.43	71.96	180.90
(V_1)	(12.15)	(19.68)	(23.19)	(18.34)	(14.71)	(22.37)
BR-40	44.27	59.83	83.33	62.48	41.67	83.29
(V_2)	(9.32)	(13.76)	(15.34)	(12.81)	(10.01)	(15.60)
Tori	39.33	79.51	106.50	75.11	53.82	96.40
(V_3)	(10.70)	(15.31)	(17.55)	(14.52)	(12.46)	(16.58)
Kranti	53.52	97.54	184.63	111.90	73.09	150.70
(V_4)	(12.48)	(16.57)	(23.09)	(17.38)	(14.27)	(20.48)
CBR-13	33.33	66.02	92.83	64.06	39.53	88.60
(V_5)	(9.52)	(13.71)	(16.13)	(13.12)	(10.27)	(15.97)
BR-23	37.54	75.95	99.83	71.11	50.70	91.52
(V_6)	(10.32)	(14.80)	(16.98)	(14.03)	(11.97)	(16.10)
TR-46	47.03	86.50	144.02	92.52	55.75	129.28
(V_7)	(10.81)	(15.62)	(20.19)	(15.54)	(11.90)	(19.18)
Mean	43.93	86.08	128.67	86.23	55.22	117.24
	(10.76)	(15.64)	(18.92)	(15.11)	(12.17)	(18.04)
I_0	30.94	54.32	80.39			
	(8.60)	(12.69)	(15.21)			
I_1	56.93	117.84	176.95			
	(12.91)	(18.57)	(22.64)			

Figures in parentheses are transformed values:

CD of D = 0.8500 I at same D = 1.4110
V = 1.2714 D at same V = 2.1313
I = 0.8147 D at same I = 1.1738
V at same D = 2.2019

where D_1 = first planting date (15 October); D_2 = second planting date (1 November); D_3 = third planting date (15 November); I_1 = inoculation with the toxigenic strain (BUML-183) of *A. flavus*; I_0 = uninoculation (control).

Ds are main plots represented by D_1, D_2, and D_3; Vs are subplots represented by V_1, V_2, V_3, V_4, V_5, V_6, and V_7; Is are sub-sub blocks represented by I_0 and I_1.

completely resistant to aflatoxin elaboration. Mustard varieties also showed significant response in relation to aflatoxin contamination during 1989–90. However, interaction of planting dates X varieties had nonsignificant impact. Inoculation with the toxigenic strain of *A. flavus*, however, had significant impact with respect to planting dates (Table 4).

Further, optimum plant populations (plant density) could be effective in minimizing *A. flavus* infection, particularly in groundnut crops, since too dense population may lead to severe drought stress where the rainfall is suboptimal in a growing season [107]. Excessive weed growth may also deplete available soil moisture, and effective weed control by use of herbicides may be one of the preventive approaches. Also, planting date may be adjusted so that the crop matures at the end of the rainy season and postharvest conditions favor rapid and effective drying of the crop. Individual plants that die from attack by pests and diseases should be lifted, as their produce is likely to contain toxin [142]. Irrigation to ensure adequate soil moisture during last 4 to 6 weeks of crop growth may be effective prevention. It is equally important to harvest the crop at optimum maturity, as excessive numbers of overmatured or immatured pods at harvest may contain high levels of aflatoxin.

III. PREHARVEST CONTAMINATION OF FUSARIAL TOXINS

Like Aspergilli, the Fusaria are also widespread in nature occurring both as facultative saprophytes and parasites on a variety of plants. Most of the fusarial species are capable of elaborating toxins of varying chemical composition and dimensions. Involvement of fusarial toxins in plant disorders was first highlighted in 1945 by Plattner and Clausonkass [143]. They observed that wilting of tomato was due to lycomarasmin produced by *Fusarium lycopersici*. The role of fusarial toxins in ailments of animals and human beings came in light due to widespread occurrence of alimentary toxic aleukia (ATA) in western Siberia during World War II. The disease was attributed to be caused due to trichothecene group of mycotoxins elaborated by *F. sporotrichioides* and *F. poae*. These days, about 20 species of Fusaria are known to be toxigenic [144]. The common fusarial toxins encountered as natural contaminants of food and feed substrates include zearalenon (ZEA) and zearalenols, some trichothecene derivatives, and moniliformin. Recently a new class of mycotoxins with cancer-promoting activity, named as fumonisin B_1 and B_2, were isolated and characterized from cultures of *F. moniliforme* (MRC 826) grown on corn [145].

Several species of *Fusarium* are important pathogens of corn, wheat, and barley, causing root, stem, or ear rot, with substantial reduction in crop yield. Besides being pathogenic certain isolates are capable of producing mycotoxins which can be accumulated in standing crops. *Fusarium* species and toxins are also a problem in rye, oats, and triticales [146–148]. We will discuss fusarial toxins under two sub-headings viz. (1) corn and (2) wheat, triticale, barley, and rye.

A. In Corn

Gibberella zeae (sexual stage of *F. graminearum*) and *F. culmorum* cause gibberella ear rot in corn which is also called as pink ear rot. These fungi also contaminate corn by deoxynivalenol, zearalenone, and T-2 toxin and also by diacetoxyscirpenol if *F. sporotrichioides* and *F. poae* are involved. This disease usually spreads from the ear tip down and develops around channels and tunnels made by insects, leaving pink to red mold on the kernels [149]. The disease is prevalent in northern temperate climates, especially in wet years. In outbreak of this disease, *F. graminearum* is predominant in Canada and northern Europe, while *F. culmorum* tends to be dominant species in north-central and eastern Europe [146,150]. A few recent reports from India also indicate the occurrence of ZEA in freshly harvested maize [151].

The second disease of *Fusarium* is *Fusarium* ear rot caused mainly by *F. moniliforme* and related fungi, *F. subglutinans* and *F. proliferatum*. These fungi cause contamination by producing moniliformin or fumonisins [152]. This disease occurs during dry weather and is facilitated by insect damage. *F. moniliforme* and *F. subglutinans* are very common in corn kernels in Canada, USA, and Europe. In southern USA and tropical countries, *F. moniliforme* is the main cause of *Fusarium* kernel rot [153,154]. Sharman et al. [155] have found moniliformin in samples of corn from USA, Great Britain, Netherlands, France, Italy, Thailand, and several African countries. Fumonisin group of mycotoxins occur naturally in corn in USA [156,157], Poland [158], South Africa [159], Taiwan [160], and other countries [161].

A third type of *Fusarium* disease in corn is called red ear rot, or red fusariosis [146]. Though less common in occurrence, this disease is associated with *F. tricinctum*, *F. equiseti*, *F. sporotrichioides*, *F. crookwellense*, *F. avenaceum*, and trichothecene of the A-type including T-2 toxin [162].

1. Agroclimatic Conditions Influencing Infection and Fusarial Toxin Production

Monitoring the growth of *F. graminearum* in experimentally infected ears showed that the growth rate of the fungus was sensitive to temperature. A period in which only about 10 days had average growing degrees days greater than 5°C virtually halted the growth. An average value of approximately 15 growing degree days greater than 5°C resulted in rapid growth [163]. Weather factors favorable for production and dispersal of spores on the host surface and infection must coincide with the time that the corn is very receptive to *F. graminearum*. Optimum temperatures are between 25 and 32°C [164]. Ear rot epidemics are usually completed in one cycle because corn ears remain highly receptive for only 10 to 20 days following silking [149].

Infection is favored by warmth and prolonged surface wetness. Splashing or wind-driven rain aids in dispersal of spores, and persistent wetness (> 48 to 60 hours) of corn ears favors mold infection [149]. Rainfall shortly after silk emergence promotes epidemic development of *F. graminearum* and subsequent accumulation of zearalenone. As kernels develop and mature in standing crop, their moisture declines from 80% to 15% to 30% at harvest. Growth of the fungus is vigorous at above 35% MC, but it can also occur, though sparsely, at as little as 20% MC.

Corn is susceptible to *F. graminearum* infection from airborne spores for a period of 7 to 10 days after silking begins. Inoculum sources are conidia or ascospores, chlamydospores, and hyphal fragments, which survive on host debris such as old corn stalks and ears and small grain straw stubble near the soil surface [165]. Susceptibility of corn to infection by *Fusarium* varies with inbred lines and hybrids, their stage of development, and environmental conditions. Late-maturing cultivars in which grain moisture content decreases slowly below 30% are most susceptible [166]. Infection is thought to be favored by upright ears and tight husks [167], high lysine, thin pericarps, and soft kernels [168,169]. Susceptibility is greatest shortly after silk emergence [149]. Choline in corn pollen is likely to stimulate conidial germination, germ tube growth, and colonization of silks [149].

Physical damage to ears increases susceptibility to infection. This physical damage is caused by birds or insects [170]. Insects such as picnic beetles and European corn borer commonly act as spore vectors and may facilitate infestation by damaging the plant. An experiment demonstrated that spores carried by the nitidulid beetle *Glischrochlis quadrisignatus* to physically damaged ears caused infection [171]. Without physical injury to the ears there was little or no infection. Bird damage predisposes corn to infection with *F. graminearum*. The accumulation of zearalenone was shown to be related to the degree of bird damage [170].

Sometimes crop rotation favors fungal infection. Both host and nonhost plants can serve as sources of *F. graminearum* inoculum. The main nonhost plants include soybeans, sorghum, wild oats, and other kinds of weeds [172,173]. Repeated planting of corn, wheat, and other cereal crops in the same or in nearby fields increases inoculum and increases insect population that attack corn plants and spread mold inoculum. It is thus advisable to break the life cycle of the pathogen to control insect-mold infestation. In field trials, a corn/soybean rotation yielded a less extensive disease outbreak than corn/corn [174]. Weed control is important to eliminate nonhost crop of *Fusarium*. If trash is left in the fields after harvesting, the infection may be easily carried to the next season. Overwintered crops in the fields should be avoided.

Certain fields and even areas within a field with poor drainage may be more prone to fungal infections. Effective land preparation, which permits early planting [175], is helpful to control fusarial diseases. Late-maturing varieties of corn may give higher yields, but adverse summer and autumn conditions may allow higher fungal infection and mycotoxin production.

Fungicides used either as seed treatment or as foliar spray have not been effective in control of ear rot [176]. Four seed treatments, with Furadan, Captan, Benlate, and water (control), and three foliar treatments, Mertict, Benlate, and water (control), were used on corn in a field that had been in corn production for 4 years. No significant difference attributed to seed or foliar treatments or a combination of both was observed.

F. moniliforme disease incidence is affected by moisture during silk emergence, and its prevalence is considerably increased with wet weather later in the season [177,178]. The significance of *F. moniliforme* in healthy tissue remains poorly understood [179]. For many years, *F. moniliforme* has been known to occur systematically in leaves, stems, roots, and kernels [180]. This leads to the suggestion that some strains of *F. moniliforme* produce disease symptoms and others do not [181]. Generally, moderate rain at silk emergence alternating with periods of warm dry weather appears to be required for *Fusarium* kernel rot of corn. The relationship of insect damage to *F. moniliforme* appears more certain. A field survey showed that the incidence of the European corn borer increased. *F. moniliforme* disease and fumonisin concentration [158]. The application of insecticides reduced thrip (*Frankliniella occidentalis*) population and disease incidence. Hybrids with increased propensity of kernel splitting have higher disease frequency [182].

A third type of *Fusarium* disease of corn, called red ear rot or red fusariosis, is associated with *F. sporotrichioides*. The growth of this fungus occurs over an extended period of wet, cold weather. This species produces low level of T-2 toxins.

B. In Wheat, Triticale, Barley, and Rye

Fusarium head blight of wheat, triticale, barley, and rye came in light due to the epidemics of disease in Japan in 1962 [183] and in Canada and USA during 1980–82 [184,185]. Head fusariosis of wheat is also named scab, wheat scab, head scab, *Fusarium* scab, *Gibberrella* scab, pink scab, head blight, *Fusarium* blight, *Fusarium* head blight, *Fusarium* ear scab, white heads, tombstone scab, etc.

F. graminearum, *F. culmorum*, *F. crookwellense*, *F. avenaceum*, *F. sporotrichioides*, *F. poae*, and *F. nivale* are the most common pathogens of wheat and barley. Among these, *F. graminearum* and *F. culmorum* are the most pathogenic. These two fungi have also been proved to be the most aggressive species toward wheat ears during pathogenicity test and caused severe blighting in inoculated wheat heads, while other species only caused damage to a number of inoculated spikelets [186–188]. Blighted heads from fields where *F. graminearum* or *F. culmorum* was dominating tended to show severe blight.

Ability to produce mycotoxins by Fusaria pathogenic to wheat, triticale, barley, and rye has been well documented [144,189,190]. Isolates of *F. graminearum* and *F. crookwellense* produce zearalenone and deoxynivalenol along with related 8-ketotricho thecenes, 15-acetyl deoxynivalenol, and nivalenol. Isolates from some of the geographical regions (Australia, New Zealand, and China) produce 3-acetyl deoxynivalenol instead of 15-acetyl deoxynivalenol [191]. In wheat of durum variety, *F. avenaceum* may produce significant level of moniliformin [192]. *F. sporotrichioides* and *F. poae* produce mainly type A trichothecene.

Wheat heads are infected shortly before, during, or shortly after anthesis when weather is humid for a few days [193]. Spores deposited on wheat spikes germinate and grow initially on anthers, but in due course the fungus grows on the spikelets, which become straw-colored. Choline and betaine in anthers have been reported to stimulate fungal growth [194]. However, cultivars vary in their response to infection with respect to inoculation [195]. DON accumulation begins about 3 days after infection, peaks about 6 weeks later, then begins to decline and reaches a constant level before harvest maturity. Concentration of DON is detected highest in chaff, less in the bran, and least in the endosperm.

Scab disease causes reduction in yield, shrivels kernels, and often results in the contamination of grains with DON. There exist differences between wheat and rye susceptibility to infection by Fusaria. Planting of *Fusarium*-infected seed may result in weak seedlings and also give rise to pour strands [172]. Germination capacity of the seed is significantly reduced. Scab is well recognized on emerged immature heads where one or more spikelets or entire heads appear prematurely bleached. If rachis is infected, all tissues above that point are faded. Bleached spikelets are usually sterile or contain partially bleached seed [196].

Infected host tissue on or near the soil surface is the potential site of the survival of the fungus and the source of inoculum dissemination [149]. This inoculum remains effective for well over a year.

1. Agroclimatic Conditions Influencing Infection and Fusarial Toxin Production

Species frequency is dependent on weather conditions not only during flowering but also during winter and early spring. Weather condition is particularly important for the formation of pathogen inoculum. During seasons of abundant rainfalls with high moisture of soil, *F. nivale* dominantes, damaging seedlings of wheat and triticale [197]. However, at seasons with lower humidity the same role in seedling infection is played by *F. culmorum* and *F. avenaceum* [197].

Epidemics, where an appreciable percentage of heads in a field are infected, appear to require moisture for some time after anthesis. Moisture is required for the fungus to sporulate on the heads and infects adjacent plants and late-flowering tillers. However, the most important factor in determining severity of infection remains moisture during the susceptible period at anthesis [198].

As with *Fusarium* head blight of wheat, there is some evidence that ear Fusaria interact with other fungi. *F. moniliforme* is reported to suppress the growth of other ear fungi [58,199]. Coinoculation of *F. poae* and *F. sporotrichioides* with *F. graminearum* leads to the conclusion that simultaneous infection by other Fusaria might be conducive to invasion by *F. graminearum* [177]. Other biotic factors like insect damage apparently facilitate entry of *F. graminearum* and other species that cause head blight [149].

Scab frequencies among wheat genotypes vary from isolate of the fungus because of the variation in virulence [200]. Cultivars resistant to seedling blight might also be resistant to scab [201]. There are several mechanisms for genetic resistance like date of flowering, proportion of extruded vs. retained anthers, and the time the flowers remain open. A resistant cultivar may have a shorter period of susceptibility, increased resistance at milk stage, and/or delay in full disease development. Another mechanism of resistance is the ability of a cultivar to prevent synthesis and/or to promote degradation of DON [201].

Limiting the seedling rate to that required for maximum yield is advocated as a scab control measure. Dense strand has been linked with increased severity because of the higher humidity within the crop [203]. Most of the *Fusarium* species incited scab in irrigated fields but not in dryland fields [204]. High moisture at anthesis plays a key role in infection. It is therefore advisable for the growers to schedule irrigation in order that it should not concide with heading and anthesis. Wheat fertilized with urea tended to have less scab than wheat fertilized with ammonium nitrate [205]. Soil phosphorus, potassium, and pH may also affect wheat scab frequencies. The fungicide propiconazole applied at or after heading significantly ($P < .05$) reduced scab severity to about 50% as compared to untreated control, but the level of DON was not reduced significantly [172]. Crop rotation to avoid planting wheat after a susceptible crop such as corn is a common and effective control measure. Plowing between two host crops is an effective means of reducing the amount of inoculum present during the second crop. Burying host crop residues by tillage with a moldboard plow is effective in reducing or eliminating this source of inoculum [206]. Weed control tends to reduce scab severity. A field surveyed with excessive weeds had twice as many scabbed heads as fields free of weeds [207].

IV. PREHARVEST CONTAMINATION OF *ALTERNARIA* TOXINS

Species of *Alternaria* are also among the widely distributed fungi in soil and on aerial plant parts, and several species of this genus are pathogenic to crop plants. They cause extensive spoilage of crops in fields. Since *Alternaria* requires relatively high moisture level (28% to 34%) for growth, their infection generally occurs on seeds and grains when the seed moisture is high or when crops are exposed to frequent showers before and after harvest.

Alternaria species elaborate about 70 secondary metabolites belonging to diverse chemical groups including dibenzopyrones, perilequinones, tetramic acid, lactones, anthraquinones, cyclic peptides, etc. [208]. However, only five of them are common as natural contaminants [17]: tenuazonic acid (TeA), altertoxin-I (ATX-I), alternariol (AOH), alternariol methyl ether (AME), and altenuene (ALT). When ingested, these mycotoxins can have severe repercussions in animals and human beings.

More than 20 species of *Alternaria* are known to elaborate toxic metabolites. These species have been isolated mostly from consumable items like fruits, vegetables, cereals, and other plant parts. It is interesting that some species can produce several toxic metabolites of diverse chemical configurations whereas same toxin can be produced by different species of this fungus depending on the substrate and environment. The most common species of *Alternaria, Alternaria alternata* (formerly known as *A. kikuchiana*), is capable of elaborating all important mycotoxins, viz., AOH; AME; ALT; TeA; ATX-I, II, and III; altenuisol; altenuic acid; altenusin; tentoxin; dehydroaltenusin; etc.

Natural occurrence of alternariol (AOH), alternariol methyl ether (AME), and altenuene (ALT) is restricted to some specific foods only. Among the cereals, sorghum constitutes the best source for the development of alternaria toxins. Seitz et al. [209] and Saur et al. [210] recorded high levels of AOH and AME (up to 7.9 μg/g) in 9 out of 20 sorghum samples [16]. They also recorded these toxins in two samples of ragi (*Eleusine coracona*). Besides sorghum, AOH and AME were also recorded in barley, oat, and wheat in concentration below 0.16 μg/g [211]. Grabarkiewicz-Szczesna et al. [212] recorded AOH in 4 out of 21 samples of wheat chaff and 1 out of 10 samples of rye chaff. Natural occurrence of AOH and AME has also been recorded in apples [213], olives [214], and sunflower and peppers [215].

Although the ability to produce altertoxins (ATX-I) has been shown by several species of *Alternaria*, natural occurrence of this toxin is not recorded on a large scale. Reports are available only for weathered sorghum and apple [210, 213]. Natural occurrence of TeA at levels below 6 μg/g was recorded in sorghum and ragi samples collected from different parts of Bihar [16]. Visconti et al. [216] reported up to 7.2 mg/kg of TeA in tomato fruits affected by the species of *Alternaria* in the field.

Virtually no work has been done with regard to the use of agronomic practices and ecological attributes for the control of these toxins. The lack of data may be related to the limited number of laboratories searching for these toxins.

V. CONCLUSIONS

Various factors that affect aflatoxin formation under field conditions include water activity, temperature, time, insect damage, morphological nature of the substrate, prevalence of toxigenic strains, spore load, fungal frequency, microbial interactions, and agronomic practices. It is essential to monitor the effect of each of these factors individually and collectively, as their interactions may be synergistic or antagonistic under field conditions. In-depth studies of this nature are required for developing preventive and precautionary measures to minimize the invasion of *A. flavus* and production of aflatoxin under field conditions. Though a volume of information has been collected on the aflatoxin problem at preharvest stage, so far no definite strategy could be worked out for tackling this problem. The preharvest problem of *Alternaria* and fusarial toxins seems to be much more acute and almost of equal dimension to that of aflatoxin. But information on this aspect is definitely wanting from various ecogeographical regions of the world. Under field conditions, once the crop becomes infected, fungal growth continues, usually with increasing concentrations of mycotoxins being produced. However, proper management and control of mycotoxins under field conditions will also be helpful in tackling this problem during transit and storage of food grains.

REFERENCES

1. RJ Cole, RI Cox, eds. Handbook of Toxic Fungal Metabolites. New York: Academic Press, 1981, p 1850.
2. KS Bilgrami. Fungal toxins: their biological effects and ecological significance. Presidential address, 74th Indian Science Congress Association, Bangalore, 1987, p 16.
3. EB Lillehoj. Effect of environmental and cultural factors on aflatoxin contamination of developing corn kernel. Aflatoxin and *Aspergillus flavus* in corn. In: UL Diener, RL Asquith, JW Dickens, eds. Southern Co-operative series bulletin 279, Alabama Agricultural Experiment Station, Auburn, AL, 1983, p 27.
4. UL Diener. Preharvest aflatoxin contamination of peanuts, corn, and cottonseed. A review. In: GC Llewellyn, CE O'Rear, eds. Biodeterioration Research 2. New York: Plenum Press, 1989, p 697.
5. KS Bilgrami, RS Misra, T Prasad, KK Sinha. Mycotoxin problem in standing maize crops in Bihar. Nat Acad Sci Lett 1:124, 1978.
6. HW Anderson, EW Nehring, WR Wischer. Aflatoxin contamination of corn in the field. J Agric Food Chem 23:775, 1975.
7. KS Bilgrami. Investigations on prevention, elimination and inactivation of aflatoxin and other mycotoxins from cereals and oilseeds with special reference to maize and mustard. Final Technical Report. US-India Project, Bhagalpur University, Bhagalpur, 1990, p 69.

8. VK Mehan, D McDonald. Mycotoxin contamination in groundnut. Prevention and control. In: KS Bilgrami, T Prasad, KK Sinha, eds. Mycotoxins in Food and Feed. Bhagalpur: Allied Press, 1983, p 237.
9. UL Diener, RE Pettit, RJ Cole. Aflatoxins and other mycotoxins in peanuts. In: HE Pattee, CT Young, eds. Peanut Science and Technology. Am Peanut Res Edu Soc, 1982, p 486.
10. ME Simpson, LR Batra. Ecological relations in respect to a boll rot of cotton caused by *Aspergillus flavus*. In: H Kurata, Y Ueno, eds. Toxigenic Fungi: Their Toxins and Health Hazards. Amsterdam: Elsevier, 1984, p 44.
11. RJ Cole, RA Hill, PD Blankenship, TH Sanders, KH Gareen. Influence of irrigation and drought stress on invasion by *Aspergillus flavus* of corn kernels and peanut pods. Dev Indn Microbiol 23:229, 1982.
12. KS Bilgrami, AK Choudhary, A Masood. Aflatoxin contamination in mustard (*Brassica juncea*) in relation to agronomic practices. J Sci Food Agric 54:221, 1990.
13. AK Sinha, KS Ranjan. Preharvest aflatoxin problem in linseed (*Linum usitatisimum*). J Indian Bot Soc 68:19, 1989.
14. RK Tripathi. Aflatoxin in sorghum grains infected with head moulds. Indian J Exp Biol 11:362, 1973.
15. RV Bhat, RB Shashidhar, Y Ramakrishna, KL Munsi. Outbreak of trichothecene mycotoxicosis associated with consumption of mould damaged wheat products in Kashmir Valley in India. Lancet 35, 1989.
16. AA Ansari, AK Shrivastava. Natural occurrence of *Alternaria* mycotoxins in sorghum and ragi from North Bihar, India. Food Addit Contam 7:815, 1990.
17. K Jewers, AE John. *Alternaria* mycotoxins—possible contaminants of few sorghum varieties. Trop Sci 30:397, 1990.
18. CP Kurtzman, BW Horn, CW Hesseltine. *Aspergillus nomius*—a new aflatoxin producing species related to *Aspergillus flavus* and *Aspergillus tamarii*. Antonie Van Leeuwenhoek 53:147, 1987.
19. PJ Cotty, D Bhatnagar. Variability among atoxigenic *Aspergillus flavus* strains in ability to prevent aflatoxin contamination and production of aflatoxin biosynthetic pathway enzymes. Appl Environ Microbiol 60:2248, 1994.
20. KS Bilgrami, AK Choudhary. Impact of habitats on toxigenic potential of *Aspergillus flavus* strains. J Stored Prod Res 29(4):351, 1993.
21. JJ Taubenhaus. A study of the black and yellow molds of ear corn. Tex Agric Exp Stn Bull 270:38, 1920.
22. LL Ilag. *Aspergillus flavus* infection of preharvest corn, drying corn and stored corn in Philippines. Philipp Phytopathol 9:37, 1975.
23. EB Lillehoj, WF Kwolek, ES Horner, et al. Aflatoxin contamination of preharvest corn: role of *Aspergillus flavus* inoculum and insect damage. Cereal Chem 57:255, 1980.
24. KS Bilgrami, T Prasad, RS Misra, KK Sinha. Survey and study of mycotoxin producing fungi associated with grains in the standing maize crops. Final Technical Report, Indian Council of Agricultural Research, Bhagalpur University, Bhagalpur, 1980, p 98.
25. RK Jones, HE Duncan, GA Payne, KJ Leonard. Factors influencing infection by *Aspergillus flavus* in silk inoculated corn. Plant Dis 64:859, 1980.

26. SF Marsh, GA Payne. Scanning EM studies on the colonization of dent corn by *Aspergillus flavus*. Phytopathology 74:557, 1984.
27. SF Marsh, GA Payne, Preharvest infection of corn silks and kernels by *Aspergillus flavus*. Phytopathology 74:1284, 1984.
28. CW Hesseltine, RJ Bothast. Mold development in ears of corn from tesseling to harvest. Mycologia 69:328, 1977.
29. GW Rambo, J Tuite, P Crane. Preharvest inoculation and infection of dent corn with *Aspergillus flavus* and *Aspergillus parasiticus*. Phytopathology 64:797, 1974.
30. UL Diener, RJ Cole, TH Sanders, GA Payne, LS Lee, MA Klich. Epidemiology of aflatoxin formation by *Aspergillus flavus*. Annu Rev Phytopathol 25:249, 1987.
31. N Zummo, GE Scott. Cob and kernel infection by *Aspergillus flavus* and *Fusarium moniliforme* in inoculated, field-grown maize ears. Plant Dis 74:627, 1990.
32. RS Misra, RK Tripathi. Effect of aflatoxin B_1 in germination, respiration and α-amylase. Plant Dis 87:155, 1980.
33. R Schoental, AF White. Aflatoxin and albinism in plants. Nature 205:57, 1965.
34. I Slowatizky, AM Mayer, A Poljakoff Mayber. The effect of aflatoxin on greening of etiolated leaves. Isr J Bot 18:31, 1969.
35. SM Kelly. Systemic infection of maize plants by *Aspergillus flavus*. In: MS Zuber, EB Lillehoj, BL Renfro, eds. Aflatoxin in Maize: A Proceeding of Workshop. Mexico City: CIMMYT, 1987, p 187.
36. L Stoloff. Aflatoxin—an overview. In: JV Rodricks, CW Hesseltine, MA Mehlman, eds. Mycotoxins in Human and Animal Health. Park Forest South, Illinois: Pathotox Publishers, Inc., 1977, p 8.
37. WW McMillian, DM Wilson, CJ Mirocha, NW Widstrom. Mycotoxin contamination in grain sorghum in fields from Georgia and Mississippi. Cereal Chem 60:226, 1983.
38. WM Hagler Jr, M Babadoost, SP Swanson, DT Bowman. Aflatoxin, zearalenone, and deoxynivalenol in North Carolina grain sorghum, 1981–1985. Crop Sci 27: 1273, 1987.
39. B Bhadraiah, P Ramarao. Aflatoxin production in preharvest, freshly harvest and stored sorghum. In: KS Bilgrami, T Prasad, KK Sinha, eds. Mycotoxins in Food and Feed. Bhagalpur: Allied Press, 1983, p 49.
40. KS Bilgrami, AK Choudhary. Incidence of *Aspergillus flavus* in the aerosphere of maize fields at Bhagalpur. Indian Phytopathol 43(1):38, 1990.
41. LS Lee, WR Goynes, PE Lacey. Aflatoxin in Arizona cottonseed: simulation of insect vectored infection of cotton bolls by *Aspergillus flavus*. J Am Oil Chem Soc 63:468, 1986. Abstract.
42. RK Jones, HE Duncan, PB Hamilton. Planting date, harvest date, and irrigation effects on infection and aflatoxin production by *Aspergillus flavus* in field corn. Phytopathology 71:8710, 1981.
43. DT Wicklow. Taxonomic features and ecological significance of sclerotia. In: UL Diener, RL Asquith, JW Dickens, eds. Aflatoxin and *Aspergillus flavus* in Corn. South Coop Ser Bull 279. Auburn University, AL: Agric Exp Stn, 1983, p 6.
44. GA Payne. Nature of field infection of corn by *Aspergillus flavus*. In: UL Diener, RL Asquith, JW Dickens, eds. Aflatoxin and *Aspergillus flavus* in Corn. South Coop Ser Bull 279. Auburn University, AL: Agric Exp Stn, 1983, p 6.

45. GA Payne, DL Thompson, EB Lillehoj. Effect of silk age and incubation temperature on kernel infection by *Aspergillus flavus* in silk inoculated corn. Phytopathology 71:898, 1981. Abstract.
46. EB Lillehoj, WF Kwolek, EE Vandegraft, MS Zuber, OH Calvert. Aflatoxin production in *Aspergillus flavus* inoculated ears of corn grown at diverse locations. Crop Sci 15:267, 1975.
47. ND Davis, CG Currier, UL Diener. Aflatoxin contamination of corn hybrids in Alabama. Cereal Chem 63:467, 1986.
48. AK Choudhary. Ecotoxicological studies on fungi. PhD thesis, Bhagalpur University, Bhagalpur, 1991.
49. KS Bilgrami. Mycotoxins in food. J Indian Bot Soc 63:109, 1984.
50. KK Sinha. Incidence of mycotoxins in maize grains in Bihar State, India. Food Addit Contam 7(1):55, 1990.
51. PD Blankenship, RJ Cole, TH Sanders, RA Hill. Effect of geocarposphere temperature on preharvest colonization of drought-stressed peanuts by *Aspergillus flavus* and subsequent aflatoxin contamination. Mycopathologia 85:69, 1984.
52. RJ Cole, PD Blankenship, RA Hill, TH Sanders. Effect of geocarposphere temperature on preharvest colonization of drought stressed peanuts by *Aspergillus flavus* and subsequent aflatoxin contamination. In: H Kurata, Y Ueno, eds. Toxigenic Fungi: Their Toxins and Human Health. Amsterdam: Elsevier, 1984, p 44.
53. RA Hill, PD Blankenship, RJ Cole, TH Sanders. Effects of soil moisture and temperature on preharvest invasion of peanuts by the *Aspergillus flavus* group and subsequent aflatoxin development. Appl Environ Microbiol 45:628, 1983.
54. DI Fennell, EB Lillehoj, WF Kwolek. *Aspergillus flavus* and other fungi associated with insect damaged field corn. Cereal Chem 52:314, 1975.
55. WD Guthrie, EB Lillehoj, D Barry, et al. Aflatoxin contamination in preharvest corn: interaction of European corn borer and *Aspergillus flavus* group isolates. J Eco Entomol 75(2):265, 1982.
56. D Barry, MS Zuber, EB Lillehoj, et al. Evaluation of two arthropod vectors as inoculators of developing maize ears with *A. flavus*. Environ Entomol 14(5):634, 1985.
57. EB Lillehoj. Secondary metabolites as chemical signals between species in an ecological niche. Sixth International Fermentation Symposium, London, 1980. In: M Moo Young, C Vezina, K Singh, eds. Advances in Biotechnology. Vol. 3. Elmsford NY: Pergamon Press, 1980, p 397.
58. KS Bilgrami, AK Choudhary. Competing mycoflora with *Aspergillus flavus* in kernels of Rabi and Kharif maize crops of Bhagalpur. Indian Phytopathol 43:547, 1990.
59. NW Widstrom, WW McMillian, DM Wilson, JL Richard, N Zummo, RW Beaver. Preharvest aflatoxin contamination of maize inoculated with *Aspergillus flavus* and *Fusarium moniliforme*. Mycopathologia 128:119, 1994.
60. KS Bilgrami, AK Choudhary. Effect of climatological factors on the incidence of *Aspergillus flavus* in relation to co-existing mycoflora on maize kernels. Indian Phytopathol 44:527, 1991.
61. AK Choudhary. Influence of microbial co-inhabitants on aflatoxin synthesis of *Aspergillus flavus* on maize kernels. Lett Appl Microbiol 14:143, 1992.

62. AW Horn, DT Wicklow. Factors influencing the inhibition of aflatoxin production in corn by *Aspergillus niger*. Can J Microbiol 29:1087, 1983.
63. KS Bilgrami, A Masood, KS Ranjan, AK Sinha. Impact of cob husk and maturity period on aflatoxin contamination in preharvested Kharif maize. Indian Phytopathol 43:508, 1990.
64. NW Widstrom, DM Wilson, WW McMillian. Aflatoxin contamination of preharvest corn as influenced by timing and methods of inoculation. Appl Environ Microbiol 42:249, 1981.
65. D Barry, EB Lillehoj, NW Widstrom, et al. Effect of husk tightness and insect (Lepidoptera) infestation on aflatoxin contamination of preharvest maize. Environ Entomol 15:1116, 1986.
66. RK Jones, HE Duncan. Effect of nitrogen fertilizer, planting date and harvest date on aflatoxin production in corn inoculated with *Aspergillus flavus*. Plant Dis 65:741, 1981.
67. KS Bilgrami, KS Ranjan, A Masood. Influence of cropping pattern on aflatoxin contamination in preharvest kharif (monsoon) maize crops (*Zea mays*). J Sci Food Agric 58:101, 1992.
68. NW Widstrom, BR Wiseman, WW McMillian, et al. Evaluation of commercial and experimental three-way corn hybrids for aflatoxin B_1 production potential. Agron J 70:986, 1978.
69. NW Widstrom, MS Zuber. Prevention and control of aflatoxin in corn: sources and mechanisms of genetic control in plant. In: UL Diener, RL Asquith, JW Dickens, eds. Aflatoxin and *Aspergillus flavus* in Corn. South Coop Series Bull 279, 1983, p 72.
70. CAC Gardner, JR Wallin. Aflatoxin production in maize endosperm mutants. Trans Mo Acad Sci 14:174, 1980.
71. VK Mehan, D McDonald. Research on aflatoxin problem in groundnut at ICRISAT. Plant Soil 79:255, 1984.
72. LJ Ashworth Jr, JL McMeans, M Brown. Infection of cotton by *Aspergillus flavus*: Time of infection and the influence of fibre moisture. Phytopathology 59:383, 1969.
73. KS Bilgrami, AK Choudhary, KS Ranjan. Aflatoxin contamination in field mustard (*Brassica juncea*) cultivars. Mycotoxin Res 8:21, 1992.
74. RJ Cole, JW Dorner, PD Blankenship. Environmental conditions required to induce preharvest aflatoxin contamination of groundnuts; summary of six years research. In: Aflatoxin Contamination of Groundnuts. Patancheru, India: ICRISAT, 1989, p 279.
75. D McDonald, C Harkness. Growth of *Aspergillus flavus* and production of aflatoxin in groundnuts. Part IV. Trop Sci VI(1):12, 1964.
76. JPF Sellschop. Field observations on conditions conducive to the contamination of groundnuts with the mold *Aspergillus flavus* Link ex Fries. In: L Abrams, JPF Sellschop, CJ Rabie, eds. Symposium on Mycotoxins in Foodstuff. Agricultural Aspects. Pretoria, South Africa: Department of Agriculture, Technical Services, 1965, pp 47–52.
77. DL Lindsey. Effect of *Aspergillus flavus* on peanuts grown under gnotobiotic conditions. Phytopathology 60:208, 1970.
78. TR Wells, WA Kreutzer, DL Lindsey. Colonization of gnotobiotically grown peanuts by *Aspergillus flavus* and selected interacting fungi. Phytopathology 62:1238, 1972.

79. CH Styer, RJ Cole, RA Hill. Inoculation and infection of peanuts flowers by *Aspergillus flavus*. Proc Am Peanut Res Educ Soc 15:91, 1983.
80. TR Wells, WA Kreutzer. Aerial invasion of peanut flower tissue by *Aspergillus flavus* under gnotobiotic conditions. Phytopathology 62:797, 1972. Abstract.
81. TH Sanders, RA Hill, AJ Cole, PD Blankenship. Effect of drought on occurrence of *Aspergillus flavus* in maturing peanuts. J Am Oil Chem Soc 58:966A, 1981.
82. JI Pitt. Field studies on *Aspergillus flavus* and aflatoxins in Australian groundnuts. In: Aflatoxin Contamination of Groundnut: Proceeding of the International Workshop—87. Patancheru, India: ICRISAT, 1989, pp 223–235.
83. RJ Cole, RA Hill, PD Blankenship, TH Sanders. Color mutants of *Aspergillus flavus* and *Aspergillus parasiticus* in a study of preharvest invasion of peanuts. Appl Environ Microbiol 52:1128, 1986.
84. MA Klich, SH Thomas, JE Mellon. Field studies on the mode of entry of *Aspergillus flavus* into cotton seeds. Mycologia 76:665, 1984.
85. MA Klich, LS Lee, HE Huizar. Occurrence of *Aspergillus flavus* in vegetative tissues of cotton plants and its relation to seed infection. Mycopathologia 95:171, 1986.
86. LS Lee, LV Lee Jr, TE Russel. Aflatoxin in Arizona cotton seed: field inoculation of bolls by *Aspergillus flavus* spores in wind-driven soil. J Am Oil Chem Soc 63:530, 1986.
87. S Sun, GM Jividen, WH Wessling, ML Ervin. Cotton cultivars and boll maturity effects on aflatoxin production. Crop Sci 18:724, 1978.
88. MA Klich, MA Chmielewski. Nectaries as entry sites for *Aspergillus flavus* in developing cotton bolls. Appl Environ Microbiol 50:602, 1985.
89. JN Ihle, LS Dure. The developmental biochemistry of cottonseed embryogenesis and germination. III. Regulation of the biosynthesis of enzymes utilized in germination. J Biol Chem 247:5048, 1972.
90. KS Bilgrami, KS Ranjan, AK Choudhary. Aflatoxin contamination in monocropping and mixedcropping mustard crops of Bihar. Indian Phytopathol 44:529, 1991.
91. BJ Blaney. Mycotoxins in crops grown in different climatic regions of Queensland. In: J Lacey, ed. Trichothecenes and Other Mycotoxins. London: John Wiley & Sons Ltd., 1985, p 97.
92. VK Mehan. The aflatoxin contamination problem in groundnut-control with emphasis on host plant resistance. Proceedings of the First Regional Groundnut Plant Protection Group Meeting and Tour '87. Harare, Zimbabwe, 1987, pp 63–92.
93. D McDonald, C Harkness. Aflatoxin in the groundnut crop at harvest in northern Nigeria. Trop Sci IX(3):148, 1967.
94. JW Dickens, JB Satterwhite, RE Sneed. Aflatoxin-contaminated peanuts produced on North Carolina farms in 1968. Proc Am Peanut Res Edu Soc 5:48–58, 1973.
95. RE Pettit, RA Taber, HW Schroeder, AL Harrison. Influence of fungicides and irrigation practice on aflatoxin in peanuts before digging. Appl Microbiol 22:629, 1971.
96. VK Mehan, D McDonald, N Ramakrishna, JH Williams. Effect of genotype and date of harvest on infection of peanut seed by *Aspergillus flavus* and subsequent contamination with aflatoxin. Peanut Sci 13:46, 1986.
97. VK Mehan, RCN Rao, D McDonald, JH Williams. Management of drought stress to improve field screening of peanuts for resistance to *Aspergillus flavus*. Phytopathology 78:659, 1988.

98. JI Davidson Jr, RA Hill, RJ Cole, AC Mixon, RJ Henning. Field performance of two peanut cultivars relative to aflatoxin contamination. Plant Sci 10:43, 1983.
99. DM Wilson, JR Stansel. Effect of irrigation regimes on aflatoxin contamination of peanut pods. Peanut Sci 10:54, 1983.
100. PD Blankenship, TH Sanders, JW Dorner, RJ Cole, BW Mitchell. Engineering aspects of aflatoxin research in groundnuts: evolution of an environmental control plot facility. In: Aflatoxin Contamination of Groundnut: Proceedings of the International Workshop '87. Patancheru, India: ICRISAT, 1989, pp 269–278.
101. TH Sanders, RJ Cole, PD Blankenship, RA Hill. Relation of environmental stress duration to *Aspergillus flavus* invasion and aflatoxin production in preharvest peanuts. Plant Sci 12:90, 1985.
102. NJ Burrell, JK Grundey, C Harkness. Growth of *Aspergillus flavus* and production of aflatoxin in groundnuts. Trop Sci 6:74, 1964.
103. AM Hussein, NF Sommer, RJ Fortlarge. Suppression of *Aspergillus flavus* in raisins by solar heating during sun drying. Phytopathology 76:335, 1986.
104. TH Sanders, RJ Cole, PD Blankenship, RA Hill. Drought soil temperature range for aflatoxin production in preharvest peanuts. Proc Am Peanut Res Edu Soc 15:90, 1983.
105. RJ Cole, TH Sanders, RA Hill, PD Blankenship. Mean geocarposphere temperatures that induce preharvest aflatoxin contamination of peanuts under drought stress. Mycopathologia 91:41, 1985.
106. RJ Cole, TH Sanders, PD Blankenship, RA Hill. Mode of formation of aflatoxin in various nut fruits and gross and histologic effects of aflatoxins in animals. In: JW Finley, DE Schwass, eds. Xenobiotics in Foods and Feeds. ACS Symp Ser 234. Washngton, DC: American Chemical Society, 1983, p 223.
107. J Graham. Aflatoxin in peanuts: occurrence and control. Queensl Agric J 108:119, 1982.
108. RG Gilbert, JL McMeans, RL McDonald. Influence of diurnal temperature cycles on infection of cotton bolls by *Aspergillus flavus*. Phytopathology 65:1043, 1975.
109. TE Russell. Aflatoxin contamination of cottonseed—climate, insects, cultural practices and segregation. Proceedings of Bettwide Cotton Prod Conf, St. Louis, 1980, pp 53–55.
110. JL McMeans, CM Brown. Aflatoxin in cottonseed as affected by the pink bollworm. Crop Sci 15:865, 1975.
111. PB Marsh, ME Simpson, GO Craig, J Donuso, HH Ramey Jr. Occurrence of aflatoxin in cotton seeds at harvest in relation to location of growth and field temperature. J Environ Qual 2:176, 1973.
112. TE Russell, TF Watson, GF Ryan. Field accumulation of aflatoxin in cottonseed as influenced by irrigation termination dates and pink bollworm infestation. Appl Environ Microbiol 31:711, 1976.
113. LS Lee, TE Russell. Distribution of aflatoxin-containing cottonseed within intact locks. J Am Oil Chem Soc 58:27, 1981.
114. AK Sinha, KS Bilgrami, KS Ranjan, T Prasad. Incidence of aflatoxin in mustard crop in Bihar. Indian Phytopathol 41:434, 1988.
115. SS Bampton. Growth of *Aspergillus flavus* and production of aflatoxin in groundnuts. Part I. Trop Sci V(2):74, 1963.

116. LJ Ashworth Jr, BC Langley. The relationship of pod damage to kernel damage by 1molds in Spanish peanuts. Plant Dis Rep 48:875, 1964.
117. JW Dickens. Aflatoxin occurrence and control during growth, harvest and storage of peanuts. In: JV Rodrick, CW Hesseltine, MA Mehlman, eds. Mycotoxins in Human and Animal Health. Park Forest South, Illinois: Pathotox Publishers, 1977, p 99.
118. DM Wilson, RE Lynch. Effect of lesser cornstalk borer peanut damage on colonization by a mutant of *Aspergillus parasiticus*. Proc Am Peanut Res Educ Soc 16:47, 1984.
119. RA Johnson, MH Gumel. Termite damage and crop loss studies in Nigeria—the incidence of termite scarified groundnut pods and resulting kernel contamination in the field and market samples. Trop Pest Manage 27:343, 1981.
120. TL Aucamp. The role of mite vectors in the development of aflatoxin in groundnuts. J Stored Prod Res 5:245, 1969.
121. DK Bell, NA Minton, B Doupnik Jr. Effect of *Meloidogyne arenaria, Aspergillus flavus* and curing time on infection of peanut pods by *Aspergillus flavus*. Phytopathology 61:1038, 1971.
122. LJ Ashworth Jr, RE Rice, JL McMeans, CM Brown. The relationship of insects to infection of cotton bolls by *Aspergillus flavus*. Phytopathology 61:488, 1971.
123. TJ Henneberry, LA Bariola, T Russell. Pink bollworm; chemical control in Arizona and relationships to infestations, lint yield, seed damage, and aflatoxin in cottonseed. J Econ Entomol 71:440, 1978.
124. DHC Bakhetia, KS Brar. Pest control in rapeseed and mustard. Indian Farming 32: 60, 1982.
125. AZ Joffe, N Lisker. Effects of crop sequence and soil types on the mycoflora of groundnut kernels. Plant Soil 32:531, 1970.
126. P Subrahmanyam, AS Rao. Occurrence of aflatoxin and citrinin in groundnut (*Arachis hypogaea* L.) at harvest in relation to pod condition and kernel moisture content. Curr Sci 43:707, 1974.
127. GJ Griffin, KH Garron, JD Taylor. Influence of crop rotation and minimum tillage on the population of *Aspergillus flavus* group in peanut field soil. Plant Dis 65:898, 1981.
128. DM Wilson, ME Walker, TP Gaines, AS Csinos, T Win, BG Mullinix Jr. Effects of *Aspergillus parasiticus* inoculation, calcium rates and irrigation on peanuts. Proc Am Peanut Res Edu Soc 17:72, 1985.
129. LS Weckbach, EH Marth. Aflatoxin production by *Aspergillus parasiticus* in competitive environment. Mycopathologia 62:39, 1977.
130. RT Hanlin. Invasion of peanut fruits by *Aspergillus flavus* and other fungi. Mycopath Mycol Appl 40:341, 1970.
131. D McDonald. Fungal invasion of groundnut fruit before harvest. Trans Br Mycol Soc 54:453, 1970.
132. HW Schroeder, LJ Ashworth Jr. Aflatoxin in Spanish peanuts in relation to pod and kernel condition. Phytopathology 55:464, 1965.
133. UL Diener. Deterioration of peanut quality caused by fungi. In: Peanuts: Culture and Uses. Stillwater, OK: American Peanut Research and Education Association, 1973, pp 523–557.
134. AZ Joffe. The mycoflora of fresh and stored groundnut kernels in Isreal. Mycopath Mycol Appl 39:255, 1969.

135. UL Diener, CR Jackson, WE Cooper, RJ Stipes, MD Davis. Invasion of peanut pods in the soil by *Aspergillus flavus*. Plant Dis Rep 49:931, 1965.
136. JW Dickens, HE Pattee. The effects of time, temperature and moisture on aflatoxin production in peanuts inoculated with a toxic strain of *Aspergillus flavus*. Trop Sci 8:11, 1966.
137. CT Kisyombe, MK Beute, GA Payne. Field evaluation of peanut genotypes for resistance to infection by *Aspergillus parasiticus*. Peanut Sci 12:12, 1985.
138. VK Mehan, D McDonald. Screening for resistance to *Aspergillus flavus* invasion and aflatoxin production in groundnuts. ICRISAT, Groundnut Improvement Program Occasional Paper 2, 1980, p 15.
139. AC Mixon. Reducing *Aspergillus* species infection of peanut seed using resistant genotypes. J Environ Qual 15:101, 1986.
140. VK Mehan, D McDonald, K Rajagopalan. Resistance of peanut genotypes to seed infection by *Aspergillus flavus* in field trials in India. Peanut Sci 14:17, 1987.
141. C Zambettakis, F Waliyar, A Bockelee-Morvan, O de Pins. Results of four years of research on resistance of groundnut varieties to *Aspergillus flavus*. Olegineux 36: 377, 1981.
142. D McDonald. *Aspergillus flavus* on groundnut (*Arachis hypogaea* L.) and its control in Nigeria. J Stored Prod Res 5:275, 1969.
143. PA Plattner, N Clausonkass. Uber ein walke erzeugendes staffwechsel product non *Fusarium lycopersici* Sacc. Helv Ch Acta 28:188, 1945.
144. WFO Marasas, PE Nelson, TA Toussoun. Toxigenic Fusarium Species: Identity and Mycotoxicology. University Park, PA: Pennsylvania State University Press, 1984.
145. WCA Gelderblom, K Jasiewicz, WFO Marasas, et al. Fumonisins—novel mycotoxins with cancer—promoting activity produced by *Fusarium moniliforme*. Appl Environ Microbiol 45:1806, 1988.
146. J Chelkowski. Formation of mycotoxins produced by *Fusarium* in heads of wheat, triticale and rye. In: J Chelkowski, ed. Fusarium Mycotoxins, Taxonomy and Pathogenicity. Amsterdam: Elsevier, 1989, p 63.
147. PM Scott. Fumonisins. Int J Food Microbiol 18:257, 1993.
148. J Perkowski, T Meidaner, HH Geiger, HM Muller, J Chelkowski. Occurrence of deoxynivalenol (DON), 3 acetyl-DON, zearalenone and ergosterol in winter rye inoculated with *Fusarium culmorum*. Cereal Chem 72:205, 1995.
149. JC Sutton. Epidemiology of wheat head blight and maize ear rot caused by *Fusarium graminearum*. Can J Plant Pathol 4:195, 1982.
150. HK Abbas, CJ Mirocha, JD Pokorney, SL Gould, T Kommedanl. Mycotoxins of *Fusarium* spp. associated with infected ears of corn in Minnesota. Appl Environ Microbiol 54:1039, 1988.
151. KK Sinha. Contamination of freshly harvested maize kernels with *Fusarium* mycotoxins in Bihar state, India. Mycotoxin Res 7A(Part II):178, 1991.
152. PG Thiel, WFO Marasas, EW Sydenham, GS Shephard, WCA Gelderblom, JJ Nieuwenhuis. Survey of fumonisin production by *Fusarium* species. Appl Environ Microbiol 57:1089, 1991.
153. TE Ochor, LE Trevathan, SB King. Relationships of harvest date and host genotype to infection of maize kernels by *Fusarium moniliforme*. Plant Dis 71:311, 1987.

154. C de Leon, S Pandey. Improvement of resistance to ear and stalk rots and agronomic traits in tropical maize gene pools. Crop Sci 29:12, 1989.
155. M Sharman, J Gilbert, J Chelkowski. A survey of the occurrence of the mycotoxin moniliformin in cereal samples from sources worldwide. Food Addit Contam 8:449, 1991.
156. RF Ross, LG Rice, RD Plattner, et al. Concentrations of fumonisin B_1 in feeds associated with animal health problems. Mycopathologia 114:129, 1990.
157. RD Plattner, WP Neffed, CW Bacon, et al. A method of detection of fumonisins in corn samples associated with field cases of equine leukoencephalomalacia. Mycologia 82:698, 1990.
158. H Lew, A Alder, W Edinger. Moniliformin and the European corn borer (*Ostrinia nubilalis*). Mycotoxin Res 7:71, 1991.
159. EW Sydenham, PG Thiel, WFO Marasas, GS Shepherd, DJ Van Schalkwyk, KR Koch. Natural occurrence of some *Fusarium* mycotoxins in corn from low and high esophageal cancer prevalence areas of the Transkei, Southern Africa. J Agric Food Chem 38:1900, 1990.
160. TC Tseng, KL Lee, TS Deng, CY Liu, JW Huang. Production of fumonisins by *Fusarium* species of Taiwan. Mycopathologia 130:117, 1995.
161. LB Bullerman, WY Tsai. Incidence and levels of *Fusarium moniliforme*, *Fusarium proliferatum* and fumonisins in corn and corn-based foods and feeds. J Food Prct 57:541, 1994.
162. J Chelkowski, P Zajkowski, H Kwasna, A Visconti, A Bottalico. *Fusarium sporotrichioides* Sherb, and trichothecenes associated with *Fusarium* ear rot of corn before harvest. Mycotoxin Res 3:111, 1987.
163. JD Miller, JC Young, HL Trenholm. *Fusarium* toxins in field corn. I. Time course of fungal growth and production of deoxynivalenol and other mycotoxins. Can J Bot 61:3080, 1983.
164. RB Hunter, JC Sutton. Plant resistance and environmental conditons. In: PM Scott, HL Trenholm, MD Sutton, eds. Mycotoxins: A Canadian Perspective. Ottawa: NRCC No. 22848, 1985, p 62.
165. EB Khonga, JC Sutton. Survival and inoculum production of *Gibberella zeae* in wheat and corn residues. Can J Plant Pathol 8:531, 1986.
166. I Manninger. Resistance of maize in ear rot on the basis of natural infection and inoculation. In: Proc 10th Meeting, Eucarpia, Maize, Sorghum Sec. Varna, Bulgaria, 1979, pp 181–184.
167. PM Emerson, RB Hunter. Response of maize hybrids to artifically inoculated ear mold incited by *Gibberella zeae*. Can J Plant Sci 60:1463, 1980.
168. TM Georgiev, FJ Bery, PJ Loesch, AV Paez. Relationship among several physical characteristics of normal and opaque-2 maize kernels and reaction to fungal diseases. In: Proc 9th Meeting, Eucarpia, Maize, Sorghum Sec. Krasnodar, USSR, 1977, pp 72–73.
169. B Palabersic, K Stastny, I Milatovic. Study on the resistance of opaque-2 maize lines and hybrids and their normal counterparts to ear rot under conditions of artifical infection with *Fusarium graminearum (Gibberella zeae)* and *Fusarium moniliforme*. In: Proc 10th Meeting, Eucarpia, Maize, Sorghum Sec. Varna, Bulgaria, 1979, p 214–218.

170. JC Sutton, W Balico, HS Funnel. Relation of weather variables to incidence of zearalenone in corn in southern Ontario. Can J Plant Sci 60:149, 1980.
171. WA Attwater, IV Busch. Role of sap beetle *Glischrochilus quadrisignatus* in the epidemiology of *Gibberella* ear rot. Can J Plant Pathol 5:158, 1983.
172. RA Martin, HW Johnston. Effect and control of *Fusarium* disease of cereal grains in the Atlantic provinces. Can J Plant Pathol 4:210, 1982.
173. MR Fernandez. Recovery of *Cochliobolus sativus* and *Fusarium graminearum* from living and dead wheat and non-gramineous winter crops in southern Brazil. Can J Bot 69:1900, 1991.
174. PE Lipps, IW Deep. Influence of tillage and crop rotation in yield, stalk sot and recovery of *Fusarium* and *Trichoderma* spp. from corn. Plant Dis 75:828, 1991.
175. WL Seaman. Epidemiology and control of mycotoxigenic *Fusarium* on cereal grains. Can J Plant Pathol 4:187, 1982.
176. VH Lengkeek. Efficacy of chemical control of stalk and ear rot of corn in South West Kansas. Fungicide and nematicide test. Field Cereal Dis Rep 36:83, 1980.
177. AA Al-Heeti. Pathological, toxicological and biological evaluations of *Fusarium* species associated with ear rot of maize. PhD thesis, Wisconsin, Madison University. Microfilms Int Diss Inf Ser 8727220, 1987.
178. J Tuite, G Shaner, G Rambo, J Foster, RW Caldwell. The *Gibberella* ear rot epidemics of corn in Indiana in 1965 and 1972. Cereal Sci Today 19:238, 1974.
179. SB King, GE Scott. Genotypic differences of maize to kernel infection by *Fusarium moniliforme*. Phytopathology 71:1245, 1981.
180. DC Foley. Systemic infection of corn by *Fusarium moniliforme*. Phytopathology 68:1331, 1962.
181. CW Becon, JW Williamson. Interaction of *Fusarium moniliforme*, its metabolites and bacteria with corn. Mycopathologia 117:65, 1992.
182. GN Odovody, JC Remmers, NM Spencer. Association of kernel splitting with kernel and ear rots of corn in a commercial hybrid grown in coastal belt of Texas. Phytopathol 80:1045, 1990.
183. T Yoshizawa. Red mold diseases and natural occurrence in Japan. In: Y Ueno, ed. Trichothecenes—Chemical, Biological and Toxicological Aspects. Amsterdam: Elsevier and Kodansha Ltd., 1983, p 195.
184. PM Scott, PY Lau, SR Kanhere. Gas chromatography with electron capture and mass spectrometric detection of deoxynivalenol in wheat and other grains. J Assoc Off Anal Chem 64:1364, 1981.
185. BC Williams. Mycotoxins in foods and foodstuffs. In: PM Scott, HL Trenholm, MD Sutton, eds. Mycotoxins: A Canadian Perspective. Ottawa: NRCC Publ. 22848, 1985, p 49.
186. A Mesterhazy. Comparative analysis of artificial inoculation methods with *Fusarium* spp. on winter wheat varieties. Phytopath Z 93:12, 1978.
187. RW Stack, MP McMullen. Head blighting potential of *Fusarium* species associated with spring wheat heads. Can J Plant Pathol 7:79, 1985.
188. I Keicana, J Perkowski, J Chelkowski, A Visconti. Trichothecene mycotoxins in kernels and head fusariosis susceptibility in winter triticale. In: Mycotoxin Research, Special Edition, European Seminar on *Fusarium*—Mycotoxins, Taxonomy, Pathogenicity. Warsaw, September 8–10, 1987, p 53.

189. WFO Marasas, NPJ Kriek, VM Wiggins, PS Steyn, DK Towers, TJ Hostie. Incidence, geographic distribution and toxigenicity of *Fusarium* species in South African corn. Phytopath 69:1181, 1979.
190. A Bottalico, P Lerario, A Visconti. Mycotoxins occurring in *Fusarium* infected maize ears in the field, in some European countries. Int Symp Mycotoxins (Pub. Sci. Dept. NIDOC, ed.), Cairo, 1983, pp 375–382.
191. CJ Mirocha, HK Abbas, CE Windels, W Xie. Variations in deoxynivalenol, 15-acetyldeoxynivalenol, 3-acetyldeoxynivalenol and zearalenone production by *Fusarium graminearum* isolates. Appl Environ Microbiol 55:1315, 1989.
192. A Alder, H Lew, W Brodacz, W Edinger, M Oberfoster. Occurrence of moniliformin, deoxynivalenol, and zearalenone in durum wheat (*Triticum durum* Desf.). Mycotoxin Res 11:9, 1995.
193. DS Tu. Factors affecting the reaction of wheat blight infection caused by *Gibberella zeae*. PhD thesis, Ohio State Univ, Columbus, 1953.
194. RN Strange, H Smith. Specificity of choline and betaine as stimulants of *Fusarium graminearum*. Trans Br Mycol Soc 70:201, 1978.
195. AH Teich. Epidemiology of wheat (*Triticum aestivum* L.) scab caused by *Fusarium* spp. In: J Chelkowski, ed. *Fusarium*—Mycotoxins, Taxonomy and Pathogenicity. New York: Elsevier, 1989, p 269.
196. MW Wiese. Compendium of Wheat Diseases. St. Paul, MN: American Phytopathological Society, 1977.
197. B Lacicowa, A Wagner, I Kiecana. Fuzariozy pszenicy uprawianej na Lubelszczyznie, Rocz Nauk Roln Ser E 15:67, 1985.
198. RM Clear, SK Patrick. *Fusarium* species isolated from wheat samples containing tombstone (scab) kernels from Ontario, Manitoba and Saskatchewan. Can J Plant Sci 70:1057, 1990.
199. JP Rheeder, WFO Marasas, PS van Wyk, DJ Schalkwyk. Reaction of South African maize cultivars to ear inoculation with *Fusarium moniliforme*, *F. graminearum*, and *Diplodia maydis*. Phytophylactica 22:213, 1990.
200. M Lacroix, D Dostaler. Influence du cultivar de ble et de desoxynivalenol par des souches du *Fusarium graminearum* (in French). Phytoprotection 66:174, 1985.
201. A Mesterhazy. Selection of head blight resistant wheats through improved seedling resistance. Plant Breeding 98:25, 1987.
202. JD Miller, JC Young, DR Sampson. Deoxynivalenol and *Fusarium* head blight resistance in spring cereals. Phytopathol Z 113:359, 1985.
203. M Hatmanu. Epiphytotics on wheat caused by *Fusarium graminearum* Schw. and factors favouring them (in Romanian). *Lucrari stiintifice, I Iasi Rumania, Inst Agron I Ionescu de la Brad*, 1972, p 157.
204. CA Strausbaugh, OC Maloy. *Fusarium* scab of irrigated wheat in central Washington, USA. Plant Dis 70:1104, 1986.
205. AH Teich, DR Sampson, L Shugar, A Smid, WE Curnoe, C Kennema. Yield, quality and disease response of soft white winter wheat cultivars to nitrogen fertilization in Ontario, Canada. Cereal Res Comm 15:265, 1987.
206. AH Teich, JR Hamilton. Effect of cultural practices, soil phosphorus, potassium and pH on the incidence of *Fusarium* head blight and deoxynivalenol levels in wheat. Appl Environ Microbiol 49:1429, 1985.

207. AH Teich, K Nelson. Survey of *Fusarium* head blight and possible effects of cultural practices in wheat fields in Lambton country in 1983. Can Plant Dis Surv 6:11, 1984.
208. N Montemurro, A Visconti. *Alternaria* metabolites—chemical and biological data. In: J Chelkowski, A Visconti, eds. *Alternaria*—Biology, Plant Disease and Metabolites. Amsterdam: Elsevier, 1992, p 451.
209. LM Seitz, DB Sauer, HE Mohr, R Burrough. Weathered grain sorghum: natural occurrence of alternariols and storability of the grain. Phytopath 65:1259, 1975.
210. DB Sauer, LM Seitz, R Burrough, et al. Toxicity of *Alternaria* metabolites found in weathered sorghum grain at harvest. J Agric Food Chem 26:1380, 1978.
211. S Gruber-Schley, A Thalman. The occurrence of *Alternaria* spp. and their toxins in grain and possible connection with illness in farm animals. Landwirsts-chaftche Forschung 41:11, 1988.
212. J Grabarkiewicz-Szczesna, J Chelkowski, P Zajkowski. Natural occurrence of *Alternaria* mycotoxins in the grain and chaff of cereals. Mycotoxin Res 5:77, 1989.
213. EE Stinson, SF Osman, EG Heisler, J Siciliano, DD Bills. Mycotoxin production in whole tomatoes, apples, oranges and lemons. J Agric Food Chem 29:790, 1981.
214. A Visconti, A Logrieco, A Bottalico. Natural occurrence of *Alternaria* mycotoxins in olives—their production and possible transfer into the oil. Food Addit Contam 3:323, 1986.
215. A Logrieco, A Bottalico, A Visconti, M Vurro. Natural occurrence of *Alternaria* mycotoxin in some plant products. Microbiol Ailment Nutr 6:13, 1988.
216. A Visconti, A Logrieco, M Vurro, A Bottalico. Tenuazonic acid in Blackmold tomatoes: Occurrence, production by associated *Alternaria* species and phytotoxic properties. Phytopathol Medit 26:125, 1987.

2

Distribution of *Fusarium* Species and Their Mycotoxins in Cereal Grains

J. Chelkowski
Institute of Plant Genetics, Polish Academy of Sciences, Poznań, Poland

I. INTRODUCTION

Five toxic fungal secondary metabolites (mycotoxins) are considered to be economically and toxicologically important worldwide: aflatoxin, ochratoxin, deoxynivalenol (DON = vomitoxin) and derivatives, zearalenone (ZON) and derivatives, and fumonisins (FB_1, FB_2). The last three of these, DON, ZON and fumonisins, are produced by various *Fusarium* species [1–3]. Mycotoxin moniliformin has been recently reported to be frequently occurring contaminant of cereal grains [4–7]. A significant increase in research on *F. moniliforme* metabolites is observed during the last 5 years, since the discovering of fumonisins B_1 and B_2 in 1988 [8–13].

Experts of FAO estimate that 25% of the world's food crops are affected by mycotoxins each year with substantial impact of *Fusarium* species to food contamination [14]. *Fusarium* species infection of cereal ears causes three undesired effects: (1) reduction of grain yield and quality; (2) economic losses in livestock fed with contaminated cereals and the consequent reduced animal production; and (3) mycotoxins carry-over to food products resulting in potential toxicity to human. For the reasons mentioned above a challenge is provided to scientific teams putting more and more attention to the development of prevention methods for the elimination of infection of cereals by *Fusarium* species and, consequently, reduction of mycotoxin accumulation. Numerous details and references on *Fusa-*

rium pathogens and *Fusarium* head (= ear) blight of cereals, both small-grain cereals and maize, have been reported on this subject [2,15,16] as well as in a recently published review [17].

II. TOXIGENIC *FUSARIUM* SPECIES COLONIZING CEREAL GRAIN

The literature on toxigenic abilities of *Fusarium* species contains significant number of confusions, caused by usage of several taxonomic systems, wrong identification of toxigenic isolates, or incorrect identification of mycotoxins [15]. The nomenclature of Nelson et al. [18] is followed in this chapter to avoid confusion. A clear relationship between currently used taxonomic systems has been reviewed by Nirenberg [19].

The number of *Fusarium* species reported to be present in cereal grain samples reaches about 20. Their list is presented in Table 1. However, only four species are significant because of: (1) their high pathogenicity to both small grain cereals and maize; (2) frequent occurrence; and (3) toxigenic abilities [2,15,17]. Those that contribute significantly to the contamination of cereal grains with trichothecene mycotoxins (DON and its derivatives), ZON and moniliformin (MON) worldwide, are the following: *F. culmorum* (DON, ZON, nivalenol [NIV]); *F. graminearum* (DON, ZON, NIV); *F. avenaceum* (MON); and *F. poae* (NIV).

The fifth species, *Microdochium nivale* (syn. *Fusarium nivale*), belongs also to important ear pathogens; however, it was definitely excluded from *Fusarium* genus in 1984 [15]. Confusing and untrue information is still present in the literature on ability of *F. nivale* to produce nivalenol and other trichothecene mycotoxins. This confusion is caused by an incorrect identification of the early isolates that were found to produce nivalenol [19].

F. culmorum and *F. graminearum* are most important species worldwide; they are able to produce DON and several derivatives (Table 1) and are most aggressive in the infection of cereal ears. Infection of small-grain cereal heads and maize ears by these two species leads to accumulation in kernels, mostly DON and ZON [2,15]. Two main chemotypes of both species occur in various parts of the world: DON type and NIV type [20–26].

F. avenaceum contributes to the contamination of grain with the mycotoxin moniliformin [4,6]. The species is of medium pathogenicity and infects few spikelets in an ear, but it can affect a high percentage of spikelets as well [4,15,27].

F. poae is reported to be dominant in countries of northern Europe. The isolates of the species are able to produce nivalenol and fusarenone and cause grain contamination with nivalenol [17,27–31].

F. sporotrichioides has been found in our studies to be both toxigenic and pathogenic to cereal ear. It produces trichothecene mycotoxins the most toxic to animals—T-2 toxin, HT-2 toxin, neosolaniol, and others. Fortunately, the species

TABLE 1 *Fusarium* Species, According to Nelson et al. [18] Nomenclature, and Their Mycotoxins Recorded in Cereal Grain Worldwide

Fusarium species	Frequency in cereals	Aggressiveness to ear (m—maize only)	Produced metabolites[a]
F. graminearum	++++	High	DON, 3AcDON, 15AcDON, NIV, ZON, 4.7DeDON
F. culmorum	+++++	High	DON, 3AcDON, NIV, ZON
F. avenaceum	+++++	Moderate	MON
F. crookwellense	++	High	NIV, FUS, ZON
F. poae	+++	Moderate	NIV, FUS
F. sporotrichioides	++	Moderate	T-2, HT-2, NEO
F. tricinctum	++	Low	MON
F. chlamydosporum	+	Low[m]	MON
F. semitectum	+	Saprophyte	BEA
F. equiseti	++	Saprophyte	FUCH, ZON
F. acuminatum	+	Low	T-2, MON
F. moniliforme	++++	Moderate[m]	FB_1, FB_2, FB_3, FU-C
F. subglutinans	+++++	Moderate[m]	MON, BEA, FUP
F. proliferatum	++	Low[m]	FB_1, FB_2, MON, BEA, FUS
F. anthophilum	+	Saprophyte	MON
F. solani	++	Saprophyte	MON
F. oxysporum	++	Saprophyte	MON
Microdochium nivale (formerly *F. nivale*)	+++++	Moderate	None

Frequency concern, European countries: +++++, very frequent; ++++, frequent; +++, medium frequency; ++, rare frequency; +, extremely rare.
DON, deoxynivalenol (vomitoxin); AcDON, acetyl deoxynivalenol; NIV, nivalenol; ZON, zearalenone; DeDON, dideoxynivalenol; MON, moniliformin; FUS, fusarenone; NEO, neosolaniol; BEA, beauvericin; FUCH, fusarochromanone; FB, fumonisin B; FU-C, fusarin C; FUP, fusaproliferin.
[a]Underline indicates the most important *Fusarium* mycotoxins worldwide.

is not found very frequently in an agricultural environment, and we have usually found *F. sporotrichioides* isolates as only 1% of the total *Fusarium* isolates from cereals [15,31,32].

The above-mentioned toxigenic species of *Fusarium* may infect small-grain cereals as well as maize ears. However, the list of maize ear pathogens is longer and encompasses, in addition to the species mentioned above, members of *Liseola* section of *Fusarium*: *F. moniliforme*, *F. subglutinans*, and *F. proliferatum*. Species *F. moniliforme* has been proven to be an endophyte in maize [9,33]. It is able to produce the mycotoxin fumonisin—probably the most widely distributed mycotoxin in maize grain, intensively studied mycotoxin in the last few years [9,34–

40]. High amounts of FB_1 and FB_2 are also produced by *F. proliferatum*; however, *F. subglutinans* is not able to produce those metabolites [9,11,40].

Recently *F. subglutinans* (as well as *F. proliferatum*) has been found to be able to produce significant amounts of beauvericin and fusaproliferin—mycotoxins whose toxicity is not well known [41–45].

III. IMPORTANT AND MINOR *FUSARIUM* MYCOTOXINS CONTAMINATING CEREAL GRAINS

Three *Fusarium* mycotoxins are considered to be economically and toxicologically most important: DON group, ZON group, and fumonisins. Two groups of cereal products should be taken into consideration, to find primary source of mycotoxins contamination [2–4]: (1) domestic cereal grains, (2) imported cereal grains.

In cereal grain samples originating from Poland we identified 16 metabolites of *Fusarium* listed in Table 1 [31,46,47]. The same metabolites have been found in numerous other countries [2,3,15,26,48]. Additionally, fusarochromanone has been detected in cereal feed associated with poultry intoxication in Denmark [49]. This mycotoxin is produced by *Fusarium equiseti*. Three categories of samples are analyzed for mycotoxins presence:

1. Samples taken at random from farms, country elevators, commercial channels, or international trade
2. Samples collected from fields with typical ears symptoms of *Fusarium* head blight (FHB) or *Fusarium* ear rot (FER)
3. Samples obtained after artificial inoculation

Results of the first category samples would provide information about the health risk in a given area as a result of grain contamination with mycotoxins. For example, when the amount of DON in wheat samples is about 100 μg/kg or lower, the health risk is low or even zero. Samples containing $\geqslant$ 1 mg DON/kg represent significant toxicity. Toxicity, health hazards, risk evaluation, and food safety regulations of various mycotoxins are reviewed in other chapters in this volume.

Results of numerous surveys published in the literature show great variation of DON concentration in cereal grains and cereal products [3–6,26,48,50,51]. In countries with low incidence of FHB the concentrations of DON and other mycotoxins are on the level 10 to 50 μg/kg. DON may be omnipresent in cereal grains, and this low amount of toxin is produced by few hyphae of toxigenic *Fusarium* species, which are present as a typical saprophytes also in quite sound and valuable bulks of grain. On the other hand, grain in fields with FHB epidemics may contain 1 to 10 mg DON/kg and sometimes more, up to about 70 mg/kg [3, 15,48,50].

In addition to the three most important *Fusarium* toxins, field samples (or samples originating from artificially inoculated ears) contain other metabolites as well, i.e., derivatives of DON or newly described metabolites (Table 1). DON, if present in higher amount, is accompanied by several derivatives—mostly 15 acetyl deoxynivalenol (AcDON) and 3 AcDON, or nivalenol and fusarenone, and sometimes with ZON. Nivalenol should get more attention; this mycotoxin is significantly more toxic than DON and is usually co-occurring with DON [3,48,50]. Gareis et al. [50] list in their review occurrence of the following *Fusarium* mycotoxins in European countries in cereals: DON up to 43.8 mg/kg in wheat, up to 6.3 mg/kg in barley, up to 6.3 mg/kg in oats, up to 67.0 mg/kg in maize, up to 1.3 mg/kg in mixed feeds; T-2 toxin up to 5.8 mg/kg in mixed feeds; and ZON up to 275.8 mg/kg in maize. Jelinek et al. [3] report the following data on DON occurrence in wheat and wheat products in nearly 2000 samples examined in Canada and USA (1980–1985): average in wheat, 41 to 3410 μg/kg, but could be up to 18,400 μg/kg, and the average in wheat products, 85 to 177 μg/kg, sometimes up to 530 μg/kg. On the other hand in Japan DON occurred in amounts up to 49600 μg/kg with NIV cocontaminating, up to 22,900 μg/kg [3]. The following *Fusarium* metabolites have been found in grain examined in New Zealand (1981–1989) in maize: moniliformin—up to 280 μg/kg, DON—up to 3500 μg/kg, mean 120 to 1000 μg/kg, NIV—up to 3600 μg/kg, mean 250 to 1890 μg/kg, ZON—up to 500 μg/kg, mean 210 to 290 μg/kg. In wheat: DON—up to 11,950 μg/kg, mean 160 to 2420 μg/kg, NIV—up to 270 μg/kg, mean 0.14 to 0.40 μg/kg [51].

The occurrence of NIV-type trichothecenes as a general contamination and frequently as the main contaminant in New Zealand is unusual compared to other regions of the world. Samples of maize naturally infected by *F. sporotrichioides* contained significant amounts of T-2 toxin, HT-2 toxin, and neosolaniol [32]. *Fusarium* damaged (scabby) kernels of wheat and other small-grain cereals contain about 10 times lower concentrations of DON than maize kernels, and rather low amounts of ZON. Grain bulks frequently contain low percentages of *Fusarium*-damaged kernels (FDK); however, in this low amount of kernels high concentration of DON and other mycotoxins are present.

Frequent occurrence of moniliformin has been shown in several countries in maize kernels, as well as in small-grain cereals [4–7]. Occurrence of moniliformin in small-grain cereals is a consequence of ear infection by *F. avenaceum* (Fig. 1). Concentration of this mycotoxin in wheat was found up to 17.1 mg/kg, up to 15.7 mg/kg in triticale, up to 12.3 mg/kg in rye, and up to 38.3 mg/kg in oats [4,6,7]. Maize contamination with moniliformin is caused by *F. subglutinans* infection, leading to kernel rot and ear rot [4,5,41,42]. Several new metabolites have been recorded in maize kernels like beauvericin, fusaproliferin. Their significance is not yet elucidated, and any health risk they cause is not yet known [41–45].

F. moniliforme is an omnipresent species in corn, in particular in subtropical and tropical parts of the world. The species colonizes stalk and ear systemically

(a)

FIGURE 1 (a and b) Wheat kernels originating from wheat ears inoculated with *F. avenaceum*. In three rows are (from top) *Fusarium*-damaged kernels, black pointed, and healthy-looking kernels.

and causes accumulation of fumonisin B_1 and its derivatives FB_2 and FB_3 in kernels. There are few reports on the occurrence in corn of fusarin C, another metabolite of *F. moniliforme* [15].

It could be summarized that the three most important *Fusarium* mycotoxins—DON, ZON, and FB_1—may be accompanied by number of metabolites which may be occasionally present in some bulks of cereal grains. However, those minor metabolites are present in grain at lower frequencies and in lower amounts than the most significant toxins.

IV. IMPACT OF *FUSARIUM* HEAD BLIGHT OF SMALL-GRAIN CEREALS ON GRAIN CONTAMINATION

Numerous surveys have demonstrated that high contamination (above 1 mg/kg) of wheat grain with DON is caused by FHB, when *F. culmorum* or *F. graminearum* is the infecting species [2,3,15,17,27]. Epidemics of FHB in United States and Canada in the early 1980s caused significant contamination of wheat grain. Levels above 1 mg/kg were quite common in wheat grain in both countries. DON was also reported in the United Kingdom in 31% of samples at the levels higher than 500 μg/kg [27]. The situation stimulated survey of wheat grain, and more data have been developed on the level of DON in wheat and wheat products than in

(b)

any other food commodity. Warm and humid weather conditions during anthesis of small-grain cereals are conducive to the infection of spikelets by the mentioned *Fusarium* species (Fig. 2). Once *Fusarium* infection is onset in the spikelet at anthesis, the pathogen grows slowly both intra- and intercellulary throughout the entire kernels. Kernels originating from infected spikelets contain abundant mycelium well visible in cross section under optical microscope (Fig. 3) and on a surface.

We observed a low percentage (usually below 1%) of scab in wheat, triticale, rye, and barley each year since 1984–1995 in Poland. In some years and some fields, a higher percentage of FHB was observed in wheat, triticale, and barley. An epidemic of scab was noticed in 1974; however, it was not possible to analyze *Fusarium* mycotoxins at that time. For the next 20 years epidemic was not observed in Poland. In the countries surrounding, the epidemics have also been infrequent [15,17]. In some regions epidemics occur rather frequently, and this causes significant difficulties to farmers and significant yield losses, as has been

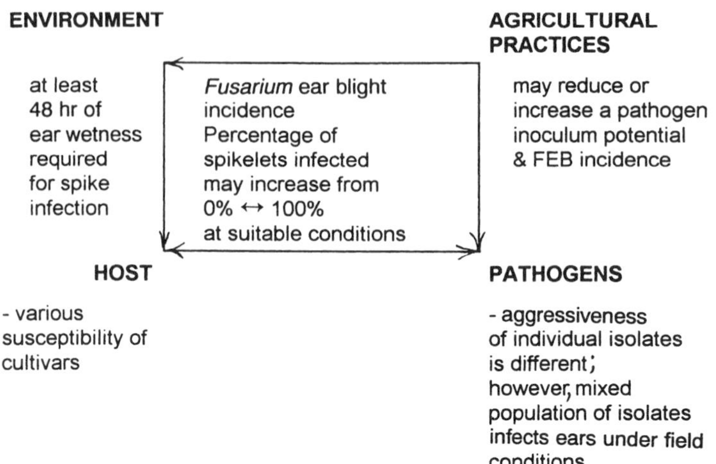

FIGURE 2 Schematic presentation of conditions important to development of *Fusarium* head blight and ear rot epidemics. FEB, *Fusarium* ear blight.

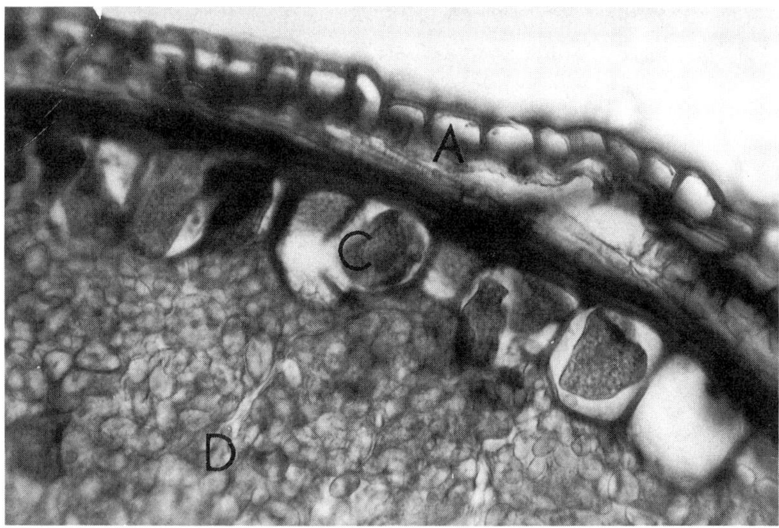

FIGURE 3 Transversel section of wheat kernels infected by *Fusarium avenaceum*. Mycelial growth, visible mostly between pericarp and aleurone layer, splits those two layers and their cell walls. Hyphae are also present inside cells and between endosperm cells.

reviewed recently [17,27,52,53]. Heavy epidemics were recorded since 1884 in England and 1890 in USA and actually still occur in Canada, USA, Japan, China, Korea, and India. In European countries epidemics of FHB are not so extensive and not so frequent. Plant protection services in some countries maintain records of FHB for longer periods. The good examples are records from the Netherlands [54], U.K. [55], and Canada [52,56]. Parry et al. [17] recently collected reports on FHB and associated species for 21 countries and four continents.

Concentration of DON in *Fusarium*-damaged (scabbed) kernels has been found in Poland, and the Netherlands in wheat and triticale on an average of 30 mg/kg [57,58]. Approximate amounts of mycotoxin may be calculated in a given sample during grading if percentage of FDK is known. The amounts of DON in naturally contaminated wheat and triticale samples correlated positively with the percentage of FDK (Fig. 4) [57–59]. Low amounts of DON are present also in kernels that are healthy looking, without any visible symptoms of scab if they originate from infected ears [15,57,58,60].

Concentration of DON in scabbed kernels has been found strongly dependent on cultivar (genotype) of wheat [58–60], as well as other cereals. Some reports present the possibility of DON production in glume and its transport to kernels [60].

The presence of DON and other trichothecenes in cereal grains has been recently reported to be the cause of several human mycotoxicoses and significant number of inhabitants of some regions have been adversely affected by the toxicity of DON [61]. Some examples are presented below:

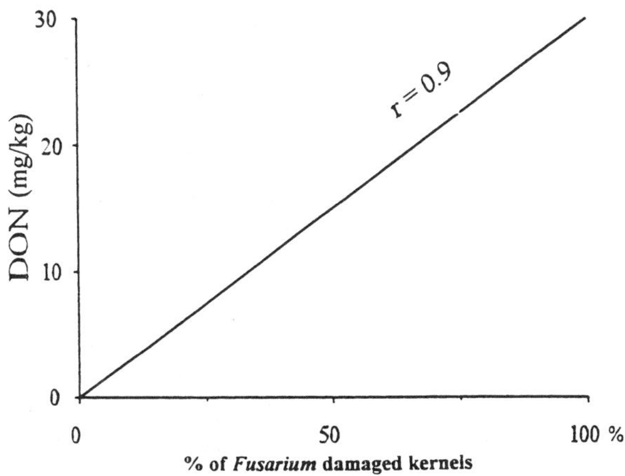

FIGURE 4 A schematic presentation of the relationship between percentage of *Fusarium*-damaged kernels and concentration of deoxynivalenol (DON) in grain sample.

1. Outbreaks of toxicosis associated with consumption of food products containing DON from moldy wheat and maize reported in China since 1961 (7000 people affected).

2. "Red mold disease" in Japan, where wheat, barley, and rice were implicated. Studies on metabolites produced by *Fusarium* species associated with cereal disease ended with the discovery of nivalenol, deoxynivalenol, and their derivatives.

3. Deoxynivalenol toxicosis in India in 1987, when over 50,000 people were estimated to have been intoxicated.

The following agricultural practices may reduce scab incidence and the DON level in cereal grain [62]:

1. Proper soil tillage—burying host crop residues prior to planting.
2. Proper crop rotation to avoid wheat (or barley) cultivation after a susceptible crop (maize, wheat), and planting wheat after nonhost crop.
3. Limiting the seeding rate to that required for maximum yield, avoiding dense stands.
4. Using urea as a nitrogen fertilizer instead of ammonium nitrate.
5. Proper weeds control.
6. Selecting and planting cereal cultivars of lowest susceptibility to scab.

Cultivars widely produced with a high yield capacity are susceptible to scab. Scab incidence may be reduced by proper agricultural practices at nonepidemic year at moderate scab outbreak. Highly resistant cultivars are still not available, and intensive work briefly summarized in recent reviews [2,15,17,62] is being done in several breeding centers to develop resistant cultivars. At this time during the years when weather stimulates heavy epiphytotic actually a severe scab is observed on even moderately and very susceptible cultivars. Recently significant concentration of DON have been found in Norwegian oat grain [63,64]. In particular, high concentrations of DON were present in samples originating from marginal areas.

Deoxynivalenol content in rye kernels inoculated with *F. culmorum* ranged from 0.72 to 28.0 mg/kg in 29 single cross hybrids [65]. FDK of rye are less discolored and less shriveled due to the narrower shape and smaller size of rye relative to wheat kernels. Mycotoxin concentrations (mg/kg) in barley kernels of 12 cultivars, inoculated with *F. culmorum*, were as follows: DON, 0.6 to 12.0; 3AcDON, 0.1 to 1.0; NIV, 0.1 to 0.7; and ZON, 0.1 to 0.5 [66]. These data show similar concentrations of DON in various small-grain cereals and significant differences between cultivars in susceptibility to accumulate DON after ear infection with *F. culmorum* and/or *F. graminearum*. *Microdochium nivale* is also a frequently occurring pathogen of ears, causing scab of small-grain cereals—mostly wheat, triticale, and rye [15,17,27,67].

Kernels damaged by *M. nivale* did not contain any known *Fusarium* metabolite, and their extract did not exhibit toxicity to *Artemia salina* larvae [67].

Examined cultures of *M. nivale* also did not produce toxic metabolites occurring in cultures of *Fusarium* species, pathogenic to cereal heads. Contamination of small-grain cereals with mycotoxins seem to be unlikely when *M. nivale* affects ears. On the other hand, high concentrations of T-2 toxin (2.4 to 13.9 mg/kg) were found in barley kernels after inoculation of 12 cultivars with an isolate of *F. sporotrichiodes* [66]. Besides toxin accumulation, mentioned species caused yield reduction at the level, similar to *F. graminearum*. Low percentage (about 1%) of this species between total *Fusarium* propagules results in usually infrequent occurrence of T-2 toxin in cereal grains [50]. Cultivars of low susceptibility to DON accumulation will be beneficial in reducing the final concentration of mycotoxins in harvested grain.

V. *FUSARIUM* EAR ROT OF MAIZE AND MYCOTOXIN ACCUMULATION IN KERNELS

Maize grain is the most contaminated with *Fusarium* mycotoxins of all agricultural commodities. The number of *Fusarium* metabolites occurring in maize is significantly higher than the number encountered in wheat and other small-grain cereals. It is a consequence of higher number of toxigenic *Fusarium* species, able to infect maize ears [31].

Amounts of *Fusarium* metabolites (like DON, ZON, MON) accumulating in maize kernels are significantly (about 10 times) higher than amounts present in kernels of wheat or barley infected by the same fungal species (Fig. 5). High contamination of maize grain with *Fusarium* mycotoxins is observed when intensity of ear rot is significant and percentage of FDK in a given bulk is high. Since the discovery of fumonisins 10 years ago, FB_1 is stated to be probably the most commonly identified mycotoxin in maize grain in all continents.

Maize ear rot caused by *Fusarium* species is a worldwide problem, with consequent contamination of grain with mycotoxins—trichothecenes, zearalenone, fumonisins, moniliformin, and others [2,15]. Numerous field outbreaks of mycotoxicoses are recorded in the literature as a consequence of livestock feeding with maize grain affected by *Fusarium* [37–39,68–70]. Several forms of the disease are differentiated: pink ear rot, red ear rot, kernel rot. The infection usually spreads from the ear tip down and develops around channels and tunnels made by insects, or injuries made by birds. Rainfall and persistent wetness for at least 48 h shortly after silk emergence promotes development of ear rot epidemics and subsequent accumulation of DON, ZON, and other mycotoxins in kernels [71].

Three zones of visibly distinct kernel rot may be identifiable: (1) severe rot (kernels brown red or pink with no bright yellow layer); (2) light rot (kernels with a portion of bright layer present); (3) healthy looking (no rot, all kernels sound and bright yellow). High amounts of mycotoxins accumulated in kernels are present in zone I of ear with visibly moldy front, moving down the ear when

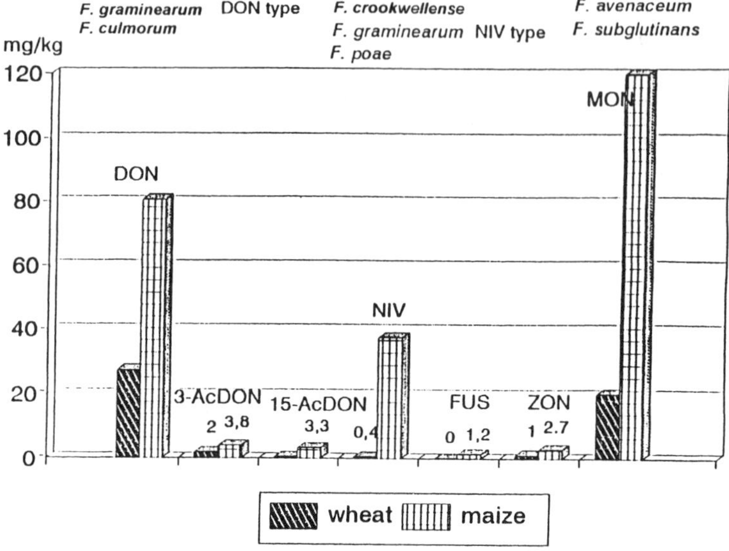

FIGURE 5 Comparison of average concentrations of deoxynivalenol (DON), zearalenone (ZON), and moniliformin (MON) in wheat and maize kernels. AcDON; acetyl deoxynivalenol; NIV, nivalenol; FUS, fusarenone.

initiated at the tip (Table 2). Significantly lower amounts of DON and other mycotoxins are present in zone II. Kernels of zone III are practically free of mycotoxins [72]. Frequently only a few kernels are moldy, in particular when *F. moniliforme* and *F. subglutinans* are the infecting species (causing the kernel rot).

F. graminearum and *F. moniliforme* are the main *Fusarium* species on maize in warmer regions such as the U.S. Corn Belt and California, Australia, South Africa, southern Europe (Italy, Spain, Yugoslavia, Ukraine, Bulgaria, France, Hungary), China, India, southern Japan. On the other hand, *F. culmorum* and *F. subglutinans* prevail in temperature regions like northern Europe (U.K., Denmark, Sweden, Finland, Poland, Netherlands, Germany), northern states of the U.S., the Prairie and Maritime provinces of Canada, and Russia [15,71].

We have observed that *F. crookwellense* is as aggressive as *F. culmorum* and occurs in regions typical for *F. graminearum* and *F. culmorum*. However, FDK of ears infected by *F. crookwellense* accumulated not DON, but NIV up to 42.5 μg/g and fusarenone up to 3.6 μg/g [72,73].

Recently we also found *F. poae* to be contributing to ear rot and contamination of grain with nivalenol. The amounts of nivalenol in FDK of corn reached up to 111.0 μg/g [29,30]. In artificially inoculated ears, NIV accumulated at harvest in amounts up to 23 μg/g and this amount was hybrid-dependent [72]. Gareis et al. [50] in their review report the following maximal amounts of *Fusarium* myco-

TABLE 2 Amounts of Mycotoxins in *Fusarium*-Damaged Kernels of Maize from Ears Naturally Infected or Artificially Inoculated

Infecting species		Amounts of mycotoxins in FDK mg/kg (= μg/g)			References
		DON	ZON	NIV	
F. graminearum	n	77.0[a]	1.0	0	73
	i	62.0–162.3	2.5–4.8	0	72
F. crookwellense	n	0	2.8–6.0	33.2–42.5	74
F. poae	i			1.7–23.2	30
	n	0	0	1.8–32.5	29
	n			0.2–111.0	30
		T-2	HT-2	NEO	
F. sporotrichioides	n	47–992	53–645	0–27.2	32
		FB_1	FB_2		75
F. moniliforme	n	0.9–273.2	0.2–102.6		
	i	5.1–196.0	1.4–75		
		MON		BEA	
F. subglutinans	n	17–425		5–60	41
	n	0.5–8.5		1.8–36.9	42
	n	4.2–399.3 av. 120		nt	4
	i	5.0–155.5		25.6–55.2	76

[a]The amount of DON in sample was highly correlated (r = .94) with percentage of FDK in the sample [73]. DON was usually accompanied by about 10 times lower amounts of 3Ac DON, 15Ac DON, or both.

n, natural infection; i, after inoculation; FDK, *Fusarium*-damaged kernels; DON, deoxynivalenol; ZON, zearalenone; NIV, nivalenol; NEO, neosolaniol; FB, fumonisin B; MON, moniliformin; BEA, beauvericin.

toxins (excluding fumonisins) in maize in European countries: DAS—2.1 mg/kg; T-2—5.0 mg/kg; DON—67.0 mg/kg; ZEA—275.0 mg/kg. Contamination of maize kernels with mycotoxins produced by *Liseola* section *Fusarium* species is characteristic and is not encountered in small-grain cereals [31]. Recent surveys of fumonisin occurrence showed a high frequency and in some regions high contamination of maize grains with FB_1, FB_2, and FB_3 (Table 2).

Visconti et al. [12] found the contamination of maize to be dependent on geographic region. A group of high-contamination countries included Italy, Portugal, Zambia, and Benin with FB_1 + FB_2 incidence at 82% to 100% with contamination levels in the range 1710 to 4450 μg/g. The second group, including Croatia, Poland, and Romania, showed very low incidence and levels of contamination ≤70 μg/g. FDK naturally colonized by *F. moniliforme* contained 0.9 to 273.2 μg FB_1/g and 0.2 to 102.6 μg FB_2/g [74].

Similar amounts of fumonisins were present in FDK from ears inoculated with *F. moniliforme*—5.1 to 196 μg FB_1/g and 1.4 to 75 μg FB_2/g. However, highly susceptible hybrids accumulated in kernels at about 10 times higher concentration of fumonisins than hybrids of lower susceptibility. Fumonisin B_3 was present in all naturally infected maize FDK samples in ratio 0.2 to 0.7 FB_3/FB_2 [75].

Fusarium subglutinans is a species frequently infecting maize ears and replaces *F. moniliforme* in regions of moderate, cooler climate. Kernels colonized by *F. subglutinans* contain at least two toxic metabolites: moniliformin and beauvericin, which were usually co-occurring. Moniliformin has been reported as a mycotoxin occurring worldwide in cereal samples, originating from 11 countries (U.S., France, U.K., South Africa, Thailand, Nigeria, Gambia, Netherlands, Kenya, Malawi, Zimbabwe, and Poland). A U.K. survey of 36 samples of maize products generally showed delectable but low levels of contamination ranging from 0.05 to 0.25 mg/kg. Highest levels of contamination were found in samples from Gambia and South Africa up to 3.16 and 2.72 mg/kg, respectively. FDK from maize samples in Poland contained 4.2 to 399.3 mg MON/kg [4]. Another two surveys of beauvericin and moniliformin in FDK samples proved presence of high levels of beauvericin (BEA) in kernels infected by *F. subglutinans*—1.8 to 36.9 mg/kg in 1993 [42] and 5 to 60 mg/kg in 1991/1992 [41,76].

Isolates of *F. subglutinans*, except two mentioned metabolites MON + BEA, are able to produce a third, toxic metabolite—fusaproliferin. This toxin has not yet been shown to be natural contaminant under field conditions. It is very likely that all three of these metabolites may be present in maize grain bulks and samples as accompanying fumonisins. Schaafsma et al. [77] detected fumonisins in all samples tested, including ears inoculated with *F. graminearum* and *F. subglutinans*, not just with *F. moniliforme*. According to these authors, the severity of ear rot may be a useful indicator of contamination of ears by mycotoxins.

Fusarin C, mycotoxin of mutagenic abilities, is produced under laboratory conditions by numerous *Fusarium* species, in particular high amounts by *F. moniliforme* [78]. This mycotoxin was detected in moldy maize samples from South Africa, U.S., and China [79]. More surveys should be performed to know the extent and significance of this metabolite contamination in cereal grain. All data mentioned above strongly suggest frequent cocontamination of maize kernels with several fusarial mycotoxins.

VI. DISTRIBUTION OF *FUSARIUM* MYCOTOXINS IN GRAIN BULK

Fusarium mycotoxins are distributed unevenly in individual grain samples as well as in grain bulks. Each sample originating from heads with scab symptoms may be separated into two fractions: (1) healthy-looking kernels, and (2) *Fusarium*-damaged (scabbed) kernels [80,81]. In the first fraction the amount of DON and other trichothecenes is about 10 times lower than in FDK fraction. A sample of FDK when was separated using sieves into four fractions: > 2.8, 2.5 to 2.8, 2.2 to 2.5, and < 2.2 mm; about 25% of total sample kernels were found in each fraction. The result documented that FDK, containing DON and other mycotoxins, vary in their size and are present not only in a fraction of small and shriveled kernels [80,81]. Concentration of DON was found to be dependent on the size of kernels in wheat, rye, and barley. Highest amounts of DON were present usually in a fraction of smallest kernels, passing sieve 2.2 mm. However, in some samples the highest level of DON was present in fraction of largest kernels > 2.8 mm [80–83]. DON (and fungal mycelium) is present mostly in pericarp, embryo, and aleurone layer, but also in endosperm [84]. In a bulk of grain containing about 5% of FDK, most of deoxynivalenol present in few percentage of kernels. However, it will not be possible to eliminate contaminated kernels from a bulk. Several papers report only partial possibility to remove DON during grain-cleaning procedure and later milling [85–87]. About 20% of total DON content may be removed from a bulk of grain using cleaning procedure and discarding the fraction < 2.2 mm. The remaining 80% of DON present in grain will remain after cleaning and will pass to cereal products [80,81].

REFERENCES

1. JD Miller. Chematotaxonomy and the toxigenic Fusaria. Abstract of 7th Intern. Congress of Mycology Division, Prague, Czech Rep., July 3–8, 1994, no. MC-10/1.
2. JD Miller, HL Trenholm (eds). Mycotoxins in Grains. St. Paul, MN: Eagan Press, 1994.
3. CF Jelinek, AE Pohland, GE Wood. Worldwide occurrence of mycotoxins in foods and feeds. J Assoc Off Anal Chem 72:223, 1989.
4. M Sharman, J Gilbert, J Chełkowski. A survey of the mycotoxin moniliformin in cereal samples worldwide. Food Addit Contam 4:459, 1991.
5. H Lew, A Adler, W Edinger. Moniliformin and European corn borer (*Ostirinia nubilalis*). Mycotoxin Res 7A:71, 1991.
6. H Lew, J Chełkowski, W Wakuliński, W Edinger. Moniliformin in wheat and triticale grain. Mycotoxin Res 9:66, 1993.
7. P Goliński, J Perkowski, M Kostecki, J Szczesna, J Chełkowski. *Fusarium* species and *Fusarium* toxins in wheat in Poland—comparison with neighbour countries. Sydowia 48:12, 1996.
8. WCA Gelderblom, K Jaskiewicz, WFO Marasas, et al. Fumonisins—novel mycotoxins with cancer-promoting activity produced by *Fusarium moniliforme*. Appl Environ Microbiol 54:1806, 1988.
9. PE Nelson. Taxonomy and biology of *Fusarium moniliforme*. Mycopathologia 117: 29, 1992.

10. GC Bezuidenhout, WCA Gelderblom, CP Gorst-Allman, et al. Structure elucidation of the fumonisins, mycotoxins from *Fusarium moniliforme*. J Chem Soc Commun 743, 1988.
11. A Visconti, MB Doko. Survey of fumonisin production by *Fusarium* isolates from cereals in Europe. AOAC 77:546, 1994.
12. MB Doko, S Rapior, A Visconti, JE Schoth. Incidence and levels of fumonisin contamination in maize genotypes grown in Europe and Africa. J Agric Food Chem 43:429, 1995.
13. A Pittet, V Parisod, M Schellenberg. Occurrence of fumonisins B_1 and B_2 in corn-based products from the Swiss market. J Agric Food Chem 40:1352, 1992.
14. LL Charmley, A Rosenberg, HL Trenholm. Factors responsible for economic losses due to *Fusarium* mycotoxin contamination of grain, foods and feedstuffs. In: JD Miller, HL Trenholm, eds. Mycotoxins in Grains. St. Paul, MN: Eagan Press, 1994, p 471.
15. J Chełkowski (ed). *Fusarium*—Mycotoxins, Taxonomy and Pathogenicity, Amsterdam: Elsevier, 1989, p 63.
16. J Chełkowski. Fungal pathogens influencing cereal seed quality at harvest. In: J Chełkowski, ed. Cereal Grain—Mycotoxins. Fungi and Quality in Drying and Storage. Amsterdam: Elsevier, 1991, p 53.
17. DW Parry, P Jenkinson, L McLead. *Fusarium* ear blight (scab) in small grain cereals—a review. Plant Pathol 44:207, 1995.
18. PE Nelson, TA Toussoun, WHO Marasas. *Fusarium* Species: An Illustrated Manual for Identification. University Park: Pennsylvania State University Press, 1983.
19. HI Nirenberg. Identification of *Fusarium* occurring in Europe on cereals and potatoes. In: J Chełkowski, ed. *Fusarium*—Mycotoxins, Taxonomy and Pathogenicity. Amsterdam: Elsevier, 1989, p 179.
20. Y Sugiura, Y Watanabe, T Tanaka, S Yamamoto, Y Ueno. Occurrence of *Gibberella zeae* strains that produce both nivalenol and deoxynivalenol. Appl Environ Microbiol 56:3047, 1990.
21. M Mańka, J Chełkowski, D Brayford, A Visconti, H Kwaśna, J Perkowski. *Fusarium graminearum* Schwabe—cultural characteristics, pathogenicity towards cereal seedlings and ability to produce mycotoxins. J Phytopathol 124:143, 1989.
22. JD Miller, R Greenhalgh, YZ Wang, M Lu. Trichothecene chemotypes of three *Fusarium* species. Mycologia 83:121, 1991.
23. CJ Mirocha, HK Abbas, CE Windels, W Xie. Variation in deoxynivalenol, 15 acetylodeoxynivalenol, 3 acetylodeoxynivalenol and zearalenone production by *Fusarium graminearum* isolates. Appl Environ Microbiol 55:1315, 1989.
24. DR Lauren, T Sharon, ST Sayer, ME di Menna. Trichothecene production by *Fusarium* species isolated from grain and pasture throughout New Zealand. Mycopathologia 120:167, 1992.
25. CJ Mirocha, W Zie, Y Xu, et al. Production of trichothecene mycotoxins by *Fusarium graminearum* and *F. culmorum* on barley and wheat. Mycopatologia 12:19, 1994.
26. A Bottalico, A Logrieco, A Visconti. *Fusarium* species and their mycotoxins in infected cereals in the field and in stored grains. In: J Chełkowski, ed. *Fusarium*—Mycotoxins, Taxonomy and Pathogenicity. Amsterdam: Elsevier, 1989, p 85.

27. JD Miller. Epidemiology of *Fusarium* ear diseases of cereals. In: JD Miller, HL Trenholm, eds. Mycotoxins in Grain. St. Paul, MN: Eagan Press, 1994, p 19.
28. H Petterson. Nivalenol production by *Fusarium poae*. Mycotoxin Res 7A:26, 1991.
29. J Chełkowski, H Lew, H Petterson. *Fusarium poae* (Peck) Wollenw—occurrence in maize ears, nivalenol production and mycotoxin accumulation in cobs. Mycotoxin Res 10:116, 1994.
30. H Lew, J Chełkowski, M Prończuk, H Wiśniewska. Nivalenol and fusarenone X in kernels of maize hybrids inoculated with *Fusarium poae* (Peck) Wollenw. Genet Pol 35B:203, 1994.
31. J Chełkowski, M Prończuk, M Prończuk. Observation on *Fusarium* ear rot in maize and related *Fusarium* species in Poland. Hod Rosl Aklim Nasien 37(4):41, 1993.
32. J Chełkowski, H Kwaśna, P Zajkowski, A Visconti, A Bottalico. *Fusarium sporotrichioides* Sherb. and trichothecenes associated with *Fusarium* ear rot of corn before harvest. Mycotoxin Res 3:111, 1987.
33. DC Foley. Systemic infection of corn by *Fusarium moniliforme*. Phytopathology 68: 1331, 1962.
34. EW Sydenham, GS Shephard, PG Thiel, WHO Marasas, S Stockenstrom. Fumonisin contamination of commercial corn-based human foodstuffs. J Agric Food Chem 39: 2014, 1991.
35. PG Thiel, WHO Marasas, EW Sydenham, et al. Survey of fumonisin production by *Fusarium* species. Appl Environ Microbiol 57:1089, 1991.
36. PE Nelson, RD Plattner, DD Shackelford, AE Desjardins. Production of fumonisins by *Fusarium moniliforme* strains from various substrates and geographic areas. Appl Environ Microbiol 57:2410, 1991.
37. PG Thiel, GS Shepherd, EW Sydenham, WHO Marasas, PE Nelson, TM Wilson. Levels of fumonisin B_1 and B_2 in feeds associated with confirmed cases of equine leukoencephalomalacia. J Agric Food Chem 39:109, 1991.
38. PF Ross, PE Nelson, JL Richard, et al. Production of fumonisins by *Fusarium moniliforme* and *Fusarium proliferatum* isolates associated with equine leukoencephalomalacia and a pulmonary edema syndrome in swine. Appl Environ Microbiol 56: 3225, 1991.
39. PF Ross, LG Rice, RD Plattner, et al. Concentration of fumonisins B_1 in feeds associated with animal health problems. Mycopathologia 114:129, 1991.
40. J Chełkowski, H Lew. *Fusarium* species of Liseola section—occurrence in cereals and ability to produce fumonisins. Microbiologie-Aliments-Nutrition 10:49, 1992.
41. A Logrieco, A Moretti, A Ritieni, et al. Natural occurrence of beuvericin in preharvest *Fusarium subglutinans* infected corn ears in Poland. J Agric Food Chem 41:2149, 1993.
42. M Kostecki, J Szczęsna, J Chełkowski, H Wiśniewska. Beauvericin and moniliformin production by Polish isolates of *Fusarium subglutinans* and natural occurrence of both mycotoxins in maize samples. Microbiologie-Aliments-Nutrition 13:67, 1995.
43. A Ritieni, V Fogliano, G Randazzo, et al. Isolation and characterization of fusaproliferin, a new toxic metabolite from *Fusarium proliferatum*. Natural Toxins 3:17, 1995.
44. A Moretti, A Logrieco, A Bottalico, A Ritieni, G Randazzo, P Corda. Beauvericin

production by *Fusarium subglutinans* from different geographical areas. Mycol Res 99:282, 1995.
45. A Logrieco, A Moretti, C Altomare, A Bottalico, EC Torres. Occurrence and toxicity of *Fusarium subglutinans* from Peruvian maize. Mycopathologia 122:185, 1993.
46. J Chełkowski. Significance of *Fusarium* metabolites in interaction between cereal plant and a pathogen. The review. Genet Pol 35B:137, 1994.
47. J Perkowski, RD Plattner, P Goliński, RF Vesonder, J Chełkowski. Natural occurrence of deoxynivalenol, 3 acetyldeoxynivalenol, 15 acetyldeoxynivalenol, 4,7 dideoxynivalenol and zearalenone in Polish wheat. Mycotoxin Res 6:7, 1990.
48. T Tanaka, A Hasegava, S Yamamoto, US Lee, Y Sugiura, Y Ueno. Worldwide contamination of cereals by *Fusarium* mycotoxins nivalenol, deoxynivalenol and zearalenone. I. Survey of 19 countries. J Agric Food Chem 36:979, 1988.
49. P Krogh, DH Christensen, B Hald, C Larsen, EJ Pedersen, U Thrane. Natural occurrence of the mycotoxin fusarochromanone, a metabolite of *Fusarium equiseti*, in cereal feed associated with tibial dyschondroplasia. Appl Environ Microbiol 55:3184, 1989.
50. M Gareis, J Bauer, C Enders, B Gedek. Contamination of cereals and feed with *Fusarium* mycotoxins in European countries. In: J Chełkowski, ed. *Fusarium*—Mycotoxins, Taxonomy and Pathogenicity. Amsterdam: Elsevier, 1989, p 44.
51. DR Lauren, MP Agnew, WA Smith, ST Sayer. A survey of the natural occurrence of *Fusarium* mycotoxins in cereals grown in New Zealand in 1986–1989. Food Addit Contam 8:599, 1991.
52. JC Sutton. Epidemiology of wheat head blight and maize ear rot caused by *Fusarium graminearum*. Can J Plant Pathol 4:195, 1981.
53. G Bai, G Shaner. Scab of wheat: prospects for control. Plant Dis 78:760, 1994.
54. CHA Snijders. *Fusarium* head blight and mycotoxin contamination of wheat. Neth J Plant Pathol 96:187, 1990.
55. RP Polley, JA Turner. Surveys of stem base diseases and *Fusarium* ear diseases in winter wheat in England, Wales and Scotland. Ann Appl Biol 126:45, 1995.
56. LSL Wong, A Tekauz, D Leisle, D Abramson, RIH McKenzie. Prevalence, distribution and importance of *Fusarium* head blight in wheat in Manitoba. Can J Plant Pathol 14:233, 1992.
57. J Perkowski, J Chełkowski, P Blażczak, CHA Snijders, W Wakuliński. A study of the correlation between the amount of deoxnivalenol in grain of wheat and triticale and percentage of *Fusarium* damaged kernels. Mycotoxin Res 7A:115, 1991.
58. CHA Snijders, J Perkowski. Effects of head blight caused by *Fusarium culmorum* on toxin content and weight of wheat kernels. Phytopathology 80:566, 1990.
59. LSL Wong, D Abramson, A Tekauz, D Leisle, RIH McKenzie. Pathogenicity and mycotoxin production of *Fusarium* species causing head blight in wheat cultivars varying in resistance. Can J Plant Sci 75:261, 1995.
60. CHA Snijders, CF Krechting. Inhibition of deoxynivalenol translocation and fungal colonization in *Fusarium* head blight resistant wheat. Can J Bot 70:1570, 1992.
61. JM Beardall, JD Miller. Diseases in humans with mycotoxins as possible causes. In: JD Miller, HI Trenholm, eds. Mycotoxins in Grain. St. Paul, MN: Eagan Press, 1994, p 471.
62. AH Teich. Epidemiology of wheat (*Triticum aestivum* L.) scab caused by *Fusarium*

spp. In: J Chełkowski, ed. *Fusarium*—Mycotoxins, Taxonomy and Pathogenicity. Amsterdam: Elsevier, 1989, p 269.

63. W Langseth, R Hie, M Gullord. The influence of cultivars, location and climate on deoxynivalenol contamination in Norwegian oats 1985–1990. Acta Agric Scand Sect B Soil Plant Sci 45:63, 1995.
64. W Langeth, personal information.
65. J Perkowski, T Miedaner, HH Geiger, HM Müller, J Chełkowski. Occurrence of deoxynivalenol (DON), 3 acetylDON, zearalenone and ergosterol in winter rye inoculated with *Fusarium culmorum*. Cereal Chem 72:205, 1995.
66. J Perkowski, I Kiecana, J Chełkowski. Susceptibility of barley cultivars and lines to *Fusarium* infection and mycotoxins accumulation in kernels. J Phytopathol 143:491, 1995.
67. J Chełkowski, P Goliński, J Perkowski, A Visconti, M Rakowska, W Wakuliński. Toxigenicity of *Microdochium nivale* (*Fusarium nivale*) isolates from cereals in Poland. Mycotoxin Res 7A:140, 1991.
68. M Schuh, E Glawisching, J Leibetseder. Natural occurrence of mycotoxicoses in Austria associated with mycotoxin-contaminated feedstuff. Proc V Int IUPAC Symp: Mycotoxin and Phycotoxins. Vienna: Technical University, 1982, pp 118–121.
69. M Schuh. Clinical cases of *Fusarium* toxicoses in Austrian domestic animals in connection with *Fusarium* toxins contaminated feedstuffs. In: Mycotoxin Research, Sp. Edition, European Seminar: *Fusarium*, Mycotoxins, Taxonomy and Pathogenicity. Warsaw, September 8–10, 1987, pp 69–73.
70. WFO Marasas, PE Nelson. Mycotoxicology. University Park: Pennsylvania State University Press, 1986.
71. AH Teich. Epidemiology of corn (*Zea mays* L.) ear rot caused by *Fusarium* spp. In: J Chełkowski, ed. *Fusarium*—Mycotoxins, Taxonomy and Pathogenicity. Amsterdam: Elsevier, 1989, p 297.
72. GE Bennett, DT Wicklow, RW Coldwell, EB Smalley. Distribution of trichothecenes and zearalenone in *Fusarium graminearum* rotted corn ears in controlled environment. J Agric Food Chem 36:639, 1988.
73. J Perkowski, J Chełkowski, RD Plattner, P Goliński. Cumulation of mycotoxins in maize cobs infected with *Fusarium graminearum*. Mycotoxin Res 7A:115, 1991.
74. A Visconti, J Chełkowski, M Solfrizzo, A Bottalico. Mycotoxins in corn ears naturally infected with *Fusarium graminearum* and *F. crookwellense*. Can J Plant Pathol 12:187, 1990.
75. M Pascale, A Visconti, M Prończuk, H Wisniewska, J Chełkowski. Accumulation of fumonisins in maize hybrids inoculated under field conditions with *Fusarium moniliforme* Sheldon. J Sci Food Agric 74:1, 1997.
76. M Kostecki, A Logrieco, J Szczęsna, et al. Susceptibility of maize hybrids to infection by *Fusarium subglutinans* and accumulation mycotoxins *moniliformin* and *beauvericin* in cobs. Genet Pol 35B:23, 1994.
77. AW Schaafsma, JD Miller ME Savard, RJ Ewing. Ear rot development and mycotoxin production in corn in relation to inoculation method, corn hybrid and species of *Fusarium*. Can J Plant Pathol 155:185, 1993.
78. P Goliński, J Chełkowski. Biosynthesis of fusarin C by *Fusarium* species. Microbiologie-Aliments-Nutrition 10:55, 1992.

79. JM Farber, PM Scott. Fusarin C. In: J Chełkowski, ed. *Fusarium*—Mycotoxins, Taxonomy and Pathogenicity. Amsterdam: Elsevier, 1989, p 41.
80. J Chełkowski, J Perkowski. Mycotoxins in cereal grain. Part 15. Distribution of deoxynivalenol in naturally contaminated wheat samples. Mycotoxin Res 8:27, 1992.
81. J Perkowski, J Chełkowski. Deoxynivalenol content dependence on kernel size in wheat grain naturally infected by *Fusarium*. Microbiologie-Ailments-Nutrition 9: 217, 1991.
82. J Perkowski, T Mieadaner. Association among deoxynivalenol, 3-acetylodeoxynivalenol, ergosterol content and kernel size in winter rye grain infected by *Fusarium culmorum*. Genet Pol 35B:317, 1994.
83. J Perkowski. Distribution of deoxynivalenol in barley kernels infected by *Fusarium*. Book of Abstracts, IV Int Sem *Fusarium*—Mycotoxins, Taxonomy and Pathogenicity. Martina Franca, Italy, May 9–13, 1995, p 17.
84. J Chełkowski, A Cierniewska, W Wakuliński. Mycotoxins in cereal grain. Part 14. Histochemical examination of *Fusarium* damaged wheat kernels. Die Nahrung 34: 357, 1990.
85. PM Scott, SR Kanhere, JE Dexter, PW Brennan, HL Trenholm. Distribution of trichothecene mycotoxin deoxynivalenol (vomitoxin) during milling of naturally contaminated hard red spring wheat and its fate in baked products. Food Addit Contam 1:313, 1984.
86. HK Abbas, CJ Mirocha, RJ Pawlowski, DJ Pusch. Effect of cleaning, milling and baking on deoxynivalenol in wheat. Appl Environ Microbiol 50:482, 1985.
87. AL Patey, J Gilbert. Fate of *Fusarium* mycotoxins in cereals during food processing and methods for their detoxification. In: J Chełkowski, ed. *Fusarium*—Mycotoxins, Taxonomy and Pathogenicity. Amsterdam: Elsevier, 1989, p 399.

3
Toxigenic *Alternaria* Species of Economic Importance

Antonio Bottalico
Institute of Plant Pathology, University of Sassari, Sassari, Italy

Antonio Logrieco
Institute of Toxins and Mycotoxins of Plant Parasites, Italian National Research Council, Bari, Italy

I. INTRODUCTION

The genus *Alternaria* is ubiquitous and abundant in the atmosphere as well as in soil, in seeds, and in agricultural commodities. It includes both plant pathogenic and saprophytic species that may damage crops in the field or cause postharvest decay of plant products in storage.

Species of *Alternaria* are known to produce many metabolites, mostly phytotoxins, which play an important role in the pathogenesis of plants. But certain species, in particular *A. alternata* (Fr.) Keissler, are capable of producing several toxic metabolites in infected plants and/or in agricultural commodities, which can contaminate foods and feeds and elicit adverse effects in animals.

A number of reviews on *Alternaria* species and their toxins dealing with different aspects such as taxonomy and ecology [1–4]; epidemiology and plant diseases [5,6]; phytotoxic effects and host-specific toxins [7,8]; mycotoxin biosynthesis [9]; chemical properties and analytical methods [10–12]; assessment of food contamination [13–15]; and toxicity [16–18] have been published recently.

The most relevant aspects of the biology, ecology, and toxicology of *Alternaria* species of economic importance are reviewed, with particular emphasis on *A. alternata* and on the most important *Alternaria* mycotoxins, including tenuazonic acid (TA), alternariol (AOH), alternariol monomethyl ether (AME), altenuene (ALT), and altertoxin I, II, and III (ATX-I, ATX-II, ATX-III).

II. TAXONOMY AND NOMENCLATURE

The genus *Alternaria*, established by Nees in 1816, is a phaeodictyosporic fungal genus belonging to the *Dematiaceae* family, order Hyphomycetales, class Hyphomycetes, of the Mitosporic fungi (= *Deuteromycotina*). Some representatives of *Alternaria* were reported as anamorphic state of *Pleospora* Rabenh. ex Ces De Not (1863), but in 1986 were recognized as *Lewia* E.M. Barr & E.G. Simmons teleomorphic state. The genus *Alternaria* is usually characterized by gray, dark blackish brown, or black colonies with a felty mycelium. The conidial shape is usually ovate or obclavate with a final beak, and they are mostly proliferated in acropetal chains or solitary. They are septate and muriform and are developed at a restricted site at the apex of distinctive pale brown or brown conidiophores.

Although *Alternaria* morphological traits are often variable and easily degenerate in laboratory conditions, they are still considered the main keys for taxonomic *Alternaria* identification, based on the presence or absence of a beak, conidial and beak shape, and conidial arrangement (solitary or in chain, branched). Three different sections of *Alternaria* were developed by Neergard [19] on the basis of distinguishing characteristics of conidial chain formation:

1. Longicatenatae, characterized by long chains of about 10 or more conidia
2. Brevicatenatae, with short chains of about three to five conidia
3. Noncatenatae, with conidia formed singly and only seldom producing a secondary conidium.

Although various morphoclassification systems are available [1,20–22], they often prove to be incongruent with phytopathological and toxicological observations. For instance, specific populations of *Alternaria* from tomato, apple, pear, citrus, and strawberry, producing different host-specific toxins, were named both as distinct species and as different *formae specialis* or pathotypes of *A. alternata* [7,8]. On the other hand, two populations of *Alternaria* isolated from mandarin fruit rot, showing some peculiarities in the fungal colony but classified as *A. alternata*, displayed significant differences both in postharvest pathology and in toxin production [23]. Therefore, molecular systematic and evolutionary studies are urgently needed for a better knowledge of this genus and other closely related fungi.

The genera most likely to present problems of identification with *Alternaria* are *Ulocladium* and *Stemphylium*. But *Ulocladium*, with the same holoblastic sympodial proliferation, and *Stemphylium*, with apically percurrent enteroblastic proliferation, lack the conidial beak and rarely form conidia in chains.

Key to Toxigenic *Alternaria* Species

Key of the species	(Preferential host)

1—Conidia solitary
  1a—Conidia beakless
    —*A. helianthi* (Hansford) Tubaki & Nishihara (*Helianthus annuus* L.)
    —*A. radicina* Meier, Drechsler & E. Rocz (*Daucus* spp.)
  1b—Conidia beaked (beak taper narrow to filiform)
    —*A. bataticola* Ikada (*Ipomea* spp.)
    —*A. brassicae* (Berk.) Sacc. (Cruciferae: as *Brassica* spp., *Raphanus* spp.)
    —*A. chartami* Chowdhury (*Carthamus* spp.)
    —*A. cucumerina* (Ellis & Everh) Elliott (Cucurbitaceae: as *Cucumis* spp., *Cucurbita* spp.)
    —*A. dauci* (Kühn) Groves & Skolko (Umbelliferae: *Daucus* spp., *Petroselinum* spp.)
    —*A. macrospora* Zimmerman (*Gossypium* spp.)
    —*A. porri* (Ellis) Ciferri (Liliaceae: as *Allium* spp.)
    —*A. solani* Sorauer (Solanaceae: as *Capsicum* spp., *Lycopersicon* spp., *Solanum* spp.)
    —*A. sesami* (Kawamura) Mohanty & Behera (*Sesamum* spp.)
    —*A. zinniae* M.B. Ellis (*Zinnia* spp.)
2—Conidia catenulatae
  2a—Conidia in short chains (two to five conidia)
    —*A. dianthi* Stevens & Hall (*Dianthus* spp.)
    —*A. japonica* Yoshii (*Brassica* spp., *Matthiola* spp., *Raphanus* spp.)
    —*A. raphani* Groves & Skolko (Cruciferae: as *Brassica* spp., *Matthiola* spp., *Raphanus* spp.)
    —*A. sesamicola* Kawamara (*Sesamum* spp.)
    —*A. tenuissima* (Kunze ex Pers) Wiltshire (no preferential host)
    —*A. triticina* Prasada & Prabhu (*Triticum aestivum* L.) (wheat)
  2b—Conidia in long chains (more than five conidia)
    —*A. alternata* (Fr.) Keissler (= *A. tenuis* Auct) (no preferential host)
      —*A. alternata* f. sp. *lycopersici* (*Lycopersicon esculentum* Mill) (tomato)
    —*A. brassicicola* (Schw:) Wiltshire (Cruciferae: as *Brassica* spp., *Matthiola* spp., *Raphanus* spp.)
    —*A. citri* Ellis & Pierce (*Citrus* spp.)
      —*A. citri* (= *A. alternata* Rough lemon pathotype) (Rough lemon, Rangpure lime)
      —*A. citri* (= *A. alternata* tangerine pathotype) (Emperor mandarin)
    —*A. fragariae* Dingley
      —*A. fragariae* (= *A. alternata* strawberry pathotype) (strawberry)

Key of the species	(Preferential host)
2—Conidia catenulatae; 2b—Conidia in long chains (cont.)	
—*A. kikuchiana* Tanaka (*Pyrus communis* L.)	
—*A. kikuchiana* (= *A. alternata* Japanese pear pathotype) (pear)	
—*A. longipes* (Ellis & Everhart) Mason (*Nicotiana tabacum* L.)	
—*A. longipes* (= *A. alternata* tobacco pathotype) (tobacco)	
—*A. longissima* Deighton & Macgarvie (*Sesamum* spp.)	
—*A. mali* Roberts (*Pyrus malus* L.)	
—*A. mali* (= *A. alternata* apple pathotype) (apple)	
—*A. tomato* (Cooke) Jones (*Lycopersicon esculentum* Mill) (tomato)	
—*A. capsici-annui* Savul & Sandu (*Capsicum annuum* L.) (pepper)	

III. BIOLOGY AND PLANT DISEASES

A. Plant Diseases by *Alternaria*

Alternaria contains numerous species that are both saprophytic on organic materials and are pathogenic on many plants, including several plants used for food. Mostly *Alternaria* species are specific pathogens to one or a few plants and are usually identified according to their host. However, a few species are cosmopolitan, very frequently seedborne and airborne, including spoilage isolates of freshly harvested crops. Among these few ubiquitous, agriculturally significant species, the most common species is *A. alternata* [20,24].

Species of *Alternaria* affect primarily the leaves, stems, flowers, ears, and fruits of annual plants, in particular vegetables and ornamentals, but also trees such as citrus, pear, and apple. *Alternaria* diseases usually appear as spots on leaves, stems, fruits, sometimes surrounded by a discoloured halo, as tuber and fruit rots, or as black heads of cereal plants; but signs of disease may also be seedling damping off, collar rot, and blights of several kinds of crop plants. During the colonization of the hosts, the pathogen can produce toxic metabolites in the infected plants that can play a role in the pathogenesis of plants (phytotoxins); when they accumulate in plant part used as food or feed, they can be a hazard to human and animal health (mycotoxins).

Some of the most common *Alternaria* species causing diseases on crops are listed below:

A. alternata (= *A. tenuis*) is a very common saprophyte and opportunistic parasite found on a wide range of hosts, associated with leaf, stem, and fruit spots; fruit rots, black heads of cereal plants. *A. alternata* may also be found in substrata such as foods, feeds, and rotting materials. Together with various groups of true saprophytes and opportunistic pathogens, this species also includes several virulent host-specific pathotypes.

The virulence of the several pathotypes of *A. alternata* to different host plants does not appear to be correlated with any morphological traits [25], but can be explained as an evolutionary adaptation based on the ability to produce host-specific toxins. As opposed to the host-specific pathotypes, the facultative or opportunistic pathogenic strains may affect fruits and senile or stressed tissues, especially close to pre- and postharvest time, under favorable conditions.

A. alternata f. sp. *lycopersici* represents a race or a pathotype of *A. alternata* specialized on tomato, causing stem canker disease on only a few cultivars of tomato, and producing host-specific toxins named ALL-toxins (AL-toxins) (T_A, T_B, T_C, T_D, and T_E) [7,8,26].

A. bataticola, causing black spots on leaves, petioles and vines of sweet potato.

A. brassicae, *A. brassicicola*, *A. japonica* (= *A. raphani*), depending on hosts and geographical areas, may be either significant or mild pathogens on cruciferous plants, causing seedling damping off, and pale brown to dark brown circular and zonate leaf and pod spots on cabbage, radish, turnip, rape, broccoli, cauliflower. *A. brassicae* includes a race of pathotype which produces a toxin with host selectivity at the species level, characterized by the same structure as destructin-B [27].

A. capsici-annui, causing black spots on pepper.

A. chartami, causing leaf spot on safflower.

A. citri, growing on lemons, oranges, and other *Citrus* fruits. This species, also classified as *A. alternata*, includes two pathotypes, *A. alternata* Rough lemon pathotype, the causal agent of brown leaf spots on young leaves only on Rough lemon and Rangpure lime, producers of ACR host-specific toxins (four compounds, from ACR-I to ACR-IV); and *A. alternata* tangerine pathotype, causing brown leaf spots on the young leaves of only Dancy tangerine and Emperor mandarin, and producing two groups of host-specific toxins—ACTG-toxins (six compounds, from ACTG-toxinA to ACTG-toxin F), and ACT-toxins (two compounds, ACT-toxin I and ACT-toxin II) [7,8].

A. cucumerina, causing leaf spot and blight on cucurbits, cantaloupes, and other melons.

A. dauci, and *A. radicina*, pathogenic to carrot seeds and plants, causing leaf blight, black rot of roots and seedlings, and damping off of seedlings.

A. dianthi, causing leaf spots, necrosis on stems, and blights of carnation.

A. fragariae, formerly *A. alternata* (= *A. tenuis*), includes *A. alternata* f. sp. *fragariae* Dingley [28], and *A. alternata* strawberry pathotype, the causal agent of black spot disease on strawberry, producing host-specific toxins named AF-toxins (AF-toxins I, II, and III) [7,8].

A. helianthi, causing leaf spots, defoliation, and blights of sunflower.

A. japonica, causing leaf spots on various cruciferous plants such as radish and Chinese cabbage.

A. kikuchiana, now known as *A. alternata* Japanese pear pathotype, is the causal agent of the black spot disease of the very few susceptible Japanese pear cultivars (Nijisseiki cvs), producing host-specific toxins (AK-I and AK-II) [7,8].

A. longipes, formerly *A. alternata* (= *A. tenuis*), includes *A. alternata* tobacco pathotype, the causal agent of brown spot disease on tobacco, producing host-specific toxin (AT-toxin) [7,8].

A. longissima, a seedborne pathogen causing zonate leaf spot, stem necrosis, and blights of sesame.

A. mali, formerly *A. alternata* (= *A. tenuis*), includes *A. alternata* apple pathotype, the causal agent of *Alternaria* blotch on apple, producing host-specific toxins (AM-I, AM-II and AM-III) [7,8].

A. macrospora, causing necrotic spots surrounded by a purplish halo on the leaves and balls of cotton.

A. porri, causing purple leaf blotch on onion, leek, and garlic.

A. sesami and *A. sesamicola* Kawamura, seedborne pathogens, causing seedling damping off, dark brown round or irregular leaf spots, necrotic lesions on stems and capsules, and blights of sesame.

A. solani, a well-known pathogen causing leaf spots, stem necrosis, fruit and tuber rots, and blights in potato, tomato, eggplant, and pepper.

A. tenuissima, causes leaf spots and blights on a wide range of hosts, both dicots and monocots. This species is very common and has been recorded on a wide range of plants, where it usually occurs as a secondary invader [29,30]. In India it is the second most common species of *Alternaria* after *A. alternata* [31]. *A. tenuissima* has not been frequently recorded on food, but its worldwide distribution allows us to presume a more relevant importance as a food spoilage agent, especially for cereals [24]. *A. tenuissima* has been reported on wheat, barley, and rye [32,33], and has been isolated from tomato fruits affected by blackmold at ripening in the field [34]. For its vegetative growth *A. tenuissima* requires a minimum water content of 0.85 a_w and a pH of about 5, whereas the optimum temperature for conidium germination is 28°C and the germination is stimulated by light [35–37].

A. tomato, causing leaf spots or nail-head spots on tomato.

A. triticina, causing leaf spots and blights on monocots, particularly on wheat and triticale (wheat/rye hybrid) [38].

A. zinniae, causing spots on leaves, stems, and flowers, and blights of many Compositae, especially on *Zinnia*.

B. Biology of *A. alternata*

Distinguishing environmental requirements and detailed information on the natural occurrence of the cosmopolitan and very multiform species *A. alternata*, which appears to be the only *Alternaria* species of particular concern from the mycotoxicological point of view, are described below. In fact, most *A. alternata* forms are saprophytes on several substrata, and are potential spoilage agents of fresh agricultural commodities, including foodstuffs and feedstuffs and can accumulate toxic metabolites in colonized products. Other *A. alternata* forms are low in virulence except on senile or stressed tissues, and are able to colonize fruits and other edible plant parts during ripening, forming toxic metabolites in the colonized tissues that can be toxic to man and animals. Moreover, toxic metabolites can be produced and accumulated in infected fruits, vegetables, and other edible plant parts also by facultative or opportunistic pathogenic forms of *A. alternata*, as well as by other *Alternaria* species. Finally, metabolites toxic to animals such as TA and alternariols (AOH, AME) can also be produced and accumulated in infected edible tissues, besides the host-specific toxins, by the several races or pathotypes of host-selective *A. alternata*.

1. Effect of Environment on Growth

Optimum growth for *A. alternata* is near 25°C, with a range of 22 to 30°C; the maximum temperature for growth is near 36°C; and the minimum is frequently reported in the range of 2.5 to 6.5°C, except for isolates adapted to the cold with minimum temperature requirements of 0°C, −2°C, or −5°C, respectively. However, no growth of *Alternaria* occurs at ⩾ 36°C and −5°C [30,39–42].

For vegetative growth *A. alternata* requires a minimum water content of 0.84 to 0.88 a_w; but optimum growth is at 0.98 to 1.00 a_w [43].

The optimum pH for the growth of *A. alternata* ranges from 4 to 6.6, but growth is possible between pH 2.7 and 8.0 [39,44,45].

The growth of *A. alternata* is linearly reduced by decreasing concentration of O_2, independent of the a_w of the medium, but vegetative growth is increasing with CO_2 concentration at least at some a_w. No growth was observed in a total absence of O_2, while at 0.25% O_2 in the atmosphere (v/v), growth was reduced to a quarter, and above 21%, O_2 concentration and growth response was linear. The maximum ratio of CO_2:O_2 tolerated by *A. alternata* is 3:1 [46–49].

2. Effect of Environment on Sporulation

The temperature range allowing sporulation appeared to be narrower than that for vegetative growth. Thus, optimal sporulation occurs at 25 to 27°C, with a minimum at 15°C, and a maximum at 33°C [50,51].

In general it appears that *Alternaria* species require more water for sporulation than for growth, and a minimum of 0.90 a_w is necessary for *A. alternata* [50]. But a heavy and profuse conidium discharge was observed when sporulating substrates were transferred from humid to dry conditions [52].

No sporulation was observed in absence of O_2, but a concentration of 0.14% O_2 was enough to allow spores to form, and concentrations of CO_2 up to 15% were ineffective on sporulation rates [47,50].

Sporulation in many *Alternaria* species, including *A. alternata*, may be induced by sunlight and near-ultraviolet light under alternating periods of illumination, with darkness being more favorable to spore release. Thus, in *A. alternata*, sporulation was better in the dark or at night than in full white light [53,54], stimulated under an infrared light spectrum [53,55,56], and was particularly improved by the near-UV range [57].

3. Effect of Environment on Spore Germination

The temperature range required by *A. alternata* for spore germination appeared to be narrower than that for vegetative growth. Thus, optimal sporulation occurs at 25 to 30°C, with a minimum of <5°C, and a maximum near 35°C [58,59].

Conidia germination of most *Alternaria* species is very sensitive to relative humidity or to a_w of the medium, and water requirement may interact with temperature. Only at 1.00 a_w does conidia germination of *A. alternata* appear insensitive to temperature change. But the range of a_w allowing germination increases as incubation temperatures depart from the optimum. In particular, the minimum water requirement, which is 0.84 to 0.85 a_w under optimal temperatures, rapidly increases at lower temperatures with 0.94 a_w at 15°C than at higher ones with 0.89 a_w [58–60].

Reducing pH from 6.5 to 4.0 increased both the minimum a_w allowing germination and the time taken for the germination, with a more pronounced influence at marginal temperatures [59,60].

Decreasing O_2 concentration from 21% to 0.14% increased the duration of the lag phase for growth initiation in *A. alternata* at all a_w tested. Increasing CO_2 concentrations from 0% to 15% increased the lag phase at all a_w and temperatures tested [49].

In storage under dry conditions, more than 40% of the conidia can survive for almost 1 year, and some remained viable for more than 10 years on dried wheat kernel. However, the amount of viable conidia dropped considerably during the first year of storage [61,62].

4. Occurrence on Food and Food Products

A. alternata has also been reported under the name of *A. tenuis* on a wide variety of foods and food products and may possibly have been confused with other pathogenic *Alternaria* species, such as spoilage agents of vegetables and fruits during both the pre- and postharvest periods. Furthermore, *A. alternata* has been frequently recorded on a wide range of cereals, both at preharvest as black heads agent or postharvest during storage, as spoilage or biodeteriogen agents.

A. alternata is extremely common on fruits and vegetables, on which it causes molding and rotting, either separately or in association with other fungi belonging to *Aspergillus*, *Botrytis*, *Cladosporium*, *Fusarium*, *Penicillium*, and *Geotrichum*, starting at the harvest time and continuing during handling and storage. Under favorable conditions, *A. alternata* can be responsible for considerable economic losses.

The colonization of seed material as well as of plants and plant products is of particular importance, and extensive lists have been compiled [19,29,63–65].

The reports on the occurrence of *A. alternata* in fresh or stored fruits are particularly numerous: *A. alternata* was isolated from molded grapes [66]; pears [67]; apples [68–70]; lemons [71]; mandarin [23]; blueberries [66]; and olives [72]. From mandarin affected by heart rot two strains were isolated, named gray and black, with different morphologies, postharvest pathologies, and toxigenicity, indicating some substantial variations within the *A. alternata* population [23]. Preliminary observations on the mycoflora of ripe olive fruits, especially on samples collected at harvest time from the ground under the olive trees after the natural falling of fruits, or obtained from retail markets or oil mills before processing, showed the presence of *Alternaria* isolates, mostly *A. alternata* [73].

A. alternata was very frequently found in ripe vegetables at harvest time, or close to postharvest time during handling and transport, in processing plant products waiting to be processed, or during storage also at low temperatures [19,24,39,74]. In particular, *A. alternata* was recorded on fresh vegetables [75]; potatoes [76]; tomatoes [76–79]; peppers [79]; chilli [80]; carrot [81]; melons [79]; bean, pea, and faba bean [76]; and soybeans [82].

Of particular concern are the attacks of *A. alternata* to tomato fruits during ripening [51,74,83]. From a survey carried out in Apulia, Italy, during the 1985 crop season, *A. alternata* turned out to be the main fungal species occurring on tomatoes affected by blackmold in the field, and a rather low incidence of *A. tenuissima* was also observed. The disease was characterized by flattened or slightly sunken fruit lesions, varying in size from 2–3 mm diameter to one-third or more fruit surface, and mycelium often developed in the internal tissues invading either the pericarp or the placenta [74].

Among stored products there are important reports concerning small grains [30,80–85]. In this regard, *A. alternata* (= *A. tenuis*) has been frequently recorded on a wide range of cereals, particularly on barley [32,33,86–91] and on soft and durum wheat [32,33,87,89,92–94]; but also on rice [95]; discolored grain sorghum [96,97]; oats [33,87]; rye [32,33]; triticale [76]; and corn [93,98,99].

Prolonged and unusually wet field conditions during ripening of small grains induced the colonization of ears by black saprophytic mycoflora responsible for disease called "black heads" which lead to kernel "black point" or "kernel smudge," resulting from epiphytic growth of mycelium and spore mass [100–102]. *Alternaria* species, especially *A. alternata*, are the predominant species of the mycoflora causing ear blackening, in association with species of *Cladosporium*, *Helminthosporium*, *Diplodia*, *Epicoccum*, and *Fusarium*.

Such black fungi occurring on kernels and on derived meals certainly contribute to the depreciation of cereals due to the occurrence of sooty kernels in cereal lots and of black particles in meals. *A. alternata* was the prevalent black point agent in kernels of durum wheat at ripening in Italy [103], USA [102,104], Canada [105], as well as in Germany, where *A. alternata* proved to be the predominant blackening mold of durum wheat kernels at the preharvest stage, in association with *Epicoccum*, *Cladosporium*, and pathogenic strains of *Drechslera* [106]. In Australia, *A. alternata* (*A. tenuis*) was responsible for severe crop losses, spoilage, and mycotoxin production in wheat and sorghum [24]. In Poland an exceptional colonization of small grain cereal heads by *A. alternata*, with the presence of copious black mycelium on brushes and kernels and a significant incidence of "smudged kernels" at harvest, was recorded in the 1990 crop year, characterized by unusually rainy weather conditions before harvest [107]. Under such conditions, the presence of *A. alternata* on "black heads" of wheat, rye, triticale, oats and barley, monitored during the 1986–1990 seasons, reached 100% [24,76]. In northern European climate, strains of *A. alternata* were also found among the prevalent fungi in growing silage maize, mainly colonizing leaves and husks [108].

A. alternata has been reported also on some kinds of fodder, such as clover and grasses [40]; maize silage and hay [30,108,109]; and coarse fodder [30].

A. alternata has been isolated from molded sunflower seeds at ripening in the field as well as during storage [79,110]. Other records on oilseed contaminations are related to the occurrence of *A. alternata* in linseed [111]; cottonseed [112]; and rapeseed and mustard [113–115]. *A. alternata* was the predominant species among the fungi colonizing siliquaes and seeds of oilseed rape (*Brassica napus* L. subsp. *oleifera* DC), found in the field in Italy, in association with *Stemphylium* spp., *Cladosporium* spp., and *Pleospora infectoria* Fuckel (*Alternaria* sp.) [116].

A. alternata has also frequently been recorded on nuts, including peanuts [117]; hazelnuts [118]; and pecans [119,120]. Other records include spices [121], marjoram [122], biltong [123], and refrigerated stored meat [124].

IV. MYCOTOXINS OF *ALTERNARIA*

A. Toxins Produced

Species of *Alternaria* are known to produce at least 70 secondary metabolites, most of which are formed by *A. alternata* and its pathotypes [7,8,13]. These metabolites can be conveniently grouped into structural classes as follows: dibenzopyrones, tetramic acids, quinones, peptides, and miscellaneous lactones. However, only seven major toxins belonging to three different structural classes are known as possible food contaminants with a potential toxicological risk [16,125,126]. These are tenuazonic acid (TA), a tetramic acid derivative; alternariol (AOH), alternariol methyl ether (AME), and altenuene (ALT), which are dibenzopyrone derivatives; and altertoxin I, II, and III (ATX-I, ATX-II, ATX-III), which are perylene derivatives (Table 1, Figs. 1–4). Moreover, several host-selective phytotoxins were mentioned in Section III, whereas other major metabolites also possessing phytotoxic but not host-selective properties (aspecific phytotoxins) are mentioned at the end of this section, and reported with more others in Table 1.

1. Tenuazonic Acid

Tenuazonic acid ($C_{10}H_{15}NO_3$, MW 197.1051), appears as a colorless viscous oil whereas Ca and Mg tenuazonate occurs as an anamorphous powder. It shows dark quenched spots on fluorescent TLC plates under UV light at 254 nm; it turns red-brown after reaction with iron chloride, and yellow with 2,4-dinitrophenylhydrazine [127,128] (Fig. 1).

TABLE 1 Toxins Produced by *Alternaria* spp.

Alternaria species	Toxin	Reference
	Alternariols and Altenuenes	
A. alternata (= *A. tenuis*)	Alternariol	138
	Alternariol monomethyl ether	138
	Altenuene	145
	Altenuisol	149
	Altenusin	148
	Dehydroaltenusin	227
	Isoaltenuene	147
	5′-Epialtenuene, Neoaltenuene	228
A. brassicae	Alternariol	76
	Alternariol monomethyl ether	76
A. brassicicola	Alternariol	32
	Alternariol monomethyl ether	32
A. capsici-annui	Alternariol	130
	Alternariol monomethyl ether	130
	Altenuene	130
A. cheiranthi	Alternariol	32
	Alternariol monomethyl ether	32
A. citri	Alternariol	138
	Alternariol monomethyl ether	138
	Altenuene	130
A. cucumerina	Alternariol	141
	Alternariol monomethyl ether	141

Table 1 Continued

Alternaria species	Toxin	Reference
	Alternariols and Altenuenes (cont.)	
A. dauci	Alternariol	138
	Alternariol monomethyl ether	138
A. kikuchiana (= *A. alternata* Japanese pear pathotype)	Alternariol	142
	Alternariol monomethyl ether	142
	Altenusin, Dehydroaltenusin	142
A. longipes (= *A. alternata* tobacco pathotype	Alternariol monomethyl ether	144
A. porri	Alternariol monomethyl ether	130
	Altenuene	130
A. radicina	Altenuene	130
A. raphani	Alternariol	32
	Alternariol monomethyl ether	32
A. solani	Alternariol	143
	Alternariol monomethyl ether	143
A. tenuissima	Alternariol	13
	Alternariol monomethyl ether	32
	Altenuene	76
A. tomato	Alternariol	135
	Alternariol monomethyl ether	135
	Altenuene	130
Pleospora infectoria (*Alternaria* sp.)	Alternariol	32
	Alternariol monomethyl ether	32
	Tetramic acids	
A. alternata (= *A. tenuis*)	Tenuazonic acid	129, 148
A. brassicae	Tenuazonic acid	32
A. brassicicola	Tenuazonic acid	32
A. capsici-annui	Tenuazonic acid	130
A. cheiranthi	Tenuazonic acid	32
A. citri	Tenuazonic acid	131
A. japonica	Tenuazonic acid	131
A. kikuchiana	Tenuazonic acid	131
A. longipes	Tenuazonic acid	132
A. porri	Tenuazonic acid	130
A. radicina	Tenuazonic acid	130
A. raphani	Tenuazonic acid	32
A. tenuissima	Tenuazonic acid	134
A. tomato	Tenuazonic acid	135
Pleospora infectoria (*Alternaria* sp.)	Tenuazonic acid	32
	Altertoxins	
A. alternata (= *A. tenuis*)	Altertoxins I, II, and III	16, 126
A. capsici-annui	Altertoxin I	130

TABLE 1 Continued

Alternaria species	Toxin	Reference
	Altertoxins (cont.)	
A. mali (= *A. alternata* apple pathotype)	Altertoxins I, II, and III	16
A. radicina	Altertoxin I	130
A. tenuissima	Altertoxin I	130
A. tomato	Altertoxins I and II	135
A. alternata (= *A. tenuis*)	Stemphyltoxin III	153
	Other major metabolites (phytotoxins)	
A. alternata (= *A. tenuis*)	Altenuic acids	148
	Tentoxin	159
	Maculosin	229
	Alterlosins	230
A. alternata (cotton pathotype)	Tentoxin	158
A. bataticola	Macrosporin	231
A. brassicae	Destruxins	27
A. carthami	Zinniol	232
	Brefeldins	233
A. chrysanthemi	Radicinins	234
A. cichorii	Zinniols	235
A. citri (= *A. alternata* citrus pathotypes)	Tentoxin	7
A. cucumerina	Macrosporin	231
A. dauci	Zinniol	236
A. helianthii	Deoxyradicinin	237
	Pyrenocines	238
A. kikuchiana (= *A. alternata* Japanese pear pathotype)	Altenin	239
A. macrospora	Zinniol	232
A. mali (= *A. alternata* apple pathotype)	Tentoxin	7
A. porri	Altersolanols, Tentoxin	240
	Porritoxin, Porriolide	240, 241
	Zinniol	232
	Macrosporin	231
	Alterporriols	242
A. radicina	Radicinins	13, 205
A. solani	Alternaric acid	154, 155
	Altersolanols, Alterporriols	231
	Solanapyrones	243
	Zinniol	232
	Zinnolide	244
	Macrosporin	231
A. tagetica	Zinniol	232
A. zinniae	Zinniol	245

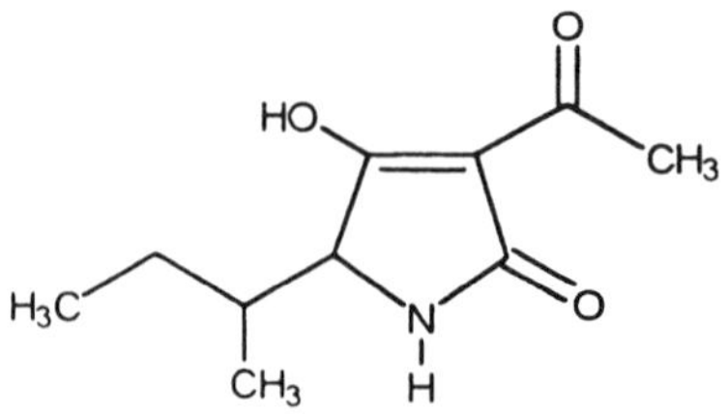

FIGURE 1 Tenuazonic acid.

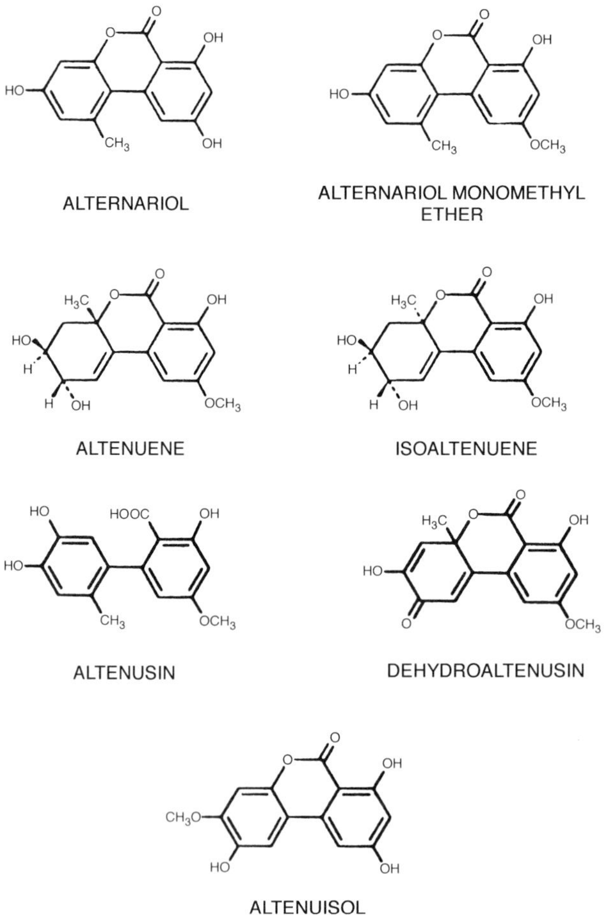

ALTERNARIOL

ALTERNARIOL MONOMETHYL ETHER

ALTENUENE

ISOALTENUENE

ALTENUSIN

DEHYDROALTENUSIN

ALTENUISOL

FIGURE 2 Alternariols and altenuenes.

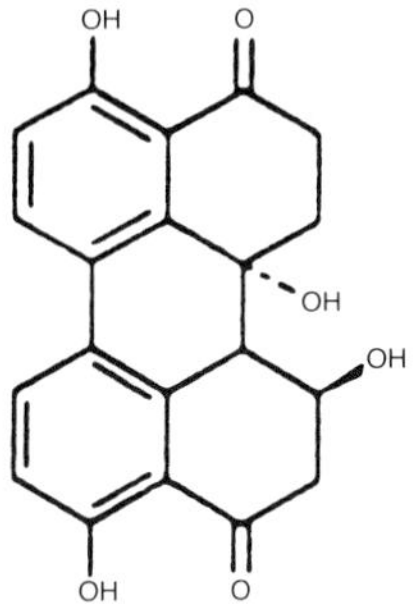

ATX-I

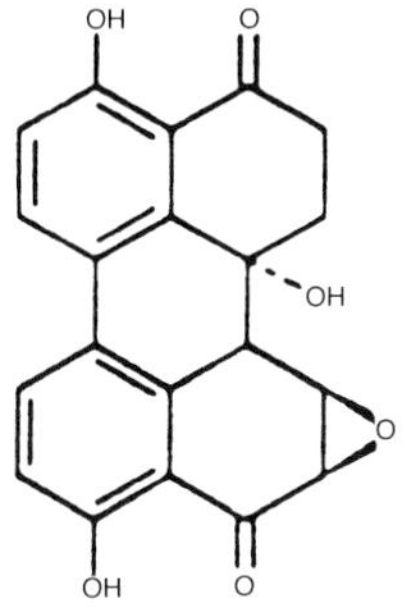

ATX-II
(STEMPHYLTOXIN-II)

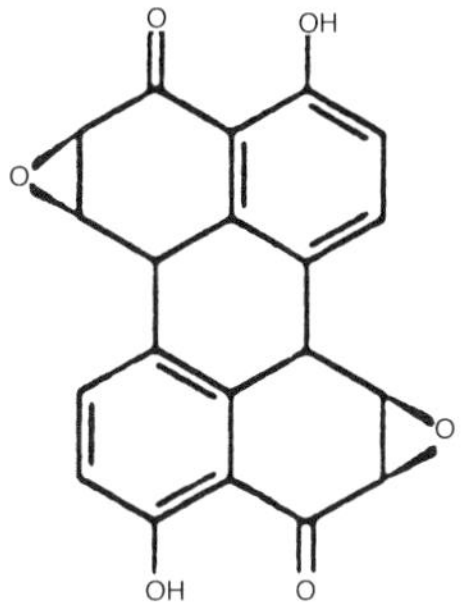

ATX-III

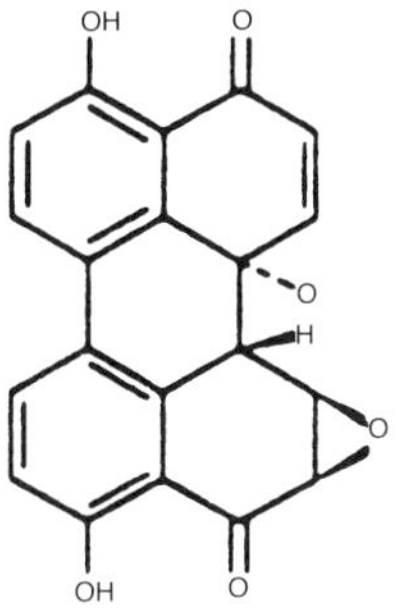

STEMPHYLTOXIN-III

FIGURE 3 Altertoxins.

Tenuazonic acid is a well-known mycotoxin as well as a phytotoxin [17], produced mainly by *A. alternata* (= *A. tenuis*) [127,129], but also by various other *Alternaria* species: *A. capsici-annui* [130]; *A. citri* [131]; *A. kikuchiana* (= *A. alternata* Japanese pear pathotype) [131]; *A. japonica* [131]; *A. longipes* (= *A. tenuis*) [132]; *A. longipes* (= *A. alternata* tobacco pathotype) [133]; *A. porri* [130]; *A. radicina* [130]; *A. tenuissima* [134]; and *A. tomato* [135].

Tenuazonic acid is also produced by species of fungi other than *Alternaria*, i.e., *Pyricularia oryzae* Briosi *et* Cavara [136], and by strains of *Phoma sorghina* (Sacc.) Boerema et al. associated to the etiology of the human hematological disorder known as Onyalay [128,137].

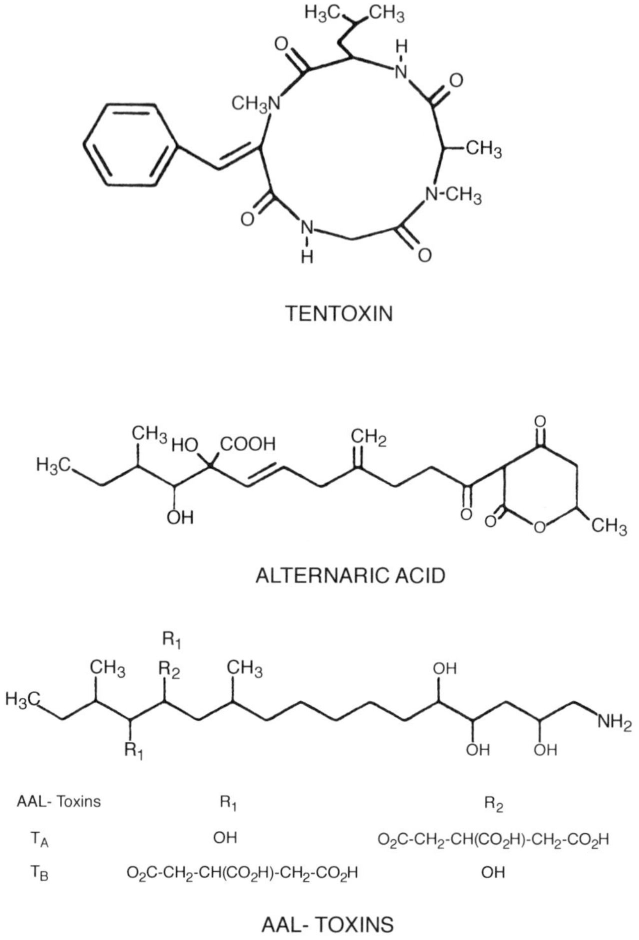

TENTOXIN

ALTERNARIC ACID

AAL- Toxins	R_1	R_2
T_A	OH	O_2C-CH_2-$CH(CO_2H)$-CH_2-CO_2H
T_B	O_2C-CH_2-$CH(CO_2H)$-CH_2-CO_2H	OH

AAL- TOXINS

FIGURE 4 Tentoxin, alternaric acid, and AAL-toxins.

2. Alternariols and Altenuenes

Alternariol ($C_{14}H_{10}O_5$, MW = 258.0528), and AME ($C_{15}H_{12}O_5$, MW = 272.0684), crystallize from ethanol as colorless needles (mp 350°C and 267°C, respectively), and sublime without decomposition in a high vacuum at 250°C and 180 to 200°C, respectively (Fig. 2). They show a blue fluorescence under UV light and a purple color after reaction with ethanolic iron chloride [138,139]. AOH and AME have been isolated from *A. alternata* (= *A. tenuis*) [138]; *A. brassicae* [76]; *A. capsici-annui* [130]; *A. citri* [140]; *A. cucumerina* [141]; *A. dauci* [140]; *A. kikuchiana*

(= *A. alternata*, Japanese pear pathotype) [142]; *A. tenuissima* [13]; *A. tomato* [135]; and *A. solani* [143]. Alternariol monomethyl ether, but not AOH, was also reported for *A. longipes* (*A. alternata* tomato pathotype) [144] and *A. porri* [130].

Altenuene ($C_{15}H_{16}O_6$, MW = 292.0946) crystallizes from acetone-hexane as colorless prisms (mp 190 to 191°C) and shows a yellow fluorescence under UV light [145,146] (Fig. 2). ALT has been isolated from *A. alternata* (= *A. tenuis*) [141] and from *A. capsici-annui*, *A. citri*, *A. porri*, *A. radicina*, *A. tenuissima*, and *A. tomato* [130].

Isoaltenuene ($C_{15}H_{16}O_6$, MW = 292.0946), isolated from *A. alternata*, shows fluorescence and UV spectral characteristics similar to those of ALT [147] (Fig. 2).

Altenusin ($C_{15}H_{14}O_6$, MW 290.0790) crystallizes from chloroform as colorless prisms (mp 202 to 203°C) (Fig. 2). Altenusin has been isolated from *A. alternata* (= *A. tenuis*) [148] and *A. kikuchiana* (= *A. alternata*, Japanese pear pathotype) [142].

Dehydroaltenusin ($C_{15}H_{12}O_6$, MW 288.0633) crystallizes from acetone as yellow plate crystals (mp 190 to 193°C) (Fig. 2). This metabolite has been isolated from *A. alternata* (= *A. tenuis*) [148] and *A. kikuchiana* (= *A. alternata*, Japanese pear pathotype) [142].

Altenuisol (altertenuol) ($C_{14}H_{10}O_6$, MW 274.0476) crystallizes from hot acetic acid as slender needles (mp 277 to 282°C) (Fig. 2). Altenuisol has been isolated from *A. alternata* (= *A. tenuis*) [149].

3. Altertoxins

Of increasing toxicological interest among the perylenequinones produced by *A. alternata* are altertoxins and stemphyltoxins (Fig. 3). Altertoxin-I (dihydroalterperylenol) ($C_{20}H_{16}O_6$, MW 352.0946) crystallizes as a yellow amorphous solid (mp >180°C), and shows a bright yellow fluorescence under UV light [150,151]. ATX-I has been produced by *A. alternata* (= *A. tenuis*) [126], *A. capsici-annui* [130], *A. mali* (= *A. alternata* apple pathotype) [16], *A. radicina* [130], *A. tenuissima* [130], and *A. tomato* [135].

Altertoxin II (stemphyltoxin II) ($C_{20}H_{14}O_6$, MW 350.0790) crystallizes as an orange crystalline solid (mp 245 to 250°C) and shows an orange-yellow fluorescence under UV light [151]. ATX-II has been produced by *A. alternata* (= *A. tenuis*) [16,126], *A. mali* (= *A. alternata* apple pathotype) [16], and *A. tomato* [135].

Altertoxin-III ($C_{20}H_{12}O_6$, MW 348.0633) has a melting point of 175 to 230°C (sublimes) [151]. This metabolite has been produced by *A. alternata* (= *A. tenuis*) [16,126], and *A. mali* (= *A. alternata* apple pathotype) [16].

Stemphyltoxin III ($C_{20}H_{12}O_6$, MW 348.0636) appears as a brown solid (mp >300°C) and was first isolated from *Stemphylium botryosum* var. *lactucum*, and then from a strain of *A. alternata* [152,153].

4. Other Major Metabolites (Phytotoxins)

Alternaric acid ($C_{21}H_{30}O_8$, MW 40) crystallizes as colorless needles (mp 138°C). It has been isolated from *A. solani* [154,155], *A. mali* (= *A. alternata*, apple pathotype), and *A. oryzae* [156] (Fig. 4).

Tentoxin ($C_{22}H_3ON_4O_4$, MW 414.2270) crystallizes as colorless prisms (mp 172 to 175°C) (Fig. 4). Tentoxin, a metabolite that inhibits chlorophyll production [157,158], is mainly produced by *A. alternata* (= *A. tenuis*) [159], especially by certain opportunistic pathotypes [7], i.e., *A. alternata*, cotton pathotype [158]; *A. citri* (= *A. alternata*, citrus pathotypes) [7]; and *A. mali* (= *A. alternata* apple pathotype) [7].

Among the several host-specific toxins produced by at least eight *A. alternata* pathotypes [7,8], the AAL-toxins are here mentioned for their structural and toxicological similarities to the fumonisins, the cancerogenic mycotoxins produced by *Fusarium moniliforme* Sheld. [225]. As already reported, AAL-toxins (AL-toxins) include five compounds indicated as T_A ($C_{13}H_{53}NO_{15}$, MW 679), T_B ($C_{13}H_{53}NO_{13}$, MW 647), [160,161], and T_C, T_D, and T_E [26] (Fig. 4). All AAL-toxins are biosynthetic products of strains of *A. alternata* f. sp. *lycopersici* causing stem canker of tomato; they are able to induce the genotype-specific necrosis characteristic of AAL-toxin in tomato bioassays, but with a markedly different relative toxicity [226].

B. Toxin Formation and Natural Occurrence

1. Effect of the Environment on Mycotoxin Production

A. alternata produces almost all of the most toxic and important secondary metabolites described for the genus *Alternaria* as a whole. In fact, the toxic secondary metabolites produced by *A. alternaria* include AOH, AME, ALT, isoALT, altenuisol, altenusin, dehydroaltenusin, TA, ATX-I, ATX-II, and ATX-III. The effects of environmental factors on the production of these mycotoxins have rarely been studied in detail.

The optimal temperature for the production of AOH, AME, and ALT by *A. alternata*, both on wheat extract agar and on wheat grain, is near 25°C, with a range of 5 to 30°C. The optimal water activity level for the production of all the three above-mentioned toxins is 0.98 a_w. The potential capability for AOH was better expressed on agar culture, while the highest amounts of alternariol methyl ether were obtained on wheat grain [162].

The optimal temperatures and water activity conditions for the production of AOH and AME by *A. alternata* in sunflower seeds were 25°C and 0.90 a_w, respectively [163]. Similar optimal conditions of 25°C and 0.90 a_w were also reported for the production of TA by *A. alternata* in sunflower seed [164].

In rotted tomato, the optimal temperature for toxin production by *A. alter-*

nata (IMI 89344) reported by Hasan [83] was as follows: 28°C for AOH and AME, 21°C for TA, and 14°C for ATX-I and ATX-II. It was also noted that both the micelial growth and the toxin production were significant at 7°C and dropped at 35°C [83].

In *A. tenuis*-decayed apples, AOH and AME were the major metabolites produced not only at 25°C but also at 2°C, indicating the occurrence of toxigenic strains adapted to cold storage [70].

The accumulation of AOH and AME by *A. alternata* was inhibited in cultures maintained under continuous white light or when the white light was supplied only during the exponential growth phase. However, mycelial growth was not affected by white light treatment even when mycotoxin accumulation was almost completely inhibited [165]. It would be of interest to investigate the inhibitory effect of light on alterariols accumulations in infected plants and plant products under field conditions, both in exposed infected surfaces and inside large rotted organs, such as in tomato, in cucurbits and melons, as well as in several kinds of fruits and citrus affected by *Alternaria* blackmold and heart rot.

Alternaria mycotoxins are easily produced both on artificial or on natural media, as well as on whole healthy host fruits. Alternariols (AOH and AME) and ALT were produced on wheat, rice, barley, maize, oats, and rye as well as on substrates of origin—i.e., whole-wheat bread, olives, apples, oranges, mandarins, lemons, blueberries, and tomatoes, sterilized and inoculated with toxigenic strains of *A. alternata* [71,77,78,166–169].

The strong toxigenic potential of *A. alternata* strains under different conditions is exemplified by the following reports.

Forty-six out of 59 strains of *A. alternata* from small grains, cultured on autoclaved rice (cultures of 50 g rice kernels and 50 ml tap water in 300-ml flasks, inoculated with a multispore suspension, incubated at 22 to 25°C for 20 days), were toxigenic; i.e., 20 isolates produced AOH, AME, and TA; 10 isolates produced only AOH and AME; while 16 produced only TA [32].

Assays on the production of mycotoxins by strains of *A. alternata* isolated from olives and grown either on rice (10 isolates) or on whole healthy olives (seven isolates) showed that the toxigenic potential of *A. alternata* on rice was much higher than that on olives. In particular, the production of TA, AME, and AOH on rice was, respectively, 1034, 147, and 59 times higher than that on olives. Moreover, ALT and ATX-I which were produced on rice at average levels of 5.9 and 5 mg/kg, respectively, were not detected in olive cultures. It is worthwhile to underline the very high amount of TA (up to 9800 mg/kg) produced on rice by these strains of *A. alternata*, with an average over 6000 mg/kg [72].

Strains of *A. alternata* isolated from several samples of black moldy tomatoes collected at harvest time in the field in Italy were comparatively cultured on rice (100 g rice kernel and 45 ml distilled water in 500-ml flasks, inoculated with a spore suspension, incubated at 25°C for 4 weeks), on whole healthy tomato fruits as described by Stinson et al. [167], and in a liquid medium of yeast extract (2%)/sucrose (4%), for 1 week at 25°C. Sixteen out of 17 strains produced the following mycotoxins on autoclaved rice and whole tomatoes: TA (up to 4750 mg/kg), AOH (up to 600 mg/kg), alternariol methyl ether (up to 100 mg/kg), ALT (up to 30 mg/kg), and ATX-I (up to 12.5 mg/kg). The production of TA by *A. alternata* on autoclaved whole tomato fruits was much lower than on rice (less than one tenth) [74].

All of the 14 strains of *A. alternata* from wheat, collected in several Mediterranean countries, cultured on autoclaved rice (100 g rice kernel and 45 ml distilled water in 500-ml flasks, inoculated with a spore suspension, incubated at 25°C for 4 weeks), were toxigenic; i.e., 14/14 produced TA (up to 6007 mg/kg); 13/14 produced AOH (up to 118 mg/kg) and AME (up to 59 mg/kg); 13/14 produced ALT (up to 37 mg/kg); 14/14 produced ATX-I (up to 32 mg/kg); and 13/14 produced ATX-II (up to 105 mg/kg) [33]. Additional investigations carried out with some of the above strains grown on wheat and rice showed that TA production was better expressed on wheat (up to 8750 mg/kg) than on rice, while the opposite was observed for the benzopyrenes and perylene derivatives [73].

A high incidence of toxigenic strains was found among 176 strains of *A. alternata* from barley collected in Spain; i.e., 88.6% produced TA, 15.3% produced AOH, and 9% produced AME [91].

In whole mandarin fruits the mycotoxin formation by strains of two morphologically different biotypes of *A. alternata* from mandarin fruits affected by heart rot in southern Italy was as follows: *A. alternata* gray strains produced TA (up to 84.8 mg/kg), AOH (up to 0.92 mg/kg), AME (up to 0.17 mg/kg), and ATX-1 (up to 0.2 mg/kg); the *A. alternata* black strain produced TA (up to 93.1 mg/kg), AOH (up to 8.7 mg/kg), AME (up to 6.3 mg/kg), but neither ATX-I nor ALT. The toxin formation was much better expressed on the rice substrate than in whole mandarin fruits. In fact, on autoclaved rice (100 g of rice kernels and 45 ml distilled water in 500-ml flasks, inoculated with a spore suspension, incubated at 25°C for 4 weeks), the gray strain of *A. alternata* from moldy mandarin produced TA (up to 6850 mg/kg), AOH (up to 12.5 mg/kg), AME (up to 7.5 mg/kg), ATX-I (up to 83.3 mg/kg), and ALT (up to 25.5 mg/kg) while the *A. alternata* black strain produced TA (up to 177.0 mg/kg), AOH (up to 20.0 mg/kg), AME (up to 20.0 mg/kg), ATX-I (up to 7.5 mg/kg), and ALT (up to 50.0 mg/kg) [23].

All of the 11 strains of *A. alternata* from oilseed rape (nine from seeds and four from siliquae), cultured on autoclaved rice (100 g of rice kernels and 45 ml distilled water in 500-ml flasks, inoculated with a spore suspension, incubated at 25°C for 4 weeks), were able to produce TA (up to 12,000 mg/kg), AOH (up to 200 mg/kg), AME (up to 200 mg/kg), ATX-I (up to 250 mg/kg), and ATX-II (up to 70 mg/kg) [116].

Finally, it seems of interest to note that AOH and AME can occur in asymptomatic tissues from inoculated whole fresh apples, and such eventuality could represent an additional risk for the moldy core occurrence in apples [170].

2. Potential Mycotoxin Formation in Field and Storage Conditions

As potential spoilage or biodeteriogen agents of fresh plant products and grains, the *Alternaria* representatives as well as those of other similar genera—i.e., *Botrytis*, *Cladosporium*, *Colletotrichum*, *Diplodia*, *Epicoccum*, *Fusarium*, *Geotrichum*, *Helminthosporium*, and *Rhizopus*—have been classified as "field fungi." They require a very high moisture content in the product in order to grow (> 22% to 25%), as opposed to "storage" fungi, i.e., essentially *Aspergillus* and *Penicillium*, which also grow at a lower moisture content (< 20% to 22%).

Therefore, *Alternaria* isolates are unable to grow well in grains and other seeds after harvest because they are usually stored at a moisture content of 12% to 14%. However, as long as the moisture content of the contaminated plant products is above 25% to 30%, *A. alternata* isolates can grow actively and accumulate mycotoxins in infected tissues.

Such conditions are very common for vegetables, for fruits close to harvest time, and for small cereals at mealy ripeness, especially when the opportunistic or saprophytic fungal colonizations are favored by rainy weather and delayed harvest. A prolonged delay of processing fruits such as tomato and other vegetables to make juices, mashes, sauces, and tomato paste, increased the *A. alternata*'s possibilities for spreading and accumulating mycotoxins. Moreover, prolonged handling or transit under environmental conditions of plant products that are normally refrigerated can increase quality losses and mycotoxin formation.

3. Natural Occurrence

Early reports on the occurrence of *Alternaria* mycotoxins in plant products are related to the presence of AOH and AME in discolored pecans [119], in moldy apples [68], and in part-per-million levels in samples of weathered and discolored grain sorghum [96,97,171], also together with ALT and ATX-I [172]. In weathered grain sorghum the concentration of mycotoxins was correlated to the degree of *Alternaria* infections in the field and to the number of rainy days at ripening time. Such evidences stressed the importance of field infections on the final crop contamination at harvest. In fact, in the same survey, AOH and AME were not found in lots of good-quality sorghum grains, while in heavily infected grain sorghum ALT was also found [96,172]. The occurrence of AME in sorghum feeds, together with zearalenone, was also reported by Sydenham et al. [173]. Other early reports on the occurrence of *Alternaria* mycotoxins in food are related to the presence of AOH, AME, and ALT, together with TA, in naturally infected tomatoes and apples (five of eight apples also contained ATX-I) [132], and to the occurrence of TA in processed products, such as tomato paste [174,175] (Table 2). In recent years more examples of foods or colonized agricultural products contaminated by *Alternaria* mycotoxins have been reported in the literature (Table 2).

TABLE 2 Natural Occurrence of *Alternaria* Toxins

Country	Product	Year	Positive sample/ total no.	Content (mg/kg)	Reference
Tenuazonic acid (TA)					
India	Sorghum			<6	97
India	Ragi[a]			<6	97
Italy	Olive[b]	1985	2/13	0.109–0.262	72
Italy	Tomato[c]	1985	2/2	0.024–7.21	34
Italy	Melon[b]	1986	1/1	0.08	79
Italy	Pepper[b]	1986	1/1	0.054	79
Italy	Mandarin[d]	1987	3/3	21.0–173.9	79
Italy	Mandarin[e]	1987	2/2	21.0–87.2	23
Italy	Mandarin[f]	1987	1/1	173.9	23
Italy	Oilseed rape	1991	0/16[g]	—	116
Japan	Rice plants				220
Japan	Tobacco leaves				133
USA	Tomato paste		8	Trace	175
USA	Tomato[h]		73/142	Trace	174
Alternariol (AOH)					
Argentina	Sunflower	1992	37/50	0.035–0.792	110
Italy	Tomato[d]	1985	1/2	1.274	34
Italy	Olive[b]	1985	4/13	0.109–2.32	72
Italy	Melon[b]	1986	0/1	—	79
Italy	Pepper[b]	1986	1/1	0.64	79
Italy	Sunflower[b]	1986	2/2	0.357–1.84	79
Italy	Mandarin[d]	1987	2/3	1.0–5.2	79
Italy	Mandarin[e]	1987	2/2	1.0–5.2	23
Italy	Mandarin[f]	1987	0/1	—	23
Italy	Oilseed rape	1991	0/16[g]	—	116
Poland	Wheat heads	1986–1987	4/21		107
Poland	Rye heads	1986–1987	1/10		107
Poland	Wheat chaff	1986	3/3	0.27–1.80	107
Poland	Wheat kernels	1986	1/3	0.59	107
Poland	Wheat chaff	1987	1/1	0.27	107
Poland	Rye chaff	1987	1/1	1.15	107
Poland	Wheat kernel black heads	1990	1/3	0.02	76
Poland	Wheat chaff black heads	1990	2/3	0.19–0.72	76
USA	Sorghum[i]				96
USA	Pecans[j]				119
USA	Tomato			Up to 1.3	78
USA	Apple				78

TABLE 2 Continued

Country	Product	Year	Positive sample/ total no.	Content (mg/kg)	Reference
Alternariol (AOH) (cont.)					
—	Fruits & Veg.		2/70[k]	<1	68
Alternariol monomethyl ether (AME)					
Argentina	Sunflower	1992	31/50	0.090–0.630	110
India	Sorghum[i]		9/20	Up to 1.8	97
India	Ragi[a]		2/20		97
Italy	Tomato[c]	1985	2/2	0.037–0.268	34
Italy	Olive[b]	1985	4/13	0.03–2.87	72
Italy	Melon[b]	1986	0/1	0.051	79
Italy	Pepper[b]	1986	1/1	0.049	79
Italy	Sunflower[b]	1986	1/2	0.129	79
Italy	Mandarin[d]	1987	2/3	0.55–1.4	79
Italy	Mandarin[e]	1987	2/2	0.55–1.4	23
Italy	Mandarin[f]	1987	0/1	—	23
Italy	Oilseed rape	1991	0/16[g]	—	116
Poland	Wheat heads	1986–1987	1/21		107
Poland	Wheat chaff	1987	1/1	0.51	107
Poland	Wheat kernel black heads	1990	1/3	0.68	76
Poland	Wheat chaff black heads	1990	0/3	—	76
USA	Sorghum[i]		3/12		96
USA	Pecans[j]				119
USA	Tomato			Up to 1.3	78
USA	Apple				78
—	Fruits & Veg.		2/70[k]	<1	68
Altenuene (ALT)					
India	Sorghum		9/20	Up to 0.7	97
India	Ragi[a]		2/20		97
Italy	Olive[b]	1985	1/13	1.4	72
Italy	Melon[b]	1986	0/1	—	79
Italy	Pepper[b]	1986	0/1	—	79
Italy	Mandarin[d]	1987	0/3	—	79
USA	Sorghum[i]		3/12	0.1–1.5	172
USA	Apple				78
Altertoxins (ATX)					
Italy	Olive[b]	1985	0/13	—	72
Italy	Mandarin[e]	1987	0/2	—	23
Italy	Melon[b]	1986	0/1	—	79
Italy	Mandarin[f]	1987	0/1	—	23

Table 2 Continued

Country	Product	Year	Positive sample/ total no.	Content (mg/kg)	Reference
Altertoxins (ATX) (cont.)					
Italy	Oilseed rape	1991	0/16[g]	—	116
USA	Sorghum[i]		3/12	Trace	172
USA	Apple		5/8	Trace	78

[a]*Eleusine coracana* Gaertner.
[b]Moldy samples at harvest time.
[c]Preharvest-infected tomato fruits.
[d]Fruit rot of mandarin (*Citrus reticulata* Blanco) at harvest time.
[e]Black fruit rot of mandarin (*C. reticulata*) at harvest time.
[f]Gray fruit rot of mandarin (*C. reticulata*) at harvest time.
[g]Samples of oilseed rape (*Brassica napus* L. subsp. *oleifera* DC) from processing factories.
[h]Samples of tomatoes from commercial processing lines.
[i]Weathered and discolored samples at harvest.
[j]Discolored pecan samples.
[k]Two infected apple samples.

Preliminary data on the natural occurrence of AOH and AME in wheat, barley, and oat colonized by *A. alternata* reported by Gruber-Schley and Thalmann [176] were then confirmed in field samples of several small cereals affected by "black heads." Investigations carried out in Poland [107] clearly showed that the level of *Alternaria* mycotoxins in small grains depends on the incidence of black heads in the fields at harvest time, caused by *A. alternata* in association with species of *Cladosporium*, *Helminthosporium*, *Drechslera*, *Epicoccum*, and *Cochliobolus*. Severe infections occur when the weather at harvest time is rainy and mature ears stay in the fields for several days. Under such conditions, AOH was found in samples of whole darkened ears of rye (10%) and wheat (15%), and AME was also found in whole darkened ears of wheat (15%). Of the five infected wheat samples, AOH was found both in the chaff (5/5; 0.27 to 1.80 mg/kg) and kernels (1/5; 0.59), whereas AME was only found in the chaff (1/5; 0.51 mg/kg) [107].

A limited survey on the natural occurrence of the major *Alternaria* mycotoxins, AOH, AME, ALT, ATX-I, and TA, in molded or damaged olive samples collected in Apulia (Italy), showed that four out of 13 samples were contaminated by two to four *Alternaria* mycotoxins. The highest contamination was found in a badly damaged sample containing 2870, 2320, 1400, and 262.5 μg/kg of AME, AOH, ALT, and TA, respectively. No mycotoxins were detected in olive oil for human consumption (six samples) or olive husks (three samples) collected from oil mills after the first pressing [72].

Alternaria mycotoxins were found in Italy in samples of sunflower seeds, peppers, and melons visibly affected by *Alternaria* black rot at harvest time [79]. Two infected sunflower seed samples were found contaminated by AOH (up to 1840 μg/kg), and AME was only observed in one (129 μg/kg); TA and ALT were not found. Peppers and melons proved to be slightly contaminated: in fact, only very low concentrations of TA (54 μg/kg), AME (49 μg/kg), and alternariol (640 μg/kg) were found in infected peppers, but only traces of TA (8 μg/kg) and AME (5.1 μg/kg) were found in melon [79].

Two samples of tomato fruits affected by *Alternaria* blackmold in the field were found to be contaminated with 24 and 7210 μg/kg of TA, respectively. Together with TA other *Alternaria* toxins were found, i.e., AME, 37 and 268 μg/kg, respectively; and AOH, traces and 1274 μg/kg, respectively [34,74].

Mandarin samples affected by two kinds of *A. alternata* heart rot were found to be contaminated by mycotoxins in different ways. In black rot samples TA, AME, and AOH (up to 87.2, 1.4, and 5.2 mg/kg, respectively) were found; TA (up to 173.9 mg/kg) was the only detectable mycotoxin in mandarin gray rot samples. None of the natural *Alternaria* rot mandarin samples contained ALT or ATX-I [23].

Alternaria mycotoxins, i.e., TA, AOH, AME, ANE, ATX-I, ATX-II, ATX-III, were not found in several samples of oilseed rape collected in the field [114] or in oil-processing plants [116], although the samples were contaminated by strains of *A. alternata* capable to produce high amounts of these mycotoxins on rice [116].

A more targeted investigation carried out in Argentina on the occurrence of *Alternaria* mycotoxins in sunflower seeds showed that of the 50 sunflower seed samples collected in seed oil plants, 38 (76%) were found to be contaminated, i.e., 30 samples (60%) by AOH (0.035 to 0.792 mg/kg) and AME (0.090 to 0.630 mg/kg), seven samples (14%) only by AOH (0.099 to 0.792 mg/kg), and one sample only by AME (0.090 mg/kg) [110]. However, the high moisture content of 0.90 a_w required by *A. alternata* for the in vitro formation of AOH, AME, and TA in sunflower seeds, which is equivalent to a 27% moisture content in in vivo sunflower seed, clearly indicates that the formation of these mycotoxins is potentially possible only in the field where the moisture content of sunflower seeds through the various stages of maturity ranges from 11% to 83%. In fact, much less probable or not at all likely appears the growth of *A. alternata* and TA production during storage, because the moisture content under storage conditions ranged from 6% to 11% [164]. However, the optimum a_w for the formation of TA in sunflower seed seems much lower than that required for the best production in cottonseed (a_w = 1.0) [177] and culture media (a_w = 0.98) [60], and these findings suggest a comparatively higher toxin risk for sunflower seeds [164].

Tenuazonic acid was also found in the culture of a strain of *Phoma sorghina* isolated from millet and sorghum in South West Africa, and associated with the etiology of a human hematologic disorder known as "Onyalai" [137].

Finally, *Alternaria* mycotoxins were not found in stored small cereals collected in Mediterranean countries, even though they were contaminated by toxigenic *A. alternaria* strains [73].

C. Toxicology of *Alternaria* Toxins

Since the early reports regarding the toxicity of *Alternaria* isolated from overwintered Russian cereals, supposedly involved in alimentary toxic aleukia [178,179], many other observations have pointed out the toxicity of *Alternaria* representatives. In particular, it was reported that the corn cultures of 65 out of 83 *Alternaria* strains from peanuts and corn were lethal when fed to rats [180,181], and that 31 out of 96 cracked corn cultures of different *A. longipes* (= *A. alternata*) strains were lethal when fed to chickens [182,183]. Subsequent studies confirmed that 57 out of 85 strains of *Alternaria* from grains, seeds, and plant materials were toxic to rats, and this led to the identification of TA [129]. Moreover, investigations on the toxicity of the culture extracts of *A. alternata* from weathered sorghum grain when fed to broiler chicks led to the identification of TA and ATX-I together with AME, AOH, and ALT [172].

Studies on the mycotoxicoses of farm animals evidenced that *Alternaria* mycotoxins might be involved in an hemorrhagic syndrome in poultry [184,185], and further investigations on the toxic mycoflora associated with cattle fescue toxicosis led to the identification of an *Alternaria* culture highly toxic to mice [187]. Observations on the cultural extracts from several strains of *A. alternata* (= *A. tenuis*) isolated from tobacco showed that they were lethal when administered intraperitoneally to mice [192]. Additional investigations on the toxicity of the *Alternaria* mycelial extracts showed their capability to cause lesions in the mucous membranes of horses [189], to enhance the antibody production in experimental animals [194], and to be toxic to brine shrimps, chick embryos, and rats [190].

Studies on the antimicrobial activities of *A. alternata* (= *A. tenuis*) culture extracts led to the chemical characterization of several metabolites toxic to mammalian [138,148], but the principles highly toxic to bacteria and fungi [191], as well as to mice [192], occurring in culture extracts of wild type and mutants of *A. mali* remained unidentified.

Moreover, extracts from mycelia or from culture filtrates of *Alternaria* strains have been reported to be mutagenic to the *Salmonella typhimurium* strains used in the Ames test. It was frequently noted that the mutagenicity of the extracts was greater than that of the estimated toxin content [77,193,194].

Finally, extracts from culture filtrates of a strain of *A. alternata* isolated from grains collected in Linxian areas of China, characterized by a high incidence of human esophageal cancer, were reported to be mutagenic and led to the identification of AME as a mutagenic metabolite [99,195,196].

Although only very limited investigations have been carried out on purified compounds, *Alternaria* metabolites have been shown to have various biological activities, i.e., antiviral, antibacterial, antifungal, insecticidal, and were toxic to several plant and animal systems.

1. Antiviral and Antimicrobial Activities

The antiviral activity of TA was reported for a wide spectrum of viruses, including enteroviruses, herpes simplex HF, measles virus, respiratory viruses, vaccinia, "B" virus. Comparatively, a lower activity was also observed for sodium isotenuazonic salt [197].

Studies on the antibacterial activities of various dibenzo-α-pyrones showed that AOH and altenuisol were the most toxic. Early studies showed that AOH was moderately toxic to *Staphylococcus aureus* (Mig.) Cast. et Chalm. and *Escherichia coli* Rosenb. [138]; successively, investigations on *Bacillus mycoides* Flügge showed that concentrations causing a zone of inhibition by AOH and altenuisol were 60 and 5 μg per disk, respectively, whereas the same effect was observed with 500 μg per disk of AME. Of interest was the strong synergism of AME and AOH, which were active at very low concentration of 0.25 μg per disk when supplied in mixture (1:1) [139]. TA has been reported to have antibacterial activity against at least nine species of bacteria, but not its sodium salt, which was not active on 31 species of bacteria [198].

Antagonistic effects of *A. alternata* have been observed on numerous bacteria and fungi in vitro [109,199–201] and against *Cochliobolus sativus* (Ito and Kurib.) Dreshs. ex Dastur on wheat roots [202], whereas the germination of spores of dwarf and covered smuts on rye has been enhanced by *A. alternata* [203].

In respect to antifungal activity, ATX-II and TA were the most toxic metabolites of *A. alternata*, with a minimal inhibitory dose versus *Geotrichum. candidum* Link ex Pers. of 10 and 20 μg per disk, respectively, followed by ATX-I with a lower minimal inhibitory dose of 200 μg per disk. Other dibenzopyrone derivatives, i.e., AOH, AME, ALT, iso-ALT, and altenuisol, were not active at these levels [204]. The different antifungal activity shown by the two altertoxins is probably related to the presence of the epoxide in ATX-II. In fact, a similar difference has previously been reported for the mutagenicity of these compounds on *Salmonella typhimurium* (Loef.) Cast. et Chalm. [18,151,205].

Preliminary observations on the antifungal activity of the sodium salt of TA showed that this mycotoxin was inactive on 48 strains of yeast and on some other fungal species [197].

2. Toxicity to Animals

Cultures of phytopathogenic isolates of *A. alternata*, supplied to tobacco budworm (*Heliothis virescens* F.), reduced the efficiency of conversion of ingested food to body tissues, decreased the larvae and pupal weight, and increased the pupation time [206]. Insecticidal activities by pure compounds of *Alternaria* are only related to sodium salt of TA and its larvicidal activity (LC_{50} = 120 μg/ml) on *Lucilia sericata* [207].

Studies on mice showed that ALT was the most toxic dibenzo-α-pyrones when given orally or administered intraperitoneally. The LD_{50} determined on mice was found to be of 50, 400, and 400 mg/kg for ALT, AOH, and AME, respectively, but when given by intraperitoneal injection the most toxic was ALT while AME was almost inactive [16].

A teratogenic/fetotoxic effect was observed to be induced by AOH, but not by AME, in pregnant mice (100 mg/kg); AME (200 mg/kg) was active in pregnant Syrian golden hamsters. Both alternariols caused a significant increase in dead and resorbed fetuses, as well as a decrease in mean fetal weight and fetus malformations. A synergistic teratogenic/fetotoxic effect caused by AOH and AME was observed in mice, with an active dose only of 25 mg/kg when administered in mixture (1:1) [16].

Alternariol, AME, and ALT assayed on 7-day-old chicken embryos by injecting a single dose through the yolk sac caused no teratogenicity or mortality at doses up to 1.0, 0.5, and 1.0 mg per egg, respectively [125]. Furthermore, feeding studies on day-old chicks with a diet supplemented with pure AME at levels up to 100 mg/kg showed no loss of weight gain or mortality, confirming the very low toxicity of this mycotoxin in poultry [125].

Some aspects of the frog skin physiology can be considered homologous to mammalian kidney nephron. In particular, frog skin actively transports sodium ions from the external environment of the animal into the blood, and this creates a potential difference between the external and internal surface [208]. Studies carried out with AOH showed that this mycotoxin decreases the short circuit current and sodium transport through frog skin. Such inhibition caused by AOH, which is partially removed by ADH, clearly suggests an impairment of the sodium pump and a toxic activity of the mycotoxin on cell metabolism [209].

Tenuazonic acid administered to mice caused diarrhea, muscle tremor, and convulsions. The LD_{50} determined on mice by different routes of dosing was observed to be between 125–225 and 81–115 mg/kg for males and females, respectively. TA is acutely toxic also against other animals. To this regard, the following LD_{50} have been reported: day-old chicks 37.5 mg/kg (orally); dogs 2.5 to 10 mg/kg (orally); rats 168 mg/kg (orally) or 157 mg/kg (intravenously). Monkeys dosed orally up to 50 mg/kg were not affected, but at higher doses, emesis, bloody diarrhea, and gastrointestinal hemorrhages were observed. TA sodium salt appeared less active and had no effect when injected intramuscularly into rabbits, while it caused mild inflammation and necrosis when given to guinea pigs by subcutaneous injections [18,210].

Tenuazonic acid has also been reported toxic to chick embryos with an LD_{50} leading to death in 10 or 18 days estimated, respectively, in 5 or 50 mg/kg. The LD_{50} calculated on 7-day-old chicken embryos by a single dose injection through the yolk sac was 0.55 mg per egg [125].

Tenuazonic acid has also been reported toxic to Leghorn chickens, and LD_{50} determined by esophageal intubation on day-old chicks was 37.5 mg/kg. In particular, increasing TA doses in chickens to sublethal and lethal levels progressively caused a poor feed efficiency, the suppression of weight gain, and the appearance of extensive internal hemorrhages [211].

3. Cytotoxicity

The *Alternaria* metabolite reported most toxic to HeLa cells was ATX-II (ID_{50} = 0.5 mg/ml), followed by ATX-I (ID_{50} = 20 mg/ml). A similar relative toxicity was reported for Chinese hamster lung fibroblast (V79). In this assay, the toxic activity of altertoxins was almost 100-fold higher than that on HeLa, and the noncytotoxic levels were found to be < 0.02, 0.2 and 5 μg/ml for ATX-II, ATX-III, and ATX-I, respectively [16,212].

Dibenzo-α-pyrones were reported less cytotoxic on HeLa cell than ATX-II, with ID_{50} values of 6, 8–14, and 28 μg/ml for AOH, AME, and ALT, respectively [16]. The same degree of toxicity (6 μg/ml) was reported for AOH on mouse lymphoma cells (L5178Y) [213].

A significant antitumoral activity was reported for TA when assayed on human adenocarcinoma-1 (Had-1) grown in embryonated eggs. Activity was higher than that shown by known antitumor agents [214].

4. Mutagenicity and Carcinogenicity

The mutagenic activity of several *Alternaria* metabolites determined on *S. typhimurium* strains by the Ames test was reported by Scott and Stoltz [194] in the following order: AOH, TA, and iso-TA as nonmutagenics; AME as weak mutagenic; and ATX-I and ATX-II with significant mutagenic activities. These early findings have been confirmed by several successive studies.

Woody and Chu [18] have reported the mutagenic activity of ATX-I to TA98 and the mutagenic activity of ATX-II also on TA100, without toxicity to bacteria in any test. Moreover, they recognized a broader activity for dibenzo-α-pyrones, and found altenuisol and AOH mutagenic to TA100, and AME mutagenic also to TA98 without metabolic activation. Altenuisol, AOH, and AME were, in decreasing degrees, toxic to the bacteria. Finally, they confirmed TA and derived products as not mutagenic and toxic to TA100.

In respect to mutagenic activity of altertoxins, Stack et al. [151] have compared the activity of ATX-I, ATX-II, and ATX-III to that of aflatoxin B_1, the well-known mutagenic and carcinogenic mycotoxins of *Aspergillus flavus* Link. The results indicated that all altertoxins were mutagenic in the following, decreasing order: ATX-III, ATX-II, and ATX-I. However, the activity of ATX-III, the most active altertoxin, was at least 10-fold lower than that of aflatoxin B_1. In fact, in TA100 system in the presence of S9, ATX-III generated only 0.7 revertant, compared with seven revertants generated by aflatoxin B_1 [126,151]. Davis and

Stack [215] have found stemphyltoxin III highly mutagenic on T98 and T1537 *S. typhimurium* systems, but the minimum active dose was higher than that of ATX-III and ATX-II.

The mutagenic activity of altertoxins and the most active dibenzo-α-pyrone AME was also confirmed in other systems, also used to assess the carcinogenic potentiality. In particular the activities of ATX-I and ATX-II were assayed by Osborne et al. [216] in the Raji cell Epstein-Barr virus early antigen (EBV-EA) induction system, and in murine fibroblast (C3H/10T-1/2) cell transformation system. Both altertoxins not only significantly enhanced fibroblast transformation, but also increased the activation of EBV-EA expression by 8- to 10-fold, strongly suggesting a mutagenic/carcinogenic role of these metabolites.

A similar mutagenic/carcinogenic activity was reported in China for culture extracts of strains of *A. alternata*, containing AME as the main active metabolite, collected from the high-esophageal cancer areas of the Linxian region [195]. Successively AME was shown to induce 6-thioguanine-resistant mutants in V79 *S. typhimurium* cells and to cause transformation of NIH/3T3 mouse fibroblast cells [99]. Furthermore, AME was also reported to be mutagenic to *E. coli*, causing from 6- to 10-fold increase in revertants as compared to the control [196].

The occurrence of higher incidence of mutagenic *Alternaria* strains in grains in high-esophageal cancer region of Linxian led to the hypothesis that such strains might be responsible for esophageal human cancer. To this regard An et al. [196] reported that the mutagenic/carcinogenic activities exhibited by the culture filtrate of strains of *A. alternata* associated to the esophageal cancer in Linxian region could be explained by those of AME alone. However, other toxicological studies strongly suggest an involvement of other metabolites with more specific mutagenic activity, such as altertoxins that are as common as alternariols in cultures of *A. alternata* [126,215,216]. Moreover, more recent, targeted studies strongly suggest that an involvement of *Fusarium moniliforme* Sheld. and their carcinogenic fumonisins, which are frequently isolated from maize in the same epidemic Linxian areas, must be primarily considered [217,218].

5. Phytotoxicity

In this section is briefly discussed only the phytotoxic activity of the major *Alternaria* metabolites considered as potential mycotoxins.

Tenuazonic acid showed a severe growth-inhibiting activity, on both the roots and shoots of germinating seeds of tomato. The phytotoxic effect was highly correlated to the TA content in the tested solutions. The 50% root growth-inhibiting concentration (ED_{50}) for TA was calculated to be 22 μg/ml. When tested on tomato cuttings by the usual methods [219], TA induced severe chlorosis and necrosis on the leaves, with a limited water loss in the leaves. The "minimal dose" was established at 130 mg/kg fresh plant weight [74].

When assayed on tomato leaves by leaf puncture assay, TA was observed to be the most toxic *Alternaria* metabolite, showing activity at a concentration of 2 μg per spot, followed by ATX-I and ATX-II with a minimum active dose of 5 μg per spot. In this assay, ALT and isoALT showed a very low activity, up to 20 μg per spot; AOH, AME, altenuisol, and tentoxin were not active at these levels [204].

Tenuazonic acid has been proved to be the major vivotoxin in naturally *P. oryzae*-infected rice plants [220] and *Datura innoxia* Mill. [221], and it has been isolated as the halo-inducing toxin in tobacco brown-spot disease [133]. Moreover, TA has been reported to reduce protein and chlorophyll content in leaves of *D. innoxia*, and also to inhibit protein and nucleic acid syntheses in cultured soybean cells [221,222]. Preliminary observations on the phytotoxic effects of AOH and AME showed the ability of these toxins to induce chlorosis in tobacco [144].

Phytotoxic activities have also been reported for altertoxins. The mode of action of altertoxins could be similar to that of other fungal perylenequinone metabolites like alteichin, from *Alternaria eichorniae* Nag Ray *et* Ponappa [223], which causes structural changes in plant membranes.

V. CONCLUSIONS

The toxicological aspects of the *Alternata* representatives are almost entirely restricted to the cosmopolitan and very multiform species *A. alternata*, also known under the name of *A. tenuis*, which includes saprophytic and opportunistic pathogenic forms, causing diseases on numerous plant species. In colonized tissues most strains of *A. alternata* are able to accumulate metabolites toxic to plants (phytotoxins) and/or animals (mycotoxins).

Although information on the incidence and levels of *Alternaria* mycotoxins in foods is still inadequate, the data clearly indicate that the frequency and contamination levels observed in specific surveys can be quite high, especially in freshly harvested plant products affected by *A. alternata* rots and blackmolding, i.e., dry or soft black rot of vegetables and fruits, heart rot of citrus, black heads, and black point in cereals. Particular toxicological risks are posed by relatively few plant products: tomatoes, apples, citrus, and related processed products; and by black heads and weathered cereals close to harvest time. However, there is an urgent need for more surveys of foodstuffs in order to evaluate human exposure to *Alternaria* toxins and their related toxicological importance.

As pointed out by several reports on the toxicity of *Alternaria*, the crude culture extracts show a degree of toxicity to rats, mice, or chicken, surprisingly higher than that shown by the more common *Alternaria* toxins. This strongly suggests the need for further studies to assess the synergistic toxic effects exhibited by AOH and AME, as well as other synergisms [17,139], in order to better clarify the toxicity of *Alternaria* mycotoxins. In view thereof, it is easy to presume also the occurrence of unidentified mycotoxins in extracts from cultures, which could explain the high toxicity of crude extracts, not corresponding to the presence or concentrations of known mycotoxins [73,224].

Moreover, the mutagenic activity of altertoxins, exhibited to a lesser degree also by AME [126,196,216], strongly suggests more investigations on the mutagenic/carcinogenic metabolites of *A. alternata* strains colonizing plants and food products, also to assess the impact of the many *Alternaria* metabolites classified as phytotoxins on the human and animal health, and in particular of AAL-toxins characterized by a basic chemical structure and physiologic activities surprisingly similar to those of the carcinogenic fumonisins [159,218,225,226].

REFERENCES

1. EG Simmons. *Alternaria* taxonomy: current status, viewpoint, challenge. In: J Chelkowski, A Visconti, eds. *Alternaria*—Biology, Plant Disease and Metabolites. Amsterdam: Elsevier, 1992, p 1.
2. SH Yu. Occurrence of *Alternaria* species in countries of the far East and their taxonomy. In: J Chelkowski, A Visconti, eds. *Alternaria*—Biology, Plant Disease and Metabolites. Amsterdam: Elsevier, 1992, p 37.
3. K Kwasna. Ecology and nomenclature of *Alternaria*. In: J Chelkowski, A Visconti, eds. *Alternaria*—Biology, Plant Disease and Metabolites. Amsterdam: Elsevier, 1992, p 63.
4. K Kwasna. Occurrence of *Alternaria* species in Poland. In: J Chelkowski, A Visconti, eds. *Alternaria*—Biology, Plant Disease and Metabolites. Amsterdam: Elsevier, 1992, p 301.
5. JO Strandberg. *Alternaria* species that attack vegetable crops: biology and options for disease management. In: J Chelkowski, A Visconti, eds. *Alternaria*—Biology, Plant Disease and Metabolites. Amsterdam: Elsevier, 1992, p 175.
6. J Rotem. The Genus *Alternaria*—Biology, Epidemiology and Pathogenicity. St. Paul, MN: APS Press, American Phytopathological Society, 1994, p 326.
7. RP Scheffer. Ecological and evolutionary roles of toxins from *Alternaria* species pathogenic to plants. In: J Chelkowski, A Visconti, eds. *Alternaria*—Biology, Plant Disease and Metabolites. Amsterdam: Elsevier, 1992, p 101.
8. H Otani, K Kohmoto. Host-specific toxins of *Alternaria* species. In: J Chelkowski, A Visconti, eds. *Alternaria*—Biology, Plant Disease and Metabolites. Amsterdam: Elsevier, 1992, p 123.
9. EE Stinson. Mycotoxins—their biosynthesis in *Alternaria*. J Food Production 48:80, 1985.
10. LM Seitz. *Alternaria* metabolites. In: V Betina, ed. Mycotoxins—Production, Isolation, Separation and Purification. Amsterdam: Elsevier, 1984, p 443.
11. N Montemurro, A Visconti. *Alternaria* metabolites—chemical and biological data. In: J Chelkowski, A Visconti, eds. *Alternaria*—Biology, Plant Disease and Metabolites. Amsterdam: Elsevier, 1992, p 449.

12. A Visconti, A Sibilia. *Alternaria* toxins. In: JD Miller, HL Trenholm, eds. Mycotoxins in Grain—Compounds Other Than Aflatoxin. St. Paul, MN: Eagan Press, 1994, p 315.
13. AD King Jr, JE Schade. *Alternaria* toxins and their importance in food. J Food Protection 47:886, 1984.
14. DH Watson. An assessment of food contamination by toxic products of *Alternaria*. J Food Protection 47:485, 1983.
15. RA Coulombe JR. *Alternaria* toxins. In: RP Sharma, DK Salunkhe, eds. Mycotoxins and Phytoalexins. Boca Raton: CRC Press, 1991, p 425.
16. RW Pero, H Posner, M Blois, D Harvan, JW Spalding. Toxicity of metabolites produced by the "Alternaria." Environ Health Perspect 6:87, 1973.
17. DJ Harvan, RV Pero. The structure and toxicity of the *Alternaria* metabolites. In: JV Rodricks, ed. Mycotoxins and Other Related Food Problems. Adv Chem Ser 149: 344, 1976.
18. MA Woody, FS Chu. Toxicology of *Alternaria* mycotoxins. In: J Chelkowski, A Visconti, eds. *Alternaria*—Biology, Plant Disease and Metabolites. Amsterdam: Elsevier, 1992, p 409.
19. P Neergard. Danish species of *Alternaria* and *Stemphylium*, Copenhagen: Communs Phytopath Lab JE Ohlens Enke, 1945, p 560.
20. MB Ellis. Dematiaceous Hyphomycetes. Kew, UK: Commonwealth Mycological Institute, 1971, p 608.
21. MB Ellis. More Dematiaceous Hyphomycetes. Kew, UK: Commonwealth Mycological Institute, 1976, p 507.
22. EG Simmons. Typification of *Alternaria*, *Stemphylium*, and *Ulocladium*. Mycologia 59:67, 1967.
23. A Logrieco, A Visconti, A Bottalico. Mandarin fruit rot caused by *Alternaria alternata* and associated mycotoxins. Plant Dis 74:415, 1990.
24. JI Pitt, AD Hocking. Fungi and Food Spoilage. Sydney: Academic Press, 1985, p 413.
25. HL Lloyd. *Alternaria* leaf spot of Tobacco. 2. Independent segregation of morphological and virulence traits in conidial populations of *Alternaria tenuis* Nees. Mycopathol Mycol Appl 47:317, 1972.
26. ED Caldas, AD Jones, B Ward, CK Winter, DG Gilchrist. Structural characterization of three new AAL toxins produced by *Alternaria alternata* f. sp. *lycopersici*. J Agric Food Chem 42:327, 1994.
27. WA Ayer, LM Pena-Rodriguez. Metabolites produced by *Alternaria brassicae*, the black spot pathogen of canola. Part I. The phytotoxic components. J Nat Prod 50: 400, 1987.
28. JM Dingley. Records of fungi parasitic on plants in New Zealand 1966–68. NZ J Agric Res 13:325, 1970.
29. DF Farr, GF Bills, GP Chamuris, AY Rossman. Fungi on Plant and Plant Products in the United States. St Paul, MN: APS Press, American Phytopathological Society, 1989, p 1252.
30. KH Domsh, W Gams, TH Anderson. Alternaria Nees ex Fr. 1821. Compendium of Soil Fungi, IHV Verlag 1:34, 1993.
31. VG Rao. The genus *Alternaria* from India. Nova Hedwigia 17:219, 1969.

32. VR Bruce, ME Stack, PB Mislivec. Incidence of toxic *Alternaria* in small grains from the USA. J Food Sci 49:1626, 1984.
33. A Logrieco, A Bottalico, M Solfrizzo, G Mulè. Incidence of *Alternaria* species in grains from Mediterranean countries and their ability to produce mycotoxins. Mycologia 82:501, 1990.
34. A Visconti, A Logrieco, M Vurro, A Bottalico. Tenuazonic acid in black-mold tomatoes: occurrence, production by associated *Alternaria* species, and phytotoxic properties. Phytopathol Medit 26:125, 1987.
35. J Pelhate. Recherche des besoins eu eau chez quelques moisissures des grains. Mycopathol Mycol Appl 36:117, 1968.
36. J Mouchacca, P Joly. Essais d'application de methodes de traitment numerique des informations systematiques. 3. Etude de l'action des variations du pH sur le developpement de quelques champignons des sols desertiques. Bull Trimest Soc Mycol Fr 85:503, 1969.
37. RD Milholland. A leaf spot disease of highbush blueberry caused by *Alternaria tenuissima*. Phytopathology 63:1395, 1973.
38. KH Anahosur. *Alternaria triticina*. In: CMI Descr Pathog Fungi and Bacteria, No. 583, London, 1978.
39. SK Hasija. Physiological studies of *Alternaria citri* and *A. tenuis*. Mycologia 62:289, 1970.
40. K Arsvoll. Fungi causing winter damage on cultivated grasses in Norway. Meld Norg Landbrhoeisk 54:49, 1975.
41. AZ Joffe. Biological properties of some toxic fungi isolated from overwintered cereals. Mycopathol Mycol Appl 16:201, 1962.
42. H Von Pechmann. Der Einfluss der Temperatur auf das Wachstum von Blaeuepilzen. In: G Becker, W Liese, eds. Holz und Organismen. Beih Mater Org 1:237, 1966.
43. VT Panasenko. Ecology of microfungi. Bot Rev 33:189, 1967.
44. V Verma. Effect of temperature and hydrogen concentration on three pathogenic fungi. Sydowia 23:164, 1969.
45. SO Orynbaev, BD Ermekova. Effect of pH of the medium on the development of *Helminthosporium sativum* and *Alternaria tenuis*. Mikol Fitopatol 7:539, 1973.
46. DM Griffin. Soil physical factors and the ecology of fungi. 4. Influence of the soil atmosphere. Trans Br Mycol Soc 49:115, 1966.
47. MN Follstad. Mycelial growth rate and sporulation of *Alternaria tenuis*, *Botrytis cinerea*, *Cladosporium herbarum*, and *Rhizopus stolonifer* in low-oxygen atmospheres. Phytopathology 56:1098, 1966.
48. JM Wells, M Uota. Germination and growth of five fungi in low-oxygen and high carbon dioxide atmospheres. Phytopathology 60:50, 1970.
49. N Magan, J Lacey. Effects of gas composition and water activity on growth of field and storage fungi and their interactions. Trans Br Mycol Soc 82:305, 1984.
50. J Lacey. Effects of environment on growth and mycotoxin production by *Alternaria* species. In: J Chelkowski, A Visconti, eds. *Alternaria*—Biology, Plant Disease and Metabolites. Amsterdam: Elsevier, 1992, p 381.
51. RC Pearson, DH Hall. Factors affecting the occurrence and severity of blackmold of ripe tomato fruits caused by *Alternaria alternata*. Phytopathology 65:1352, 1975.
52. DS Meredith. Violent spore release in some fungi imperfecti. Ann Bot 27:39, 1963.

53. M Firpi, O Verona. Influenza della luce sullo sviluppo di *Alternaria tenuis* ed *Ulocladium chartarum*, agenti di deterioramento di carta e cartoni. Cellulosa Carta 1971:25, 1971.
54. PK Bhargava, MN Khare. Epidemiology of *Alternaria* blight of chickpea. Indian Phytopathol 41:195, 1988.
55. H Etzold. Die wirkung des lichtes auf einige Pilze und ihre Spektrale Grenze zum Langwelligen hin. Arch Mikrobiol 37:226, 1960.
56. CM Leach. Influence of relative humidity and red-infrared radiation on violent spore release by *Drechslera turcica* and other fungi. Phytopathology 65:1303, 1975.
57. CM Leach. Sporulation of diverse species of fungi under near-ultraviolet radiation. Can J Bot 40:151, 1962.
58. CH Dickinson, D Bottomley. Germination and growth of *Alternaria* and *Cladosporium* in relation to their activity in the phylloplane. Trans Br Mycol Soc 74:309, 1980.
59. N Magan, J Lacey. Effect of temperature and pH on water relations of field and storage fungi. Trans Br Mycol Soc 82:71, 1984.
60. N Magan, J Lacey. The effect of water activity and temperature on mycotoxin production by *Alternaria alternata* in culture and wheat grain. In: J Lacey, ed. Trichothecenes and other mycotoxins. Chichester: John Wiley, 1985, p 243.
61. JE Machacek, HAH Wallace. Longevity of some common fungi in cereal seed. Can J Bot 30:164, 1952.
62. RC Russel. Longevity studies with wheat seed and certain seed-borne fungi. Can J Plant Sci 38:29, 1958.
63. P Joly. Le Genre *Alternaria*, Recherches Physiologiques, Biologiques et Systematiques. Paris: Lechevalier, 1964, p 250.
64. B Lacicowa. Interaction between some fungi living on cereal seeding material. Acta Mycol 9:7, 1973.
65. M Dawood. Seed-borne fungi, especially pathogens of spring wheat. Acta Mycol 18: 83, 1982.
66. JM Harvey, WT Pentzer. Market diseases of grapes and other small fruits. Washington: US Department of Agriculture, 1960, p 1.
67. H Udagawa, T Kohguchi, H Otani, K Kohmoto. A decade of transition of polyoxin-tolerant strains of *Alternata alternata* Japanese pear pathotype in the field ecosystem. J Fac Agric Tottori Univ 18:9, 1983.
68. M Wittkowski, W Baltes, W Kronert, R Weber. Determination of *Alternaria* toxins in fruit and vegetable products. Z Lebensm Unters Forsch 177:447, 1983.
69. S Maeno, K Kohmoto, H Otani, S Nishimura. Different sensitivities among apple and pear cultivars to AM-toxin produced by *Alternaria alternata* apple pathotype. J Fac Agric Tottori Univ 19:8, 1984.
70. I Vinas, J Bonet, V Sanchis. Incidence and mycotoxin production by *Alternaria tenuis* in decayed apples. Lett Appl Microbiol 14:284, 1992.
71. ALE Mahmoud, SA Omar. Enzymatic activity and mycotoxin-producing potential of fungi isolated from rotted lemon. Cryptogamie Mycol 15:117, 1994.
72. A Visconti, A Logrieco, A Bottalico. Natural occurrence of *Alternaria* mycotoxins in olives—their production and possible transfer into the oil. Food Addit Contam 3:323, 1986.

73. A Bottalico, A Logrieco. *Alternaria* plant diseases in Mediterranean countries and associated mycotoxins. In: J Chelkowski, A Visconti, eds. *Alternaria*—Biology, Plant Disease and Metabolites. Amsterdam: Elsevier, 1992, p 209.
74. A Bottalico, A Logrieco, A Visconti. Presenza di isolati tossigeni e di micotossine di *Alternaria alternata* nei frutti di pomodoro e di peperone affetti da nerume. Difesa Piante 12:163, 1989.
75. TA Webb, JO Mundt. Molds on vegetables at the time of the harvest. Appl Environ Microbiol 35:655, 1978.
76. J Grabarkiewicz-Szczesna, J Chelkowski. Metabolites produced by *Alternaria* species and their natural occurrence in Poland. In: J Chelkowski, A Visconti, eds. *Alternaria*—Biology, Plant Disease and Metabolites. Amsterdam: Elsevier, 1992, p 363.
77. J Harwig, PM Scott, DR Stoltz, BJ Blanchfield. Toxins of molds from decaying tomato fruit. Appl Environ Microbiol 38:267, 1979.
78. EE Stinson, SF Osman, EG Heisler, J Siciliano, DD Bills. Mycotoxin production in whole tomatoes, apples, oranges, and lemon. J Agric Food Chem 29:790, 1981.
79. A Logrieco, A Bottalico, A Visconti, M Vurro. Natural occurrence of *Alternaria*-mycotoxins in some plant products. Microbiol Ailment Nutr 6:13, 1988.
80. SW Khodke, KB Gahukar. Fruit rot disease of chilli caused by *Alternaria alternata* (Fr.) Keissler in Maharashtra. Pkv Res J 17:206, 1993.
81. K Tylkowska. Significant of carrot seeds infection with *Alternaria radicina* M.D. *et* E. Rocz. AR Poznan, 1988, p 189.
82. BJ Jacobsen, KS Harlin, SP Swanson, et al. Occurrence of fungi and mycotoxins associated with field mold damaged soybean in the midwest. Plant Dis 79:86, 1995.
83. HAH Hasan. *Alternaria* mycotoxins in black rot lesion of tomato fruit: conditions and regulation of their production. Mycopathologia 130:171, 1995.
84. CM Christensen, HH Kaufmann. Deterioration of stored grains by fungi. Annu Rev Phytopathol 3:69, 1965.
85. CM Christensen, HH Kaufmann. Grain Storage. The Role of Fungi in Quality Loss. Minneapolis: University of Minnesota Press, 1969, p 153.
86. B Flannigan. Mycoflora of dried barley grain. Trans Br Mycol Soc 53:371, 1969.
87. B Flannigan. Comparison of seed-borne mycofloras of barley, oats and wheat. Trans Br Mycol Soc 55:267, 1970.
88. MIA Abdel-Kader, AH Moubasher, SII Abdel-Hafez. Survey of the mycoflora of barley grains in Egypt. Mycopathologia 69:143, 1979.
89. JT Mills, HAH Wallace. Mycoflora and condition of cereal seeds after a wet harvest. Can J Plant Sci 59:645, 1979.
90. JH Clarke, ST Hill. Mycofloras of moist barley during sealed storage in farm and laboratory silos. Trans Br Mycol Soc 77:557, 1981.
91. S Sanchis, A Sanclemente, J Usall, I Vinas. Incidence of mycotoxigenic *Alternaria alternata* and *Aspergillus flavus* in barley. J Food Prot 56:246, 1993.
92. J Pelhate. Inventaire de la mycoflore des bles de conservation. Bull Trimest Soc Mycol Fr 84:127, 1968.
93. AH Moubasher, MA Elnaghy, SI Abdel-Hafez. Studies on the fungus flora of three grains in Egypt. Mycopathol Mycol Appl 47:261, 1972.
94. HAH Wallace, RN Sinha, JT Mills. Fungi associated with small wheat bulks during prolonged storage in Manitoba. Can J Bot 54:1332, 1976.

95. M Saito, K Ohtsubo, M Umeda, et al. Screening test using HeLa cells and mice for detection of mycotoxin-producing fungi isolated from foodstuffs. Jpn J Exp Med 41:1, 1971.
96. LM Seitz, DB Sauer, HE Mohr, R Burroughs. Weathered grain sorghum: natural occurrence of alternariols and storability of the grain. Phytopathology 65:1259, 1975.
97. AA Ansari, AK Shrivastava. Natural occurrence of *Alternaria* mycotoxins in sorghum and ragi from North Bihar. Food Addit Contam 7:815, 1990.
98. RW Lichtwardt, GL Barron, LW Tiffany. Mold flora associated with shelled corn in Iowa. Iowa State Coll J Sci 33:1, 1958.
99. GT Liu, J Miao, YZ Zhen, et al. Studies of mutagenicity of *Alternaria alternata* in grain from the areas with high incidence of esophageal cancer. Proc Jpn Assoc Mycotoxicol Special Issue No 1:131, 1988.
100. JE Machacek, FJ Greaney. The "black point" of kernel smudge disease of cereals. Can J Res Sect C16:84, 1938.
101. TP Bhowmik. *Alternaria* seed infection of wheat. Plant Dis Rep 53:77, 1969.
102. RJ Southwell, JF Brown, PTW Wong. Effect of inoculum density, stage of plant growth and dew period on the incidence of black point caused by *Alternaria alternata* in durum wheat. Ann Appl Biol 96:29, 1980.
103. C Orsi, G Chiusa, V Rossi. Ulteriori indagini sulla micoflora delle cariossidi di frumento duro, in rapporto alla volpatura. Petria 4:225, 1994.
104. RA Kilpatrik. Factors affecting black point of wheat in Texas 1964–67. Texas Agric Exp Stn Misc Publ No. 884, 1968, p 11.
105. RL Conner, JB Thomas. Genetic variation and screening techniques for resistance to black point in soft white spring wheat. Can J Plant Pathol 7:402, 1985.
106. HI Nirenberg, H Schmitz-Elsherif, CI Kling. Auftreten von Fusarien und Schwärzepilzen an Durumweizen in Deutschland. II. Befall mit Schwärzepilzen. Z Pfanzenkrankh Pfanzensch 102:164, 1995.
107. J Grabarkiewicz-Szczesna, J Chelkowski, P Zajkowski. Natural occurrence of *Alternaria* mycotoxins in the grain and chaff of cereals. Mycotoxin Res 5:77, 1989.
108. M Mueller. Untersuchungen zum *Alternaria*-Befall von Silomais und Heu. Zentralbl Mikrobiol 146:481, 1991.
109. A Codignola, M Gallino. Su alcuni micromiceti isolati da insilati di mais. 1. Attività antibiotica. Allionia 20:43, 1974/1975.
110. A Torres, S Chulze, E Varsavsky, MI Rodriguez. *Alternaria* metabolites in sunflower seeds. Incidence and effect of pesticides in their production. Mycopathologia 121:17, 1993.
111. BDL Fitt, I Vloutoglou. *Alternaria* diseases of linseed. In: J Chelkowski, A Visconti, eds. *Alternaria*—Biology, Plant Disease and Metabolites. Amsterdam: Elsevier, 1992, p 289.
112. Y Bashan, Hernandes-Saavedra. In: J Chelkowski, A Visconti, eds. *Alternaria*-blight of cotton: epidemiology and transmission. *Alternaria*—Biology, Plant Disease and Metabolites. Amsterdam: Elsevier, 1992, p 233.
113. M Wiewiorowska, B Godlewska. *Alternaria* and *Ulocladium* species of rapeseed seeds in Poland. Abstracts of Second European Seminar "*Fusarium*—mycotoxins, taxonomy and pathogenicity, and *Alternaria* metabolites," Poznan, 1990, p 61.
114. I Vinas, J Palma, S Garza, A Sibilia, V Sanchis, A Visconti. Natural occurrence of

aflatoxin and *Alternaria* mycotoxins in oilseed rape from Catalonia (Spain): incidence of toxigenic strains. Mycopathologia 128:175, 1994.
115. R Kumar, PP Gupta, RD Parashar, PR Kumar. Detection of seedborne microorganisms on different media in rapeseed and mustard. Indian J Mycol Plant Pathol 23: 169, 1993.
116. A Visconti, A Sibilia, A Sabia. *Alternaria alternata* from oilseed rape: mycotoxin production and toxicity to *Artemia salina* larvae and rape seedlings. Mycotoxin Res 8:9, 1992.
117. AZ Joffe. The mycoflora of fresh and stored groundnut kernels in Israel. Mycopathol Mycol Appl 39:255, 1969.
118. F Senser. Untersuchungen zum Aflatoxingehalt in Haselnüssen. Gordian 79:117, 1979.
119. HW Schroeder, RJ Cole. Natural occurrence of alternariols in discolored pecans. J Agric Food Chem 25:204, 1977.
120. LH Huang, RT Hanlin. Fungi occurring in freshly harvested and in-market pecans. Mycologia 67:689, 1975.
121. N Misra. Influence of temperature and relative humidity on fungal flora of some species in storage. Z Lebensmittelunters Forsch 172:30, 1981.
122. T Grzybowska. The effect of fungicides on marjoram (*Origanum majorana* L.) and pathogenicity of four fungus species isolated from it. Roczn Nauk Roln Ser E 17: 159, 1988.
123. WB Van der Riet. Studies on the mycoflora of biltong. S Afr Food Rev 3:105, 1976.
124. N Inagaki. On some fungi isolated from foods. Trans Mycol Soc Jpn 4:1, 1962.
125. GF Griffin, FS Chu. Toxicity of *Alternaria* metabolites alternariol, alternariol methyl ether, altenuene, and tenuazonic acid in the chicken embryo assay. Appl Environ Microbiol 46:1420, 1983.
126. ME Stack, MJ Prival. Mutagenicity of the *Alternaria* metabolites altertoxins I, II, and III. Appl Environ Microbiol 52:718, 1986.
127. CE Stickings. Studies in biochemistry of micro-organisms. 106. Metabolites of *Alternaria tenuis* Auct.: the structure of tenuazonic acid. Biochem J 72:332, 1959.
128. PS Steyn, CJ Rabie. Characterization of magnesium and calcium tenuazonate from *Phoma sorghina*. Phytochemistry 15:1977, 1976.
129. RA Meronuck, JA Steele, CJ Mirocha, CM Christensen. Tenuazonic acid, a toxin produced by *Alternaria alternata*. Appl Microbiol 23:613, 1972.
130. KS Bilgrami, AA Ansari, AK Sinha, AK Shrivastava, KK Sinha. Mycotoxin production by some Indian *Alternaria* species. Mycotoxin Res 10:56, 1994.
131. T Kinoshita, Y Renbutsu, ID Khan, K Kohmoto, S Nishimura. Distribution of tenuazonic acid production in the genus *Alternaria* and its pathological evaluation. Ann Phytopathol Soc Jpn 38:397, 1972.
132. C Ramm, GB Lucas. Production of enzymes and antibiotic substances by *Alternaria longipes*. Tobacco Sci 7:81, 1963.
133. Y Mikami, Y Nishijima, H Iimura, A Suzuki, S Tamura. Chemical studies on brown-spot disease of tobacco plants. I. Tenuazonic acid as a vivotoxin of *Alternaria longipes*. Agric Biol Chem Tokyo 35:611, 1971.
134. ND Davis, UL Diener, G Morgan-Jones. Tenuazonic acid production by *Alternaria alternata* and *Alternaria tenuissima* isolated from cotton. Appl Environ Microbiol 34:155, 1977.

135. PB Mislivec, VR Bruce, ME Stack, R Bandler. Moulds and tenuazonic acid in fresh tomatoes used for catsup production. J Food Prot 50:38, 1987.
136. N Umetsu, J Kaji, K Tamari. Investigation on the toxin production by several blast fungus strains and isolation of tenuazonic acid as a novel toxin. Agric Biol Chem Tokyo 36:859, 1972.
137. CJ Rabie, SJ Van Rensberg, JJ Van Der Watt, A Lubben. Onyalai—the possible involvement of a mycotoxin produced by *Phoma sorghina* in the aetiology. S Afr Med J (Nutr Suppl) 49:1647, 1975.
138. H Raistrick, CE Stickings, R Thomas. Studies on the biochemistry of microorganisms. 90. Alternariol and alternariol monomethyl ether, metabolic products of *Alternaria tenuis*. Biochem J 55:421, 1953.
139. RW Pero, D Harvan. Simultaneous detection of metabolites from several toxigenic fungi. J Chromatogr 80:255, 1973.
140. GG Freeman. Isolation of alternariol and alternariol monomethyl ether from *Alternaria dauci* (Kühn) Groves and Skolo. Phytochemistry 5:719, 1965.
141. AN Starrat, GA White. Identification of some metabolites of *Alternaria cucumerina* (E. & E.) Ell. Phytochemistry 7:1883, 1968.
142. K Kameda, H Akoi, H Tanaka, M Namiki. Studies on metabolites of *Alternaria kikuchiana* Tanaka, a phytopathogenic fungus of japanese pear. Agric Biol Chem 37: 2137, 1973.
143. GA Pollock, CE DiSabotino, RC Heimsch, DR Hilbelink. The subchronic toxicity and teratogenicity of alternariol monomethyl ether produced by *Alternaria solani*. Food Chem Toxicol 20:899, 1982.
144. RW Pero, CE Main. Chlorosis of tobacco induced by alternariol monomethyl ether produced by *Alternaria tenuis*. Phytopathology 60:1570, 1970.
145. RW Pero, RG Owens, SW Dale, D Harvan. Isolation and identification of a new toxin, altenuene, from the fungus *Alternaria tenuis*. Biochim Biophys Acta 230:170, 1971.
146. AT McPhail, RW Miller, D Harvan, RW Pero. X-ray crystal structure revision for the fungal metabolite (±)-altenuene. J Chem Soc Chem Commun p. 882, 1973.
147. A Visconti, A Bottalico, M Solfrizzo, F Palmisano. Isolation and structure elucidation of isoaltenuene, a new metabolite of *Alternaria alternata*. Mycotoxin Res 5:69, 1989.
148. T Rosett, RH Sankhala, CE Stickings, MEU Taylor, R Thomas. Studies on the biochemistry of micro-organisms. 103. Metabolites of *Alternaria tenuis* Auct: culture filtrate products. Biochem J 67:390, 1957.
149. RW Pero, D Harvan, MC Blois. Isolation of the toxin, altenuisol, from the fungus *Alternaria tenuis* Auct. Tetrahedron Lett 12:945, 1973.
150. EE Stinson, SF Osman, PE Pfeffer. Structure of Altertoxin I, a mycotoxin from *Alternaria*. J Org Chem 47:4110, 1982.
151. ME Stack, EP Mazzola, SW Page, et al. Mutagenic perylenequinone metabolites of *Alternaria alternata*: altertoxins I, II, and III. J Nat Prod 49:866, 1986.
152. A Arnone, G Nasini, L Merlini, G Assante. Secondary mould metabolites. Part 16. Stemphyltoxins. New reduced perylenequinone metabolites from *Stemphylium botryosum* var. *lactucum*. J Chem Soc Perkin Trans 1:525, 1986.
153. ME Stack, EP Mazzola. Stemphyltoxin III from *Alternaria alternata*. J Nat Prod 52:426, 1989.

154. PW Brian, PJ Curtis, HG Hemming, CH Unwin, JM Wright. Alternaric acid, a biologically active metabolic product of the fungus *Alternaria solani*. Nature 164:534, 1949.
155. JF Grove. Alternaric acid. Part I. Purification and characterization. J Chem Soc p. 4056, 1952.
156. CC Reilly, D Gottlieb. The mode of action of alternaric acid on *Myrothecium verrucaria*. Phytopathology 69:550, 1979.
157. ND Fulton, K Bollenbacher, GE Templeton. A metabolite from *Alternaria tenuis* that inhibits chlorophyll production. Phytopathology 55:49, 1965.
158. WL Meyer, GE Templeton, CI Grable, et al. Use of ^{1}H nuclear magnetic resonance spectroscopy for sequence and configuration analysis of cyclic tetrapeptides. The structure of tentoxin. J Am Chem Soc 97:3802, 1975.
159. GE Templeton. *Alternaria* toxins related to pathogenesis in plants. In: S Kadis, A Ciegler e SJ Aji, eds. Microbial Toxins. Vol. VIII. Fungal Toxins. New York: Academic Press, 1972, p 169.
160. DG Gilchrist, RG Grogan. Production and nature of a host-specific toxin from *Alternaria alternata* f. sp. *lycopersici*. Phytopathology 66:165, 1976.
161. AT Bottini, JR Bowen, DJ Gilchrist. Phytotoxins. II. Characterization of a phytotoxic fraction from *Alternaria alternata* f. sp. *lycopersici*. Tetrahedron Lett p. 2723, 1981.
162. N Magan, GR Cayley, J Lacey. Effect of water activity and temperature on mycotoxin production by *Alternaria alternata* in culture and on wheat grain. Appl Environ Microbiol 47:1113, 1984.
163. A Torres, S Chulze, E Varsavsky, A Dalcero, M Etcheverry, C Farnochi. Influencia de la temperatura y la actividad acuosa en la production de micotoxinas de *Alternaria* (*Hyphomycetales*) en girasol. Bol Soc Argent Bot 28:175, 1992.
164. M Etcheverry, S Chulze, A Dalcero, E Varsavsky, C Magnoli. Effect of water activity and temperature on tenuazonic acid production by *Alternaria alternata* on sunflower seeds. Mycopathologia 126:179, 1994.
165. P Häggblom, M Hiltunen. Regulation of mycotoxin biosynthesis in *Alternaria*. In: J Chelkowski, A Visconti, eds. *Alternaria*—Biology, Plant Disease and Metabolites. Amsterdam: Elsevier, 1992, p 435.
166. R Burroughs, LM Seitz, DB Sauer, HE Mohr. Effect of substrate on metabolite production by *Alternaria alternata*. Appl Environ Microbiol 31:685, 1976.
167. EE Stinson, DD Bills, SF Osman, J Siciliano, MJ Ceponis, EG Heisler. Mycotoxin production by *Alternaria* species grown on apples, tomatoes, and blueberries. J Agric Food Chem 28:960, 1980.
168. J Reiss. *Alternaria* mycotoxin in grain and bread. XVII. Mycotoxins in foodstuffs. Z Lebensm Unters Forsch 176:36, 1983.
169. S Ozcelik, N Ozcelik, LR Beuchat. Toxin production by *Alternaria alternata* in tomatoes and apples stored under various conditions and quantitation of the toxins by high-performance liquid chromatography. Int J Food Microbiol 11:187, 1990.
170. AL Robiglio, SE Lopez. Mycotoxin production by *Alternaria alternata* strains isolated from red delicious apples in Argentina. Int J Food Microbiol 24:413, 1995.
171. LM Seitz, DB Sauer, HE Mohr, R Burroughs, JV Paukstelis. Metabolites of *Alternaria* in grain sorghum. Compounds which could be mistaken for zearalenone and aflatoxin. J Agric Food Chem 23:1, 1975.

172. DB Sauer, LM Seitz, R Burroughs, et al. Toxicity of *Alternaria* metabolites found in weathered sorghum grain at harvest. J Agric Food Chem 26:1380, 1978.
173. EW Sydenham, PG Thiel, WFO Marasas. Occurrence and chemical determination of zearalenone and alternariol monomethyl ester in sorghum-based mixed feeds associated with a new metabolite from *Alternaria porri*. Biosci Biotechnol Biochem 57:334, 1988.
174. ME Stack, PB Mislivec, JAG Roach, AE Pohland. Liquid chromatographic determination of tenuazonic acid and alternariol methyl ether in tomatoes and tomato products. J Assoc Off Anal Chem 68:640, 1985.
175. PM Scott, SR Kanhere. Liquid chromatographic determination of tenuazonic acids in tomato paste. J Assoc Off Anal Chem 63:612, 1980.
176. S Gruber-Schley, A Thalmann. Zum Vorkommen von *Alternaria* spp. und deren Toxine in Getreide und moegliche Zusammenhaenge mit Leistungsminderungen landwirtschaftlicher Nutziere. Landwirtsch Forsch 41:11, 1988.
177. AB Young, ND Davis, UL Diener. Effect of temperature and moisture on tenuazonic acid production by *Alternaria tenuissima*. Phytopathology 7:607, 1980.
178. AZ Joffe. The mycoflora of overwintered cereals and its toxicity. Bull Res Counc Isr Sect D 9:101, 1960.
179. AZ Joffe. Toxin production of cereal fungi causing toxic alimentary aleukia in man. In: GN Wogan, ed. Mycotoxins in Foodstuffs. Cambridge: M.I.T. Press, 1965, p 77.
180. CM Christensen, GH Nelson, CJ Mirocha, F Bates. Toxicity to experimental animals of 943 isolates of fungi. Cancer Res 28:2293, 1968.
181. CJ Mirocha, CM Christensen, GH Nelson. Toxic metabolites produced by fungi implicated in mycotoxicoses. Biotechnol Bioengineer 10:469, 1968.
182. B Doupnik Jr, EK Sobers. Mycotoxicosis: toxicity to chicken of *Alternaria longipes* isolated from tobacco. Appl Environ Microbiol 16:1596, 1968.
183. EK Sobers, B Doupnik Jr. Relationship of pathogenicity to tobacco leaves and toxicity to chicks of isolates of *Alternaria longipes*. Appl Microbiol 23:313, 1972.
184. J Forgacs, H Koch, WT Carll, RH White-Stevens. Additional studies on the relationship of mycotoxicoses to the poultry hemorrhagic syndrome. Am J Vet Res 19:744, 1953.
185. J Forgacs, H Koch, WT Carll, RH White-Stevens. Mycotoxicoses. I. Relationship of toxic fungi to moldy feed toxicosis in poultry. Avian Dis 6:363, 1962.
186. SG Yates. Toxin-producing fungi from fescue pasture. In: S Kadis, A Ciegler, S Ajl, eds. Microbial Toxins. Vol. VII. Algal and Fungal Toxins. New York: Academic Press, 1971, p 194.
187. P Hamilton, G Lucas, R Weltz. Mouse toxicity of fungi of tobacco. Appl Microbiol 18:570, 1968.
188. MV Gorlenko. The toxins of molds. Am Rev Soviet Med 5:163, 1948.
189. BD Reddy, DC Kelley, HC Minocha, HD Anthony. Pathogenicity of *Alternaria alternata* and its antibody production in experimental animals. Mycopathol Mycol Appl 54:385, 1974.
190. UL Diener, RE Wagener, G Morgan-Jones, ND Davis. Toxigenic fungi from cotton. Phytopathology 66:514, 1976.
191. MK Slifkin, J Spalding. Studies on the toxicity of *Alternaria mali*. Toxicol Appl Pharmacol 17:375, 1970.

192. MK Slifkin, A Ottolenghi, H Brown. Influence of fungicides on mycotoxin production by *Alternaria mali*. Mycopathol Mycol Appl 50:241, 1973.
193. LF Bjeldanes, GW Chang, SV Thomson. Detection of mutagens produced by fungi with the *Salmonella typhimurium* assay. Appl Environ Microbiol 35:1150, 1978.
194. PM Scott, DR Stoltz. Mutagens produced by *Alternaria alternata*. Mutat Res 78:33, 1980.
195. ZG Dong, GT Liu, ZM Dong, et al. Induction of mutagenesis and transformation by the extract of *Alternaria alternata* isolated from grains in Linxian, China. Carcinogenesis 8:989, 1987.
196. YH An, TZ Zhao, J Miao, et al. Isolation, identification, and mutagenicity of alternariol monomethyl ether. J Agr Food Chem 37:1341, 1989.
197. FA Miller, WA Rightsel, BJ Sloan, J Ehrlich, JC French, QR Barlz. Antiviral activity of tenuazonic acid. Nature Lond 200:1338, 1963.
198. CO Gitterman. Antitumor, cytotoxic, and antibacterial activities of tenuazonic acid and congeneric tetramic acids. J Med Chem 8:483, 1965.
199. BS Drabkin, AZ Joffe. Protistocidal activity of some moulds. Mikrobiologiya 21: 700, 1952.
200. I Levisohn. Antagonistic effects of *Alternaria tenuis* on certain root-fungi of forest trees. Nature London 179:1143, 1957.
201. Z Jeziorska. The influence of fungi isolated from the rhizosphere of chosen species. Pam Pulawski 60:187, 1974.
202. E Csuti, JM Lemaire, J Ponchet, F Rapilly. Exemples d'interactions fongiques au niveau de plantules de ble contaminees par l'*Helminthosporium sativum*. Ann Epiphyt 16:37, 1965.
203. G Gassner, E Niemann. Synergistische und Antagonistische Wirkung von Pilzen und Bakterien auf die Sporenkeimung Verschiedener *Tilletia*-Arten. Phytopath Z 23:395, 1955.
204. A Visconti, A Bottalico, M Solfrizzo. Activity of *Alternata alternata* metabolites on tomato leaves and *Geotrichum candidum*. In: A Graniti, R Durbin, A Ballio, eds. Phytotoxins and Plant Pathogenesis. Berlin: Springer-Verlag, 1989, p 457.
205. JF Grove. Metabolic products of *Stemphylium radicinum*. Part I. Radicinin. J Chem Soc (C) p. 3234, 1964.
206. HK Abbas, JE Mulrooney. Effect of some phytopathogenic fungi and their metabolites on growth of *Heliothis virescens* F. and its host plants. Biocontrol Sci Technol 4:77, 1994.
207. M Cole, GN Rolinson. Microbial metabolites with insecticidal properties. Appl Microbiol 24:660, 1972.
208. W Koefoed-Johnsen, HH Ussing. The nature of the frog skin potential. Acta Physiol Scand 42:298, 1958.
209. L Barbarossa, E Gallucci, A Bottalico, S Micelli. The effect of alternariol mycotoxin (AOH) on active sodium transport across frog skin (*Rana esculenta*). Boll Soc Ital Biol Sper 64:825, 1988.
210. ER Smith, TN Fredrickson, Z Hadidian. Toxic effects of the sodium and the N′-dibenzylethylenediamine salts of tenuazonic acid (NSC-525816 and NSC-82260). Cancer Chemother Rep 52:579, 1968.

211. JJ Giambrone, ND Davis, UL Diener. Effect of tenuazonic acid on young chicken. Poult Sci 57:1554, 1978.
212. BK Boutin, JT Peeler, RM Twedt. Effects of purified altertoxins I, II, and III in the metabolic communication V79 system. Toxicol Environ Health 26:75, 1989.
213. JW Spalding, RW Pero, RG Owens. Inhibition of the G_2 phase of the mammalian cell cycle by the mycotoxin alternariol. J Cell Biol 47:199a, 1970.
214. EA Kaczka, CO Gittermann, EL Dulaney, et al. Discovery of inhibitory activity of tenuazonic acid for growth of human adenocarcinoma-1. Biochem Biophys Res Commun 14:54, 1964.
215. DM Davis, ME Stack. Mutagenicity of stemphyltoxin III, a metabolite of *Alternaria alternata*. Appl Environ Microbiol 57:180, 1991.
216. LC Osborne, VI Jones, JT Peeler, EP Larkin. Transformation of C3H/10T1/2 cells and induction of EBV-early antigen in Raji cells by altertoxins I and III. Toxicol In Vitro 2:97, 1988.
217. FS Chu, GY Li. Simultaneous occurrence of fumonisin B_1 and other mycotoxins in moldy corn collected from the People's Republic of China in regions with high incidences of esophageal cancer. Appl Environ Microbiol 60:847, 1994.
218. WFO Marasas. Fumonisins: their implications for human and animal health. Natural Toxins 3:193, 1995.
219. A Bottalico. Valutazione degli effetti tossici della fusicoccina sulle piante con metodo ponderale. Phytopathol Medit 11:77, 1972.
220. N Umetsu, J Kaji, K Tamari. Isolation of tenuazonic acid from blast-diseased rice plants. Agric Biol Chem Tokyo 37:451, 1973.
221. KK Janardhanan, A Hussain. Phytotoxic activity of tenuazonic acid isolated from *Alternaria alternata* (Fr.) Keissler causing leaf blight of *Datura innoxia* Mill. and its effect on host metabolism. Phytopathol Z 111:305, 1984.
222. N Umetsu, T Muramatsu, H Honda, K Tamari. Studies on the effect of tenuazonic acid on plant cells and seedlings. Agric Biol Chem Tokyo 38:791, 1974.
223. D Robeson, G Strobel, GK Matusumoto, EL Fisher, MH Chen, J Clardy. An unusual phytotoxin from *Alternaria eichorniae*, a fungal pathogen of water hyacinth. Experientia 40:1248, 1984.
224. DN Mortimer, MV Howell, MJ Shepherd. Investigation of toxic *Alternaria* metabolites potentially present in UK-produced foods. Int Biodeterior 24:409, 1988.
225. WT Shier, HK Abbas, CJ Mirocha. Toxicity of the mycotoxins fumonisins B_1 and B_2 and *Alternaria alternata* f. sp. *lycopersici* toxin (AAL) in cultured mammalian cells. Mycopathologia 116:97, 1991.
226. DG Gilchrist, B Ward, V Moussato, CJ Mirocha. Genetic and physiologic response for fumonisin and AAL-toxin by intact tissue of a higher plant. Mycopathologia 117: 57, 1992.
227. RG Coombe, JJ Jacobs, TR Watson. Metabolites of some *Alternaria* species. The structures of altenusin and dehydroaltenusin. Aust J Chem 23:2343, 1970.
228. N Brandburn, RD Coker, G Blunden, CH Turner, TA Crabb. 5′-Epialtenuene and neoaltenuene, dibenzo-α-pyrones from *Alternaria alternata* cultured on rice. Phytochemistry 35:665, 1994.
229. AC Stierle, JH Cardellina II, GA Strobel. Maculosin, a host-specific phytotoxin of

spotted knapweed from *Alternaria alternata*. Proc Natl Acad Sci USA 85:8008, 1988.
230. AC Stierle, JH Cardellina II, GA Strobel. Phytotoxins from *Alternaria alternata*, a pathogen of spotted knapweed. J Natl Prod 52:42, 1989.
231. N Okamura, H Haraguchi, K Hashimoto, A Yagi. Altersolanol-related antimicrobial compounds from a strain of *Alternaria solani*. Phytochemistry 34:1005, 1993.
232. PJ Cotty, IJ Misaghi. Zinniol production by *Alternaria* species. Phytopathology 74:785, 1984.
233. KG Tietjen, E Schaller, U Matern. Phytotoxins from *Alternaria chartami* Chowdhury: structural identification and physiological significance. Physiol Plant Pathol 23:387, 1983.
234. DJ Robeson, GR Gray, GA Strobel. Production of the phytotoxins radicinin and radicinol by *Alternaria chrysanthemi*. Phytochemistry 21:2359, 1982.
235. A Stierle, J Hershenhorn, G Strobel. Zinniol-related phytotoxins from *Alternaria cichorii*. Phytochemistry 32:1145, 1993.
236. I Barash, H Mor, D Netzer, Y Kashman. Production of zinniol by *Alternaria dauci* and its phytotoxic effect on carrot. Physiol Plant Pathol 19:7, 1981.
237. DJ Robeson, GA Strobel. Deoxyradicinin, a novel phytotoxin from *Alternaria helianthi*. Phytochemistry 21:1821, 1982.
238. B Tal, DJ Robeson. The production of pyrenocine A and B by a novel *Alternaria* species. Z Naturforsch 41c:1032, 1986.
239. N Sugiyama, C Kashima, M Yamamoto, R Mohri. Altenin, a new phytopathologically-toxic metabolite from *Alternaria kikuchiana*. Bull Chem Soc Jpn 38:2028, 1965.
240. R Suemitsu, K Ohnishi, M Horiuchi, A Kitaguchi, K Odamura. Porritoxin, a phytotoxin of *Alternaria porri*. Phytochemistry 31:2325, 1992.
241. R Suemitsu, K Ohnishi, M Horiuchi, Y Morikawa, Y Sakaki, Y Matsumoto. Structure of porriolide, a new metabolite from *Alternaria porri*. Biosci Biotechnol Biochem 57:334, 1993.
242. K Ohnishi, R Suemitsu, M Kubota, H Matano, Y Yamada. Biosintheses of alterporriol D and E by *Alternaria porri*. Phytochemistry 30:2593, 1991.
243. A Ichihara, H Tazaki, S Sakamura. Solanapyrones A, B and C phytotoxic metabolites from the fungus *Alternaria solani*. Tetrahedron Lett 24:5373, 1983.
244. A Ichihara, H Tazaki, S Sakamura. The structure of zinnolide, a new phytotoxin from *Alternaria solani*. Agr Biol Chem 49:2811, 1985.
245. AN Starratt. Zinniol: a major metabolite of *Alternaria zinniae*. Can J Chem 46:767, 1968.

4

Design of Sampling Plans for Determination of Mycotoxins in Foods and Feeds

Raymond D. Coker
Natural Resources Institute, University of Greenwich, Kent, England

I. INTRODUCTION

The control of mycotoxins in foods and feeds requires a combination of surveillance and regulatory and quality assurance procedures.

Surveillance provides valuable information on the nature and extent of mycotoxin contamination. However, a "mycotoxin problem" should not be considered in isolation but as an integral component of a broader study, involving interdisciplinary teams, which aims to identify the constraints and opportunities associated with the production and utilization of a particular commodity. In some instances, an evaluation of a complete commodity system (a "plough to the plate" approach) will be required, whereas in other instances a specific component of the system (e.g., a processing subsystem) will be evaluated. The adoption of a holistic, systems approach to the surveillance of mycotoxins facilitates an analysis of the many interacting components of the system (or subsystem) and, subsequently, the identification of those constraints within the system that are leading to the onset of spoilage and, ultimately, to the production of mycotoxins. Consequently, sampling plans are required that ensure that the analytical data generated identifies those key points ("critical points") which are contributing to the spoilage of the commodity and to the production of mycotoxins. Once these critical points have been identified, a suitable intervention program can be implemented and/or the critical points can be integrated into quality assurance schemes (includ-

Figure 1 Sampling a 20-ton lot of unprocessed edible peanuts.

ing Hazard Analysis Critical Control Points [HACCP] management systems), and/or regulatory programs.

Currently, more than 50 countries *regulate* the occurrence of mycotoxins (especially the aflatoxins) in foods and feeds. Commonly, there are specific regulations for both unprocessed and processed material. In the U.K., for example, a lot of unprocessed edible peanuts is acceptable if it contains 10 μg/kg or less of total aflatoxins, whereas a retail pack (a small unit of food purchased by the consumer) of peanuts offered for sale must not contain more than 4 μg/kg of aflatoxins. In this example, attention is focused both upon a *lot* of unprocessed peanuts (typically of the order of 20 tons in weight; Fig. 1) and upon a small, *retail pack* (of around 100 g) of processed nuts. Consequently, sampling plans are required by the regulators that can be applied to both these situations.

Those responsible for the production, processing, and marketing of foods and feeds need to constantly monitor the quality of the commodity they are handling either through the implementation of a traditional quality assurance program or, ideally, through an HACCP management system. Again, a variety of contamination levels and quantities of materials will need to be addressed from, for example, a 50,000-ton shipment of animal feed (Fig. 2) to a 50-g packet of roasted peanuts, and sampling plans will be required which can be applied to a wide variety of situations.

II. SAMPLING SEQUENCE

The *sampling sequence* describes the activities involved during the sequence of events which leads to the provision of a laboratory sample which can be analyzed for the selected mycotoxins. It is summarized in Figure 3.

Before embarking upon any sampling activity it is essential that the *aim* of the activity be clearly defined, especially when undertaking a surveillance study. Frequently, reports of surveys do not include a clear description of the aim of the study and an explanation of how this aim has influenced the design of the surveillance program.

Similarly, the aim of the sampling activity will determine the location of the *sampling points*. Any given commodity can occur in a number of "locations" and "situations." For example, it may be *located* within a warehouse where it is *situated* in bulk or in a variety of containers including sacks, tins, plastic bags, or bottles. Once the locations of the commodity have been carefully identified, the precise location of the sampling points can then be determined. When planning a surveillance program, for example, the identification of the sampling points will require inputs from all members of the interdisciplinary team involved in the study including, of course, a statistician. Ideally, those *critical points*, which are identified by the survey as significantly contributing to the production of mycotoxins,

Figure 2 Discharging a large shipment of animal feed.

(1) Aim of sampling

⇓

(2) Identify sampling point

⇓

(3) Identify lot

⇓

(4) Select sampling method

⇓

(5) Collect sample

⇓

(6) Perform sample preparation

⇓

(7) Laboratory sample

FIGURE 3 Sampling sequence.

can then be utilized as sampling points when designing subsequent regulatory and quality assurance procedures.

Once the sampling points have been identified, *lots* of food or feed that are representative of the system under evaluation have to be selected. Again, if a survey is being undertaken, the selection should represent the views of the interdisciplinary team associated with the study. Similarly, if lots are being selected for regulatory purposes, it is important that a broad range of views be canvassed during the selection process. In the latter case, however, it may be appropriate to occasionally select lots that are specifically suspected of being contaminated with mycotoxins, as opposed to the selection of representative material. It can be especially difficult for regulators to define the lot when samples, in the form of individual retail packs, are collected from supermarket shelves. Lots selected for quality assurance purposes can take a variety of forms. If, for example, the production of jars of peanut butter is under evaluation, a lot may constitute the output from a single shift. Alternatively, if the quality of stored grain is being monitored, a truck load of grain may constitute a lot during the loading and

FIGURE 4 Maize sack-stack under construction in Zambia.

unloading of the store (Fig. 4); considerably larger lots, often involving thousands of tons, may be the focus of attention during storage (Fig. 5).

After the lot(s) have been selected, the next step in the sampling sequence is the collection of samples that are representative of the lot. Most attempts to develop an effective procedure for the collection of representative samples from a lot of food or feed have focused upon the aflatoxins, since the majority of current regulations are concerned specifically with this group of mycotoxins. However, the design of effective sampling procedures has been seriously hampered by the highly positively skewed distribution of the aflatoxins, by difficulties experienced with the withdrawal of the samples (e.g., when sampling an oilseed cake in sacks), and by the significant difficulties experienced when attempting to collect samples from throughout a lot. By observing Figure 5, the problems associated with the sampling of large lots of stored material are self-evident. Wherever possible, samples should be collected when the lot is mobile and, consequently, readily accessible. The large lot shown in Figure 5, for example, can only be properly sampled when the stack is being constructed or dismantled. Ideally, of course, samples should be collected automatically from a moving stream of material; an automatic cross-cut sampler [1], for example, will collect a sample as the material falls from the end of a conveyor belt. The design and execution of effective sampling plans are discussed further in Section III.

Representative samples of foods and feeds are typically between 3 and 20 kg in weight (see Sect. III). Consequently, it is essential that sample preparation

FIGURE 5 Large sack-stack of maize in Zambia.

procedures be available that allow the production of laboratory samples which are representative of the original sample. Laboratory samples may be produced by successively comminuting and dividing the sample, using either static or rotary dividers (Fig. 6). Ideally, the comminution and sample-division steps should be performed simultaneously using a subsampling mill (Fig. 7). The device illustrated converts large samples of particulate materials into comminuted, representative subsamples in a single operation.

III. DEVELOPMENT OF SAMPLING PLANS

A. Sampling, Sample Preparation, and Analytical Variances

The discussion in this section will focus upon steps 4 and 5 of the sampling sequence (see Fig. 3) and will include the development of procedures for the sampling of corn (maize) and peanuts.

The distribution of aflatoxin among peanut and corn kernels is highly positively skewed. It has been reported [2], for example, that only 0.03% of the kernels are contaminated at a mean lot concentration of 5 μg/kg aflatoxin, and that a single contaminated kernel can contain as much as 1100 μg aflatoxin [3].

The skewed distribution of aflatoxin in processed edible peanuts, FAQ peanuts, and peanut cake is illustrated in Table 1. For each commodity, a large

FIGURE 6 Rotary cascade sample divider.

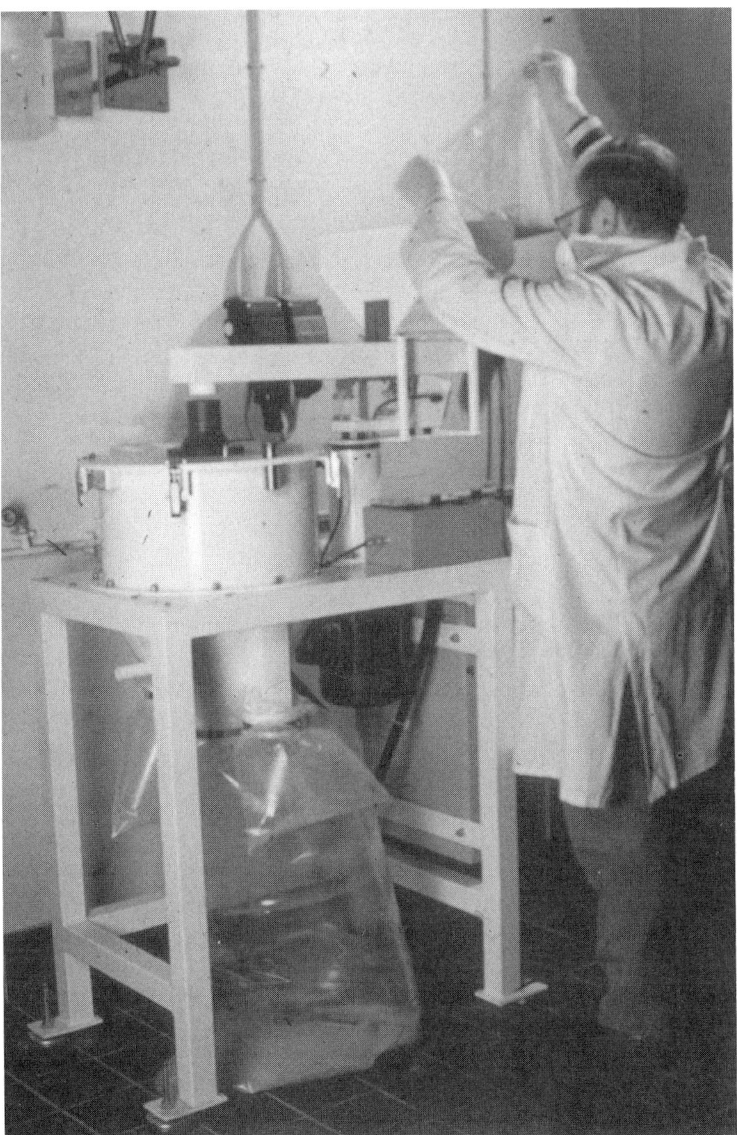

FIGURE 7 Subsampling mill in operation.

TABLE 1 Distribution of Aflatoxin in Edible Peanuts, FAQ Peanuts, and Peanut Cake

Commodity	Lot size/ samples	Contaminated samples	Aflatoxin content (μg/kg)	Lot average (μg/kg)
Processed edible peanut kernels	20 tons; 200 × 100 g	1/200 (0.5%)	10	0.05
	15 t; 200 × 100 g	3/200 (1.5%)	31, 92, 109	1.00
	15 t; 200 × 100 g	2/200 (1.0%)	7, 609	3.00
	20 t; 200 × 100 g	7/200 (3.5%)	8–1938	10.00
FAQ peanut kernels	10 t; 84 × 100 g	13/84 (15.5%)	<1–4235	209.0
Peanut cake	5.6 t; 204 × 100 g	204/204 (100%)	87–249	163.0

Source: Ref. 4.

number (84 to 204) of 100-g incremental samples was systematically collected [4] from each lot, and each sample was analyzed for aflatoxin. The lot means for the processed (blanched, sorted, and roasted) edible peanuts varied from 0.05 to 10.0 μg/kg aflatoxin, whereas the corresponding percentage of contaminated samples varied from 0.5% to 3.5%. It is noteworthy that the highest levels of aflatoxin in the contaminated samples were as much as 200 times greater than the composite sample mean (i.e., the mean aflatoxin content of the incremental samples). All lots contained samples which were contaminated at levels of aflatoxin that were above the 4.0 μg/kg level currently tolerated in the U.K. [5]. The task of designing a practicable, cost-effective quality assurance sampling plan which allows manufacturers to confidently determine the acceptability of lots of edible peanuts is a real challenge for the statisticians!

Table 1 also presents data describing the distribution of aflatoxin in a 10-ton lot of FAQ peanuts and in peanut cake manufactured from the same lot. It is evident that, when compared with the edible peanuts, the distribution of aflatoxin is less skewed in the more highly contaminated FAQ kernels, the level of contamination in the most highly contaminated sample being about 20 times greater than the composite sample mean. After the kernels had been crushed, the aflatoxin in the resulting 5.6-ton lot of peanut cake was far more uniformly distributed than in the original kernels. Every incremental sample contained aflatoxin, and the highest level of contamination in the samples was less than twice the composite sample mean.

Given the skewed distribution of aflatoxin, particularly in whole kernels, it is extremely difficult to collect a sample which accurately represents the mean lot concentration of aflatoxin, and, as a result, the sampling step is usually the largest source of variation in the *analytical sequence* (i.e., the sequence of sampling, sample preparation, and analysis). However, it is essential that the sample be reasonably representative of the lot from which it has been collected if meaningful conclusions are to be drawn from the analytical data. Consequently, the total variance (S_t^2) associated with the analytical sequence should be reduced to an acceptable level.

The total variance (S_t^2) is equal to the sum of the sampling (S_s^2), sample preparation (S_{sp}^2), and analytical (S_a^2) variances (i.e., $S_t^2 = S_s^2 + S_{sp}^2 + S_a^2$). In general, for a given lot concentration, sampling variance will decrease as the sample size increases; whereas, for a given sample size, the relative sampling variance (coefficient of variation, CV) will decrease as the mean contamination level of the lot increases. The variance associated with the sample preparation step can also be reduced by increasing the degree of comminution and/or by increasing the size of the subsample. Finally, the analytical variance can be reduced by increasing the number of analyses.

The sampling, sample preparation, and analytical variances associated with the testing of shelled peanut kernels have been estimated empirically by Whitaker and co-workers [6–8] and are shown in Eqs. 1 to 3, below:

$$S_s^2 = (49.0546\ M_L^{1.3955} - 0.39M_L^{1.7867})/W_s \tag{1}$$

where S_s^2 is the sampling variance; M_L is the aflatoxin concentration in the lot in μg/kg, and W_s is mass of peanuts in the sample in kilograms (with a kernel count per gram of 1.95).

$$S_{sp}^2 = (0.978M_s^{1.7867} - 0.0178M_s^{1.9339})/W_{sp} \tag{2}$$

where S_{sp}^2 is the sample preparation variance when using a subsampling mill with a 3.18-mm screen; M_s is the aflatoxin concentration in the sample in μg/kg; and W_{sp} is the mass of comminuted peanuts in the subsample in kilograms.

$$S_a^2 = (1/na)*(0.00482*M_{ss}^{1.7518}) \tag{3}$$

where S_a^2 is the analytical variance for HPLC analysis; na is the number of replicate analyses performed; and M_{ss} is the aflatoxin concentration in the subsample in μg/kg.

Similar equations have been developed for farmer's stock peanuts, shelled corn, and cottonseed. For example, the effect of the sample size on sampling variance has been estimated (Table 2) by Whitaker and co-workers [8] for a lot of cottonseed containing 100 μg/kg aflatoxin. It is evident that the variance is halved each time the sample size is doubled. Similarly, the sample preparation and analytical variances can be halved by doubling the subsample size or the number of replicate analyses, respectively.

To reduce the variability of the analytical sequence to an acceptable level, it is necessary to achieve a cost-effective balance between the inputs into the

TABLE 2 Effect of Sample Size on the Range of Aflatoxin Test Results for 95% Confidence Limits When Testing a Cottonseed Lot with 100 μg/kg Aflatoxin

Sample size (kg)	Standard deviation[a]	Concentration of aflatoxin in sample (μg/kg)		
		Low[b]	High[c]	Range
1	87	0	271	271
2	62	0	222	222
4	45	13	187	174
8	32	37	163	126
16	24	53	147	94
32	19	64	136	72

Source: Ref. 1.
[a]Reflects the effect of the sample size shown in the table, a 200-g subsample and a single analysis.
[b]Low calculated as 100 − 1.96* (standard deviation); a value of 0 was recorded if a negative result was obtained.
[c]High calculated as 100 + 1.96* (standard deviation).

sampling, sample preparation, and analytical steps. For example, increasing the sample size beyond a certain point may be less cost-effective than increasing the size of the subsample and/or the number of replicate analyses.

B. Design and Evaluation of Sampling Plans

The following factors should be considered when designing a sampling plan:

Type of commodity
Maximum concentration of aflatoxin permitted in the lot (where the "permitted" level may be defined by a regulation or a quality assurance specification)
Maximum concentration of aflatoxin permitted in the sample (the "critical concentration")
Sample size
Number of incremental samples (which are combined to produce the sample)
Number of samples
Sample preparation method
Analytical method
Producer and consumer risks

There are two types of "risk" associated with any sampling plan: the *producer risk*, and the *consumer risk*. The producer risk (PR) reflects the probability that the sampling plan will reject a "good" lot (i.e., a lot that does not exceed the permitted level of aflatoxin), whereas the consumer risk (CR) reflects the probability that a "bad" lot (i.e., a lot that does exceed the permitted level of aflatoxin) will be accepted. A well-designed sampling plan will attain an acceptable balance between the two types of risk, which in turn will be determined by the relative economic and "social" consequences associated with the CR and PR.

For any given sampling plan, the magnitude of the producer and consumer risks can be estimated by constructing a unique *operating characteristic* (OC) curve, which is a plot of the probability [P(M)] of the sampling plan accepting a lot against the mean concentration (M) of aflatoxin in the lot. The shape of the OC curve will be determined by the sample size, the size and degree of comminution of the subsample, the type and number of analyses performed, and the maximum concentration of aflatoxin permitted in the sample (the critical concentration) if the lot is to be accepted by the plan. The acceptance probability (PM) may be expressed as follows:

$$P(M) = \text{prob}\,(\overline{X} < \overline{X}c|M) \tag{4}$$

where $\overline{X}$ is the concentration of aflatoxin in the sample, $\overline{X}c$ is the critical concentration of the sample, ($\overline{X}c$ is frequently equivalent to Mc, the concentration of aflatoxin which is permitted in the lot), and M is the concentration of aflatoxin in the lot.

It can be seen from Eq. 4 that, as expected, P(M) approaches 1 as M approaches 0, and that P(M) approaches 0 as M becomes large. A typical OC curve is illustrated in Figure 8, where M_c is the concentration of aflatoxin that is permitted in the lot. The producer risk refers to those lots with an aflatoxin content which is less than M_c but which are rejected by the plan; the consumer risk is determined by those lots with an aflatoxin level which is greater than M_c but which are accepted by the plan.

1. Use of Theoretical Probability Models

To construct an OC curve, a method is required that predicts the distribution of the analytical results obtained when a specific sampling plan is applied to a lot of aflatoxin-contaminated material. Several theoretical distributions for aflatoxin in peanuts have been proposed [9–13] by Waibel (a modified gamma distribution), Brown (log normal), Jewers et al. (Weibull), and Whitaker et al. (negative binomial); an FAO Technical Consultation [2] has recommended the use of the negative binomial distribution. The negative binomial density function f(x) is:

$$f(x) = (\Gamma(x + K)/(x!\Gamma(K)))\,(K/(K + M))^{k}\,(M/(K + M))^{x} \tag{5}$$

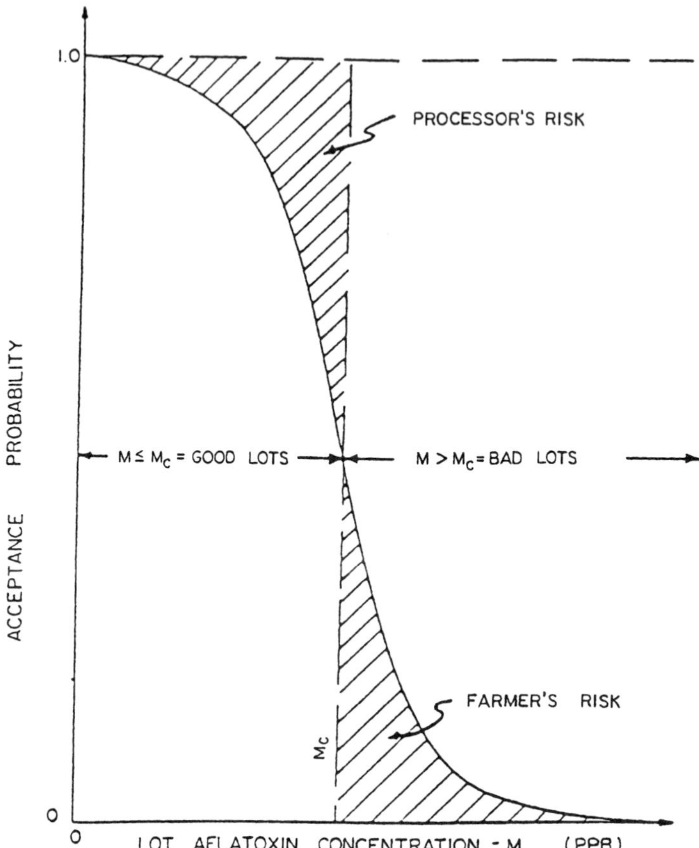

FIGURE 8 Typical operating characteristic curve. (From Ref. 1.)

where x is the quantity of aflatoxin in a single kernel, Γ is the gamma function, M is the average quantity of aflatoxin within all kernels in the lot, and K is the shape parameter.

Using the estimated total aflatoxin testing variance (S^2_t) and the method of moments [13], the shape parameter (K) of the negative binomial distribution can be calculated from Eq. 6; the negative binomial distribution can be used to compute the probability that a specific sampling plan will accept a lot of contaminated peanuts, for a given lot concentration (M):

$$K = (M*M)/((ns* S_t^2) - M) \quad (6)$$

where M is the concentration of aflatoxin in the lot in μg/kg, ns is the number of kernels in the sample, and S_t^2 = total aflatoxin testing variance = $S_s^2 + S_{sp}^2 + S_a^2$.

The effect of sample size, and of the critical concentration ($\overline{X}c$), on the shape of the OC curve has been illustrated [1] by Whitaker and co-workers for the sampling of cottonseed, where $\overline{X}c$ is equivalent to Mc (40 μg/kg aflatoxin) for each of the plans. Figure 9 illustrates the effect of sample sizes of 2.3, 6.8, and 20.4 kg on the OC curve. Inspection shows that both the PR and the CR decrease as the sample size increases, with the CR decreasing at a faster rate than the PR; and that balanced, low values for the PR and CR are obtained with a sample size of 20.4 kg. However, it is also evident that a reasonable balance between the producer and consumer risks will be retained when the sample size lies between 6.8 and 20.4 kg. The effect of the critical concentration ($\overline{X}c$) on the OC curve has also been studied [1] and is illustrated in Figure 10, for a permitted aflatoxin concentration (M_c) of 40 μg/kg and a sample size of 9.1 kg. Each increase in the critical concentration causes a decrease in the PR and an increase in the CR. If $\overline{X}c$ is equivalent to 30 μg/kg, the PR is significantly greater than the CR; whereas, if $\overline{X}c$ is increased to 50

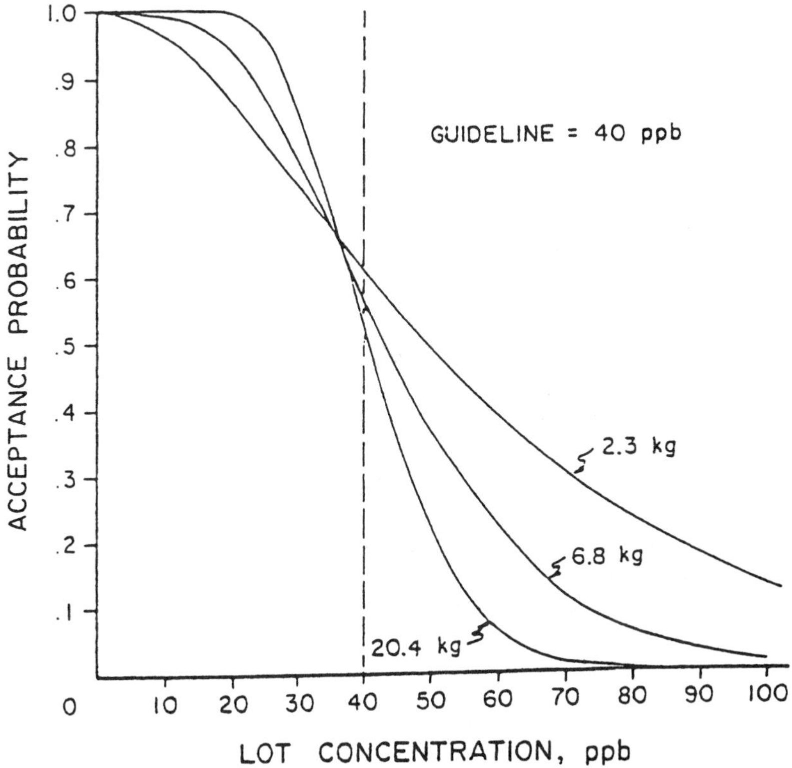

FIGURE 9 Effect of sample size on the shape of the operating characteristic curve. (From Ref. 1.)

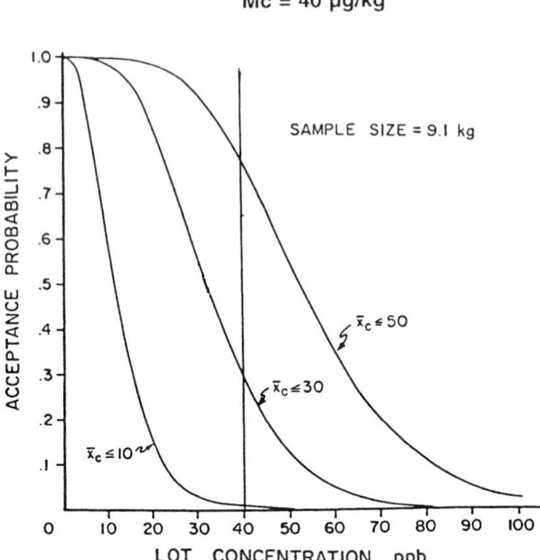

FIGURE 10 Effect of the critical concentration on the shape of the operating characteristic curve. (From Ref. 1.)

μg/kg, the CR is significantly greater than the PR. Consequently, for a sample size of 9.1 kg, a reasonable balance between the producer and consumer risks is obtained with a critical concentration of approximately 40 μg/kg. If the conclusions drawn from Figures 9 and 10 are considered together, it can be anticipated that an effective sampling plan for cottonseed, where the maximum permitted concentration of aflatoxin in a lot is 40 μg/kg, would involve the collection of a 10-kg sample and acceptance of the lot if the sample contained 40 μg/kg or less of aflatoxin.

An FAO Technical Consultation [2] constructed OC curves for 35 sampling plans for peanut and corn products. For raw shelled peanuts, it was assumed that comminution was performed using a hammer mill, and that 100-g subsamples were produced and submitted for analysis by thin-layer chromatography (TLC). Figure 11, for example, illustrates the effect of the critical concentration ($\overline{Xc}$) on the shape of the OC curve for 20-kg samples of peanuts, if $\overline{Xc}$ is equivalent to Mc. In the case of corn, a combination of comminution by hammer mill, a 50-g subsample and TLC analysis was assumed. Figure 12 shows the effect of $\overline{Xc}$ on the OC curve for 10-kg samples of corn if, again, $\overline{Xc}$ is equivalent to Mc. The PRs and CRs associated with an Mc of 20μg/kg, and a variety of critical concentrations, can be deduced from Figures 11 and 12.

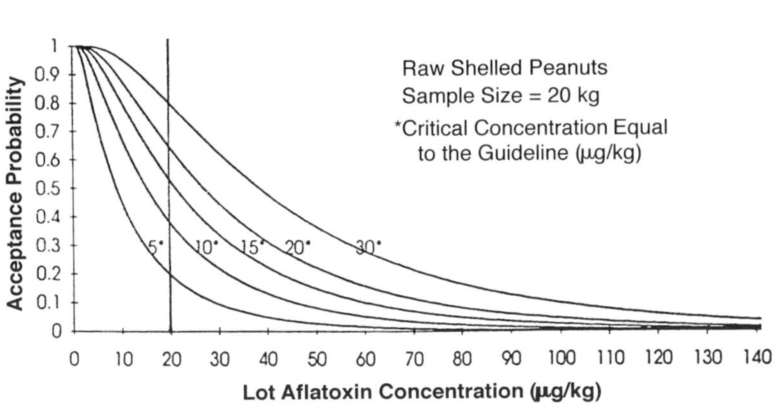

FIGURE 11 Effect of the critical concentration on the shape of the operating characteristic curve when collecting 20-kg samples of peanuts. (From Ref. 2.)

Whitaker and co-workers have also constructed [14] OC curves for the sampling plans currently used (or under development) by the regulatory authorities and/or traders in the U.S., U.K., and Netherlands, and described in Table 3. The corresponding OC curves are illustrated in Figure 13. It can be seen that the Dutch plan (Mc = 9 µg/kg total aflatoxin) has a PR that is substantially greater

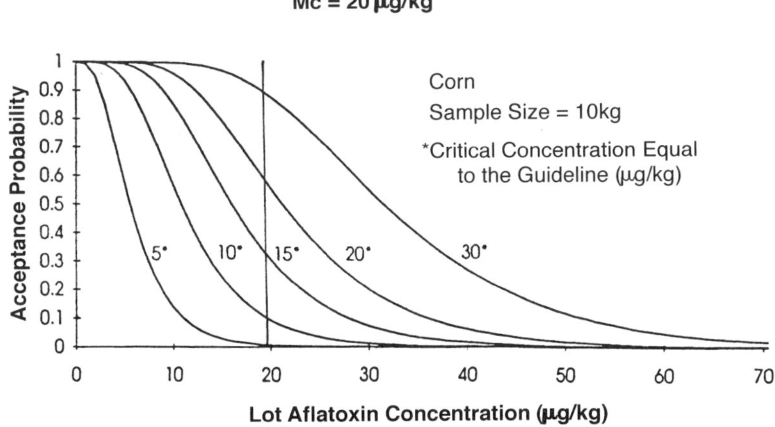

FIGURE 12 Effect of the critical concentration on the shape of the operating characteristic curve when collecting 10-kg samples of corn. (From Ref. 2.)

Table 3 The U.S., U.K., and Dutch Sampling Plans

Country	Mc[a] (μg/kg)	$\overline{Xc}$[b] (μg/kg)	No. samples	Sample size, kg	Mill[c]	Subsample size, g	Analytical method[d]
U.S.	20	15	3	21.8	Hammer	1100	TLC
U.K.	10	10	1	10.0	VCM	50	HPLC
Netherlands	9	5	4	7.5	VCM	50	HPLC

Source: Ref. 14.

[a](Mc) is the maximum permitted level of total aflatoxins; for the Dutch plan, a Mc of 5 μg/kg aflatoxin B_1 is assumed to be equivalent to 9 μg/kg total aflatoxins.

[b]($\overline{Xc}$) is the critical concentration of aflatoxin in the sample; for the Dutch plan, a $\overline{Xc}$ of 3 μg/kg of aflatoxin B_1 is assumed to be equivalent to 5 μg/kg total aflatoxins.

[c]VCM is a vertical cutter mill.

[d]TLC is thin-layer chromatography; HPLC is high-performance liquid chromatography.

than the CR and that virtually no lots in excess of approximately 15 μg/kg are accepted. On the other hand, with the U.K. plan (Mc = 10), the CR is considerably greater than the PR; most lots in excess of 60 μg/kg aflatoxin will be rejected. The U.S. plan (Mc = 20) affords reasonably equally balanced producer and consumer risks and rejects all lots that exceed about 45 μg/kg aflatoxin.

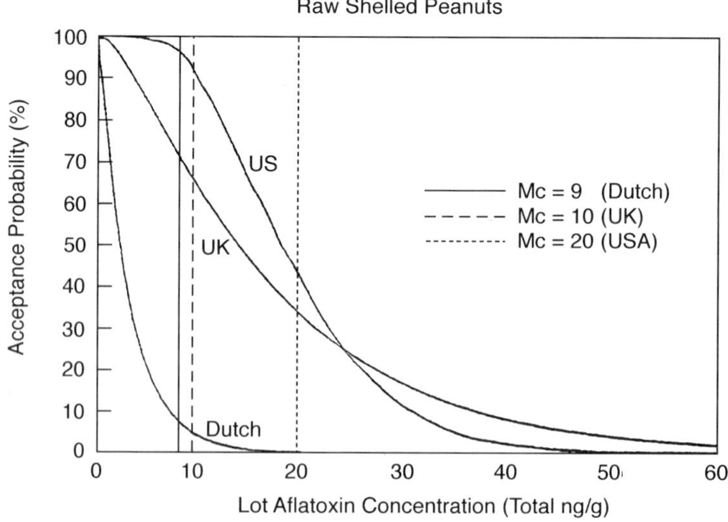

Figure 13 Operating characteristic curves for the U.S., U.K., and Dutch sampling plans. (From Ref. 14.)

The efficacy of a sampling plan can be further described by simulating the performance of the plan as applied to a collection of lots with a specific distribution of aflatoxin. Whitaker and co-workers have constructed [14] a *lot distribution* by averaging the observed distribution of aflatoxin test results in lots of raw peanut kernels produced in the U.S. during 1976–1985 (Table 4). In this 10-year period, 50% of the lots were uncontaminated, only 0.25% of the lots exceeded an aflatoxin level of 100 μg/kg, and the mean contamination level was 5.5 μg/kg aflatoxin. The effect of applying the U.S., U.K., and Dutch sampling plans to the lot distribution described in Table 4 is shown in Table 5. For example, it is evident that the U.S. plan results in the greatest number of correct decisions (95.6%), followed by the U.K. (91.1%) and Dutch (82.4%) plans. The Dutch plan accepted the smallest number of bad lots (0.2%), whereas the U.K. plan accepted the greatest number of bad lots (5.1%). However, the Dutch plan also rejected a large number (17.4%) of good lots compared to the U.S. (3.2%) and U.K. (3.8%) plans.

2. Use of Statistical Models to Develop and Evaluate Sampling Plans

When using the negative binomial model to simulate the performance of sampling plans, it is assumed that each sample has been collected in a statistically random manner.

Statistical models can be used to evaluate the effect on the efficacy of the plan when sample selection is not truly random and additional errors are intro-

TABLE 4 Cumulative Distribution Among Lot Aflatoxin Concentrations for Shelled Peanuts in the U.S. from 1976 to 1985

Lot aflatoxin concentration (μg/kg)	Cumulative distribution (%)	Lot aflatoxin concentration (μg/kg)	Cumulative distribution (%)
0	50.00	10	84.00
1	60.00	15	89.60
2	65.00	20	93.10
3	69.00	25	95.10
4	72.33	30	96.40
5	75.00	40	97.60
6	77.33	50	98.33
7	79.33	70	99.16
8	81.03	100	99.75
9	82.57		

Source: Ref. 14.

TABLE 5 A Comparison of the Performance of the U.S., U.K., and Dutch Sampling Plans

Evaluation factor (per 100 lots sampled)	U.S.	U.K.	Netherlands
Maximum permitted level; Mc, μg/kg	20	10	9 ($5B_1$)
Critical sample concentration: $\overline{X}c$, μg/kg	15	10	3 ($3B_1$)
Total lots tested	100	100	100
Good lots tested[a]	93.1	84.0	82.6
Bad lots tested[b]	6.9	16.0	17.4
Average concentration among lots, μg/kg	5.5	5.5	5.5
Lots accepted by plan	91.0	85.3	65.4
Lots rejected by plan	9.0	14.7	34.6
Good lots accepted by plan	89.8	80.2	65.2
Bad lots rejected by plan	5.8	10.9	17.2
Good lots rejected by plan	3.2	3.8	17.4
Bad lots accepted by plan	1.2	5.1	0.2
Correct decisions	95.6	91.1	82.4
Average concentration among accepted lots, μg/kg	2.7	2.2	0.5
Average concentration among rejected lots, μg/kg	33.8	24.5	14.8

Source: Ref. 14.
[a]Good lots are those contaminated with ≤ Mc μg/kg aflatoxins.
[b]Bad lots are those contaminated with > Mc μg/kg aflatoxins.

duced during the sample collection process, when incremental samples are combined to give a composite sample. The (relative) variance of the *true* aflatoxin concentration of incremental samples of a given size may be expressed by the model [4]:

$$S_{is}^2 = A + B/M_{is} \tag{7}$$

where S^2_{is} is the true sampling variance of the incremental samples, M_{is} is the incremental sample weight, A and B are constants, A reflects the segregation variance, and B/M_{is} reflects the fundamental variance associated with particulate materials.

From Eq. 7 the (relative) variance of the true aflatoxin concentration of a sample can be expressed as:

$$S_s^2 = (A + B/M_{is})/N \tag{8}$$

where S_s^2 is the true sampling variance of the composite sample, and N is the number of incremental samples combined to produce the sample.

As discussed before, the measured aflatoxin concentration of a sample will differ from the true value because of the preparation (S^2_{sp}) and analytical (S^2_a) errors. Therefore, the total variance of the analytical sequence (sampling, sample preparation, and analysis) used to determine the aflatoxin concentration of a lot may be expressed as:

$$S^2_t = (A + B/M_{is})/N + S^2_{sp} + S^2_a \quad (9)$$

The values of the constants A and B may be estimated from analytical data derived from the analysis of a large number of incremental samples, of a variety of sizes, collected from lots with a range of aflatoxin contents. Once A and B have been estimated, the variance associated with sampling plans involving a variety of combinations of incremental sample size (M_{is}) and number (N) can then be determined.

3. Sampling Plans for Problematic Commodities

A "problematic" commodity is one upon which it is difficult to perform sampling and/or sample preparation procedures. It is difficult, for example, to sample oilseed cakes in a packaged form without having to open and empty each sack. On the other hand, palm kernels are easy to sample but, because of their extreme hardness, are very difficult to comminute. Such problematic commodities, and/or those which are of a high value, require a sampling plan which involves the collection (and preparation) of a sample of minimal weight.

Sharkey et al. [15] have developed a sequential acceptance test that involves the collection of 1-kg samples, minimal sample preparation, and a small number of analyses. The continuous three-parameter Weibull function (Eq. 10) was adopted as a model for the distribution of aflatoxin, and the values of γ, η, and β were estimated using the method of Dubey [16]:

$$F(x) = 1 - \exp[-\{(X - \gamma)/\eta\}^{\beta}] \quad (10)$$

where γ, η, and β are constants.

The Weibull function was used to generate 1000 simulated aflatoxin test results for 100-g incremental samples of three problematic commodities (palm kernels, groundnut cake, and peanut butter) which could in turn be used to simulate the collection of 1-kg samples. The simulated 1-kg samples were then used to evaluate the following plan (the "omega" plan):

1. Use an empirically derived statistic, $\omega = (\bar{x} - M_t)/(s/\sqrt{n})$, where, $\bar{x}$ is the mean aflatoxin content of n 1-kg samples, with a standard deviation s and a deviation of $(\bar{x} - M_t)$ from the "target" aflatoxin concentration (M_t) of the lot under test.

2. Define the target aflatoxin concentration (M_t) of the lot under test (where the target concentration may be equivalent to the maximum permitted level of aflatoxin or some other assigned value; see Fig. 14).

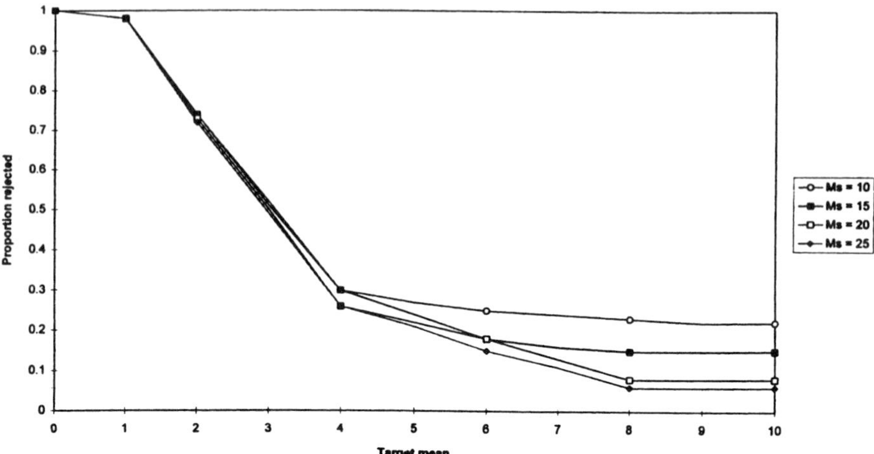

FIGURE 14 Application of the omega sampling plan to peanut butter. Proportion of sampling plan decisions which rejected the lot plotted against the target lot mean (M_c) for various values of M_s. (From Ref. 15.)

3. Define the maximum concentration of aflatoxin permitted in any sample ($\overline{X}c$).

4. Collect six 1-kg samples (each composed of 10 100-g incremental samples).

5. Analyze the first three samples and if any 1-kg sample has an aflatoxin concentration in excess of $\overline{X}c$, immediately reject the lot.

6. After the analysis of the third sample, compute the statistic, ω, and proceed as follows: if $\omega > +2$, immediately reject the lot; if $\omega < -2$, immediately accept the lot; if $-2 \leqslant \omega \leqslant +2$, analyze the fourth sample and repeat the calculation of ω until a decision is made to accept or reject the lot (the value 2 in these decision rules was obtained as an empirical value).

7. If any of the subsequently analyzed 1-kg samples has an aflatoxin concentration in excess of $\overline{X}c$, immediately reject the lot.

8. If no decision has been made after the analysis of the sixth sample: if $\bar{x} \leqslant M_t$, accept the batch; if $\bar{x} > M_t$, reject the batch.

Sharkey and co-workers applied [15] the above approach to the sampling of peanut butter where the analysis of 498 100-g samples of peanut butter (systematically collected, as retail packs, from a production line) afforded the distribution data shown in Table 6. Most of the samples of peanut butter contained less than 4 μg/kg aflatoxin with a mean value of 7 μg/kg. Using the Weibull distribution model, 1000 aflatoxin concentration values were simulated for 100-g samples which, in turn, were utilized in the simulation of values for 1-kg samples of peanut

TABLE 6 Distribution of Aflatoxin in 100-g Samples of Peanut Butter

Aflatoxin concentration (μg/kg)	Frequency	Aflatoxin concentration (μg/kg)	Frequency
<4	368	15–	7
4–	54	20–	4
6–	29	25–	3
8–	14	⩾30	5
10–	14		

Source: Ref. 15.

butter. The test procedure described above was then applied to all the 1-kg sample values in order to compute (1) the number of decisions (accept or reject) reached, and (2) the number of times the decision was to reject the lot. The fraction (2)/(1) was then plotted against the target mean for the concentration of aflatoxin in the lot (Fig. 14). The target means evaluated ranged from 0 to 10 μg/kg, whereas the M_s (the maximum aflatoxin value permitted in any sample) values were 10, 15, 20, and 25 μg/kg aflatoxin. From Table 6, it is evident that this particular lot of peanut butter would only be considered acceptable if a distribution were tolerated where approximately one-third of the 100-g samples, in the form of jars of peanut butter, contained in excess of 4 μg/kg aflatoxin (assuming that the distribution of aflatoxin in the samples is equivalent to the lot distribution), and that aflatoxin concentrations in excess of 30 μg/kg were also tolerated in the 100-g samples. It can be seen from Figure 14 that if the values of M_t and M_s are set at 1 and 10 μg/kg, respectively, the lot would be rejected by the plan on most occasions. Alternatively, the plan would accept the lot with a very high frequency if M_t and M_s were set at 10 and 25 μg/kg. When the test utilized M_t and M_s values of 1 and 10 μg/kg, an average of 3.5 samples (i.e., 3.5 kg peanut butter) required analysis before a decision was reached. For M_t and M_s values of 10 and 25 μg/kg, the average number of analyses required was also about 3.5.

IV. CONCLUSIONS

Most activities associated with the control of mycotoxins involve, at some stage, the collection and analysis of a sample of food or feed, and subsequent actions determined by the analysis results. Consequently, if appropriate action is to be taken, it is essential that the errors (variances) associated with the analytical sequence be minimized.

Although considerable progress has been made with the design of sampling plans for surveillance, regulatory, and quality assurance purposes, much work still remains to be done. We can look forward to the day when regulators automatically attach details of appropriate sampling and analysis methods to the quality specification for a specific commodity. Furthermore, it is important that cost-effective, practicable methods be developed which allow producers and regulators to accurately estimate the distribution of mycotoxins within retail packs of foods, both at the manufacturing plant and at the point of sale.

ACKNOWLEDGMENTS

I gratefully acknowledge the very helpful comments of Dr. Thomas Whitaker (U.S. Department of Agriculture, North Carolina State University), Peter Defize (TNO Institute of Applied Physics, Netherlands), and Angela Sharkey (University of Portsmouth, U.K.) during the preparation of this chapter.

REFERENCES

1. TB Whitaker, JW Dickens, FG Giesbrecht. Testing animal feedstuffs for mycotoxins: sampling, subsampling, and analysis. In: JE Smith, RS Henderson, eds. Mycotoxins and Animal Foods. Boca Raton: CRC Press, 1991, p 153.
2. Sampling Plans for Aflatoxin Analysis in Peanuts and Corn. Report of an FAO Technical Consultation. FAO Food and Nutrition Paper 55, Rome, Italy, 1993.
3. AF Cucullu, LS Lee, RY Mayne, LA Goldblatt. Determination of aflatoxin in individual peanuts and peanut sections. J Am Oil Chem Soc 43:89, 1966.
4. RD Coker, MJ Nagler, G Blunden, et al. The design of sampling plans for mycotoxins in foods and feeds. Natural Toxins 3:257, 1995.
5. Aflatoxins in nuts, nut products, dried figs and dried fig products regulations. U.K. Statutory Instruments No. 3236, 1992.
6. TB Whitaker, JW Dickens, RJ Monroe. Variability of aflatoxin test results. J Am Oil Chem Soc 49:590, 1974.
7. TB Whitaker, JW Dickens, RJ Monroe. Variability associated with testing corn for aflatoxin. J Am Oil Chem Soc 56:789, 1979.
8. TB Whitaker, JW Dickens, RJ Monroe. Variability associated with testing cottonseed for aflatoxin. J Am Oil Chem Soc 53:502, 1976.
9. J Waibel. Dtsh Lebensm Rundsch 73:353, 1977.
10. GH Brown. Food Technol Austr 36:128, 1984.
11. K Jewers, RD Coker, G Blunden, MJ Jazwinski, AJ Sharkey. Problems involved in the determination of aflatoxin in bulk commodities. Int Biodeterioration 22:83, 1986.
12. TB Whitaker, EH Wiser. Theoretical investigation into the accuracy of sampling shelled peanuts for aflatoxin. J Am Oil Chem Soc 46:377, 1969.
13. TB Whitaker, JW Dickens, RJ Monroe. Comparison of the observed distribution of aflatoxin in shelled peanuts to the negative binomial distribution. J Am Oil Chem Soc 49:590, 1972.

14. TB Whitaker, J Springer, PR Defize, WJ DeKoe, RD Coker. Evaluation of sampling plans used in the United States, United Kingdom, and the Netherlands to test raw shelled peanuts for aflatoxin. AOAC Int 78:1010, 1995.
15. AJ Sharkey, OG Roch, RD Coker. A case-study on the development of a sampling and testing protocol for aflatoxin levels in edible nuts and oilseeds. Statistician 43:267, 1994.
16. SD Dubey. Some percentile estimators of Weibull parameters. Technometrics 9:119, 1967.

5
Mycotoxin Analytical Techniques

David M. Wilson
University of Georgia, Tifton, Georgia

Eric W. Sydenham
Medical Research Council, Cape Town, South Africa

Gary A. Lombaert
Health Canada, Winnipeg, Manitoba, Canada

Mary W. Trucksess
Food and Drug Administration, Washington, D.C.

David Abramson
Agriculture and Agri-Food Canada, Winnipeg, Manitoba, Canada

Glenn A. Bennett
National Center for Agricultural Utilization Research, United States Department of Agriculture, Agricultural Research Service, Peoria, Illinois

I. INTRODUCTION

Mycotoxin contamination of agricultural commodities is practically unavoidable due to uncontrollable environmental conditions; however, the mycotoxin content of foods and feeds can be managed. Efforts to minimize mycotoxins include

The opinions expressed in this publication are those of the authors and do not necessarily reflect the official views of the University of Georgia, South African Medical Research Council, Health Canada, U.S. Food and Drug Administration, Agricultural and Agri-Food Canada, or U.S. Department of Agriculture. This manuscript includes U.S. government authors and therefore is not copyrightable. The authors thank their respective institutions for supporting this effort. D.M. Wilson thanks Darlene Willis for her help in typing the manuscript.

monitoring, managing, and controlling quality of product from the farm to the marketplace, establishing regulatory limits or guidelines, utilizing decontamination procedures, and implementing strategies to divert contaminated grains to feed and other industrial uses. Surveillance programs in agricultural products are necessary to reduce the risk of mycotoxin consumption by humans and animals. Chemical and immunological assays, which can routinely be used to rapidly analyze large numbers of samples, are needed in all monitoring programs. The data obtained from monitoring efforts should be useful in determining and reducing the risk of animal and human exposure.

Analytical methods for mycotoxin analyses are based on many factors. No one method or technique is suitable for all applications. The initial considerations in analytical schemes include the chemical nature of the mycotoxins of interest, the molecular weight, and the functional groups. These factors determine the volatility and solubility properties of the mycotoxin and influence the analytical approach selected for a specific toxin or group of toxins. The sample matrix to be extracted for mycotoxin assay also influences the extraction and analytical method selection. Current analytical techniques are usually based on extraction into a solvent, followed by a partial purification step (i.e., cleanup) and quantitation. Use of TLC, HPLC, GC, and other chemical methods are not limited by selection of solvents for use in the procedures. However, immunoassays used for quantitation are limited to a few solvents.

Analytical methods for aflatoxins, fumonisins, deoxynivalenol, ochratoxins, zearalenone, citrinin, and patulin are discussed in this review. This review is a collaborative effort by scientists expert in mycotoxin analysis. M.W. Trucksess and D.M. Wilson contributed portions relating to aflatoxins; E.W. Sydenham, those of fumonisins and patulin; G.A. Lombaert on deoxynivalenol; D. Abramson on ochratoxin A and citrinin; and G.A. Bennett on zearalenone. As senior author, D.M. Wilson, collated, edited and compiled the manuscript and is responsible for any omissions or errors.

This review is by no means all-inclusive but attempts to illustrate the nature and principles of current analytical techniques and problems in mycotoxin analyses. The mycotoxins were selected for this review based on their relative importance and frequency of occurrence. Other important compounds could not be covered due to space considerations.

II. EXTRACTION OF MYCOTOXINS

The seven mycotoxins we have selected to review can be used to illustrate the various approaches used in extraction, purification, separation, and determination. Sampling and sample preparation are beyond the scope of this paper and are covered in another chapter. Some of the problems associated with sampling and reliability have been recently reviewed [1–3].

A. Aflatoxins

The aflatoxins were the first mycotoxins to be regulated and can be extracted and analyzed with a plethora of techniques. These compounds are easily extracted into water-saturated chloroform, aqueous methanol, acetonitrile, acetone, and other polar solvents. Cole and Dorner [4] reported that acetonitrile-water was less efficient in extraction than methanol-water under different solvent to sample ratios in peanuts. Sometimes salt has to be added to the extraction vessel to improve the recovery, especially from animal tissues. When chloroform is the solvent choice, it is common to purify the extracts by passing the extract through a silica gel column for preliminary cleanup. The major problems in aflatoxin extraction include occasional emulsions, incomplete extractions from some matrices, and the coextraction of compounds that interfere with quantitation. For immunoassay, it is common to use an aqueous methanol or acetonitrile extraction followed by dilution with water to minimize denaturing or incomplete binding with the appropriate antibody because of solvent effects.

B. Fumonisins, Deoxynivalenol, and Zearalenone

The fumonisins are highly polar diesters of propane-1,2,3,tricarboxylic acid (TCA) and various 2-amino-12,16-dimethylpolyhydroxyeicosanes in which the C_{14} and C_{15} hydroxyl groups are esterified with the terminal carboxyl group of TCA (Fig. 1). Initial extraction of fumonisins usually involves water with either methanol or acetonitrile ranging from 1:1 [5,6] to 1:3 [6,7], respectively. The matrix influences the extraction of fumonisins [6,8] with recovery being low in corn bran flour. Because of the low recoveries, a methanol borate buffer (pH 9.2)

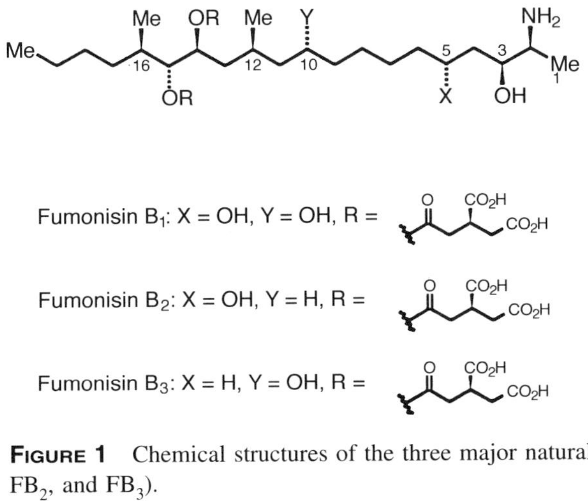

Fumonisin B_1: X = OH, Y = OH, R =

Fumonisin B_2: X = OH, Y = H, R =

Fumonisin B_3: X = H, Y = OH, R =

FIGURE 1 Chemical structures of the three major naturally occurring fumonisins (FB_1, FB_2, and FB_3).

TABLE 1 Extraction and Determination of Deoxynivalenol from Various Commodities

Commodity(s)	Extraction solvent	Cleanup	Detection method	Detection limit	Reference
Wheat, flour	Methanol/aqueous KCl	Activated carbon/celite/ alumina	GC/FID; (HFBs)	100–500 ng/g	113
Corn, other cereal grains	Methanol/water	Florisil SPE; C18 SPE	GC/ECD; (HFBs, TMSs)	3.5 ng/g	109
Barley, corn	Acetonitrile/water	n-Hexane; florisil	GC/NICI-MS (TMSs)	5 ng/g	129
Flour, bran	Acetonitrile/water	Romer	LC/UV	50 ng/g	74
Grain-based foods[a]	Methanol/ or acetonitrile/water	Nil	ELISA	1000 ng/g	231
Wheat	Methanol/water	Nil	ELISA	200 ng/g	188
Beer	Ethyl acetate	ChemElut C18 SPE	GC/MS; (HFBs)	0.1–1.5 ng/ml	123
Milk	Nil	Precipitate proteins; Extrelut activated carbon/alumina	LC/UV	5 ng/ml	75
Swine tissue	Acetonitrile/water	Activated carbon/alumina	LC/MS	1 ng/g	133

ECD, electrochemical detection; ELISA, enzyme-linked immunosorbent assay; FID, flame ionization detection; GC, gas chromatography; HFB, heptafluorabuytl; LC, liquid chromatography; MS, mass spectrometry; NICI-MS, negative ion chemical ionization detection using a mass spectrometer; SPE, solid phase extraction; TMS, trimethylsilyl derivative; UV, ultraviolet.

[a]Grain-based foods: breakfast cereals, corn meal, wheat flour.

was used for corn bran [8] and acidified methanol was used for processed corn commodities [9]. Further cleanup is usually a solid-phase extraction (SPE) column and C_{18} or SAX functional groups.

The fusarium metabolites deoxynivalenol (DON) and zearalenone can be analyzed by numerous techniques, and there are several extraction procedures in the literature. Table 1 lists some of the references and extraction solvents that have been used for DON analysis. Following the initial extraction several techniques have been used for cleanup.

C. Ochratoxin and Citrinin

Ochratoxin A and citrinin both have carboxylic acid functional groups that can be exploited in acidic or basic extraction and cleanup. Citrinin is capable of forming stable chelate complexes and is not particularly stable, so extractions have to be carefully designed. Hald et al. [10] published a collaborative study (Table 2) comparing ochratoxin A assays of blank and spiked wheat samples using various extractions and methods. One can easily see that extractions and cleanup considerations are critical in selection and application of methods.

III. THIN LAYER CHROMATOGRAPHIC METHODS

Thin-layer chromatography was the first method for detection of mycotoxins and is routinely accomplished with silica gel plates. The TLC methods remain important in many laboratories. TLC methods for alfatoxins, DON and zearalenone have been reviewed many times [11–13], so they will not be considered here.

A. Fumonisins

Qualitative methods based on solvent extraction, separation on either reversed-phase (C_{18}) or normal-phase TLC plates, and visualization with ninhydrin or *para*-anisaldehyde spray solutions were initially developed during the isolation of the fumonisins [14,15]. Improvements were made which ultimately incorporated two-stage development of TLC plates coupled with spectrodensitometry. [16].

Rottinghaus et al. [17] significantly improved both the sensitivity and selectivity characteristics of a reversed-phase TLC method for the analysis of fumonisin B_1 (FB_1) and fumonisin B_2 (FB_2) in animal feeds, by developing a prechromatography purification step using C_{18} SPE cartridges. The sensitivity was enhanced by using fluorescamine for the visualiation of the resultant chromatographic spots, which could be observed under long-wavelength UV light [17]. This method was found to be suitable for the codetermination of FB_3, in addition to FB_1 and FB_2. Although the sensitivity of the method had been reported to be 0.1

TABLE 2 Performance of Ochratoxin A Assay Methods Using Various Extraction-Cleanup Procedures and Fluorescence HPLC Final Determination[a]

Cleanup type	Mean concentration (ng/g)	Standard deviation (ng/g)	% Recovery[b]	Reference
None	5.7	0.7	72	233
Liquid/liquid	12.6	2.1	86	234
Liquid/liquid	12.9	1.7	48	235
DE	11	1.4	73	236
DE	14.3	1.3	80	21
	9.4	0.9	84	
SPE/silica	14.0	1.8	89	237
	10.8	0.5	91	
	11.1	0.7	70	
SPE/silica	15.9	0.7	85	
SPE/silica + SPE/CN	14.2	2.4	97	239
SPE/C18	13.9	1.7	85	23
	11.5	0.9	74	
IAC	15.9	7.4	38	202
	14.7	1.9	45	

Source: Ref. 232.

DE, diatomaceous earth; SPE, solid-phase extraction; CN, cyano-bonded solid phase; IAC, immunoaffinity column.

[a]A naturally contaminated sample of wheat containing 15 ng/g ochratoxin A was distributed to 33 laboratories for analysis.

[b]Corrected for recovery at the 15 ng/g level.

μg/g in animal feeds [17], the presence of matrix interferences hampered chromatographic band discernment when applied to corn samples [18]. These interferences were less noticable for those extracts purified on SAX media, and it was recommended that this latter purification technique be used for fumonisin levels in corn ≤1 μg/g[18]. The use of 0.5% vanillin solution for component visualization coupled with high-performance reversed-phase TLC separation of SAX-purified extracts has also been recommended [19].

A high-performance normal-phase TLC method was developed for the analysis of fumonisins in rice [20]. The method involves the purification of extracts of SAX media followed by single directional separation by TLC and visualization by dipping the chromatographed plate into a 0.16% acidic solution of *para*-anisaldehyde, prior to quantification by scanning fluorodensitometry. The response of the method was linear between 0 and 5 μg/g (c.v. = 15.4 to 0.9%, respectively), while recoveries averaged 81% [20].

B. Ochratoxins

The current official method of Association of Official Analytical Chemists (AOAC) International for ochratoxins A and B and their esters in barley [21] has been extensively studied and collaboratively validated. This method uses an alkaline diatomaceous-earth column cleanup, and a final determination via silica TLC with fluorescence visualization under 365-nm light. This method also qualitatively detects ochratoxin B and ochratoxin esters. Confirmation of ochratoxin A is accomplished by making the fluorescent methyl ester with boron trifluoride reagent. This TLC-based method is regarded as outdated [22]. Another method involving partitioning into bicarbonate solution, solid-phase extraction onto a C_{18} adsorbent, elution with ethyl acetate-methanol-acetic acid, and determination by HPLC with fluorescence detection has been collaboratively studied [23] and has been adapted as an AOAC official method.

C. Citrinin

Cereal samples can be screened for citrinin by TLC. Citrinin is visible on silica plates under longwave UV light as yellow fluorescent spots, which change to blue after spraying with boron trifluoride reagent. Tailing of citrinin spots on silica TLC plates has prompted improvements to the stationary phase by impregnating the silica with oxalic acid [24] or glycolic acid [25]. Chromatography of citrinin and other mycotoxins on C_{18} reverse-phase plates has been characterized, and is useful for confirmation of identity in screening procedures [26].

D. Patulin

TLC has recently been applied to the codetermination of patulin and zearalenone in corn [27]. The toxins are extracted and purified by silica-gel column chromatography, prior to sequential bisolvent normal-phase TLC using *iso*-octane followed by a chloroform-acetone mixture. Patulin levels may be quantified by observing the TLC plates at 275 nm. The detection limit has been reported to be 0.003 μg patulin per spot with recoveries of >80% (RSD = 15%) for patulin levels between 0.07 and 0.3 μg per chromatographic spot.

The use of semipermeable membranes for the diphasic dialysis extraction of patulin from apple juice, followed by TLC analysis, has been described [28,29]. Extraction by diphasic dialysis is based on the molecular weight and solubility characteristics of the target analyte in organic solvents where the compounds of interest pass through the membrane from the aqueous to an organic phase. No further purification step was reported, and patulin, as its 3-methyl-2-benzothiozolinone hydrazone hydrochloride (MBTH) derivative, was quantitatively determined by normal-phase TLC. Analyte recoveries were reported to be relatively low (between 62.5% and 65%) at a spiking level of 0.1 μg patulin/ml, with a detection limit of 0.05 μg/ml [28,29].

Abramson et al. [30] cited various TLC retention characteristics for patulin and 17 other mycotoxins on reversed-phase TLC plates, chromatographed using binary mixtures of water and methanol, acetonitrile, or tetrahydrofuran. Patulin was observed as its MBTH derivative, and it was suggested that the use of reversed-phase TLC could serve as a convenient confirmatory method for normal-phase TLC screening procedures.

IV. HIGH-PERFORMANCE LIQUID CHROMATOGRAPHIC METHODS

High-performance liquid chromatography (HPLC) can be used to separate many mycotoxins. The HPLC columns used for separations vary and include normal-phase, reversed-phase, and ion exchange columns depending on the particular method. The success of HPLC separation for mycotoxin analysis depends on factors such as extraction, cleanup, concentration, separation, detection techniques and HPLC conditions suitable for the specific compounds and matrices of concern.

A. Aflatoxins

Aflatoxin analysis using HPLC with fluorescence detection has become the method of choice for many applications. All of the common aflatoxins can be separated and quantified using HPLC. The aflatoxins exhibit fluorescence with excitation around 365 nm and emission above 400 nm. Quantitation using UV detection at 365 nm is also possible but less commonly used.

Trucksess and Wood [11] and Holcomb et al. [31] recently reviewed the literature on aflatoxin methods. Therefore, this review will not cover all HPLC methods. The official AOAC methods can be conveniently found in the AOAC Official Methods of Analysis [32]. The most recent AOAC method used a proprietary multifunctional cleanup column with TFA precolumn derivatization followed by reversed-phase HPLC [33].

Separation of aflatoxins by normal-phase HPLC has generally been replaced by reversed-phase separations. However, the fluorescence of B_1 and G_1 is quenched in the mobile phase of reversed-phase HPLC. Pre- or postcolumn derivatization of B_1 and G_1 with various reagents can enhance the fluorescence. Trifluoracetic acid (TFA) is used for precolumn derivatization to form B_{2A} and G_{2A}, respectively. TFA derivatization is effective, but the method is time-consuming and may decrease precision. Postcolumn derivatization reduces sample preparation and analysis time but may require an additional pump for the reagent.

Several methods for postcolumn derivatization have been reported. Shepherd and Gilbert [34], and Thiel et al. [35] reported the introduction of aqueous iodine to column effluent and the passing of the effluent and iodine through a heated reaction coil. Beaver et al. [36] compared this method with TLC and found the methods equivalent for aflatoxin B_1. The disadvantages of PCD using iodine are the need for an additional pump, the need for a reaction coil, and the lack of stability of the aqueous iodine solution.

Kok et al. [37,38] reported postcolumn derivatization using electrochemically generated bromine. This method requires the addition of potassium bromide and nitric acid to the mobile phase. The column effluent passes through an electrochemical cell (KOBRA cell), where the bromide is converted to bromine. Bromine reacts with B_1 and G_1 to give derivatives that fluoresce in the reverse-phase solvents. Advantages of electrochemically generated bromine over iodine derivatization include a more rapid reaction, with no need for a reaction coil and additional pumps.

Joshua [39] reported HPLC analysis using postcolumn photochemical UV irradiation and fluorescence detection for enhancement of B_1 and G_1 determination. This method eliminates the additional chemical reagents, additional pumps, and electrochemical cells. A photochemical reactor (PHRED) with a low-pressure mercury lamp and a knitted Teflon reactor coil are needed. Joshua [39] suggested that UV irradiation converts B_1 and G_1 to B_{2A} and G_{2A}, respectively. Wilson and Brock compared the response of KOBRA and PHRED cells with a reversed-phase separation and found that both performed satisfactorily (Table 3). Roch et al. [40] compared pre- and postcolumn derivatization and found that both gave over 90% recoveries with limits of detection of B_1 of 1 ng/g 0.3 ng/g for B_2. Therefore, the analyst has many acceptable choices for HPLC analysis of aflatoxins.

Cyclodextrins have also been used [41,42] to increase the fluorescence response of aflatoxins B_1 and G_1. A C-18 column with a mobile phase of methanol-water (60:40) was used with the cyclodextrins being introduced post-column. The cyclodextrins increased the fluorescence of B_1 and G_1 on the order of the 45- and 70-fold, respectively [43]. Pyridinium bromide perbromide was successfully used as a novel postcolumn derivative for determination of aflatoxins in spices [44]. Immunochemical cleanup of aflatoxins including M_1 is sometimes followed by HPLC detection [45–49]. Immunoaffinity column cleanup has been reported to be successful for HPLC of B_1, B_2, G_1, G_2, M_1, and Q_1 in urine [50]. An HPLC method has been developed for sterigmatocystin and aflatoxin B_1 [51].

Automation of aflatoxin analyses is desirable in many laboratories, and several authors have addressed this issue. Aflatoxin M_1 was determined using an automated immunoprecolumn followed by a C18 precolumn before the C-18 analytical column. Defatted milk or urine was loaded onto the immunocolumn, and several steps later the concentrated extract was injected on the analytical column. The dynamic range was 20 to 400 ng/L with a relative standard deviation (RSD) of 5% to 10% [48,49]. Van Rhijin et al. [52] developed a two-column

TABLE 3 Relative Area Counts of Aflatoxin B_1, B_2, G_1, and G_2 Peaks After Separation on a Zorbax ODS 4.6 mm × 2.5 cm C-18 Column with a Mobile Phase of Water/Acetonitrile/Methanol (60:20:20), a 1 ml/min Flow Rate, and Fluorescence Detection (Excitation 365 nm, Emission 425 nm)

	Relative area count[a]			
Postcolumn treatment	G_2	G_1	B_2	B_1
No treatment[b]	1	1	1	1
KOBRA cell[b]	0.93	46.13	0.98	23.01
PHRED cell[b]	0.96	32.35	1.03	19.7
No treatment[c]	1	1	1	1
PHRED cell[c]	0.96	19.18	0.99	14.46

[a]Values are the averages for three injections (unpublished data furnished by D.M. Wilson and J.H. Brock).
[b]Aqueous mobile phase with 1 mM KBr and 1 mM HNO_3.
[c]Aqueous mobile phase without KBr and HNO_3.

cleanup coupled with HPLC and found that the automated method was useful for screening but not quantitation. Urano et al. [53] reported that an automated online immunoaffinity column LC system using the KOBRA cell for aflatoxin in corn and peanuts could be effectively used. Niedwetzki et al. [45], Kussak et al. [47], and Carmen et al. [54] all successfully used the robotic system for automated sample cleanup and detection by HPLC. Kussak et al. [47] used C2 extraction columns while Niedwetzki et al. [45] and Carmen et al. [54] used immunoaffinity cleanup columns.

B. Fumonisins

The fumonisins do not absorb UV light or fluoresce; therefore detection by conventional spectrophotometric methods requires derivatization. Several reports have described techniques for reactions with the primary amine group (Fig. 1). The initial quantitative HPLC method, was a fumonisin detection limit of 10 μg/g, involved the formation of a maleyl derivative and detection by UV at 236 nm [55]. Enhanced sensitivity was achieved following the formation of fluorescamine derivatives coupled with fluorescence detection [5,56]. The precolumn derivatization of the fumonisins with fluorescamine can result in the formation of an equilibrium mixture of two reaction products that can be separated by HPLC [55].

The precolumn preparation of *ortho*-phthaldiaedehyde (OPA) derivatives of the fumonisins, in the presence of 2-mercapto-ethanol and borate buffer at pH 9 to 10, is both rapid and reproducible at room temperature [7,57]. The resultant derivatives of FB_1, FB_2, and FB_3, may be separated by isocratic reversed-phase HPLC coupled with fluorescence detection (Fig. 2), with a detection limit of between 0.01 and 0.05 μg analog/g [56,58]. Limited stability of the OPA reaction products requires that the derivatives be injected into the HPLC system with 2 min. of preparation, and the adoption of standardized periods of time between derivative preparation and injection have been recommended [7,57]. Rice et al. [59] cited the pH of the derivative reaction mixture as a critical factor in the optimum chromatography of OPA derivatives, and indicated that fluorescence response, chromatographic separation, and derivative stability could be enhanced by maintaining the pH between 8.0 and 8.5 [59].

A number of other fluorescent derivatives have been proposed for the analysis of the fumonisins in corn or corn-based products. Enhanced stability was observed for those derivatives prepared with 4-fluoro-7-nitrobenzofurazan (NBDF), where a detection limit of 0.1 μg/g was observed [60]. Derivatives prepared using 9-fluorenylmethyl chloroformate (FMOC) were shown to be stable for a period of 72 hours [61]. Cohen and Michaud [62] synthesized an alternative fluorescent derivatizing agent (6-amino-quinolyl-N-hydroxy-succinimidyl carba-

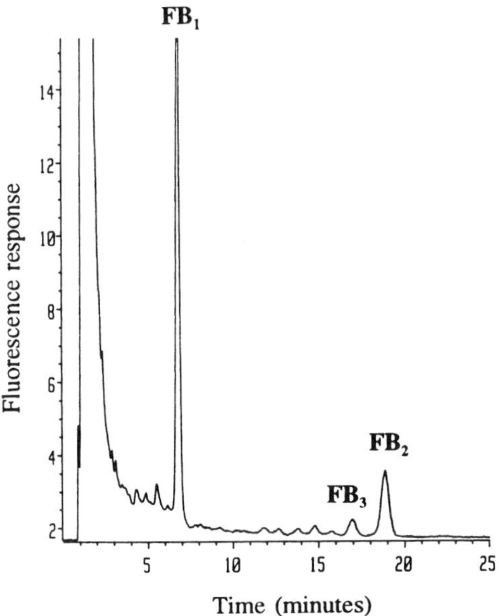

FIGURE 2 Chromatographic separation of FB_1, FB_2, and FB_3 in a naturally contaminated corn sample purified in accordance with the method of Sydenham et al. [42]. The corn sample contained 0.80, 0.34, and 0.03 μg/g of the three fumonisin analogs, respectively.

mate) similar to OPA but which exhibits greater stability when applied to the analysis of amino acids. Supplied under trade name AccQ-Fluor (Millipore, Milford, MA), the reagent has also been reported to be stable for 48 hours for analysis of fumonisins in corn. The detection limit is severalfold higher than that of OPA, at approximately 0.26 μg/g [63]. One of the most sensitive derivatives yet prepared, which also exhibited enhanced stability over 24 hours and could detect 50 pg of FB_1, is naphthalene-2,3-dicarboxaldehyde in the presence of potassium cyanide [6]. The derivative has been used in a method for the determination of fumonisins in milk, which has a sensitivity of 0.005 μg/ml [64].

Interlaboratory collaborative studies have been conducted for the determination of fumonisins in corn. The first was a study of the fluorescamine procedure, which showed acceptable agreement in the range of 4 to 1800 μg/g [65]. The HPLC method for FB_1 and FB_2 described by Shephard et al. [7], with purification of corn extracts on SAX media and derivatization with OPA, was collaboratively assessed by the Commission on Food Chemistry of IUPAC using naturally contaminated corn samples with 11 participants from six countries [66]. Between-laboratory reproducibilities ranged from 18% to 26.7% for FB_1 (at 0.2 to 2.0 μg FB_1/g) and from 28% to 45.6% for FB_2 (at 0.07 to 0.74 μg FB_2/g).

A third collaborative study was conducted under the joint auspices of IUPAC and AOAC International, in which the accuracy and reproducibility characteristics of OPA-based method [57] were assessed using both spiked blank (fumonisin-free) and naturally contaminated corn samples [67]. The study included the determination of FB_3 in addition to FB_1 and FB_2, and was conducted by 12 participants from 10 countries [67]. Between-laboratory reproducibilities ranged from 13.9% to 22.2% for FB_1 (at 0.5 to 8.0 μg/g), from 15.8% to 26.7% for FB_2 (at 0.2 to 3.2 μg/g), and from 19.5% to 24.9% for FB_3 (at 0.1 to 1.6 μg/g) [67], while the calculated HORRAT ratios ranged from 0.75 to 1.73 [67]. The HORRAT ratio is considered to be the "best single index of acceptability of method performance" with "values above 2 being considered unacceptable" [2]. Overall, recoveries ranged from 75.8% to 86.6% for the individual toxins [67], and the method has been adopted as official first action by AOAC International.

In addition to the separation of fumonisins by reversed-phase HPLC, the application of ion pair chromatography coupled to postcolumn derivatization with OPA and fluorescence detection has been reported [68].

C. Deoxynivalenol

Early application of HPLC with UV (LC/UV) detection of DON was reported by Canadian researchers [69] employing a reversed-phase chromatographic column and an aqueous methanol mobile phase. Rajakylä et al. [70] determined DON (and other mycotoxins) in grains at levels of 250 to 1000 ng/g using LC/UV detection. Researchers in India [71] and New Zealand [72] used similar systems to detect

DON in grains and foods. More recently, Stratton et al. [73] employed reversed-phase LC/UV to determine DON and other *Fusarium* toxins in grains. The mobile phase was a methanol-water gradient, and a DON detection limit of 50 ng/g was reported. In the U.S., Trucksess et al. [74] used a similar system in a survey of flour and bran. Vudathala et al. [75] used a reversed-phase chromatographic column, and an acetonitrile-water mobile phase to detect DON added to cow's milk.

Early work involving electrochemical detection (LC/EC) includes that of Sylvia et al. [76], who used reductive EC detection to determine DON levels in naturally contaminated wheat. Increased sensitivity and selectivity for DON (and other trichothecenes) were reported [77] via HPLC coupled with online post-column photolysis and amperometric detection. Wheat samples with DON in the range of 100 to 900 ng/g were analyzed and compared with good correlation to determinations by LC/UV and TLC.

D. Ochratoxins

Many HPLC methods have appeared over the years for ochratoxins, and most procedures employ fluorescence detection. HPLC methods for ochratoxin A analysis in animal tissues and fluids include those of Wilken et al. [78] and Bauer and Gareis [79]. A method involving partitioning into bicarbonate solution acidification, back-extraction, solid-phase extraction onto a silica adsorbent, elution with toluene-acetone-formic acid, and determination by HPLC with fluorescence detection has been proposed for swine urine and feces [80]. The authors give the detection limits as 0.3 ng/ml and 1.5 ng/g, respectively.

E. Zearalenone

Since several *Fusarium* mycotoxins may be produced by species that synthesize zearalenone, multitoxin screening procedures have been developed to detect zearalenone and the commonly occurring trichothecenes [81]. One such method involves extraction with acetonitrile-methanol-water followed by purification in a C_{18} solid-phase column. Zearalenone (and moniliformin) were determined by HPLC with recoveries of 72% and 78%, respectively, and a detection limit for zearalenone of 10 ng/g. A more sophisticated and sensitive method for determination and confirmation of zearalenone and its isomers, as well as zeranol and its isomers, was developed for the detection of these compounds in animal tissues [82]. Liquid chromatography with electrochemical detection was used for identification and quantitation of zearalenol, zeranol, zearalanone, and zearalenone. Capillary GC/MS was required to separate α-zearalenol from β-zearalenol and zeranol from taleranol. This method consistently found incurred levels of zeranol in beef liver tissue but did not detect any zearalenone in any tissue tested. More recently, a similar procedure for zeranols and zearalenols (normal-phase HPLC for separation and GC-MS for quantitation) was used to verify the presence of zeranols in pasture-fed animals [83]. This surprising report implies that zenobiotic anabolic compounds at physiological levels can occur in domestic animals without deliberate treatment with growth promoters.

A high-performance liquid chromatography with fluorescence detection method has been tested collaboratively [84] and is an official method of the AOAC International, the American Oil Chemists' Society, and the American Association of Cereal Chemists. An important improvement in HPLC with fluorescence detection is post-column derivatization with aluminum chloride [85]. This technique increases the fluorescence of zearalenone up to fivefold. A rapid method for isolation and determination of zearalenone in rice cultures involves simple extraction of moist culture material with acetonitrile: water, defatting with hexane, and, after diluting with water, assay by HPLC using diode array detector. The procedure is rapid and recovery of spiked rice was 76% to 94% [86]. An improvement in HPLC separation and simultaneous detection of zearalenone and ochratoxin by fluorescence detection has been accomplished by adding β-cyclodextrin to the mobile phase [87].

F. Citrinin

Estimation of citrinin by HPLC normally involves extraction, cleanup, chromatography, and fluorescence detection (88,89). HPLC procedures require exhaustively purified extracts. In the method of Lepom [88], for example, the acetonitrile-water cereal extract is defatted with *n*-hexane and extracted with chloroform. These chloroform layers are then partitioned against sodium bicarbonate. After acidification, the bicarbonate layers are back-extracted with chloroform. The chloroform layers are dried over sodium sulfate, combined, and evaporated to dryness. Recoveries of citrinin from spiked ground wheat at 100 and 1000 ng/g were 71% and 70%, with CV = 3.7% and 4.4%, respectively. In the method of Zimmerli et al. [89], a variant of the AOAC method for ochratoxin A is used to extract and purify citrinin from cereal matrices. A novel HPLC method using a normal-phase (silica) column impregnated with citric acid, and fluorescence detection, gave on average 85% recovery (SD = 7.4%; n = 9) from wheat flour spiked at 1 to 30 ng/g. However, the retention time of citrinin was erratic with continued use of the column.

G. Patulin

Patulin is a relatively polar molecule and exhibits a strong absorption spectrum. Therefore, methods based on reversed-phase HPLC coupled with UV detection have been developed and used extensively. Prieta et al. [90] optimized the diphasic dialysis extraction procedure with a purification step using silica gel solid-phase extraction (SPE) technology. The overall recoveries were between 80% and 85%, at a spiking level of 0.02 μg patulin/ml apple juice. The relative standard deviations (RSD) values were between 2% and 15%, and the detection limit was 0.001 μg/ml. Recoveries for patulin of between 80 and 88% (over a spiking range of 0.003 to 0.1 μg/ml) and a detection limit of 0.002 μg/ml, have been reported for a method involving partition of patulin from apple juice into ethyl acetate, followed by concentration and purification on silica gel SPE cartridges [91]. Separation of patulin was achieved using reversed-phase HPLC with a mixture of water: tetrahydrofuran as mobile phase and UV detection at 272 nm. Important aspects of the method included the total removal of residual water prior to purification on silica gel media, as well as selection of the correct temperature and inert gas flow rates used during the final concentration step of the method [91].

The only method subjected to collaborative assessment over the past 5 years involved extraction of patulin from apple juice with ethyl acetate followed by partitioning of the organic phase with a sodium carbonate solution, prior to reversed-phase HPLC and determination with UV detection at either 254 nm or 276 nm. The study was conducted by 22 participants from 10 countries. The 12 test samples contained levels of between 20 and 200 ng patulin/ml [92]. Recoveries ranged from 91% to 108%. This method has subsequently been adopted as official first action by the AOAC International.

During the HPLC analyses of processed apple juices, the separation of patulin from intrinsic compounds, such as 5-hydroxymethyl-2-furaldehyde (HMF), is of particular importance. HMF can be effectively separated from patulin in purified apple juice concentrates naturally contaminated with about 20 μg patulin/ml (based on single-strength fruit juice at 12° Brix). The separation, which is superior to previous reports [90, 91], was achieved using 250 × 4.6 mm i.d. HPLC column containing 5 μm Inertsil ODS2 material prepared from high-purity silica, with acetonitrile: water (10:90) as mobile phase (Fig. 3). HMF is formed during the production and storage of apple juice concentrates, by the reaction between reducing sugars and amino acids, and its separation from patulin is crucial since it exhibits a UV spectrum similar to that of patulin with a maximum absorbance that differs by less that 10 nm (i.e.,λ_{max} at 284 nm for HMF, as opposed to a λ_{max} at 275 nm for patulin; Fig. 3). Consequently, insufficient resolution of these compounds, coupled with single wavelength UV detection, may lead to component misidentification. The use of photodiode array detection coupled with reversed-phase HPLC (in order to confirm the identification of patulin) has been strongly recommended [93,94].

Bartolomé et al. [95] suggested the use of absorbance ratios at alternative wavelengths, as well as wavelength maxima in second-order derivative UV spectra, as confirmatory methods for the identification of patulin. Conversely, others

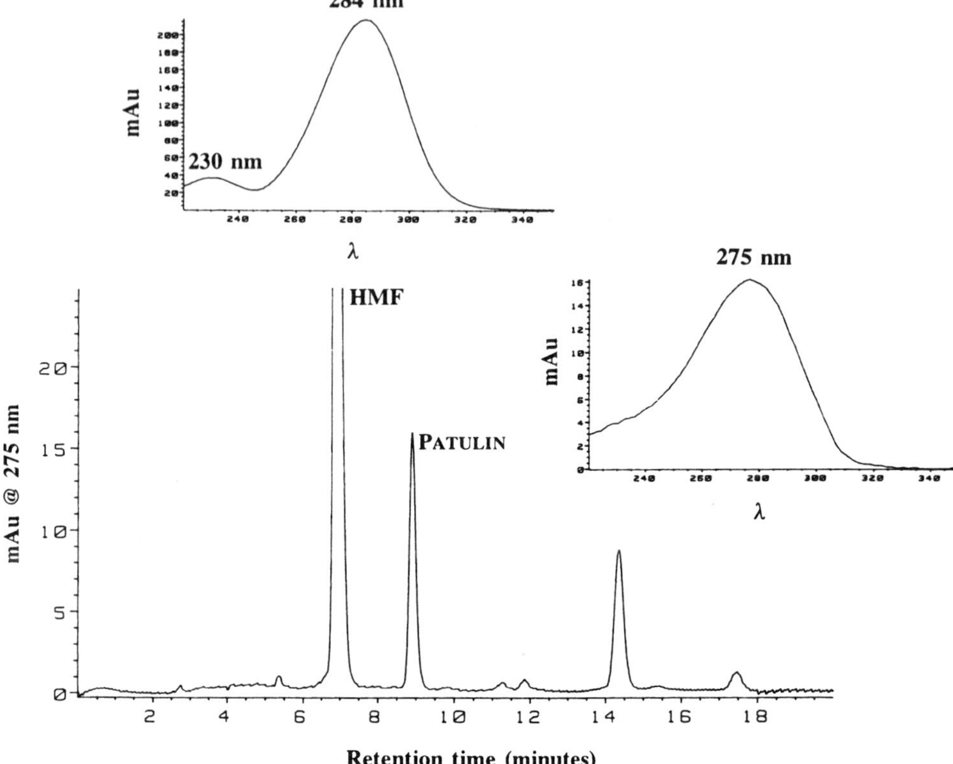

FIGURE 3 Reversed-phased HPLC separation of a purified apple juice extract naturally contaminated with ~0.02 μg patulin/ml, together with the individual UV spectra of HMF, λ_{max} at 284 nm, and patulin, λ_{max} at 275 nm, recorded in the sample using diode array detection. HMF, 5-hydroxymethyl-2-furaldehyde.

have reported that the unequivocal confirmation of secondary metabolites including patulin, based on their UV spectra in neutral solutions, could be inhibited due to spectral similarities, whereas the collection of UV spectra, derived in alkaline solutions, could distinguish between parent compounds and their metabolites [96].

V. GAS CHROMATOGRAPHIC METHODS

A. Fumonisins

Analysis of the fumonisins using conventional capillary GC has centered on the separation and detection of the hydrolysis products (i.e., the TCA and aminopolyol moieties), since volatilization of the intact molecule was difficult due to the high polarity of the compounds. Initial studies focused on the analysis of TCA (which was esterified with *iso*-butanol), although its determination yields limited information and is nonspecific [55]. Increased specificity was achieved following the formation and determination of trimethylsilyl derivatives of the aminopolyols [97]. Subsequent preparation of the trifluoroacetyl derivatives of the hyrolysates tended to improve the chromatographic separation of the individual aminopolyols [15, 98]. The development of capillary GC methods also resulted in the ability to simultaneously quantify and confirm observations, through the use of GC-MS interfaces [98]. The accuracy and precision of GC-MS methods were improved following the preparation and use of deuterium-labeled FB_1 (produced in liquid culture using methyl-D_3-labeled methionine) as an internal standard, which was added to extracts prior to hydrolysis [99].

B. Deoxynivalenol

Gas chromatography, usually employing fused silica capillary columns, remains the most common technique for the determination of DON. Derivatization is required prior to the determinative step to increase the volatility and detectablity of DON. The most common derivatization mechanisms are the production of the trimethylsilyl (TMS) ethers or fluorinated propionyl or butyryl esters from the free hydroxyl groups of the parent compound. The optimization of TMS reactions has been reported [100,101], and a solid-phase reaction catalyst has been studied for the production of heptafluorobutyryl (HFB) esters [102]. Although most GC methods employ an electron capture detector (ECD), the use of flame ionization detectors (FIDs) has also been reported. The application of GC/MS is becoming more common.

GC-ECD has been employed to detect DON (and other trichothecenes) in cereal grains by many authors. In a wheat survey, Scott et al. [103] injected HFB esters onto a DB-1701 megabore capillary column and reported detection limits of 20 to 100 ng/g. Sydenham et al. [104] employed an ECD to detect both TMS and HFB derivatives, and reported a detection limit of 20 ng/g wheat. Luo et al. [105] estimated in DON detection limit of 5 ng/g in corn and wheat using an OV-17 packed column and TMS ethers. Muñoz et al. (106) injected HFBs onto a DB-5 capillary and detected as little as 43 ng/g DON in corn. Langseth and Clasen [107] also employed an ECD to detect TMS derivatives and reported detection limits of about 20 ng/g of sample. Moller and Gustavsson [108] employed TMS derivatives, and Siedel et al. [109] employed HFB derivatives to detect DON (and other trichothecenes) in various cereal grains. Croteau et al. [110] derivatized corn extracts with heptafluorobutyryl anhydride, and reported a detection limit of 50 ng/g. GC/ECD using the TMS derivatives was the method of choice for Scott et al. [111] for the determination of DON in wort.

Some researchers have demonstrated the advantage of hydrolysis of DON to the parent alcohol prior to derivatization and analysis. For the analysis of feeds, Rood et al. [112] employed hydrolysis and then detected pentafluoropropionyl esters with an ECD, while Lauren and Agnew [72] reported that hydrolysis resulted in improved stability of grain extracts and more reliable and consistent quantitations of trifluoroacetic anhydride derivatives with an ECD. Both groups reported detection limits of about 20 ng/g.

The use of an FID for the analysis of DON (and other trichothecenes) in wheat has been reported [113]. Samples were extracted with methanol-aqueous potassium chloride, and the extracts were derivatized by heptafluorobutyryl-imidazole. Confirmation of identity was accomplished by acetylation and reduction reactions; detection limits were reported as 50 to 100 ng/g.

C. Patulin

Capillary gas chromatography has been applied to the analysis of patulin in apple juice [114]. Extracts were purified by silica gel column chromatography, and the residue was then derivatized with heptafluorobutyrylimidazole (HFB). The patulin-HFB derivative was separated on an apolar 30 m $\times$ 0.32 mm i.d. fused-silica capillary column (coated with a 0.25-μm film of 5% phenylmethylsiloxane) with electron capture detection, using hexachlorobenzene as an internal standard. The patulin-HFB derivative was relatively stable in *n*-heptane, while the derivatization procedure was highly reproducible.

VI. MASS SPECTROMETRIC METHODS

A. Aflatoxins

Kussak et al. [115] developed a liquid chromatography/electrospray tandem mass spectrometry method for aflatoxins B_1, B_2, G_1, and G_2 in airborne dust and spiked urine. The detection limits (S/N = 3) with tandem mass spectrometry were between 4 and 10 pg. A reversed-phase LC method was reported by Cappiello et al. [116]; the separation was on a reversed-phase column coupled with a modified particle beam interface. The specificity of the electron impact ionization was excellent, and the method should be useful for complex matrices.

Holcomb et al. [117] found using thermospray mass spectrometry that the iodine derivatives of B_1 and G_1 were formed by the addition of an iodine ion and a methoxy group across the double bond of the furan portion of the molecules.

B. Fumonisins

In a comparative study, Korfmacher et al. [118] evaluated thermospray (TS-MS), fast-atom bombardment (FAB-MS), and electrospray mass spectrometry (ES-MS), and concluded that FAB-MS and ES-MS were ideal as techniques for the characterization of nanogram quantities of FB_1, with both techniques providing strong molecular ions coupled with limited fragmentation. Plattner and Branham [99] applied FAB-MS to the analysis of naturally contaminated corn samples and observed a detection limit of 0.1 μg/g. Conversely, TS-MS was reported to generate multiple fragment ions, while the sensitivity was considered to be relatively limited [118], although the use of an aqueous ammonium acetate: acetonitrile phase mobile phase and negative ion detection has since resulted in a decrease of the detection limit of TS-MS to about 2 ng FB_1 [119].

In contrast to studies on GC-MS interfaces, those based on liquid chromatography (LC-MS) have the added advantages that the fumonisins need not be hydrolyzed nor derivatized, thus allowing the detection of the intact molecules. The initial ES-MS methods were based on the analysis of a FB_1 standard which was separated on a reversed-phase C_{18} column using a water:acetonitrile:acetic acid mixture as mobile phase [120]. The sensitivity of the technique was assessed to be in the low picogram range, and similar techniques have subsequently been used for the analysis of fumonisins and their hydrolyzed products in both naturally contaminated samples and culture material of *F. moniliforme* [121,122]. Particle beam interfaces have also been used for the confirmation of fumonisin production in liquid media, following the reversed-phase HPLC separation of their methyl esters [123]. An advantage of the particle beam interface (PB-MS) is its ability to provide both electron and chemical ionization spectra, which can provide greater structural information. Liquid chromatography-PB-MS has been applied to the quantitative determination of FB_1, FB_2, and FB_3 in commercial cornmeal, where detection limits for the method have been found to be approximately 0.1, 0.02, and 0.05 μg/g for the three fumonisin analogs, respectively [124].

C. Deoxynivalenol

An excellent review on the use of negative ion chemical ionization (NICI) MS for the analysis of DON (and other trichothecenes) was published by Krishnamurthy et al. [125]. Investigators in the U.S. [112] used NICI/MS to confirm the presence of DON (and other trichothecenes) in feeds and estimated a detection limit of 20 to 50 ng/g. Researchers in South Africa [104] and Japan [105] have used GC/MS to confirm the presence of DON (and other tricholthecenes) in infected wheat and corn. DON (and other trichothecenes) in cereal grains have been analyzed by GC coupled to an (ion trap) mass selective detector operated in the NICI mode [126]. The researchers prepared trifluoroacyl derivatives and used methanol as the reactant gas. Capturing full-scan spectra (400 to 650 atomic mass units), detection limits of 1 to 5 ng/g were reported. Canadian [127] and American investigators [128] have used GC/MS (in the electron impact mode) to determine the presence of DON (and other *Fusarium* mycotoxins) in beer. NICI/MS (in selected ion monitoring mode) was utilized by Kim et al. [129] to determine levels of DON (and other *Fusarium* toxins) in a survey of barley and corn, with a reported detection limit of 5 ng/g. The method of Scott et al. [103] has been updated to include six trichothecenes and employ either ECD or NICI/MS [130].

Early investigations into the use of LC/MS for DON include that of Rajakylä et al. [70], who improved upon an LC/UV method with the introduction of a thermospray MS interface and reported detection limits in the range of 3 to 40 ng/g of grain. The mass spectra of DON (and other trichothecenes) have been studied [131], and both thermospray and plasma spray introduction techniques were reported suitable for monitoring purposes. Frit-fast-atom bombardment-high resolution MS was coupled to LC with postcolumn addition of glycerol reactant to result in a sensitive and selective method for identification of DON (and other trichothecenes) [132]. An LC/MS system with thermospray interface has been used successfully [133] for the determination of DON in swine tissues.

D. Ochratoxins

Mass spectrometry provides a specific detection technique for mycotoxins after chromatographic separation by LC or GC. In LC/MS detection of ochratoxin A, the chlorine atom in the molecule affords high parent ion abundance when using negative-ion chemical ionization, and ochratoxin A has been monitored by this method in barley [134] and swine blood [135] with selected-ion monitoring. Thermospray LC/MS has also been used to detect ochratoxin A in grain with ammonium positive-ion chemical ionization and selected-ion monitoring [136]. GC/MS of the ochratoxin A methyl ester, produced by diazomethane treatment, has been used to detect ochratoxin A at 0.1 ng/g (100 pg/g) in corn and cereal bran [137].

E. Zearalenone

A definitive and sensitive procedure for zearalenone and α-zearalenol in ruminant urine is a gas chromatography–tandem mass spectrometry method [138]. The sensitivity was reported to be 1 ng/ml with a linear response between 1 and 20 ng/ml zearalenone. Tandem mass spectrometry allows for selection of appropriate parent ions and subsequent detection of unique daughter ions for confirmation. This technique minimizes interferences from the matrix, thus increasing sensitivity.

F. Patulin

The application of capillary GC has enabled the use of the online electron impact mass spectrometry for confirmation purposes. Tarter and Scott [114] reported the confirmation of the patulin-HFB derivative, based on the presence of a molecular ion at *m/z* 350 and the co-occurrence of a series of ions between *m/z* 153 and *m/z* 53, which were consistent with patulin-HFB fragment ions. Conversely, the occurrence of patulin in purified apple juice extracts, determined by reversed-phase HPLC, was confirmed by direct–loop liquid chromatography–electrospray mass spectrometry [93]. However, under the alkaline conditions used, patulin must have undergone some degree of molecular rearrangement, since it was observed as a doubly charged species with a molecular ion at ca. *m/z* 77. The electrospray mass spectrum also yielded a major fragment ion at *m/z* 123, although its occurrence could not be structurally ascribed [93].

VII. IMMUNOLOGICAL METHODS

Immunological assays (IAs) have become one the cornerstones of rapid testing of food and feeds for aflatoxins and other mycotoxins. Polyclonal and monoclonal antibodies are the binding agents. Polyclonal antibodies are produced by immunizing animals with a mycotoxin-protein conjugate. For aflatoxins, polyclonal antibodies with specificity toward the dihydrofuran portion or the cyclopentanone ring moiety of the aflatoxin molecule are generated depending on the site of conjugation [139]. Monoclonal antibodies are secreted by fusions of immunized mouse spleen cells with a myeloma cell line. Immunological recognition is based on the spatial complement of specific groups (for aflatoxin it is the dihydrofuran portion or the cyclopentanone ring) of the antigen with those of the antibody. Since the antibodies target the chemical groups and not the antigen, compounds with similar chemical groups will also interact with the antibodies. This cross-reacting phenomenon is both a negative and a positive feature. Analysts need to be very careful in selecting methods that contain antibodies that recognize only compounds of interest.

A. Aflatoxins

1. Immunological Assays

Three types of IA are commonly employed: the radioimmunoassay (RIA), the enzyme-linked immunosorbent assay (ELISA), and the immunoaffinity column assay [140]. Radiolabeled aflatoxin is used in the RIA. In the competitive RIA the unlabeled aflatoxin standard or the aflatoxins in the test solution and the labeled aflatoxin in the assay system are competing for the limited number of binding sites on the antibodies. The amount of aflatoxins in a sample is inversely related to the amount of radiolabeled aflatoxin in the solution. Only a small amount of antibody is required for RIA. However, the radioisotopes used in the assays present disposal problems. The use of nonisotopic labels such as enzymes has gradually replaced the use of RIA.

In the typical direct ELISA the binding of aflatoxin-enzyme conjugate by the immobilized antibodies is inhibited by the presence of aflatoxins in the test solution. The bound enzyme catalyzes the transformation of a substrate to a colored complex. The color intensity formed is inversely related to aflatoxin concentration. Horseradish peroxidase, which catalyzes the oxidation of the substrate (tetramethylbenzidine) to form a blue colored complex, is the most commonly used enzyme for labeling. Microtiter plates, beads, and membranes (cup, card, probe tests) have been used as solid supports for ELISA. The assays may be qualitative or quantitative.

Affinity columns are prepared by adsorption of the antibodies onto a gel material contained in a small plastic cartridge. Aflatoxins are captured from test solutions by the immunospecific antibodies. After nonbinding impurities have been washed from the cartridge, aflatoxins are desorbed with methanol, then derivatized and quantitated by solution fluorometry or transferred to the liquid chromatographic reversed-phase column for further separation and fluorescence determination [141,142]. The use of affinity columns to capture and concentrate aflatoxins has several advantages: increased selectivity, the ability to trap the aflatoxins in large volumes of test samples (biological fluids), and integration with other analytical techniques. However, large amounts of antibodies are needed to prepare the immunosorbent columns and the cost is greater than RIA and ELISA.

2. Commerical Immunoassay Kits

The status of immunoassays has changed through the years. Since the 1970s both direct and indirect IAs for aflatoxins have been developed by research institutes [143–150]; as a result, more than 10 commercially available IA kits have been developed applying the research findings. Table 4 shows many of the commercial IA kits for aflatoxins. Before using these kits the analyst must examine the ability of the kits to determine the analyte at the target level and in the matrix of interest as well as what the crossreactivities are. It is also important to perform confirmatory tests by comparing immunoassay results with those of the conventional techniques.

Some of the kits shown in Table 4 have been evaluated by comparing statistical parameters obtained by immunoassay with those by traditional methods such as thin-layer chromatography and liquid chromatography [152–158]. Organizations such as the U.S. Food and Drug Administration (FDA) and the U.S. Department of Agriculture (USDA) have set some of the criteria used to evaluate the commercial procedures [159]. AOAC International/IUPAC has sponsored several international collaborative studies to evaluate IA kits. Results of some of the studies have been published and have been approved as either official first action or official final action. The AOAC Research Institute certifies IA kits by using an independent laboratory to validate their performance claims. The Insti-

TABLE 4 Commercial Immunoassay Kits for Aflatoxins[a]

Test kit	Format	Applicable range (ng/g)	Commodities	Mode of detection
Afla 5, 10, 20 Cup[b]	Cup	5, 10, 20	Peanuts, peanut butter, corn, cottonseed	Visual
Alfatest[c]	Affinity column for B_1, B_2, G_1, G_2	10	Corn, peanuts, peanut butter, milk	LC or fluorometer
	Affinity column for M_1	0.1		
Agri-screen[d]	Microwell	5	Corn, peanuts, cottonseed, feed	Visual or ELISA reader
Biokits[e]	Microwell	9	Peanut butter	Visual or ELISA reader
Cite Probe[f]	Probe	5, 20	Corn, cottonseed	Visual
Dosage[g]	Microwell	0.5	Corn, peanuts, feed	ELISA reader
Detection[g]	Tube	1, 5	Corn, peanuts, feed	Visual or instrumental
EZ-screen[h]	Card	5, 20	Corn, peanuts	Visual
One-Step ELISA[b]	Microwell	5	Corn, peanuts, peanut butter, cottonseed, feed, milk	ELISA reader
	Microwell for M_1	0.5		
Veratox[d]	Microwell	5	Corn, peanuts, cottonseed, feed, milk	ELISA reader
	Microwell for M_1	0.25		

LC, liquid chromatography; ELISA, enzyme-linked immunosorbent assay.
[a]Information from reference 150 and manufacturers.
[b]International Diagnostic, St. Joseph, MI.
[c]Vicam, Watertown, MA.
[d]Neogen Corp., Lansing, MI.
[e]Cortecs Diagnostic Ltd., Deeside Industrial Park, Clwyd, U.K.
[f]Idexx Labs, Westbrook, ME.
[g]Transia, Lyon, France.
[h]MEDTOX Diagnostics, Inc., Burlington, NC.

tute and the USDA Federal Grain Inspection Service (FGIS) signed a Memorandum of Understanding stating that when the kits have been successfully evaluated by the Institute using the FGIS protocol, then the kits will be considered to be certified by both FGIS and the Institute.

Six IA kits have ben approved as AOAC International official methods for aflatoxins. The manufacturers and lowest level measured are given in parentheses.

1. ELISA method for aflatoxin B_1 in corn and roasted peanuts (Neogen, Agri-Screen applicable to screening aflatoxin B_1 at ⩾20 ng/g).
2. ELISA method for aflatoxin B_1 in cottonseed products and mixed feed (Neogen, Agri-Screen ⩾15ng/g).
3. Immunoaffinity column method for total aflatoxins in corn, peanuts, peanut butter (Vicam, AflatestP column, ⩾10ng/g).
4. ELISA method for total aflatoxin B_1, B_2, and G_1 in corn, cottonseed, peanuts, peanut butter (International Diagnostic Systems, ImmunoDot Screen Cup, ⩾30ng/g in corn and raw peanuts).
5. ELISA method for total aflatoxin B_1, B_2,and G_1 in corn (International Diagnostics Systems, Afla-20 Cup Test, ⩾20ng/g).
6. ELISA method for total aflatoxins in peanut butter (Biokits, range of 9 to 90 ng/g).

Two IA kits have been certified by AOAC Research Institute:

1. Veratox AST (Neogen): aflatoxin residues in grain and grain products.
2. Aflatest P (Vicam): aflatoxin residues in grain and grain products.

The steps for purification and isolation of aflatoxins using the affinity column have been automated. A commercially available robotic system enables IAC cleanup and LC separation including postcolumn derivatization with iodine or bromine and fluorescence detection automatically [160–162]. The use of reusable affinity precolumns coupled with a column switching system allows online transfer of aflatoxins from affinity columns to LC [163,164]. Even though the automated methods are mainly for food analysis, they can possibly be applied to biological fluid as well.

3. Aflatoxin Metabolites in Milk, Sera, and Urine

Epidemiological studies in the areas of high incidence of hepatoma throughout Africa and Southeast Asia have suggested that the exposure to aflatoxins may play a role in primary liver cancer. It has not been clarified whether aflatoxin alone or through an interaction with hepatitis B virus infection causes the effect [165]. In order to demonstrate the relationship between exposure to aflatoxins and the etiology of disease, methods have been developed to measure aflatoxin metabolites in milk, sera, and urine. IAC cleanup and concentration in conjunction with offline LC is the most common technique for determination of aflatoxin M_1, an aflatoxin B_1 metabolite, in milk [166–168]. The detection limit ranged from 20 to 80 pg/ml. An automated system including IAC cleanup and LC quantitation was used to determine M_1 and aflatoxin Q_1 in human urine [165]. The limit of detection was 18 pg/ml.

4. Aflatoxin-Albumin Adducts in Blood

Aflatoxin-albumin (Af-alb) adducts are found in blood after aflatoxin B_1 exposure. The half-life of Af-alb is about 20 days, which is much longer than M_1 in blood or urine. The measurement of Af-alb adducts in serum is a more useful biomarker for determining individual exposure to aflatoxins. The structure of the major Af-alb has been characterized as an Af-lysine (Aflys) residue [169]. A combination of ELISA and LC fluorescence techniques permitted the quantitation of Af-alb adducts in serum samples from fingerprick blood samples [170]. In that study Af-alb adducts were measured in various African and Asian populations. A procedure to increase the recovery to aflatoxin adducts consisted of precipitating all serum proteins with ammonium sulfate, digesting with pronase, and then quantitating by RIA [171]. This procedure avoided the isolation of albumin before analysis and thereby reduced interference in the RIA. It was used to determine the Af-alb adducts in samples collected from Chongming Island, China, a region of high liver cancer incidence, and food was found to be highly contaminated with aflatoxins. The af-alb levels ranged from 0 to 890 pmol Af/g alb.

5. Aflatoxin-DNA Adducts in Tissues and Body Fluids

Aflatoxin B_1 binds covalently with DNA to form aflatoxin-DNA (AF-DNA) adducts. Aflatoxin B_1 is enzymatically oxidized to the 2,3-epoxide which attacks the N^7 atom of guanine to produce the DNA adduct, 2,3-dihydro-2-(N^7-guanyl)-3-hydroxyaflatoxin B_1 (AF-N^7-gua). AF-N^7-gua and the ring opened form of AF-N^7-gua,2,3-dihydro-2-(N^5-formyl-2′,5′,6′-triamino-4′-oxo-N^5-pyurimidy)-3-hydroxyaflatoxin B_1(AF-FAPyr) are the major AF-DNA adducts. These DNA adducts are biomarkers that indicate that FB_1 could cause genotoxic damage to humans or animals. A high-affinity monoclonal antibody that recognizes B_1, B_2, M_1, and the two DNA adducts had been reproduced and was covalently bound to Sepharose 4B [172,173]. This preparative IAC was used to detect aflatoxin metabolites and DNA adducts in the urine of people from China and Gambia exposed to AFB_1 through dietary contamination. AF-N^7-gua, M_1, and aflatoxin P_1 were found.

AF-DNA adducts were detected in formalin-fixed human liver, kidney and lung autopsy samples from persons acutely exposed to aflatoxins in Southeast Asia [174]. DNA was isolated, acid-hydrolyzed, neutralized, and immunoconcentrated in an IAC. The acid-hydrolyzed DNA inhibited antibody binding in a competitive aflatoxin inhibition ELISA both before and after IAC purification. The ELISA inhibition data were confirmed by HPLC analysis. Adduct levels ranged from 0 to 170 adducts/10^6 nucleotides. The adducts were identified as derivatives of AF-N^7-gua.

B. Fumonisins

Murine polyclonal antibodies to FB_1 have been produced using cholera toxin as both hapten carrier protein and adjuvant [175]. The antiserum produced cross-reacted with FB_2 (87%) and FB_3 (40%), but did not crossreact with the hydrolyzed moieties of the fumonisins (i.e., the aminopolyol backbones or TCA), suggesting that the antibody recognized a section of the basic fumonisin molecule near the union of the two moieties (Fig. 1), thus enabling it to detect each of the naturally occurring analogs. These antibodies were incorporated into an indirect ELISA with a sensitivity of 0.1 μg/ml. Further development of monoclonal antibodies and their incorporation into a direct ELISA format (which was subsequently made commercially available—Neogen, Lansing, MI) increased the sensitivity of the technique to 0.05 μg/ml [176]. The performance of this direct ELISA has been the subject of comparative studies, in which the technique was evaluated against other chemical methods (GC-MS and HPLC) [177,178]. Applied to a series of naturally contaminated grain-based samples, the results obtained from the two chemical techniques were well correlated with each other ($r = .946$), although the correlations of the individual methods with the ELISA technique were relatively poor ($r = .512$ and .478, respectively) with the ELISA consistently recording higher results than the chemical methods [177]. Sydenham et al. [178] applied the same antibody-based ELISA to both naturally and artificially spiked corn samples, as well as to extracts prepared from corn-based culture material of *F. moniliforme* and compared their performance of an established HPLC procedure. In accordance with previous observations [179], the ELISA produced higher values than HPLC for the naturally contaminated samples, while a similar, higher ELISA response for the culture samples could be partially attributed to the presence of the fumonisin A analogs [121], which also crossreacted with the monoclonal antibody [178]. However, when applied to artificially spiked corn samples (over the range 8.0 to 12.8 μg/g), the ELISA and HPLC results were well correlated ($r = .996$), suggesting that the higher estimations determined in naturally contaminated samples may have resulted from the presence of fumonisin precursors or unidentified metabolites [178]. Shelby et al. [179] compared the performance of an indirect ELISA with TLC for the determination of FB_1 in a large number of corn samples. The indirect ELISA consistency reported higher fumonisin levels than TLC (in about 50% of the samples), although the authors considered that any underestimation by TLC may have been due to the presence of other fumonisin analogs (FB_2, FB_3, and possibly FB_4) in the corn samples that were not determined by

TLC but that crossreacted with the antibody incorporated into the indirect ELISA assay [179].

A highly sensitive polyclonal-based direct ELISA, which can detect about 10ng fumonisin analog per gram, has also been reported to exhibit nonspecific binding due to corn-based matrix effects, especially at low fumonisin concentrations [180]. Improved correlations between a commercially available polyclonal antibody-based direct ELISA (Neogen, Lansing, MI) and HPLC results were recorded in a limited study on 20 corn samples naturally contaminated at combined FB_{1-3} levels ranging from <0.05 to 5 μg/g [181]. Linear regression analysis between the total fumonisin results generated by HPLC and the direct ELISA provided a correlation coefficient of 0.967 and a regression slope of 1.059 [181].

Membrane-based competitive enzyme immunoassays for FB_1, in both dipstick and immunofiltration (ELIFA) formats, have been developed [182]. Both tests have visual detection limits for FB_1 in the range of 0.04 to 0.06 μg/g in corn-based foods, and the results generated by each technique compared favorably with those determined by microtiter plate ELISA and HPLC methods. It was concluded that with assay times of 60 and 10 min for the dipstick and ELIFA formats, respectively, the methods would be useful for the qualitative screening of corn-based foods [182].

Methanol eluates from the immunoaffinity columns for fumonisin may also be used for the formation of fluorescent derivatives and direct determination of total fumonisin concentrations (based on fluorescence response). The method has been developed to be applicable primarily to corn-based human foods with a detection range of between 0.1 and 5 μg/g (Vicam, Watertown, MA).

Antibodies reactive to the fumonisins have been incorporated into a commercially available immunoaffinity column (Vicam, Watertown, MA), for the purification of extracts from food commodities prior to HPLC with fluorescence detection. The column has been reported to have equal affinity for FB_1 and FB_2 and to give recoveries of 85.4% and 87.1%, respectively; the capacity is about 1.2 μg/g fumonisin per column [183]. The column has been used for analysis of fumonisins from canned and frozen sweet corn products [184]. For corn, minor alterations were made to the basic method, and recoveries for FB_1 from sweet corn ranged from 76% to 88% at spiking levels of between 0.05 and 0.2 μg FB_1/g [184].

C. Deoxynivalenol

Immunochemical techniques offer perhaps the greatest potential in the field of mycotoxin analysis, but the evolution of validated laboratory methods for DON has been arduously slow. Methods are, however, beginning to appear for the determination of DON (and other trichothecenes) in cereal grains. Initial studies of both direct and indirect ELISA for the analysis of DON [185] found direct ELISA advantageous over indirect ELISA in terms of both sensitivity and specificity.

Detection limits were reported as 10 ng/g in corn and wheat, and good correlation with radioimmunoassay and TLC techniques were found. The characteristics of direct ELISAs for DON (and 3- and 15-acetyl-DON) in buffers were studied [186], and differentiation between 3-acetyl-DON and DON was found possible through hydrolysis. This work has led to the commercial availability of an ELISA kit applicable to the analysis of DON (after acetylation) in feeds, grains, and cereal products (RIDASCREEN DON, Rbiopharm GmbH). When used as a qualitative screening procedure for the presence of DON in wheat, the AGRISCREEN (Neogen Corporation) laboratory kit showed no significant difference with a TLC procedure [187]. A microtiter well test (EZQuant, Diagnostix, Inc.) was favorably compared to a GC method [188]. As research into the production and reactivity of antibodies against specific mycotoxins continues, their application to field and laboratory methods will be increasingly common.

D. Ochratoxins

Ochratoxin A was one of the first mycotoxins for immunoassay techniques were developed. EIA procedures have appeared for analysis of ochratoxin A in barley [189], wheat [190], swine kidneys [191], and swine serum [192], and RIA methods have been described for barley and swine kidneys [193–195]. Polyclonal antisera from rabbits have proved to be 10 to 20 times more sensitive than polyclonal antisera from chickens for determining ochratoxin A by EIA in wheat samples [196]. A convenient test-strip assay using rabbit polyclonal antibodies gives rapid visual detection of ochratoxin A at 2 ng/ml in buffers and 100 ng/g or less in spiked corn sample [197]. Monoclonal cell culture technology facilitates production of antibodies of high specificity [198], and EIA and RIA procedures have been described for foods, animal blood, and swine kidneys [199–201) using monoclonal antibodies.

Using immunoaffinity column cleanup coupled with HPLC separation, and fluorescence detection, it is possible to achieve 70% to 80% recovery of ochratoxin A added at 10 ng/g to wheat, rye, barley, pig kidneys, and blood sausage [202]. The limit of detection was around 0.2 ng/g.

E. Zearalenone

Enzyme-linked immunosorbent assay procedures have been used for simultaneous screening of zearalenone and other important mycotoxins [203]. Monoclonal antibodies for zearalenone, aflatoxin B_1, and fumonisin B_1 were immobilized (as multiple lines) on nitrocellulose membrane strips. After a modified ELISA was conducted, the color intensity of lines formed by a precipitating substrate was related to mycotoxin concentration. The detection limit for zearalenone was 3 ng/ml. Line density was quantitatively assessed using a camera, video monitor, and a microcomputer with video-digitizing capability. This computer-assisted multianalytic assay system could be used to determine values for mycotoxins in extracts in less than 30 min. Another multimycotoxin ELISA procedure employs a dipstick format for the detection of up to five mycotoxins in food and feed products [204]. In artificially contaminated wheat, the detection for zearalenone was reported to be 60 ng/g. These specific and cost-effective ELISA procedures can conveniently check for the presence or absence of various mycotoxins in feedstuffs at certain threshold values within a short time frame.

Recent developments have been prompted by the need to have rapid, cost-effective screening procedures for zearalenone and other mycotoxins. These procedures (affinity columns and ELISA) are specific for the target mycotoxin and, when properly used, greatly reduce the number of negative samples which would previously be analyzed by specific methods such as HPLC, GC, GC/MS, or MS/MS [205]. Specific monoclonal antibodies for zearalenone are very sensitive (15 to 500 pg per assay) [206], and similar monoclonal antibodies have been incorporated into assay procedures and applied to cereal grains and milk [207]. The rapid detection of zearalenone and T-2 toxin has been accomplished in the same extract of cereals with monoclonal antibody-based ELISA with no cleanup of the cereal extract [208]. Zearalenone was detectable at 25 to 400 ng/g and the assay could be completed in 3 hours.

A commercially available ELISA screening procedure for zearalenone was tested collaboratively and adapted as the official method of analysis for the AOAC International [209]. This screening method was used to detect zearalenone in corn, wheat, and feed by both visual and spectrophotometric examination. The method, as tested, successfully detected zearalenone in these commodities at $\geqslant$800ng/g toxin. The detection limit for this method is higher than desired for this mycotoxin.

An important aspect of the rapid, specific ELISA procedures is that they can be adapted to measure zearalenone (and α-zearalenol, which is three times more estrogenic than zearalenone) in animal fluids such as serum or urine. For example, porcine urine can be assayed by a modified ELISA to verify that the animal has in fact been exposed to zearalenone [210]. This direct competitive ELISA of the animals' urine eliminates the uncertainty of analyses of suspected feed samples which may no longer be available when the animal exhibits zearalenone toxicosis. Since zearalenone is excreted in glucuronide form, urine samples were treated with glucuronidase prior to ELISA measurements. This technique showed that 5.3% of ingested zearalenone was excreted in urine within 8 hours following ingestion and that the average excretion rate was about 100μg zearalenone equivalents per hour.

F. Citrinin

The first ELISA procedure for citrinin has been recently published [211]. Citrinin was coupled to keyhole limpet hemocyanin carrier protein, and this antigenic conjugate was used to immunize rabbits. Citrinin was also coupled to glucose oxidase protein from *A. niger*, which was used for coating microtiter plates for determinations using the indirect competitive ELISA format. Citrinin could be detected in buffer solutions at 1 ng/ml. Recovery of citrinin from spiked wheat flour at 200 to 2000 ng/g was 89% to 104%, with CV = 9% to 13%. Except for dilution, no treatment of the extract is necessary. A direct ELISA for citrinin was published by Abramson et al. [212]. Recoveries of citrinin added to ground barley at 500 to 2000 ng/g were 108% to 111% with a CV of 8.4% to 24%.

G. Patulin

Commercial immunoassays for patulin are not yet available, but polyclonal antibodies against patulin, as its hemiglutarate derivative, have been produced [213].

VIII. SUPERCRITICAL FLUID CHROMATOGRAPHY

A. Aflatoxins

Analytical-scale supercritical fluid extractions were used to remove aflatoxin B_1 from field-inoculated corn [214]. Various pressures and temperatures were tested, and it was determined that the optimum conditions were 5000 psi at 80°C with a 15% organic modifier composed of acetonitrile/methanol at a 2:1 ratio. Additional work is needed to make this a useful analytical technique. Densities > 1.072 g/ml or pressures > 820 atm (at 40°C) are required to extract aflatoxin at reasonable yields. The addition of methanol makes the extraction less selective. The limitations of supercritical fluid extraction of aflatoxin are related to its low solubility in CO_2, adsorption on the matrix, and slow diffusion [215].

B. Deoxynivalenol and Zearalenone

Little advancement has been reported in the application of SFC to the analysis of DON in the past decade. Investigators have used carbon dioxide supercritical fluids to effect the elution of DON (and separation from other trichothecenes) on bonded-phase capillary columns prior to either FID or NICI/MS detection. Sub picogram detection quantities are reported [216].

Supercritical fluid chromatography has been investigated as an alternative method for resolution and possible analysis of some *Fusarium* mycotoxins [217]. Capillary and packed columns were used to determine retention indices and effects of modifiers as well as column temperatures and pressures for several mycotoxins. Retention times of toxins studied, including zearalenone, were eight times shorter than those on capillary columns. A new and related concept in isolation and identification of zearalenone, zeranol, and their metabolites, is analytical supercritical fluid extraction (SFE) [218]. This technology utilizes solid-phase extraction columns to preconcentrate toxins in urine on a C_{18} column followed by SFE with unmodified CO_2. The extracted zearalenones-zeranols are trapped on an alumina column. The trapped mycotoxins are then eluted with acetone/water and quantitated by HPLC with UV detection. Recoveries from spiked urine was a remarkable 97%.

IX. OTHER NOVEL METHODS

A. Aflatoxin

Micellar electrokinetic capillary chromatography was applied to high-speed analysis of aflatoxin by Cole et al. [219]. Baseline separation of aflatoxins B_1, B_2, G_1, and G_2 could be accomplished in under 30 sec with laser-based fluorescence detection of 0.05 to 0.9 fmo of underivatized aflatoxins. Holland and Sepaniak [220] resolved 10 mycotoxins in 45 min using micellar electrokinetic capillary chromatography.

B. Fumonisins

The fumonisin methods described earlier are relatively advanced and require the provision of sophisticated and expensive equipment. There is, however, a need for inexpensive rapid screening techniques for the analysis of fumonisins in food commodities. To a great extent, immunochemical methods described earlier are fulfilling these criteria. However, the development of minicolumn techniques such as chemiselective immobilization and detection (CSID) may also provide rapid inexpensive fumonisin analytical procedures. The CSID method relies upon the derivatization of sample extracts with fluorescamine and the absorption on a minicolumn, sequentially packed with various inorganic sorbents. The fluorescent fumonisin derivatives are then selectively immobilized at an interface between two sorbents, and the fluorescent intensity may then be determined under long-wave UV light [221].

Capillary zone electrophoresis (CZE) has also recently been investigated for the analysis of the fumonisins, as the fluorescein isothiocyanate (FITC) derivatives of their hydrolyzed backbones [222]. Detected by laser-induced fluorescence, the detection limit of the method has been found to be 25 fg FB_1 [222]. Thompson and Maragos [223] have also reported the development of a fiber-optic immunosensor for the measurement of FB_1 as its FITC derivative. The sensors were produced using immobilized monoclonal antibodies against FB_1, which were then placed in a capillary flow cell. Solutions of FITC-labeled FB_1 and varying amounts of unlabeled FB_1 were pumped past the optical sensor in order that they could compete for antibody binding sites on the fiber [223]. The resultant fluorescence was detected using conventional optics at $490nm_{(Ex)}$ and $520nm_{(Em)}$, with the signals from the detector being inversely proportional to the unlabeled FB_1 concentration. The detection limit of the technique was assessed to be 0.05 μg/ml [223]. Vo-Dinh [224] reviewed the development of fiber-optic immunosensors for environmental analysis.

The application of evaporative light-scattering detectors (ELSD) to the analysis of fumonisins has been investigated, although the ELSD response was shown to be nonlinear with concentration [121]. The effective detection limit of the technique has been reported to be approximately 50 ppm, which was considered adequate for the analysis of culture material [121]. Clearly, extensive sample purification procedures would be required for the detection technique to be applied to naturally contaminated corn extracts. However, the effective application of ELSD to the semipreparative purification of fumonisins from culture material of *F. moniliforme* has been demonstrated [225].

C. Ochratoxins

Spectrofluorometric methods [226,227] have long been employed for the determination of ochratoxin A in blood after hydrolyzing with carboxypeptidase A, and measuring the differences in fluorescence before and after treatment. Although it may be possible to detect ochratoxin A in blood at 6 ng/ml, this technique has been superseded by more sensitive and precise methodologies based on HPLC with fluorescence detection [228].

D. Zearalenone

Test systems for determining estrogenic activity in vitro of zearalenone have been developed [229] as alternatives to frequently used and time-consuming uterotropic assays with mice. Specific chemical analyses, such as HPLC, GC-MS, or TLC, can accurately determine "known" metabolites (i.e. zearalenone; α- and β-zearalenol) which may or may not be responsible for total estrogenic potency of food for human and animal consumption. In this test system, defined estrogen-sensitive cell lines (MCF-9 and LeC-9 cells) were used to determine estrogenic activity of purified compounds and zearalenone-contaminated wheat, barley, and oats. When compared to uterotropic assays, the relative estrogenic response obtained in LeC-9 cells was higher, indicating that the cereal matrix itself was influencing the test. When compared to HPLC analysis of cereal grains, 12 of 15 samples gave similar results in this test system. Further refinements of the in vitro tests could lead to new analytical systems to measure several environmental estrogens. A microbial assay, using *Bacillus brevis*, was developed to detect a wide range of mycotoxins, including zearalenone [230]. The lowest detectable amount of zearalenone was 10 μg per disk.

X. CONCLUSION

We have surveyed many of the current techniques for mycotoxin analysis in this review. There are many approaches that can be used to devise an analytical procedure. While there is really no one best technique, there is an abundance of methods that can be used or modified to design a protocol to meet specific analytical requirements. The choice of a method for an analytical procedure depends on the analyst's preference, the sample matrix, the mycotoxin of interest, and the availability of supplies and equipment. Many chemical and immunochemical methods are available for specific applications.

Thin-layer chromatography can be used as the separation technique for almost all mycotoxins, although derivatives sometimes have to be made to visualize the compound of interest or to confirm the identity. The major disadvantage of TLC is that the procedure is inherently variable, so the coefficient of variation will be large and the precision will be poor.

Analytical separations accomplished by HPLC and GC are more precise because there is generally less variation associated with these procedures. Therefore for most analytical tasks, HPLC or GC is preferred over TLC. HPLC is suitable for quantitation of the aflatoxins, fumonisins, ochratoxins, patulin, and citrinin. GC methods are preferred for determination of DON. Zearalenone can be determined using either HPLC or GC methods. Mass spectral techniques can be applied to many compounds and are extremely useful for trace analysis as well as chemical confirmation of mycotoxins.

Immunochemical methods are most useful for screening or quantitation within a limited concentration range. With immunochemical methods analysts must be careful to test for matrix and solvent effects, crossreactivities, and method consistency. Immunochemical techniques are very useful for studies on the metabolism, fate, and persistence of mycotoxins in animals.

AOAC International Official Methods are particularly useful for samples in commerce because these methods have been collaboratively evaluated. However, official methods should only be routinely used for the commodities included in the collaborative studies. For research and unusual samples, methods frequently have to be modified and evaluated for specific analytical requirements.

REFERENCES

1. TB Whitaker, DL Park. Problems associated with accurately measuring aflatoxin in foods and feeds: errors associated with sampling, sample preparation and analysis, In: The Toxicology of Aflatoxins: Human Health, Veterinary and Agricultural Significance. DL Eaton, JD Groopman, eds. San Diego: Academic Press, 1994, p 433.

2. W Horwitz, R Albert S Nesheim. Reliability of mycotoxin assays—an update. J AOAC Int 75:461, 1993.
3. T Whitaker, W Horwitz, R Albert, S Nesheim. Variability associated with analytical methods used to measure aflatoxin in agricultural commodities. J AOAC Int 79:476, 1996.
4. RJ Cole, JW Dorner. Extraction of aflatoxins from naturally contaminated peanuts with different solvents and solvent/peanut ratios. J AOAC Int 77:1509, 1994.
5. TM Wilson, PF Ross, LG Rice, et al. Fumonisin B_1 levels associated with an epizootic of equine leukoencephalomalacia. J Vet Diagn Invest 2:213, 1990.
6. GA Bennett, JL Richard. Liquid chromatographic method for analysis of the naphthalene dicarboxaldehyde derivative of fumonisins. J AOAC Int 77:501, 1994.
7. GS Shepard, EW Sydenham, PG Thiel, WCA Gelderblom. Quantitative determination of fumonisins B_1 and B_2 by high-performance liquid chromatography with fluorescence detection. J Liq Chromatog 13:2077, 1990.
8. PM Scott, GA Lawrence. Stability and problems in recovery of fumonisins added to corn-based foods. J AOAC Int 77:541, 1994.
9. O Zeller, F Sager, B Zimmerli. Occurrence of fumonisins in foods. Mitt Gebiete Lebensm Hyg 85:81, 1994.
10. B Hald, GM Wood, A Boenke, B Schurer and P Finglas, Ochratoxin A in wheat: an intercomparison of procedures. Food Addit Contam 10:185, 1993.
11. MW Trucksess, GE Wood. Recent methods of analysis for aflatoxins in foods and feeds, In: The Toxicology of Aflatoxins: Human Health, Veterinary and Agricultural Significance. DL Eaton, JD Groopman, eds., San Diego: Academic Press, 1994, p 409.
12. V Betina. Review: Chromatographic methods as tools in the field of mycotoxins. J Chromatog 477:187, 1989.
13. GA Bennett, JL Richard. Fate and distribution of *Fusarium* mycotoxins during processing of contaminated grains. Proceedings of Distillers Feed Research Conference, New Orleans, LA, 1994, pp 23–29.
14. WCA Gelderblom, K Jaskiewicz, WFO Marasas, et al. Fumonisins—novel mycotoxins with cancer-promoting activity produced by *Fusarium moniliforme*. Appl Environ Microbiol 54:1806, 1988.
15. PG Theil, WFO Marasas, EW Sydenham, GS Shepard, WCA Gelderblom, JJ Nieuwenhuis. Survey of fumonisin production by *Fusarium* species. Appl Environ Microbiol 57:1089, 1991.
16. J Dupuy, P Le Bars, H Boudra, J Le Bars. Thermostability of fumonisin B_1, a mycotoxin from *Fusarium moniliforme*, in corn. Appl Environ Microbiol 59:2864, 1993.
17. GE Rottinghaus, CE Coatney, HC Minor. A rapid, sensitive thin layer chromatographic procedure for the detection of fumonisin B_1 and B_2. J Vet Diagn Invest 4:326, 1992.
18. S Stockenström, EW Sydenham, PG Thiel. Determination of fumonisins in corn: evaluation of two purification procedures. Mycotoxin Res 10:9, 1994.
19. A Pittet, V Parisod, M Schellenberg. Occurrence of fumonisins B_1 and B_2 in corn-based products from the Swiss market. J Agric Food Chem 40:1352, 1992.

20. M Dawlatana, RD Coker, MJ Nagler, G Blunden. A normal phase HPTLC method for the quantitative determination of fumonisin B_1 in rice. Chromatographia 41:3, 1995.
21. Association of Official Analytical Chemists. Ochratoxins, In: K. Helrich, ed. Official Methods of Analysis, 15th ed. Arlington, VA: AOAC 1990, p 1207.
22. Scott PM. Mycotoxins. J Assoc Off Anal Chem 71:70, 1998.
23. S Nesheim, ME Stack, MW Trucksess, RM Eppley, P Krogh. Rapid solvent-efficient method for liquid chromatographic determination of ochratoxin A in corn, and ochratoxin A in wheat and barley by high-performance liquid chromatography. J Chromatog 355:335, 1992.
24. A Gimeno, ML Martins. Rapid thin layer chromatographic determination of patulin, citrinin and aflatoxin in apples and pears, and their juices and jams. J Assoc Off Anal Chem 66:85, 1983.
25. A Gimeno. Determination of citrinin in corn and barley on thin layer chromatographic plates impregnated with glycolic acid. J Assoc Off Anal Chem 67:194, 1984.
26. D Abramson, T Thorsteinson, D Forest. Chromatography of mycotoxins on precoated reverse-phase thin-layer plates. Arch Environ Contam Toxicol 18:327, 1989.
27. LM Lin, J Zhang, K Sui, WB Sung. Simultaneous thin layer chromatographic determination of zearalenone and patulin in maize. J Planar Chromatog 6:274, 1993.
28. J Prieta, MA Moreno, JL Blanco, G Suarez, L Dominguez. Determination of patulin by diphasic dialysis extraction and thin-layer chromatography. J Food Prot 55:1001, 1992.
29. L Dominguez, JL Blanco, MA Moreno, et al. Diphasic dialysis: a new membrane method for a selective and efficient extraction of low molecular weight organic compounds from aqueous solutions. J AOAC Int 75:854, 1992.
30. D Abramson, T Thorsteinson, D Forest. Chromatography of mycotoxins on precoated reverse-phase thin-layer plates. Arch Environ Contam Toxicol 18:327, 1989.
31. M Holcomb, DM Wilson, MW Trucksess, HC Thompson. Determination of aflatoxins in foods products by chromatography. J Chromatog 624:341, 1992.
32. AOAC International. Official Methods of Analysis of AOAC International, 16th ed. Arlington, VA: AOAC International, 1995, p 49.
33. MW Trucksess, ME Stack, S Nesheim, RH Albert, TR Romer. Multifunctional column coupled with liquid chromatography for determination of aflatoxins B_1,B_2, G_1 and G_2 in corn, almonds, Brazil nuts, peanuts and pistachio nuts; collaborative study. J AOAC Int 77:1512. 1994.
34. MS Shepherd, J Gilbert. An investigation of HPLC postcolumn iodination conditions for the enhancement of aflatoxin B fluorescence. Food Addit Contam 1:325, 1984.
35. PG Thiel, S Stockenström, PS Gathercole. Aflatoxin analysis by reverse phase HPLC using post-column derivatization for enhancement of fluorescence. J Liq Chromatog 9:103, 1986.
36. RW Beaver, DM Wilson, MW Trucksess. Comparison of LC with post-column derivatization with TLC for the determination of aflatoxins in naturally contaminated corn. J Assoc Off Anal Chem 73:579, 1990.
37. W Kok, UA Brinkman, RW Frei. On-line electrochemical reagent production for

detection in liquid chromatography and continuous flow systems, Anal Chim Acta 162:19, 1984.
38. W Kok, CH Van Neer, WA Traag, LGM Tuinstra. Determination of aflatoxin in cattle feed by liquid chromatography and post-column derivatization with electrochemically generated bromine. J Chromatog 367:231, 1986.
39. H Joshua. Determination of aflatoxin by reversed-phase high performance liquid chromatography with post-column in-line photochemical derivatizaiton and fluorescence detection. J Chromatog 654:247, 1993.
40. OG Roch, G Blunden, DJ Haig, RD Coker, C Gay. Determination of aflatoxins in groundnut meal by high-performance liquid chromatography: a comparison of two methods of derivatization of aflatoxins B_1. Br J Biomed Sci 52:312, 1995.
41. ML Vazquez, A Cepeda, P Prognon, A Mahuzier, J Blais. Cyclodextrins as modifiers of the luminescence characteristics of aflatoxins. Anal Chim Acta 255:343, 1991.
42. ML Vazquez, CM Franco, A Cepeda, P Prognon, G Mahuzier. Liquid chromatographic study of the interaction between aflatoxins and beta-cyclodextrin. Anal Chim Acta 269:239, 1992.
43. A Cepeda, CM Franco, CA Fente, et al. Post-column excitation of aflatoxins using cyclodextrins in liquid chromatography for food analysis. J Chromatog 721:69, 1996.
44. RC Garner, MM Whattam, PJL Taylor, MW Stow. Analysis of United Kingdom purchased spices for aflatoxins using an immunoaffinity column clean-up procedure followed by high-performance liquid-chromatographic analysis and post column derivatization with pyridinium bromide perbromide. J Chromatog 648:485, 1993.
45. G Niedwetzki, G Lach, K Geschwill. Determination of aflatoxins in food by use of an automatic workstation. J Chromatog 661:175, 1994.
46. A Kussak, B Andersson, K Andersson. Immunoaffinity column cleanup for the high-performance liquid-chromatographic determination of aflatoxins B_1, B_2, G_1, G_2, M_1, and Q_1 in urine. J Chromatog Biomed Appl 672:253, 1995.
47. A Kussak, B Andersson, K Andersson. Determination of aflatoxins in airborne dust from feed factories by automated immunoaffinity column cleanup and liquid chromatography. J Chromatog 708:55, 1995.
48. A Farjam, NC Merbel, H Lingeman, et al. Non-selective desorption of immuno precolumns coupled on-line with column liquid chromatography: determination of aflatoxins. Int J Environ Anal Chem 45:73, 1991.
49. A Farjam, R DeVries, H Lingeman, UAT Brinkman. Immuno precolumns for selective on-line sample pretreatment of aflatoxins in milk prior to column liquid chromatography. J Liq Chromatog 14:2541, 1991.
50. A Kussak, B Andersson, K Andersson. Automated sample cleanup with solid-phase extraction for the determination of aflatoxins in urine by liquid-chromatography. J Chromatog Biomed Appl 616:235, 1993.
51. C Huang-Hua, X Zhao-Zun, B Li-Yin, ZX Zhao, YB Li. Studies on the simultaneous determination of aflatoxin B_1 and sterigmatocystin in high performance liquid chromatography. Chin J Chromatog 12:299, 1994.
52. JA van Rhijn, J Viveen, GMT Tuinstra. Automated determination of aflatoxin B_1 in cattle feed by two-column solid-phase extraction with on-line high-performance liquid chromatography. J Chromatog 592:1, 1992.

53. T Urano, MW Trucksess, SW Page. Automated affinity liquid chromatography system for on-line isolation, separation and quantitation of aflatoxins in methanol-water extracts of corn or peanuts. J Agric Food Chem 41:1982, 1993.
54. AS Carmen, SS Kuan, GM Ware, PP Umrigar. Robotic automated analysis of foods for aflatoxin. J AOAC Int. 79:456, 1996.
55. EW Sydenham, WCA Gelderblom, PG Thiel, WFO Marasas. Evidence for the natural occurrence of fumonisin B_1, a mycotoxin produced by *Fusarium moniliforme*, in corn. J Agric Food Chem 38:285, 1990.
56. PF Ross, LG Rice, RD Plattner, et al. Concentrations of fumonisin B_1 in feeds associated with animal health problems. Mycopathologia 114:129, 1991.
57. EW Sydenham, GS Shephard, PG Thiel. Liquid chromatographic determination of fumonisins B_1, B_2, and B_3 in foods and feeds. J AOAC Int 75:313, 1992.
58. MB Doko, A Visconti. Occurrence of fumonisins B_1 and B_2 in corn and corn-based human foodstuffs in Italy, Food Addit Contam. 11:433, 1994.
59. LG Rice, PF Ross, J Dejong, RD Plattner, JR Coats. Evaluation of a liquid chromatographic method for the determination of fumonisins in corn, poultry feed, and *Fusarium* culture material. J AOAC Int 78:1002, 1995.
60. PM Scott, GA Lawrence. Liquid chromatographic determination of fumonisins with 4-fluoro-7-nitrobenzofurazan. J AOAC Int 75:829, 1992.
61. M Holcomb, HC Thompson Jr, LJ Hankins. Analysis of fumonisin B_1 in rodent feed by gradient elution HPLC using precolumn derivatization with FMOC and fluorescence detection. J Agric Food Chem 41:764, 1993.
62. SA Cohen, and DP Michaud, Synthesis of a fluorescent derivatizing reagent, 6-aminoquinoyl-N-hydroxysuccinimidyl carbamate, and its application for the analysis of hydrolysate amino acids via high-performance liquid chromatography. Anal Biochem 211:279, 1993.
63. C Velaquez, C Van Bloemendal, V Sanchis, R Canela. Derivation of fumonisins B_1 and B_2 with 6-aminoquinolyl-N-hydroxysuccinimidyl carbamate. J Agric Food Chem 43:1535, 1995.
64. CM Maragos, JL Richard. Quantitation and stability of fumonisins B_1 and B_2 in milk. J AOAC Int 77:1162, 1994.
65. RD Plattner, PF Ross, J Reagor, J Stedelin, LG Rice. Analysis of corn and cultured corn for fumonisin B_1 by HPLC and GC/MS by four laboratories. J Vet Diagn Invest 3:357, 1991.
66. PG Thiel, EW Sydenham, GS Shephard, DJ Van Schalkwyk. Study of the reproductibility characteristics of a liquid chromatographic method for the determination of fumonisins B_1 and B_2 in corn: IUPAC collaborative study. J AOAC Int 76:361, 1993.
67. EW Sydenham, GS Shephard, PG Thiel, S Stockenström, PW Snijman, DJ Van Schalkwyk. Liquid chromatographic determination of fumonisins B_1, B_2, and B_3 in corn: IUPAC/AOAC interlaboratory collaborative study. J AOAC Int 79:688, 1996.
68. M Miyahara, H Akiyama, M Toyoda, Y Saito. New procedure for fumonisin B_1 and B_2 in corn and corn products by ion pair chromatography with ortho-phthaldialdehyde post-column derivatization and fluorometric detection, Presented at 108th Annual AOAC Int. Meeting, Portland, OR, 1994.
69. HL Trenholm, RM Warner, DB Prelusky. Assessment of extraction procedures in the

analysis of naturally contaminated grain products for deoxynivalenol (vomitoxin). J Assoc Off Anal Chem 68:645, 1985.
70. E Rajakylä, K Laasasenaho, PJD Sakkers. Determination of mycotoxins in grain by high-performance liquid chromatography and thermospray liquid chromatography-mass spectrometry. J Chromatog 384:391, 1998.
71. Y Ramakrishna, RV Bhat, S Vasanthi. Natural occurrence of mycotoxins in staple foods in India. J Agric Food Chem 38:1857, 1990.
72. DR Lauren, MP Agnew. Multitoxin screening method for *Fusarium* mycotoxins in grains. J Agric Food Chem 39:502, 1991.
73. GW Stratton, AR Robinson, HC Smith, L Kittilsen, M Barbour. Levels of five mycotoxins in grain harvested in Atlantic Canada as measured by high performance liquid chromatography. Arch Environ Contam Toxicol 24:399, 1993.
74. MW Trucksess, DE Ready, MK Pender, CA Ligmond, GE Wood SW Page. Determination and survey of deoxynivalenol in white flour, whole wheat flour and bran. Presented at 109th AOAC Int. Annual Meeting, Nashville, TN, 1995.
75. DK Vudathala, DB Prelusky, HL Trenholm. Analysis of trace levels of deoxynivalenol in cow's milk by high pressure liquid chromatography. J Liq Chromatog 17:673, 1994.
76. VL Sylvia, TD Phillips, BA Clement, JL Green, LF Kubena, ND Heidelbaugh. Determination of deoxynivalenol (vomitoxin) by high-performance liquid chromatography with electrochemical detection. J Chromatog 362:79, 1986.
77. WL Childress, IS Krull, CM Selavka. Determination of deoxynivalenol (DON, vomitoxin) in wheat by high-performance liquid chromatography with photolysis and electrochemical detection (HPLC-hv-EC). J Chromatog 28:76, 1990.
78. C Wilken, W Baltes, I Mehlitz, R Tierbach, R Weber. Ochratoxin A in pork kidneys method for determination. Z Leben-Untersuch Forsch 180:496–497, 1985.
79. J Bauer, M Gareis. Ochratoxin A in the food chain. Z Veterinärmed B 34:613–627, 1987.
80. H Valenta, I Kühn, K Rohr. Determination of ochratoxin A in urine and feces of swine by high-performance liquid chromatography. J Chromatog 613:295-302, 1993.
81. DR Lauren, MP Agnew. Multitoxin screening method for *Fusarium* mycotoxins in grains. J Agric Food Chem 39:502, 1991.
82. JE Roybal, RK Munns, WJ Morris, JA Hurlburt, W Shimoda. J Assoc Off Anal Chem 71:263, 1988.
83. AF Erasmuson, BG Scahill, DM West. Natural zeranol (α-zearalanol) in urine of pasture-fed animals. J Agric Food Chem 42:2721, 1994.
84. GA Bennett, OL Shotwell, WF Kwolek. Liquid chromatographic determination of α-zearalenol and zearalenone in corn: collaborative study. J Assoc Off Anal Chem 68:958, 1985.
85. MT Hetmanski, KA Scudamore. Detection of zearalenone in cereal extracts using high-performance liquid chromatography with post-column derivatization. J Chromatog 588:97, 1991.
86. AK Shrivastava, AA Ansari. Isolation and determination of zearalenone in rice cultures using liquid chromatography with diode array detection. Food Addit Contam 9:331, 1992.

87. V Seidel, E Poglits, K Schiller W Lindner. Simultaneous determination of ochratoxin A and zearalenone in maize reverse-phase high-performance liquid chromatography with fluorescence detection and β-cylodextrin as mobile phase additive. J Chromatog 635:227, 1993.
88. P Lepom. Simultaneous determination of the mycotoxins citrinin and ochratoxin A in wheat and barley high performance liquid chromatography. J Chromatog 355:335, 1986.
89. B Zimmerli, R Dick, U Baumann. High-performance liquid chromatographic determination of citrinin in cereals using an acid-buffered silica gel column. J Chromatog 462:406, 1989.
90. J Prieta, MA Moreno, J Bayo, et al. Determination of patulin by reversed-phase high-performance liquid chromatography with extraction by diphasic dialysis. Analyst 118:171, 1993.
91. R Rovira, F Ribera, V Sanchis, R Canela. Improvements in the quantitation of patulin in apple juice by high-performance liquid chromatography. J Agric Food Chem 41:214, 1993.
92. AR Brause, MW Trucksess, FS Thomas, SW Page. Determination of patulin in apple juice by liquid chromatography: collaborative study. J AOAC Int 79:451, 1996.
93. EW Sydenham, HF Vismer, WFO Marasas, et al. Reduction of patulin in apple juice samples-influence of initial processing. Food Control 6:195, 1995
94. B Bartolomé, ML Bengoechea, FJ Pérez-Ilzarbe, T Hernández, I Estrella, C Gómez-Cordovés. Determination of patulin in apple juice by high-performance liquid chromatography with diode-array detection. J Chromatog A 664:39, 1994.
95. B Bartolomé, ML Benegoechea, MC Gálvez, et al. Photodiode array detection for elucidation of the structure of phenolic compounds. J Chromatog 655:119, 1993.
96. RMA Paterson, C Kemmelmeier. Neutral, alkaline and difference ultraviolet spectra of secondary metabolites from *Penicillium* and other fungi, and comparisons to published maxima from gradient high-performance liquid chromatography with diode-array detection. J Chromatog 511:195, 1990.
97. RD Plattner, WP Norred, CW Bacon, et al. A method of detection of fumonisins in corn samples associated with field cases of equine leukoencephalomalacia. Mycologia 82:698, 1990.
98. RD Plattner, D Weisleder, DD Shackelford, R Peterson RG Powell. A new fumonisin from solid cultures of *Fusarium moniliforme*. Mycopathologia 117:23, 1992.
99. RD Plattner, BE Branham. Labeled fumonisins: production and use of fumonisin B_1 containing stable isotopes. J AOAC Int 77:525, 1994.
100. J Gilbert, JR Startin, C Crews. Optimization of conditions for the trimethylsilyation of trichothecene mycotoxins. J Chromatog 319:376, 1985.
101. AF Rizzo, L Saari, E Lindfor. Derivatization of trichothecenes and water treatment of their trimethylsilyl ethers in an anhydrous apolar solvent. J Chromatog 368:381, 1986.
102. SR Kanhere, PM Scott. Heptafluorobutyrylation of trichothecenes using a solid-phase catalyst. J Chromatog 511:384, 1990.
103. PM Scott, GA Lombaert, P Pellaers, et al. Application of capillary gas chromatography to a survey of wheat for five trichothecenes. Food Add Contam 6:489, 1989.
104. EW Sydenham, PG Thiel, WFO Marassas, JJ Nieuwenhuis. Occurrence of deoxy-

nivalenol in *Fusarium graminearum* infected undergrade wheat in South Africa. J Agric Food Chem 27:921, 1989.
105. Y Luo, T Yoshizawa, T Katayama. Comparative study on the natural occurrence of *Fusarium* mycotoxins (trichothecenes and zearalenone) in corn and wheat from high- and low-risk areas for human esophageal cancer in China. App Environ Microbiol 56:3723, 1990.
106. L Muñoz, M Cardelle, M Pereiro, R Riguera. Occurrence of corn mycotoxins in Galicia (northwest Spain). J Agric Food Chem 38:1004, 1990.
107. W Langseth, PE Clasen. Automation of a cleanup procedure for determination of trichothecenes in cereals using a charcoal-alumina column. J Chromatog 603:290, 1992.
108. TE Möller, HF Gustavsson. Determination of type A and B trichothecenes in cereals by gas chromatography with electron capture detection. J AOAC Int 75:1049, 1992.
109. V Seidel, B Lang, S Fraissler, et al. Analysis of trace levels of trichothecene mycotoxins in Austrian cereals by gas chromatography with electron capture detection. Chromatographia 37:191, 1993.
110. SM Croteau, DB Prelusky, HL Trenholm. Analysis of trichothecene mycotoxins by gas chromatography with electron capture detection. J Food Chem 42:928, 1994.
111. PM Scott, SR Kanhere, EF Daley, JM Farber. Fermentation of wort containing deoxynivalenol and zearalenone. Mycotoxin Res 8:58, 1992.
112. HD Rood, BB Buck, SP Swanson. Gas chromatographic screening method for T-2 toxin, diacetoxycirpenol, deoxynivalenol and related trichothecenes in feeds. J Assoc Off Anal Chem 71:483, 1988.
113. EB Furlong, LM Valente Soares. Gas chromatographic method for quantitation and confirmation of trichothecenes in wheat. J AOAC Int 78:386, 1995.
114. EJ Tarter, PM Scott. Determination of patulin by capillary gas chromatography of the heptafluorobutyrate derivative. J Chromatog 538:441, 1991.
115. A Kussak, CA Nilsson, B Andersson, J Langridge. Determination of aflatoxins in dust and urine by liquid-chromatography electrospray-ionization tandem mass-spectrometry. Rapid Commun Mass Spectrom 9:1234, 1995.
116. A Cappielli, G Famiglini, B Tirillini. Determination of aflatoxins in peanut meal by LC/MS with a particle-beam interface. Chromatographia 40:411, 1995.
117. M Holcomb, WA Kortnacher, HC Thompson. Characterization of iodine derivatives of aflatoxin B_1 and G_1 thermospray mass spectrometry. J Anal Toxicol 15:289, 1991.
118. WA Korfmacher, MP Chiarelli, JO Lay Jr, J Bloom, M Holcomb, KT McManus. Characterization of the mycotoxin fumonisin B_1: comparison of thermospray, fast-atom bombardment and electrospray mass spectrometry. Rapid Commun Mass Spectrom 5:463, 1991.
119. RA Thakur, JS Smith. Analysis of fumonisin B_1 by negative-ion thermospray mass spectrometry. Rapid Commun Mass Spectrom 8:82, 1994.
120. CJ Mirocha, DG Gilchrist, WT Shier, HK Abbas, Y Wen, RF Vesonder. AAL toxins, fumonisins (biology and chemistry) and host-specificity concepts. Mycopathologia 117:47, 1992.
121. RD Plattner. Detection of fumonisins produced in *Fusarium moniliforme* cultures by HPLC with electrospray MS and evaporative light scattering detectors. Nat Toxins 3:294, 1995.

122. EW Sydenham, PG Thiel, GS Shephard, KR Koch, T Hutton. Preparation and isolation of the partially hydrolyzed moiety of fumonisin B_1. J Agric Food Chem 43:2400, 1995.
123. JC Young, P Lafontaine. Detection and characterization of fumonisin mycotoxins as their methyl esters by liquid chromatography/particle-beam mass spectrometry. Rapid Commun Mass Spectrom 7:352, 1993.
124. DR Doeger, PC Howard, S Bajic, S Preece. Determination of fumonisins using on-line liquid chromatography coupled to electrospray mass spectrometry. Rapid Commun Mass Spectrom 8:603, 1994.
125. T Krishnamurthy, MB Wasserman, EW Sarver. Mass spectral investigations on trichothecene mycotoxins. Biomed Environ Mass Spec 13:503, 1986.
126. K Schwadorf, HM Müller. Determination of trichothecenes in cereals by gas chromatography with ion trap detection. Chromatographia 32:137, 1991.
127. PM Scott, SR Kanhere, D Weber. Analysis of Canadian and imported beers for *Fusarium* mycotoxins by gas chromatography-mass spectrometry. Food Add Contam 10:381, 1993.
128. DJ Hastings, LE Stenroos. Determination of deoxynivalenol in barley, malt, and beer by gas chromatography-mass spectrometry. J Am Soc Brewing Chem 53:78, 1993.
129. JC Kim, HJ Kang, DH Lee, YW Lee, T Yoshizawa. Natural occurrence of *Fusarium* mycotoxins (trichothecenes and zearalenone) in barley and corn in Korea. Appl Environ Microbiol 59:3798, 1993.
130. Health Canada, Health Protection Branch. Determination of nivalenol, deoxynivalenol (DON, vomitoxin), T-2 toxin, HT–2 toxin, diacetoxyscirpenol, and monoacetoxyscirpenol in cereals by gas chromatography with electron capture detection. Lab Bull WPG-LB 13:1–13, 1993.
131. R Kostiainen. Identification of trichothecenes by thermospray, plasmaspray, and dynamic fast-atom bombardment liquid chromatography-mass spectrometry. J Chromatog 562:555, 1991.
132. R Kostiainen, K Matsuura, K Nojima. Identification of trichothecenes by frit- and fast-atom bombardment liquid chromatography-high resolution mass spectrometry. J Chromatog 538:323, 1991.
133. DB Prelusky, HL Trenholm. Tissue distribution of deoxynivalenol in swine dosed intravenously. J Agric Food Chem 39:748, 1991.
134. D Abramson. Measurement of ochratoxin A in barley extracts by liquid chromatography-mass spectrometry. J Chromatog 391:315, 1987.
135. RR Marquardt, AA Frohlich, O Sreemannarayana, D Abramson, A Bernatsky. Ochratoxin A in blood from slaughter hogs in western Canada. Can J Vet Res 52:186, 1988.
136. E Rajakylä, K Laasasenaho, PJD Sakkers. Determination of mycotoxins in grain by high-performance liquid chromatography and thermospray liquid chromatography-mass spectrometry. J Chromatog 384:391, 1987.
137. Y Jiao, W Blaas, C Ruhl, R Weber. Identification of ochratoxin A in food samples by chemical derivatization and gas chromatography-mass spectrometry. J Chromatog 595:364, 1992.
138. J Plasencia, CJ Mirocha, RJ Pawlosky, JF Smith. Analysis of zearalenone and

α-zearalenol in urine of ruminants using gas chromatography-tandem mass spectrometry. J Assoc Off Anal Chem 73:973, 1990.
139. FS Chu. Development of antibodies against aflatoxins, In: DL Eaton, JD Groopman, eds. The Toxicology of Aflatoxins: Human Health, Veterinary and Agricultural Significance, San Diego: Academic Press, 1994, p 451.
140. FS Chu. Immunoassays for mycotoxins: current state of art, commercial and epidemiological applications. Vet Hum Toxicol (suppl) 32:42, 1990.
141. MW Trucksess, ME Stack, S Nesheim, et al. Immunoaffinity column coupled with solution fluorometry or liquid chromatography post-column derivatization for determination of aflatoxins in corn, peanuts, and peanut butter: collaborative study. J Assoc Off Anal Chem 74:81, 1991.
142. FS Chu. Recent progress on analytical techniques for mycotoxins in feedstuffs. J Anim Sci 70:3950, 1992.
143. FS Chu, and I Ueno, Production of antibody against aflatoxin B_1. Appl Environ Microbiol 33:1125, 1977.
144. JJ Pestka, PK Gaur, FS Chu. Quantitation of aflatoxin B_1 and aflatoxin B_1 antibody by an enzyme-linked immunosorbent microassay. Appl Environ Microbiol 40:1027, 1980.
145. JJ Pestka, Y Li, WO Harder, FS Chu. Comparison of radioimmunoassay and enzyme-linked immunosorbent assay for determining aflatoxin M_1 in milk. J Assoc Off Anal Chem 64:294, 1981.
146. T Sun, Y Wu, S Wu. Monoclonal antibody against aflatoxin B_1 and its potential applications. Chin J Oncol 5:401, 1983.
147. MRA Morgan, AS Kang, HWS Chan. Aflatoxin determination in peanut butter by enzyme-linked immunosorbent assay. J Sci Food Agric 37:908, 1986.
148. DN Mortimer, MJ Sheperd, J Gilbert. A survey of the occurrence of aflatoxin B_1 in peanut butters by enzyme-linked immunosorbent assay. Food Add Contam 5:127, 1988.
149. C De Boevere, C Van Peteghem. Development of an immunoaffinity column and an indirect immunoassay with a biotin-streptavidin detection system for aflatoxin M_1 in milk. Anal Chim Acta 275:341, 1993.
150. Council for Agricultural Science and Technology, Mycotoxins: economic and health risks. Task Force Rep 116:61, 1989.
151. AS Pately, M Sharman, R Wood, J Gilbert. Determination of aflatoxin concentrations in peanut butter by enzyme-linked immunosorbent assay (ELISA): study of three commercial ELISA kits. J Assoc Off Anal Chem 72:965, 1989.
152. MW Trucksess, K Young, KF Donahue, DK Morris, E Lewis. Comparison of two immunochemical methods with thin-layer chromatographic methods for determination of aflatoxins. J Assoc Off Anal Chem 73:425, 1990.
153. M Carvajal, F Mulholland, RC Garner. Comparison of the EASI-EXTRACT immunoaffinity concentration procedure with the AOAC CB method for the extraction and quantitation of aflatoxin B_1 in raw ground unskinned peanuts. J Chromatog 511:379, 1990.
154. RJ Cole, JW Dorner, FE Dowell. Aflatoxin immunoassays for peanut grading. In: M Vanderland, LH Stanker, BE Watkins, DW Roberts, eds. Immunoassay for Trace Chemical Analyses: Monitoring Toxic Chemicals in Humans, Foods and the Environment. Washington: American Chemical Society, 1991, p 158.

155. DL Park, H Njapau, SM Rua Jr, KV Jorgensen. Field evaluation of immunoassays for aflatoxin contamination in agricultural commodities, In: M Vanderland, LH Stanker, BE Watkins, DW Roberts, eds. Immunoassay for Trace Chemical Analyses: Monitoring Toxic Chemicals in Humans, Foods and the Environment, Washington: American Chemical Society, 1991, p 162.
156. M Azer, C Cooper. Determination of aflatoxins in foods using HPLC and a commercial ELISA system. J Food Prot 54:291, 1991.
157. KI Hongyo, Y Itoh, E Hifumi, A Takeyasu. Comparison of monoclonal antibody-based enzyme-linked immunosorbent assay with thin-layer chromatography and liquid chromatography for aflatoxin B_1 determination in naturally contaminated corn and mixed feed. J AOAC Int 75:307, 1992.
158. JW Dorner, PD Blankenship, RJ Cole. Performance of two immunochemical assays in the analysis of peanuts for aflatoxin at 37 field laboratories. J AOAC Int 76:637, 1993.
159. MW Trucksess, DE Koeltzow. Evaluation and application of immunochemical methods for mycotoxins in food, In: JO Nelson, AE Karu, RB Wong, eds. Immunoanalysis of Agrochemicals Emerging Technologies. Washington: American Chemical Society, 1995, p 326.
160. M Sharman, J Gilbert. Automated aflatoxin analysis of foods and animal feeds using immunoaffinity column clean-up and high-performance liquid chromatographic determination. J Chromatog 543:220, 1991.
161. G Niedwetzki, G Lach, K Geschwill. Determination of aflatoxins in food by use of an automatic work station. J Chromatog A 661:175, 1994.
162. T Ureno, MW Trucksess, SW Page. Automated affinity liquid chromatography system for on-line isolation, separation and quantitation of aflatoxins in methanol-water extracts of corn or peanuts. J Agric Food Chem 41:1982, 1993.
163. A Farjam, R De Vries, H Lingeman, UAT Brinkman. Immuno precolumns for selective on-line sample pretreatment of aflatoxins in milk prior to column liquid chromatograph. Int J Environ Anal Chem 44:175, 1992.
164. A Farjam, NC Van de Merbel, H Lingeman, RW Frei, UAT Brinkman. Non-selective desorption of immuno precolumns coupled on-line with liquid column chromatography: determination of aflatoxins. Int J Environ Anal Chem 45:73, 1991.
165. R Ross, JM Yuan, M Yu, et al. Urinary aflatoxin biomarkers and risk of hepatocellular carcinoma. Lancet 339:943, 1992.
166. TJ Hansen. Affinity column cleanup and direct fluorescence measurement of aflatoxin M_1 in raw milk. J Food Prot 53:75, 1990.
167. LGMT Tuinstra, AH Roos, JMP van Trijp. Liquid chromatographic determination of aflatoxin M_1 in milk powder using immunoaffinity columns for cleanup: interlaboratory study. J AOAC Int 76:1248, 1993.
168. E Ioannou-Kakouri, M Christodoulidou, E Christou, E Constantinidou. Immunoaffinity column/HPLC determination of aflatoxin M_1 in milk. Food Agric Immunol 7:131, 1995.
169. G Sabbioni, PL Skipper, G Buchi, SR Tannenbaum. Isolation and characterization of the major serum albumin adduct formed by aflatoxin B_1 in vivo in rats. Carcinogenesis 8:819, 1987.
170. CP Wild, R Montesano. Immunological quantitation of human exposure to afla-

toxins and N-nitrosamines. In: M Vanderlaan, LH Stanker, BE Watkins, DW Roberts, eds. Immunoassays for Trace Chemical Analysis: Monitoring Toxic Chemicals in Humans, Foods and the Environment. Washington: American Chemical Society, 1991, p 215.

171. FZ Sheabar, JD Groopman, GS Qian, GN Wogan. Quantitative analysis of aflatoxin-albumin adducts. Carcinogenesis 14:1203, 1993.

172. JD Groopman, KF Donahue. Aflatoxin, a human carcinogen: determination in foods and biological samples by monoclonal antibody affinity chromatography. J Assoc Off Anal Chem 71:861, 1988.

173. JD Groopman, A Zarba. Immunoaffinity-based monitoring of human exposure to aflatoxins in China and Gambia, In: M Vanderlaan, LH Stanker, BE Watkins, DW Roberts, eds. Immunoassays for Trace Chemical Analysis: Monitoring Toxic Chemicals in Humans, Foods and the Environment. Washington: American Chemical Society, 1991 p 207.

174. JC Harrison, RC Garner. Immunological and HPLC detection of aflatoxin adducts in human tissues after an acute poisoning incident in SE Asia. Carcinogenesis 12:741, 1991.

175. JI Azcona-Olivera, MM Abouzied, RD Plattner, WP Norred JJ Pestka. Generation of antibodies reactive with fumonisins B_1, B_2, and B_3 by using cholera toxin as the carrier-adjuvant. Appl Environ Microbiol 58:169, 1992.

176. JI Azcona-Olivera, MM Abouzied, RD Plattner, JJ Pestka. Production of monoclonal antibodies to the mycotoxins fumonisins B_1, B_2, and B_3. J Agric Food Chem 40:531, 1992.

177. JJ Pestka, JI Azcona-Olivera, RD Plattner, F Minervini, MB Doko, A Visconti. Comparative assessment of fumonisin in grain-based foods by ELISA, GC-MS and HPLC. J Food Prot 57:169, 1994.

178. EW Sydenham, GS Shephard, PG Thiel, C Bird, BM Miller. Determination of fumonisins in corn; evaluation of competitive immunoassay and HPLC techniques. J Agric Food Chem 44:159, 1996.

179. RA Shelby, GE Rottinghaus, HC Minor. Comparison of thin-layer chromatography and competitive immunoassay methods for detection fumonisin on maize. J Agric Food Chem 42:2064, 1994.

180. E Usleber, M Straka, G Terplan. Enzyme immunoassay for fumonisin B_1 applied to corn-based food. J Agric Food Chem 42:1392, 1994.

181. EW Sydenham, S Stockenström, PG Thiel, et al. Polycolonal antibody-based ELISA and HPLC methods for the determination of fumonisins in corn: comparative study. J Food Prot 59:893, 1996.

182. E Schneider, E Usleber, E Märtlbauer. Rapid detection of fumonisin B_1 in corn-based food by competitive direct dipstick enzyme immunoassay/enzyme-linked immunofiltration assay with integrated negative control reaction. J Agric Food Chem 43:2548, 1995.

183. GM Ware, PP Umirgar, AS Carmen Jr, SS Kuan. Evaluation of fumonitest immunoaffinity columns. Anal Lett 27:693, 1994.

184. MW Trucksess, ME Stack, S Allen, N Barrion. Immunoaffinity column coupled with liquid chromatography for determination of fumonisin B_1 in canned and frozen sweet corn. J AOAC Int 78:705, 1995.

185. YC Xu, GS Zhang, FS Chu. Enzyme-linked immunosorbent assay for deoxynivalenol in corn and wheat. J Assoc Off Anal Chem 71:945, 1988.
186. E Usleber, E Märtlbauer, R Dietrich, G Terplan. Direct enzyme-linked immunosorbent assays for the detection of the 8-ketotrichothecene mycotoxins deoxynivalenol, 3-acetyldeoxynivalenol, and 15-acetyldeoxynivalenol in buffer solutions. J Agric Food Chem 39:2091, 1991.
187. ML Putnam, KA Binkerd. Comparison of a commercial ELISA kit and TLC for detection of deoxynivalenol in wheat. Plant Dis 76:1078, 1992.
188. R Schmidt. EZ Quant, an easy quantitative test for deoxynivalenol, Presented at 109th AOAC Int. Annual Meeting, Nashville, TN, 1995.
189. MRA Morgan, R McNerny, HWS Chan. Enzyme-linked immunosorbent assay of ochratoxin A in barley. J Assoc Off Anal Chem 66:1481, 1993.
190. SC Lee, FS Chu. Enzyme-linked immunosorbent assay of ochratoxin A in wheat. J Assoc Off Anal Chem 67:45, 1984.
191. MRA Morgan, R McNerny, HWS Chan, PH Anderson. Ochratoxin A in pig kidney determined by enzyme-linked immunosorbent assay (ELISA). J Sci Food Agric 37:475, 1986.
192. E Märtlbauer, G Terplan. Ein enzymimmunologisches verfahren von ochratoxin A in schweineseren. Archiv Lebensmittelhyg 39:143, 1988.
193. DM Roussseau, AAG Candlish, GA Slegers, CH Van Peteghem, WH Stimson, JE Smith. Detection of ochratoxin A in porcine kidneys by a monoclonal antibody-based radioimmunoassay. Appl Environ Microbiol 53:514, 1987.
194. DM Rousseau, GA Slegers, CH Van Peteghem. Radioimmunoassay of ochratoxin A in barley. Appl Environ Microbiol 50:529, 1985.
195. DM Rousseau, GA Slegers, CH Van Peteghem. Solid-phase radioimmunoassay of ochratoxin A in serum. J Agric Food Chem 34:862, 1986.
196. JR Clarke, RR Marquardt, AA Frohlich. Comparative studies on the specificity and sensitivity of rabbit and laying-hen antisera to ochratoxin A. Food Agric Immunol 7:33, 1995.
197. ER Schneider, R Dietrich, E Märtlbauer, E Usleber, G Terplan. Detection of aflatoxins, trichothecenes, ochratoxin A and zearalenone by test strip immunoassay: a rapid method for screening cereals for mycotoxins. Food Agric Immunol 3:185–193, 1991.
198. AAG Candlish, WH Stimson, JE Smith. A monoclonal antibody to ochratoxin A. Lett Appl Microbiol 3:9, 1986.
199. DM Rousseau, AAG Candlish, GA Slegers, et al. Detection of ochratoxin A in porcine kidneys by a monoclonal antibody-based radio immunoassay. Appl Environ Microbiol 53:514, 1987.
200. AAG Candlish, WH Stimson, JE Smith. Determination of ochratoxin A by monoclonal antibody-based enzyme immunoassay. J Assoc Off Anal Chem Int 71:961, 1988.
201. O Kawamura, S Sato, H Kajii, et al. A sensitive enzyme-linked immunosorbent assay of ochratoxin A based on monoclonal antibodies. Toxicon 27:887, 1989.
202. M Sharman, S MacDonald, J Gilbert. Automated liquid chromatographic determination of ochratoxin A in cereals and animal products using immunoaffinity column clean-up. J Chromatog 603:285, 1992.
203. MM Abouzied, JJ Pestka. Simultaneous screening of fumonisin B_1, aflatoxin B_1 and

zearalenone by line immunoblot: a computer-assisted multianalyte assay system. J Assoc Off Anal Chem Int 77:495, 1994.

204. E Schneider, E Usleber, E Märtlbauer, R Dietrich, G Terplan. Multimycotoxin dipstick enzyme immunoassay applied to wheat. Food Addit Contam 12:387, 1995.
205. RD Plattner, GA Bennett. Rapid detection of *Fusarium* mycotoxins in grains by quadrapole mass spectrometry/mass spectrometry. J Assoc Off Anal Chem 66:1470, 1983.
206. R Teshima, M Kawase, T Tanaka, et al. Production and characterization of a specified monoclonal antibody against mycotoxin zearalenone. J Agric Food Chem 38:1618, 1990.
207. JI Azcona, MM Abouied, JJ Pestka. Detection of zearalenone by tandem immunoaffinity-enzyme-linked immunosorbent assay and its application to milk. J Food Prot 53:577, 1990.
208. I Barna-Vetró, A Gyöngyösi, L Solti. Monoclonal antibody-based enzyme-linked immunosorbent assay of *Fusarium* T-2 and zearalenone toxins in cereals. Appl Environ Microbiol 60:729, 1994.
209. GA Bennett, TC Nelsen, BM Miller. Enzyme-linked immunosorbent assay for detection of zearalenone in corn, wheat and pig feed: collaborative study. J Assoc Off Anal Chem Int 77:1500, 1994.
210. OA MacDougald, AJ Thulin, JJ Pestka. Determination of zearalenone and related metabolites in porcine urine by modified enzyme-linked immunosorbent assay. J Assoc Off Anal Chem 73:65, 1990.
211. D Abramson, E Usleber, E Märtlbauer. An indirect enzyme immunoassay for the mycotoxin citrinin. Appl Environ Microbiol, 61:2007, 1995.
212. D Abramson, E Usleber, E Märtlbauer. Determination of citrinin in barley by indirect and direct enzyme immunoassay. J AOAC Int 79:3125, 1996.
213. LJ McElroy, CM Weiss. The production of polyclonal antibodies against the mycotoxin derivative patulin hemiglutarate. Can J Microbiol 39:861, 1993.
214. SL Taylor, JW King, JL Richard, JI Greer. Analytical-scale supercritical fluid extraction of aflatoxin B_1 from field-inoculated corn. J Agric Food Chem 41:910, 1993.
215. H Engelhardt, P Haas. Possibilities and limitations of SFE in the extraction of aflatoxin B_1 from food matrices. J Chromatog Sci 31:13, 1993.
216. RD Smith, HR Udseth, BW Wright. Rapid and high resolution capillary fluid chromatography (SFC) and SFC/MS of trichothecene mycotoxins. Chromatog Sci 23:192, 1985.
217. JC Young, DE Games. Supercritical fluid chromatography of *Fusarium* mycotoxins. J Chrom 627:247, 1992.
218. AAM Stolker, MAS Marques, RW Stephany, LA van Ginkel, RW Maxwell. Extraction of zearalenone, zeranol, and their metabolites from fortified bovine using a coupled SPE/SFE technique, Proceedings of the 7th International Symposium on Supercritical Fluid Chromatography and Extraction, Indianapolis, IN, 1996. Abstract C-15.
219. RO Cole, RD Holland, MJ Sepaniak. Factors influencing performance in the rapid separation of aflatoxins by micellar electrokinetic capillary chromatography. Talanta 39:1139, 1992.

220. RD Holland, MJ Sepaniak. Qualitative analysis of mycotoxins using micellar electrokinetic capillary chromatography. Anal Chem Wash 65:1140, 1993.
221. TD Phillips, BA Clement, AB Sarr, ZG Huang, SM Williams. Chemiselective immobilization and detection of fumonisin B_1. Proceedings of the VIII International IUPAC Symposium on Mycotoxins and Phycotoxins, Mexico City, Mexico, 1992, p 42.
222. CM Maragos. Capillary zone electrophoresis and HPLC for the analysis of fluorescein isothiocyanate-labelled fumonisin B_1. J Agric Food Chem 43:390, 1995.
223. VS Thompson, CM Maragos. A fiber-optic immunosensor for the detection of fumonisin B_1. Presented at the 109th AOAC International Annual Meeting, Nashville, TN, 1995.
224. T Vo-Dinh, GD Griffin, JP Alarie, MJ Sepanjak, JR Bowyer. Development of fiber-optic immunosensors for environmental analysis. Pollu Pre Indust Processes, ACS Symp Ser 508:270, 1992.
225. JG Wilkes, JB Sutherland, MI Churchwell, AJ Williams. Determination of fumonisins B_1, B_2, B_3 and B_4 by high-performance liquid chromatography with evaporative light-scattering detection. J Chromatog A 695:319, 1995.
226. K Hult, E Hökby, S Gatenbeck. Analysis of ochratoxin B alone and in the presence of ochratoxin A using carboxypeptidase A. Appl Environ Microbiol 33:1275, 1977.
227. K Hult, E Hökby, S Gatenbeck, L Rutqvist. Ochratoxin A in blood from slaughter hogs in Sweden: use in evaluation of toxin content of consumed feed. Appl Environ Microbiol 39:828, 1977.
228. D Baker, B Radic. Fast determination of ochratoxin A in serum by liquid chromatography: comparison with enzymic spectrofluorometric method J Chromatog 570:441, 1991.
229. U Mayr, A Butsch, S Schneider. Validation of two in vitro test systems for estrogenic activities with zearalenone, phytoestrogens and cereal extracts. Toxicol 74:135, 1992.
230. MS Madhyastha, RR Marquart, A Masi, J Borsi, AA Frohlich. Comparison of toxicity of different mycotoxins to several species of bacteria and yeasts: use of *Bacillus brevis* in a disc diffusion assay. J Food Prot 57:48, 1994.
231. MM Abouzied, JI Azcona, WE Braselton, JJ Pestka. Immunochemical assessment of mycotoxins in 1989 grain foods: evidence for deoxynivalenol (vomitoxin) contamination. Appl Environ Microbiol 57:672, 1991.
232. B Hald, GM Wood, A Boenke, B Schurer, P Finglas. Ochratoxin A in wheat: an intercomparison of procedures, Food Addit Contam 10:185, 1993.
233. H Tsubouchi, K Yamamoto, K Hisada, Y Sakabe, S Udagawa. Effect of roasting on ochratoxin A level in green coffee beans inoculated with *Aspergillus ochraceus*. Mycopathologia 97:111, 1987.
234. J Bauer, M Gareis, B Gedek. Zum nachweis und verkommen von ochratoxin A in bei schlachtschweinen. Berliner Munchener Tierarzt Wochenschr 97:279, 1984.
235. A Breitholtz-Emmanuelsson, G Dalhammar, K Hult. Immunoassay of ochratoxin A, using antibodies developed against a new ochratoxin-albumin conjugate. J AOAC Int 75:824, 1992.
236. A Visconti, A Bottalico. High levels of ochratoxin A and B in moldy bread responsible for mycotoxicosis in farm animals. J Agric Food Chem 31:1122, 1983.

237. W Langseth, Y Elingsen, U Nymoen, EM Ökland. High-performance liquid chromatographic determination of zearalenone and ochratoxin A in cereals and feed. J Chromatog 478:269, 1989.
238. BA Roberts, EM Glangy, DSP Patterson. Rapid economical method for determination of aflatoxins and ochratoxin in animal feedstuffs. J Assoc Off Anal Chem 64: 961, 1981.
239. H Cohen, M Lapointe. Determination of ochratoxin A in animal feed and cereal grains by liquid chromatography with fluorescence detection. J Assoc Off Anal Chem 69:957, 1986.

6

Determination of Human Exposure to Aflatoxins

Gabriele Sabbioni
Ludwig-Maximilians-Universität München, München, Germany

Ovnair Sepai
University of Newcastle, Newcastle upon Tyne, England

I. INTRODUCTION

The carcinogenic properties of the aflatoxins have been studied extensively in several animal species [1–3]. Primary liver tumors have been induced by administration of AFB_1 to fish, birds, rodents (five strains of rats, B6C3F1 mouse, hamster), ferrets, and subhuman primates. These experiments showed that rats are highly susceptible, mice highly resistant, and primates of intermediate susceptibility. Human exposure to high levels of aflatoxin B_1 (AFB_1) from the diet is an important risk factor for the development of liver cancer [4–6]. In several reports the association between dietary exposure and human liver cancer has been strong [3,7]. However, epidemiological studies cannot reveal information about an individual's risk to develop disease. Molecular biomarkers of aflatoxin exposures in individuals may provide tools for prevention strategies and indicate the risk for liver cancer.

The rationale for the use of biomarkers [8,9] is represented in Figure 1. This model classifies the molecular biomarkers according to the steps in the progressive nature of disease. The scientific community distinguishes between markers of (1) susceptibility, (2) markers of the internal dose, (3) markers of the biologically effective dose, (4) markers of early biological effects, (5) markers of altered function, and (6) markers of clinical disease. This chapter of the present book will

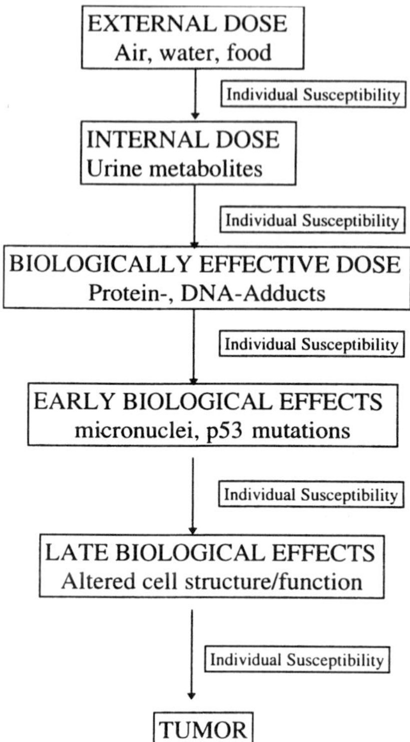

FIGURE 1 Principles of biomonitoring.

focus on the progress made on the development of markers of internal dose and biologically effective dose.

II. METABOLISM OF AFLATOXINS

The crucial step in the process of carcinogenesis is the formation of the 8,9-dihydro-8,9-epoxy-AFB_1 (AFB_1-epoxide) (Fig. 2). This intermediate could not be isolated in the past. AFB_1-epoxide was synthesized in situ with 4-chloroperbenzoic acid or with microsomes [10–14]. A few years ago, AFB_1-epoxide was isolated from the reaction of AFB_1 with dimethyldioxirane and fully characterized [15]. The reaction of dimethyldioxirane with AFB_1 gives an approximately 10:1 mixture of the *exo*-AFB_1-epoxide and the *endo*-stereoisomer. As an alternative, 8,9-dihydro-8,9-dichloro-AFB_1 or 8,9-dihydro-8,9-dibromo-AFB_1 (AFB_1-dibromo),

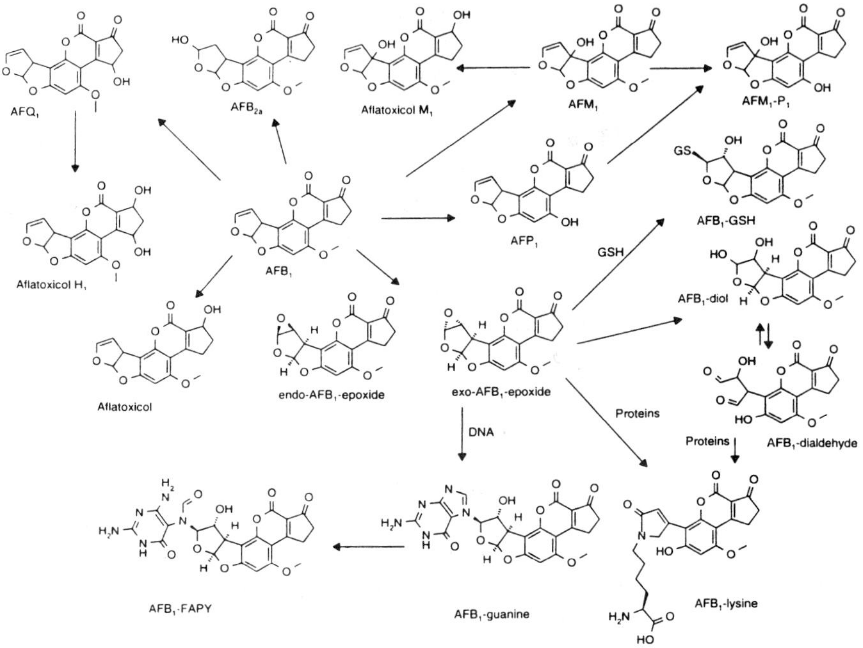

FIGURE 2 Metabolism of aflatoxin and DNA-, protein-, and GSH-adduct formation.

which can be isolated, has been used instead of the AFB_1-epoxide for in vitro reactions with macromolecules [16–18]. The primary AFB_1-DNA adduct was identified by Essigmann et al. [10,19] as 8,9-dihydro-8-(N^7-guanyl)-9-hydroxy-AFB_1 (AFB_1-guanine). The presence of AFB_1-guanine in vivo was confirmed by Croy et al. [20]. The major DNA adduct in rat liver after an IP injection of 1.28 μmol AFB_1/kg was AFB_1-guanine. At 48 h 30% to 40% of the AFB_1-guanine adducts initially present in liver DNA was excreted in the urine [21–23].

Essentially only trans-8,9-dihydro-8-(N^5-formyl-2,5,6-triamino-4-oxopyrimidin-N^5-yl)-9-hydroxy-AFB_1 (AFB_1-FAPY) was detectable after 48 h. This adduct is persistent. The formation of the AFB_1-glutathione adduct (AFB_1-GSH) is an important detoxification pathway. This process is catalyzed by species-specific glutathione S-transferases (GST). It appears that animals that are very susceptible to AFB_1 lack an efficient GST. One percent to 2% of the administered dose is excreted into urine as the glutathione (GSH) adduct [24]. In vitro experiments with rat liver microsomes, AFB_1, and reduced GSH indicated the formation of AFB_1-GSH [25]. In rats given radiolabeled AFB_1 a compound excreted in the bile (10% to 30% of the dose) was isolated and provisionally identified as 8,9-

dihydro-8-(S-glutathionyl)-9-hydroxy-AFB_1 (AFB_1-GSH) [26]. This adduct was chemically characterized by Moss et al. [27].

The metabolism of AFB_1 in animals and humans has been summarized by Busby and Wogan [2] and Eaton et al. [28] (Fig. 2). In the following paragraph only the studies relevant to human metabolism will be mentioned. Moss and Neal [29] showed a rapid NADPH-dependent metabolism of AFB_1 with human liver microsomes, especially from males. The major metabolite was AFQ_1, with some 8,9-dihydro-8,9-dihydroxy-AFB_1 (AFB_1-diol) being found and AFB_1-GSH formed by postmitochondrial supernatant from human liver.

Epoxidation of AFB_1 with human liver microsomes yields mixtures of *exo*-AFB_1-epoxide and *endo*-AFB_1-epoxide [30]. The formation of mixtures of *exo*- and *endo*-8,9-epoxides by human cytochrome P450 (CYP) mixed-function oxidases was confirmed in studies with human CYP1A2 and CYP3A4 expressed in *Escherichia coli*. CYP3A4 is more active than CYP1A2 in forming genotoxic AFB_1 oxidation products. Analysis of the AFB_1 products indicated that CYP3A4 formed AFQ_1 and the *exo*-8,9-epoxide. CYP1A2 formed AFM_1, a small amount of AFQ_1, and both the *exo*- and *endo*-8,9 epoxides. Raney et al. [31] found that in human liver microsomes AFQ_1 and AFB_1 epoxide (measured as the GSH adduct) are the major oxidative products formed from AFB_1 at all substrate concentrations. AFQ_1 is not appreciably oxidized in human liver microsomes and is not very genotoxic. The 3a-hydroxylation of AFB_1 to AFQ_1 is viewed as a potentially significant detoxification pathway.

The C8 of the *exo*-AFB_1-epoxide reacts with the N7 position of guanine to give the trans adduct described above, AFB_1-guanine (Fig. 2). Only the *exo*-AFB_1 is genotoxic [32]. The *exo* isomer is strongly mutagenic in a base-pair reversion assay using *Salmonella typhimurium*; the *endo* isomer is essentially nonmutagenic. Harris et al. [33] explained this with NMR, by studying the intercalation of AFB_1 and its derivatives in DNA. The *exo*-epoxide AFB_1 intercalates with the epoxide optimally oriented for an SN2 reaction with guanine. Intercalation of the *endo* isomer places the epoxide in an orientation that precludes reaction. Thus, while the *exo* epoxide is a potent mutagen, the *endo* epoxide fails to react with DNA.

The discovery of the *endo* and *exo* AFB_1 epoxides raised questions about the reaction of the *endo*-AFB_1 epoxide with GSH. As shown earlier the rate of conjugation of GSH with AFB_1, *exo*-epoxide is an important factor in determining species variation in risk to aflatoxins. Induction of GSTs can yield a significant protective effect. Raney et al. [34] synthesized and fully characterized the *exo*- and *endo*-epoxide-AFB_1 conjugates with GSH. The two epoxides were incubated with liver cytosol from different species. The *endo*-epoxide was a good substrate for GSH conjugate formation in rat liver cytosol, whereas mouse liver cytosol conjugated the *exo*-epoxide almost exclusively. Human liver cytosol conjugated both epoxide isomers to a much lower extent than rat or mouse liver cytosol. Purified human GST catalyzed the conjugation of the *exo*-epoxide in the order GST μ-class (M1a-1a) > GST α-class A1-1 > A2-2, while only GST M1a-1a conjugated the *endo*-epoxide. These results are important in the extrapolation from animal models to human risk assessment.

The sensitivity of humans relative to animals remains unclear. Liu et al. [35] discovered an interesting polymorphism of a GST, which plays a role in the detoxication of AFB_1-epoxide. GST μ (according to the new nomenclature, reviewed in [36]) activity was tested with trans-stilbene oxide (tSBO) in human liver and lymphocytes. GST μ has polymorphic expression in human lymphocytes. These results showed that activity of GST-tSBO in lymphocytes correlates with its activity in liver ($r = .7$). The functional significance of the GST μ polymorphism was evaluated by measuring its effect on AFB_1-DNA adduct formation in vitro. Human liver cytosols prepared from persons having low or high GST-tSBO activity were incubated with human liver microsomes, calf thymus DNA, and AFB_1. HPLC analysis of the DNA adducts, AFB_1-FAPY, and AFB_1-guanine revealed that AFB_1-DNA binding was inhibited to a greater extent in high conjugators than low conjugators. The correlation between AFB_1-DNA adduct concentrations and GST μ activity was highly significant, with a correlation coefficient of $r = .88$. This suggests that GST μ plays an important role in detoxifying DNA-reactive metabolites of AFB_1, and thus this enzyme may be a susceptibility marker for AFB_1-related liver cancer.

The GST μ catalyzed formation of AFB_1-GSH in human samples was not confirmed by two studies [37,38]. Kirby et al. compared the expression of hepatic CYP and GST [37] in liver tissues from 20 liver cancer patients from Thailand, where the incidence of liver cancer is high and where exposure to aflatoxin occurs. The expression of hepatic enzymes was compared to the in vitro metabolism of AFB_1. There was a > 10-fold interindividual variation in expression of the various CYPs: CYP3A4 (57-fold), CYP2B6 (56-fold), and CYP2A6 (120-fold). Microsomal metabolism of AFB_1 to AFB_1-epoxide and AFQ_1, which is the major metabolite, increases with CYP3A3/4 expression and, to a lesser extent, with CYP2B6 expression. Ten of 14 individuals (71%) were homozygous null when genotyped for GST M1. AFB_1-epoxide was measured as the tris-(hydroxymethyl)-methylamine (Tris) complex with the AFB_1-diol. There was no detectable conjugation of AFB_1-epoxide to GSH by cytosol from tumorous or normal liver. Therefore, ability of human cytosols to bind reactive AFB_1 metabolites to GSH resembled AFB_1-sensitive species such as rat, duck, and trout rather than resistant species such as mouse and hamster. These results show that multiple forms of human hepatic CYPs metabolize AFB_1 to both the reactive intermediate AFB_1-epoxide and the detoxification product AFQ_1. The authors concluded that variations in expression of hepatic CYP, due either to genetic polymorphisms or to modulation by environmental factors, may be more important determinants in the risk of liver cancer development in AFB_1-exposed populations, since human livers lack significant GST-mediated protection against AFB_1.

Heinonen et al. [38] confirmed the GSTM1 studies by Kirby et al. [37]. No GSH adducts of aflatoxin were found in experiments with human liver. AFB_1 metabolism and the extent of AFB_1 binding to cellular macromolecules in human liver slices were studied under experimental conditions that allow direct comparison to similar endpoints in the rat. In human liver slices, significant interindividual variations were observed in the rates of oxidative metabolite formation and in specific binding to cellular macromolecules. The rates of oxidative metabolism of AFB_1 to AFQ_1, AFP_1, and AFM_1 in the three human liver samples were similar to those previously observed in rat liver slices. AFB_1-GSH conjugate formation was not detected in any of the human liver samples. However, specific binding of AFB_1 to cellular macromolecules was considerably lower in human liver slices than in rat liver slices. AFB_1-DNA binding levels in human samples ranged from 3% to 26% of the rat samples. The AFB_1-protein binding level in the one human sample measured was 20% of the rat samples. Heinonen et al. [38] summarized that humans do not possess GST isozyme(s) with high specific activity toward AFB_1-epoxide, and that humans do not form as much AFB_1-epoxide as the rat. The former two studies by Kirby et al. [37] and Heinonen et al. [38] contrast with another study [39]. Langouët et al. [39] determined AFM_1, AFB_1-GSH, and unchanged AFB_1 in cultures derived from eight human liver donors. The hepatocytes were pretreated with or without oltipraz (OPZ), a synthetic derivative of the natural 1,2-dithiole-dithione, which is known to drastically reduce tumor incidences in rats. Parenchymal cells obtained from the three GSTM1-positive livers metabolized AFB_1 to AFM_1 and to AFB_1-GSHs from *exo*- and *endo*-AFB_1-epoxides. However, no AFB_1-GSHs were formed in the GSTM1-null cells. Pretreatment of the cells with inducers of CYP1A2 and CYP3A4 (3-methylcholanthren or rifampicin, respectively) produced significant increases in AFB_1 metabolism. Although OPZ induced GSTA2, and GSTM1, and GSTA1 to a lesser extent, OPZ decreased formation of AFM_1 and AFB_1-GSH, which involves oxidation by CYP1A2 and CYP3A4. The inhibition of CYP1A2 and CYP3A4 by OPZ was confirmed by decreased mono-oxygenase activities toward ethoxyresorufin and nifedipine, respectively. The authors concluded that only GSTM1-positive human hepatocytes formed AFB_1-GSH, and that chemoprotection with OPZ is due to an inhibition of activation of AFB_1 and to a GST-dependent inactivation of the carcinogenic *exo*-AFB_1-epoxide.

The toxicity of other aflatoxins that are present in the food has been reviewed by several authors [1]. DNA adducts of AFP_1 and AFM_1 epoxides were identified in liver extracts following treatment of rats with AFB_1 [19]. Recently, epoxides of AFG_1 and AFQ_1 [31]—the putative ultimate carcinogen of these compounds—have been synthesized and isolated. With these intermediates, in vitro experiments with calf thymus DNA and *Salmonella typhimurium* were performed [31]. 8,9-Epoxy-8,9-dihydro-AFG_1 was synthesized and found to yield lower levels of AFB_1-guanine than obtained from AFB_1-epoxide when mixed with calf thymus DNA or *Salmonella typhimurium* TA 98. However, the ratios of revertants to the corresponding guanine adducts were similar when *Salmonella typhimurium* TA 98 was treated with the (analogous) epoxides of AFB_1, AFG_1, or AFQ_1. Recently, Bujons et al. [40] performed similar experiments with AFM_1. They synthesized the 8,9-dihydro-8,9-epoxide-AFM_1 (AFM_1-epoxide) from AFM_1 and dimethyldioxirane. Interpretation of ^{1}H-NMR spectra of the above epoxide revealed that one stereoisomer was present as a major component, most likely that with the *exo* configuration. The mutagenicities in *Salmonella typhimurium* strain TA-100 of AFM_1-epoxide and AFM_1 were ca. 3 times lower than the AFB_1-epoxide.

III. BIOMONITORING OF AFLATOXINS

A. Internal Dose: AFB_1 Metabolites in Biological Fluids

Several studies have been published investigating the presence of AFB_1 and AFM_1 in urine, blood, and milk. In this section we present studies that did not include any measurements of DNA and protein adducts.

1. Urine

Campbell et al. [41] analyzed the AFB_1 content of food and urine in a group of Filipinos. Analysis of 500 food products in the Philippines showed that peanut butter was highly contaminated with a mean concentration of 1.60 nmol/g of AFB_1. In urine specimens from peanut butter consumers (0 to 200 g/24 h), AFM_1 was found by thin-layer chromatography (TLC). Approximately 1% to 4% of the ingested AFB_1 was excreted as AFM_1. A further method for the separation of aflatoxins from human urine using TLC was published by Lovelace et al. [42]. The detection limit for AFB_1, AFG_1, AFM_1, aflatoxicol, and aflatoxicol M_1 were 2.34, 1.80, 1.34, 2.19, and 1.88 μmol/100 ml, respectively. The sensitivity of these TLC approaches are not high enough to measure individual exposure for long-term studies. The detection limit of HPLC with UV or fluorescence detection is lower, but limited by the interference of contaminants. Similar analytical problems have been reported for the analysis of foods and feeds [43]. Therefore, several researchers produced antibodies against AFB_1 metabolites. The antibodies were used for the production of immunoaffinity columns or for immunoassays, such as enzyme-linked immuno assay (ELISA) and radio immunoassay (RIA) [44]. Concentration by immunoaffinity chromatography (IAC) followed by HPLC or immunoassay can regularly detect aflatoxins down to fmol/ml levels in fluids, including urine. Haugen et al. [45], Sizaret et al. [46], Garner et al. [47,48], and Groopman et al. [49,50] produced antibodies that recognize most metabolites. These studies are discussed later.

The detection limits of urine AFB_1 metabolites have been improved by several authors. The detection of AFB_1, AFB_2, AFG_1, AFG_2 [58–60], and AFQ_1 was improved by postcolumn derivatization with bromine and subsequent analysis by HPLC with fluorescence detection [61]. Kussak et al. [62] presented a method using HPLC coupled to an electrospray-ionization tandem mass spectrometer. Samples of naturally contaminated airborne dust and spiked urine were cleaned up on immunoaffinity columns and analyzed by HPLC using either mass spectrometry detection or postcolumn derivatization with bromine and fluorescence detection. With tandem mass spectrometry, detection limits calculated as the amount ejected on column were: AFB_1 12.8 fmol, AFB_2 12.7 fmol, AFG_1 15.2 fmol, and AFG_2 30.3 fmol.

Degan et al. [63] prepared Eu-labeled antibodies specific for aflatoxins. These antibodies were used for a solid-phase competitive time-resolved fluoroimmunoassay. The results correspond ($r = .97$) well with the ELISA test based on use of the unmodified antibody to aflatoxin. This procedure has been proposed as a quick, sensitive, and reliable immunoassay for use in mycotoxin screening in foods and body fluids.

The following paragraphs summarize studies performed in regions of China and Africa, where a high incidence of liver cancer has been reported. Data from a group of Danish workers exposed to AFB_1 will also be discussed.

China. Wu et al. [51] analyzed 24-h urine samples from residents of Beijing and Qidong, which are areas with a high incidence of liver cancer. Urine was concentrated by IAC and analyzed for AFM_1 by HPLC. Levels as low as 1.52 pmol AFM_1 could be detected. AFM_1 in free form was identified as the major metabolite in urine. The urinary excretion of AFM_1 was < 6.40 pmol/day in Beijing, whereas the rate was higher in Qidong. AFM_1 excretion correlated with the consumption of contaminated food. Subjects with a fluctuating pattern of serum α-fetoprotein (an indicator for hepatic hyperplasia) also showed a higher level of urinary AFM_1 excretion compared to healthy subjects living in the same area.

Sun et al. [52] analyzed samples from the same areas (Qidong and Beijing, China). For these analyses, they developed a panel of monoclonal IgG antibodies in the murine system against AFB_1 and/or AFM_1. The antibodies obtained were used for IAC and for immunoassays. In an area with high risk of liver cancer, especially during wet seasons, 10% of the local inhabitants had a urinary output of AFM_1 1 or 2 orders of magnitude higher than that of Beijing people. The ingestion of AFB_1 among these local people was estimated to exceed 3.20 μmol/yr. The major source was from contaminated corn and rice, but local alcoholic beverages also contributed to the AFB_1 intake. The increased excretion of AFM_1 was more pronounced in patients with chronic active hepatitis.

Another study in a region with high liver cancer incidence was performed by

Zhu et al. [53] in Fushui county of the Guangxi province. AFB_1 corn and peanut oil (total, 253 samples) were analyzed daily over a period of 1 week from 32 households. A total of 252 urine samples were collected simultaneously from the residents in the households that were shown to have consumed AFB_1. A good correlation was observed between dietary AFB_1 intake and total AFM_1 excretion in human urine; AFM_1 (nmol) = 0.487 + 0.0123 × AFB_1 (nmol). Between 1.23% and 2.18% of dietary AFB_1 was found to be present as AFM_1 in human urine. A good correlation was also observed between the AFB_1 concentration in corn and the AFM_1 concentration in human urine.

A further study of urine samples from the Fushui county was published by Liu et al. [54]. The method for the analysis of AFM_1 was modified. Urine samples were treated with saturated lead acetate, and AFM_1 was extracted with chloroform. After washing with water to remove impurities, the compound was derivatized with trifluoroacetic acid. The AFM_1 derivative was quantified by HPLC with fluorescence detection. Over a range of 0.15 to 1.22 pmol AFM_1, a good correlation between AFM_1 peak height and its concentration was obtained. Recovery was 87% and sensitivity 0.01 ppb. AFM_1 in urine of residents who live in the high liver cancer incidence area were slightly higher than those of residents of low liver cancer incidence area in Fushui county. AFB_1 concentrations in food (maize, rice, peanuts, and peanut oil) were also measured. However, the differences in the level of contamination were not significant in these two regions.

Africa. The largest survey of AFM_1 excretion into urine was reported by Nyathi et al. [55]. Over 1200 urine samples were collected from different areas of Zimbabwe. The urine samples were extracted with chloroform. The resultant aflatoxins were quantified by TLC and HPLC. The most commonly observed metabolite was AFM_1 at an average concentration of 12.8 pmol/ml urine. Nationally, 4.3% of all urine samples were positive for AFM_1. In some areas up to 10% of the urine samples were positive for AFM_1. These data indicate that AFM_1 is one of the most predominant aflatoxin metabolites in human urine.

Bean et al. [56] collected urine samples from 161 Nigerians residing in Lagos. The samples were screened for AFB_1 and AFG_1 and the chloroform-soluble metabolites of AFB_1, AFQ_1, AFB_{2a}, AFM_1, and aflatoxicol (Fig. 2). AFB_1 and AFG_1 and three of the metabolites—aflatoxicol, AFB_{2a} and AFM_1—were found in the urine samples. All of the aflatoxins observed were detected by HPLC. AFB_{2a} was found in 32.7%, AFG_1 in 9.9%, aflatoxicol in 9.3%, AFM_1 in 8.7%, and AFB_1 in 3.1% of the urine samples analyzed. The highest average concentration among aflatoxins was AFG_1 at approximately 0.365 pmol/ml urine. The following order of concentration was detected: $AFG_1 > AFB_1 \gg AFM_1 > AFB_{2a} >$ aflatoxicol.

Europe. Few studies have been performed in Europe. Urine samples were collected in Denmark by Dragsted et al. [57]. They analyzed the samples using a competitive enzyme immunoassay. One or more compounds were recognized by a monoclonal antibody against AFB_1. The concentration of urinary aflatoxin-like substances was equivalent to 0.0 to 20.8 pmol AFB_1/mg creatinine (80 individuals). The source of aflatoxin-like compounds was investigated to explain the seemingly high level of aflatoxin-like material in urine samples from people living in a temperate climate. The excretion of these compounds depended mainly on the food ingested 24 to 48 h before urine samples were collected. The excretion of aflatoxin-like substances was increased when diets included beer, dairy products, or meat. The structure of these aflatoxin-like compounds has not been elucidated. These data may result from food components with aflatoxin adducts, for example AFB_1-lysine adducts in proteins, or from crossreactivity of the antibody with compounds not derived from aflatoxin.

2. Milk

Aflatoxin metabolites are also found in breast milk. Several reports have described the presence of aflatoxins in human milk [64–66]. Wild et al. developed an ELISA for the detection of aflatoxin in human breast milk [67]. The assay allows the quantitation of 6.1 fmol AFM_1/ml of milk using < 10 ml of sample. A good correlation was observed between ELISA and an HPLC fluorescence technique using naturally contaminated milk at levels up to 121.8 fmol AFM_1/ml. In women from rural villages in Zimbabwe, six of 54 samples were positive. No positive samples were detected in 42 milk samples obtained from women in France.

Zarba et al. [68] studied the maternal-to-child exposure of AFM_1 in breast milk. AFM_1 in breast milk was quantified by HPLC and fluorescence after preconcentration of the samples with an aflatoxin immunoaffinity column. Of the total intake, 0.09% to 0.43% of AFB_1 was excreted in milk. In addition, in three out of five samples AFG_1 was found.

El-Nezami et al. [69] examined samples from lactating women in Australia and Thailand. AFM_1 was determined by HPLC with fluorescence detection and by ELISA. Eleven out of 73 samples from Australia (0.085 to 3.14 pmol AFM_1/ml) and five out of 11 from Thailand (0.12 to 5.29 pmol AFM_1/ml) were positive for AFM_1.

B. Biologically Effective Dose: DNA Adducts

Further determinations of AFB_1, AFP_1, and AFM_1 in urine included the determination of the excreted DNA adducts or of blood protein adducts. Therefore, these studies are presented in Sections B and C.

1. Urine

Several very good reviews have been published on biomonitoring of AFB_1 by measuring urine metabolites [2,47,70,71]. AFB_1-guanine is the major product formed by the interaction in vivo of AFB_1 with rat liver nucleic acids. After oral administration of AFB_1 to rats, AFB_1-guanine was detected in the urine. The presence of AFB_1-guanine and the relation to the AFB_1-guanine adducts in the liver were first published by Bennett et al. [22]. AFB_1-guanine was isolated from urine by preparative HPLC. Spectral and chemical analysis of mg quantities of this compound provided strong evidence that this compound was identical to authentic AFB_1-guanine. The dose-response curves for AFB_1-guanine excretion and AFB_1-guanine in rat liver DNA were similar. The levels of AFB_1-guanine excreted in urine corresponded to 30% to 40% of the levels seen initially in liver DNA.

In the following studies, researchers produced antibodies specific to AFB_1-guanine. For the analysis of biological samples by ELISA, Martin et al. [72] prepared an antibody against AFB_1 by immunization of rabbits with a bovine serum albumin-AFB_1 conjugate. The minimum detectable concentration of AFB_1 was approximately 16 to 32 fmol/ml. Anti-AFB_1 antibody was also inhibited by AFB_1-FAPY and AFB_1-guanine. Groopman et al. [49] obtained monoclonal antibodies following fusion of mouse SP-2 myeloma cells with spleen cells of mice immunized with AFB_1 covalently bound to bovine γ-globulin. The aflatoxin-modified protein used to immunize mice was produced chemically by activating AFB_1 to an AFB_1-epoxide, which then covalently bound to the protein. These antibodies are specific for AFB_1, AFB_2, AFM_1, and the major aflatoxin-DNA adducts. One of the monoclonal antibodies isolated (2B11) was a high-affinity IgM antibody for AFB_1, AFB_2, and AFM_1. In a competitive RIA using $[^3H]AFB_1$, 3.20 pmol of AFB_1, 3.18 pmol of AFB_2, or 3.05 pmol of AFM_1 caused 50% inhibition with this antibody. The antibody also had significant crossreactivity for the major aflatoxin-DNA adducts: AFB_1-guanine and AFB_1-FAPY. The antibody was covalently bound to Sepharose-4B and used in a column-based solid-phase immunosorbent assay system. Aflatoxins added in vitro to phosphate buffer, human urine, human serum, or human milk at levels expected in human samples were quantitatively recovered by applying the mixture to this immunoaffinity column purification system. In rats injected with ^{14}C-labeled AFB_1, AFB_1-guanine and the oxidative metabolites AFM_1 and AFP_1 as the major aflatoxin species were present in the urine [73]. This methodology was applied to human urine samples obtained from people from the Guangxi Province of China exposed to AFB_1 through dietary contamination. The major aflatoxin metabolites detected were AFB_1-guanine and aflatoxins AFM_1 and AFP_1.

AFB_1-DNA adducts in urine and AFB_1 in food were detected in Kenyan samples by Autrup et al. [74]. They found a putative adduct with fluorescence characteristics identical to AFB_1-guanine in human urine collected in Murang'a district, Kenya. Six out of 81 urine samples were positive. In a further study, Autrup et al. [75] analyzed samples from three different districts of Kenya. The highest number of individuals with detectable urinary levels of AFB_1-guanine lived in Murang's a district or neighboring Meru and Mebu districts. A seasonal variation in the number of positive cases was observed in Murang's district. Positive case rates of 11.5% from January to March and only 3% from July to September were observed. A relatively high number of positive cases were found among older men, and the rate of positive cases was higher in men than in women, although the difference was not significant. Autrup et al. [76] also studied possible synergistic effects of the hepatitis B virus (HBV) and AFB_1 in the induction of liver cancer in Kenya. Of all individuals tested, 12.6% were positive for aflatoxin exposure as indicated by the urinary excretion of AFB_1-guanine. The authors observed regional differences. However, they assumed no annual and seasonal variation of exposure. The incidence of HBV infection nationwide, as measured by the presence of the surface antigens, was 10.6%, but a wide regional variation was observed. The authors did not find any synergistic effect between HBV infection and AFB_1 exposure in the induction of liver cancer. This is in contrast to later studies [5,6,53]. However, a moderate degree of correlation between the exposure to aflatoxin and liver cancer was observed when the study was limited to certain ethnic groups.

Groopman et al. [77] improved their analytical methods [49]. They devised a combined monoclonal antibody immunoaffinity chromatography/HPLC method to isolate and quantify aflatoxin-DNA adducts and other metabolites from urine samples. The authors produced 11 different monoclonal antibodies recognizing AFB_1, AFQ_1, AFG_1, aflatoxicol, and AFM_1 and applied these antibodies to a multiple monoclonal antibody affinity chromatography technique. Over 90% of total aflatoxin in urine from dosed rat bound to the column using the multiple monoclonal antibody affinity column. Subsequent analyses by HPLC showed that > 55% of the aflatoxins in rat urine are AFP_1, AFB_1-guanine, AFM_1, AFQ_1, AFB_1-diol, and AFB_1, accounting for 34.5%, 9.6%, 8.0%, 1.8%, 1.5%, and 1.0% of the total aflatoxins, respectively. AFB_1-guanine excretion in rat urine correlated with the adduct formation in the liver (r = .99). In addition to the nucleic acid adduct excretion data, Groopman et al. evaluated AFM_1 and AFP_1 as molecular dosimeters in the urine. AFM_1 was an excellent marker (r = .93), whereas no linear relationship between dose and AFP_1 excretion in urine was found. This is quite remarkable, when according to the epidemiological studies by Ross et al. [5] and Qian et al. [6] the best molecular dosimetry marker for liver cancer incidence was the amount of AFP_1 present in the urine. One reason could be that AFP_1 is produced mainly by tumor microsomes rather than by normal liver microsomes [37]; i.e., the cases were metabolizing AFB_1 more efficiently to AFP_1 due the presence of a tumor [144]. This finding needs more attention in future studies.

Groopman et al. [77,78] applied the techniques developed in rat experiments to explore the relationship between dietary exposure to aflatoxins and the excretion of the major aflatoxin-DNA adduct and metabolites into the urine of chronically exposed people from Gambia, who were either HBV surface antigen-positive or negative. The diets of 20 individuals, 10 males and 10 females, from

15 to 56 years old, were monitored for 1 week. The average AFB_1 intake of total aflatoxins was 38.4 nmol for the entire study group during the 1-week collection period. However, there was considerable day-to-day variation in exposures: 0 to 94.8 nmol total aflatoxins/day. Starting on the fourth day, consecutive 24-h urines were obtained for 4 days. The subjects were generally paired for HBV status. The urine samples were purified by preparative monoclonal antibody IAC followed by HPLC and quantified by UV and fluorescence detection. ELISAs were carried out on each of the samples purified by IAC. The aflatoxin intake values were compared to the excretion (1) of total aflatoxin metabolites, and (2) of AFB_1-guanine. Analysis of total aflatoxin metabolites in the urine samples were made by competitive ELISA. The individual aflatoxin components in urine were determined by HPLC. These analyses revealed the preponderance of AFG_1 in many of the urine samples in addition to AFP_1, AFQ_1, and AFB_1-guanine. The total AFB_1 exposure in the diet correlates well with total AFB_1-guanine/mg creatinine excretion in the urine over the complete collection period ($r = .82$; $P < .0001$). The aflatoxin-DNA adduct levels in urine for a given dietary exposure were not different in HBV carriers and noncarriers. Albumin adducts were determined in the same group of people (discussed in Sect. C).

Weaver et al. [79] published an interesting new method for the determination of AFB_1-guanine in urine. They developed the procedure with urine from rats dosed with radiolabeled AFB_1. The AFB_1-guanine adduct was hydrolyzed in concentrated hydrochloric acid. The resulting product was quantified by HPLC-fluorescence or, after reaction with Tris, as the Tris-diol-AFB_1 complex by synchronous fluorescence spectrometry (SFS). The SFS method was developed several years ago by Harris et al. [80]. Conventional excitation and emission spectra of aflatoxins contain many peaks. By scanning excitation and emission synchronously, with the proper fixed wavelength difference, only one peak emerges. The sensitivity of the method is similar to the detection of AFB_1 by HPLC coupled to a fluorescence detector. However, its ability to distinguish between closely related compounds is limited. So far, these methods have not been used in field studies.

2. DNA Adducts in Tissues

DNA adducts in human tissue has been mainly determined in Europeans. Garner et al. [48] used immunological methods to analyze aflatoxin-DNA adducts in human liver. Eight liver samples obtained from Czechoslovakian patients with primary hepatocellular carcinoma were studied. Seven out of eight samples contained detectable and antiaflatoxin inhibitory material. The values ranged between 0.63 and 3.51 pmol aflatoxin/mg DNA.

Further samples were analyzed by Harrison and Garner [81] from persons acutely exposed to aflatoxins during a poisoning incident. DNA was isolated from formalin-fixed human tissues, and acid hydrolyzed. The hydrolysate inhibited antibody binding in a competitive aflatoxin inhibition ELISA both before and after immunoaffinity column purification. HPLC analysis of the DNA hydrolyzates revealed a peak with a shorter retention time than synthetic AFB_1-guanine. The major peak seen when DNA was extracted from formalin-fixed tissues from rats treated with AFB_1 was identical to that seen in the formalin-fixed human tissues. Adduct levels ranged from 0 to $170/10^6$ nucleotides depending on tissue type and individual examined.

3. DNA Adducts of Aflatoxin and Cancer Risk

Harrison et al. [82] investigated the carcinogenic risk from aflatoxins to a variety of organs from data obtained in the U.K. Human colon and rectum DNA were examined from normal and tumorous tissue obtained from cancer patients, and colon, liver, pancreas, breast, and cervix DNA from autopsy specimens. AFB_1-DNA adducts were detected in all tissue types examined and ranged from 0 to 60 adducts per 10^6 nucleotides. In samples analyzed by HPLC, the adducts had the same chromatographic properties of AFB_1-FAPY. This study demonstrates that there are either unknown sources of aflatoxins in the diet or the present permitted level in food is too high. In a further study, the same researchers found AFB_1 covalently bound to colonic DNA in 27 out of 42 human tissue samples analyzed, using immunoassay or chromatography procedures [83]. AFB_1-DNA adduct values ranged from 0.03 to 55 adducts per 10^6 nucleic acid bases, up to tens of thousands of DNA adducts per cell. This level of adducts has been correlated with cancer in experimental studies in a variety of animal species [83,84]. The steady-state DNA-adduct levels in rats and trout given a TD_{50} dose (= daily dose at which 50% of the animals developed a tumor) of AFB_1 were 850 adducts per 10^9 nucleotides and 1040 adducts per 10^9 nucleotides, respectively. In human studies the levels in liver cancer patients were between 1200 and 1700 adducts per 10^9 nucleotides [81,85]. Therefore, it appears that the levels are very similar in animals with liver tumors.

Urinary biomarkers were used in epidemiological studies. As part of a prospective study of 18,244 middle-aged men in Shanghai (China), Ross et al. determined urinary AFB_1, AFP_1, AFM_1, and AFB_1-guanine to assess the relation between aflatoxin exposure and liver cancer [5]. Subjects with liver cancer were more likely to have detectable concentrations of any of the aflatoxin metabolites in comparison to controls (relative risk 2.4, 95% confidence interval 1.0 to 5.9). The highest relative risk was for AFP_1 (6.2; 95% confidence interval 1.8 to 21.5). Ross et al. found a strong interaction between serological markers of chronic HBV infection and aflatoxin exposure in liver cancer risk. Therefore, reduction of aflatoxin exposure may be a useful intermediate goal in prevention of liver cancer, since the benefits of widescale hepatitis B vaccination will not be apparent for many years.

C. Biologically Effective Dose: Protein Adducts

1. Noncovalent Protein Binding of AFB_1

AFB_1 covalently bound to albumin and noncovalently bound AFB_1 are present in plasma. Several research groups have investigated the noncovalent binding of AFB_1 to serum proteins, especially albumin. Spectrometric investigations showed the formation of a complex with albumin [86]. AFB_1 in rat plasma was primarily bound to plasma albumin. Very little binding activity was shown by other plasma proteins [87]. AFB_1 was bound to human serum albumin at least at two sites [88]. AFB_1 bound to whole human plasma and serum albumins was stable in an alkaline solution at pH 8.6 and after overnight dialysis [89,90]. Neither charcoal treatment of rat albumin nor the presence of 0.15 M NaCl had any significant effect on the interaction [87]. However, it was completely dissociated in 1 M guanidine-HCl, suggesting that hydrogen bonds accounted for the association between AFB_1 and the protein [89].

AFB_1 metabolites as well as the parent compound are present in plasma. Wong and Hsieh [91] identified aflatoxicol as the major aflatoxin metabolite in the plasma of rats following oral or IV administration of ^{14}C-labeled AFB_1. Aflatoxicol, however, was not detected in the plasma of similarly dosed mice and monkeys.

2. Covalent Binding of AFB_1 to Albumin

Covalent binding of AFB_1 to albumin was discovered 25 years ago by Daleizios and Wogan [92,93]. It was postulated that protein adducts were derived from the reaction of amino-groups of the amino acids with the AFB_1-diol in the dialdehyde form. This was deduced from experiments performed with rat liver microsomes [13]. At physiological pH, AFB_1-diol was formed as a major metabolite of AFB_1. AFB_1 bound to microsomal proteins. This binding did not occur at pH 6.5. Amines (Tris and isopropylamine) and both oxidized and reduced glutathione competed with microsomal protein for the binding of AFB_1. Changes in the UV absorption spectrum of AFB_1 to a dialdehydic phenolate and the binding to amines and microsomal proteins indicated that binding proceeds via a Schiff's base reaction. It should be noted that the reaction of the dialdehydic phenolate with Tris has often been used as a measure of AFB_1-epoxide present in an incubation mixture. However, this Tris-diol complex has never been chemically characterized.

Nassar et al. [94] found circulating fluorescent compounds, conjugated to plasma albumin, remaining in the plasma after injection of aflatoxin. In vitro conjugation of AFB_1 and AFG_1 with plasma albumin did not occur until fresh liver or kidney slices were added. Thus, liver and kidney metabolism of AFB_1 and AFG_1 must occur before they bind to plasma albumin.

Appleton et al. [95] found that AFB_1 covalently binds to liver proteins, DNA, and RNA. The amount that binds is proportional to the dose. Two research groups further investigated the binding of AFB_1 to albumin [96–99]. Wild et al. [99] studied binding of AFB_1 to plasma protein and liver DNA in male Wistar rats dosed by oral intubation in either single or repeated doses. A constant ratio was found between levels of aflatoxin bound to plasma protein and that bound to liver DNA 24 h after a single dose (11.2 to 40.4 nmol AFB_1/kg) (Fig. 3). In total, 0.98% to 2.15% of the administered dose was bound to the plasma protein at this time point. In the chronic study, rats received two doses of 1.60 nmol AFB_1/day; groups of animals were killed on days 2 through 24. Binding of aflatoxin to plasma protein accumulated to a level threefold higher than that seen after a single dose. Levels of binding reached a plateau between days 7 and 14 of treatment and then remained stable until the end of the experiment. Binding to DNA also accumulated, 2.5-fold and in parallel to plasma protein. The binding reached a plateau between days 7 and 14 of treatment. In both the chronic and acute studies,

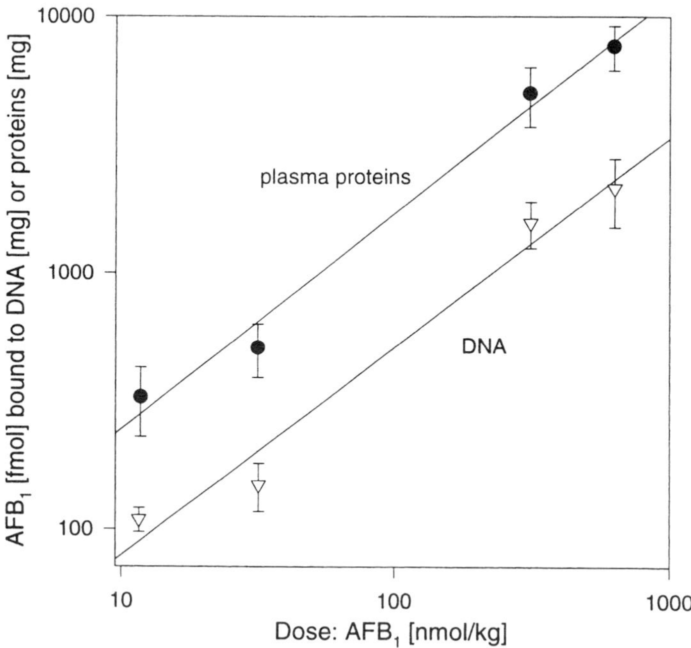

FIGURE 3 DNA- and plasma protein-adducts of $[^3H]AFB_1$ in male Wistar rats [100].

fractionation of the plasma proteins showed that all detectable bound aflatoxin was associated with a single peak corresponding to albumin. Thus, a constant ratio was observed, after chronic or single exposure, between the concentration of plasma albumin-bound aflatoxin and that bound to liver DNA, the target organ for carcinogenesis by AFB_1.

Characterization of the Main AFB_1-Albumin Adduct. Sabbioni et al. [98] dosed male Fisher rats with AFB_1. AFB_1 reacted primarily with one or more lysine residues in serum albumin. The radioactivity associated with albumin following administration of ^{14}C-AFB_1 to rats was cleared with a half-life of 2.5 days, which is not significantly different from the half-life of unmodified albumin in the normal rat. The product isolated from a pronase digest of in vivo modified albumin was identical to the synthetic product, by chromatography retention time, and by UV and mass spectroscopy. The synthetic product was obtained by the acylase-catalyzed deacetylation of the reaction product of N_α-acetyl-L-lysine with AFB_1-dibromo and characterized by UV, fluorescence, 500 MHz ^1H-NMR, ^{13}C-NMR, and fast atom bombardment mass spectrometry [100]. Based on the major adduct of AFB_1 (1) with guanine and GSH the formation of the simple substitution product 8,9-dihydro-8-(N_ε-lysyl)-9-hydroxy-AFB_1 (11) was expected; however, this structure does not fit the spectroscopic data reported in this study (Fig. 4). The spectral data strongly supported a structure in which the terminal dihydrofuran ring of AFB_1 has been converted to a pyrolinone ring. The lack of doublets above 5 ppm in the ^1H-NMR (proton in the bridge of the bis-dihydrofuran moiety in AFB_1), the smaller mass (456 instead of 474) found by FAB-MS, and the pH-dependent shift of the UV absorbance indicate a major structural change of the original bis-dihydrofuran moiety of AFB_1. In Figure 4, the proposed pathways for the AFB_1-lysine adduct formation are presented. Based on the possible reaction mechanisms the following three structures were considered: (8) = 5-oxo-3-pyrroline, (9) = 5-oxo-2-pyrroline, or (13) = 4-oxo-2-pyrroline with the coumarin unit attached to the 3 position. AFB_1-dibromo (2) or AFB_1-epoxide (3) most likely react first with water to yield the same intermediate, the AFB_1-diol (4). The dialdehyde tautomer (5) of (4) reacts with N_ε- of lysine to form a Schiff base (10) or a 2,4,5-trihydroxy-pyrrolidine (6). Compound (6) loses two water molecules to yield the 5-hydroxy-pyrrole (7) which is known to tautomerize to 5-oxo-3-pyrroline (8) and 5-oxo-2-pyrroline (9) in a ration of 9 to 1. The Amadori rearrangement of the Schiff base (10) followed by the condensation of the ε-amino group with the remaining aldehyde leads to the 4-oxo-2-pyrroline (13). The collected spectroscopic properties correspond the best to the structure with the coumarin unit attached to the 3 position of 5-oxo-3-pyrroline. In particular, the carbon NMR data fit best structure (8) (AFB_1-lysine = [8-[N-(2-amino-hexanoyl-6-yl)-5-oxo-3-pyrrolin-3-yl]-7-hydroxy-5-methoxycyclopentenone[2.3-*c*]coumarin]). AFB_1-lysine is unstable in acids and in base ($>$ pH 9.0). Incubation at 37°C and

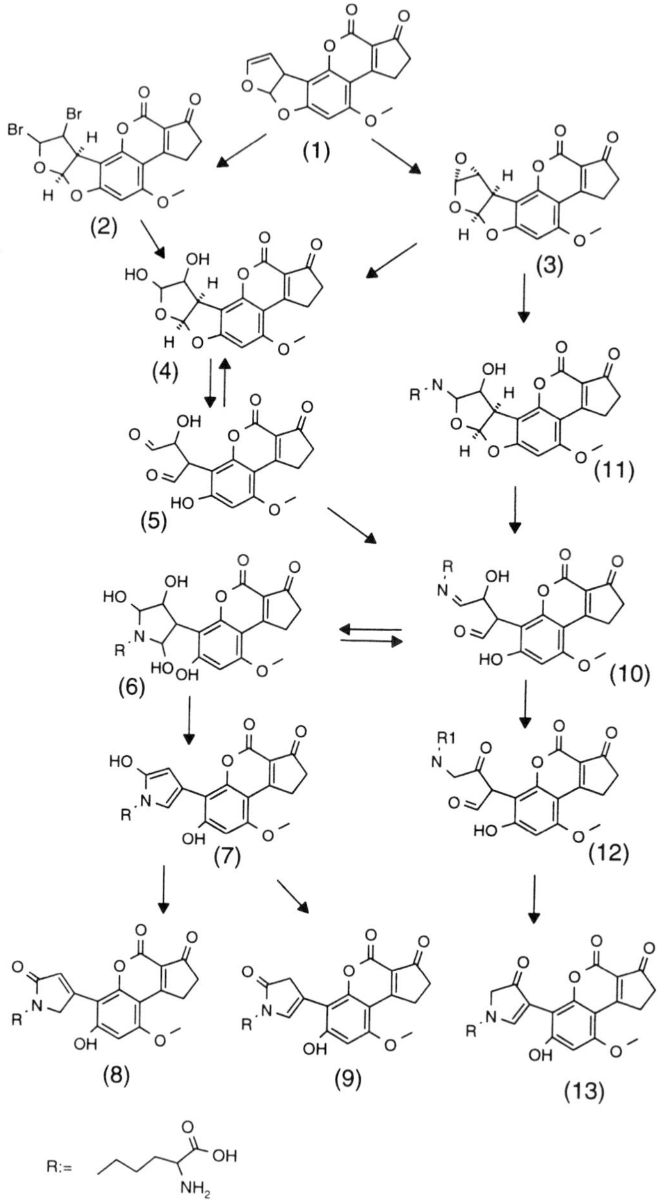

FIGURE 4 Reaction mechanism of protein adduct formation.

pH 4.0 leads to a loss of ca. 25% of AFB_1-lysine in 24 hours. Heat (> 70°C) destroys the adduct at all pH values within a few minutes. The presence of DMSO, methanol, or acetonitrile accelerates the loss of AFB_1-lysine. After chromatographic purification, the organic solvents must be evaporated immediately. AFB_1-lysine can be stored for over a year as a solid or in solution at pH 7.4 to 8.0 at −50°C. A very rapid and reliable purity control of AFB_1-lysine may be performed on a UV spectrometer. The various decomposition products have distinct absorption maxima, but have not been characterized. The ratio between the maxima at 400 nm and at 343 nm is a very good indicator of ongoing decomposition. However, the pH of the UV solutions must be adjusted exactly because the absorbance of 400 nm varies significantly between pH 7.0 and 7.4. Purity of AFB_1-lysine used as standard should be checked twice by HPLC using a methanol gradient in either phosphate buffer at pH 7.2 or ammonium formate buffer at pH 4.0.

To determine whether the liver is the sole site of aflatoxin-albumin adduct formation, preliminary experiments with isolated perfused rat liver were carried out [101]. The data showed that AFB_1 metabolites covalently react not only with albumin in the hepatocyte, but also with circulating proteins in the perfusate. This suggests that a reactive aflatoxin metabolite secreted by the liver may form serum albumin adducts in circulating blood. Okoye et al. [102] demonstrated in an experiment with rats injected simultaneously IP with $[^3H]AFB_1$ and L-$[^{14}C]$leucine that the majority (> 90%) of rat serum albumin-AFB_1 adduct was formed by a modification of the protein at the time of its synthesis in the hepatocyte. This finding further indicates that albumin adducts are a dosimeter for degree of modification of DNA in hepatocytes.

Analysis of human samples have progressed slowly as a consequence of the instability of AFB_1-lysine under some conditions of proteolyis. The reason for the unstability of AFB_1-lysine is unknown. The instability of AFB_1-lysine was noted after digestion of albumin with pronase (110000 PUK/g, Calbiochem 537088), which is no longer commercially available. After acidification to pH 4 with acetic acid, the digests were analyzed by HPLC at pH 4 [98]. Attempts to isolate the decomposition products of AFB_1-lysine were not successful. Further experiments showed that AFB_1-lysine is stable when other pronase brands (pronase E from Serva or pronase 70000 PUK/g 53702 from Calbiochem) were used, and when all stages of analysis were performed without acidification [100]. The digestion conditions were optimized to obtain the highest possible yield of AFB_1-lysine from biological samples. The yields of AFB_1-lysine were tested using different pronases, different ratios of albumin-enzyme (mg albumin/mg enzyme), and different reaction times. The optimum conditions for the digestion of albumin are 12 to 15 hours at pH 7.4 and 37°C with an albumin enzyme ratio of 3/1. The levels found remain about constant for up to 24 h. Serum albumin was digested with pronase, purified on AFB_1-specific immunoaffinity columns, and analyzed by HPLC with a fluorimeter as detector. The detection limit for AFB_1-lysine was 20 fmol. Six samples with 2 mg serum albumin were digested in the presence of synthetic AFB_1-lysine (120 fmol). The recovery of AFB_1-lysine determined by HPLC and fluorescence detection was 83% ± 6% after 12 h incubation at 37°C. In order to reuse the immunoaffinity columns more often, it is advisable to eliminate pronase by precipitation with acetone.

The hydrolysis conditions were also tested from other laboratories. The stability was tested by incubating 2 mg of albumin with 100 fmol AFB-lysine with or without proteinase K. Without proteinase K the mean recoveries were 98.8% ± 18.6%, and with proteinase K the recoveries were 62.4% ± 14.4%. These recoveries were not time-dependent [104].

Another group investigated several protocols to quantify AFB_1-albumin adducts by ELISA [103]. These authors found an incomplete release of the major adduct (AFB_1-lysine) even after prolonged hydrolysis, but the adduct was very unstable under some conditions of proteolysis for unknown reasons. A considerable fraction of aflatoxin-modified material was produced by proteolysis of in vivo-modified rat albumin or in vitro-modified bovine albumin, which is not recognized by ELISA with an antiaflatoxin polyclonal antibody having a wide spectrum of aflatoxin metabolite detection. This fraction is increased in parallel with proteolysis time.

An excellent overview about the quantification of serum albumin adducts of AFB_1 by ELISA has been published by Chapot and Wild [104]. ELISA requires considerable expertise to ensure reliable and reproducible quantitative results. The samples are quantified against a calibration curve obtained from 1 to 30 fmol AFB_1-lysine. This range corresponds to 20% to 85% inhibition. The inhibition values of the samples should range from 20% to 65% inhibition to obtain accurate quantitation. Therefore, it is advisable to run a sample at two different concentrations. The results are reported as fmol AFB_1-lysine equivalent per mg albumin, as the inhibition will not only be due to AFB_1-lysine but to other hydrolysate products as well. The precision of the assay (repeatability) was evaluated by the sixfold determination of an albumin sample [104]. The authors obtained 102.7 ± 7.4 fmol AFB_1-lysine equivalent per mg albumin, which corresponds to a relative standard deviation of 7.2%. The reproducibility of the assay was determined by spiking a human serum sample with three different amounts of serum from a rat treated with 3.84 μmol/kg AFB_1. The sera was aliquoted and analyzed in a series of eight ELISAs over a period of 4 months. Mean results (± SD) were 6.9 ± 1.6, 12.1 ± 2.3, and 4.4 ± 1.5 fmol AFB_1-lysine equivalents per mg albumin for the three samples.

Similar results were obtained for the determination by HPLC and fluorescence. Albumin purified by Reactive Blue Sepharose columns or by ammonium sulfate precipitation was digested with pronase. The digest was purified on an AFB_1-specific immunoaffinity column and analyzed by HPLC with fluorescence detection. The precision of a sixfold determination of a human sample was ± 8%

for AFB_1-lysine. The precision of AFG_1-lysine in rats dosed with AFG_1 was ± 8.4%. Therefore, albumin of rats dosed with AFG_1 might be considered as an internal standard for the analysis of human samples, since to date no such adducts have been found in vivo [100,101,109].

The results obtained from quantitation with ELISA and with HPLC-fluorescence were compared in several studies [105–108]. The methods were optimized and compared in animal experiments: for example, male BDIV rats were given 3.84, 1.02, or 0.38 μmol/kg AFB_1 (three rats per group) or vehicle alone (two rats per group) by IP injection in DMSO [109]. Binding to albumin was confirmed by purification of albumin on an anion exchange column (Mono Q, Pharmacia, 50 × 5 mm). Albumin binding was examined by HPLC-fluorescence and by ELISA following hydrolysis. Linear dose responses were obtained by HPLC-fluorescence for AFB_1-lysine (Fig. 5a), where $y = 1.19 \times +0.35$; $r = .99$. Thus, in AFB_1-treated rats, binding to albumin was measured as 1.19 pmol of AFB_1-lysine/mg albumin per μmol/kg dose of aflatoxin. In the case of the ELISA AFB_1 binding to albumin is also linear to the dose: $y = 5.4 \times +0.87$; $r = .99$. Thus, there was 5.4 pmol AFB_1-lysine equivalent/mg albumin per μmol/kg dose of AFB_1. AFB_1 binding measured by ELISA and by HPLC- fluorescence correlate with $r = .95$, y (HPLC) = 0.218 × (ELISA) + 0.175 (Fig. 5b).

A further approach for the analysis of albumin adducts of AFB_1 was published by Sheabar et al. [110]. The goal of these experiments was to devise an analysis procedure that would increase the overall recovery of aflatoxin adducts in serum albumin. The method developed consisted of the following procedures. Proteins were precipitated from serum (minimum of 100 μl) with ammonium sulfate. Following dialysis against phosphate-buffered saline, the proteins were digested with pronase (1:4.1 wt/wt enzyme:protein) for 15 h at 37°C with shaking. Enzyme and other undigested proteins were precipitated with acetone. After evaporation of the acetone, levels of AFB_1-albumin adducts were determined by RIA. This procedure eliminated the tedious isolation of albumin prior to analysis and reduced interference in the RIA. High recoveries (25%) of AFB_1 adducts were achieved together with a low limit of detection (0.5 pmol AFB_1-lysine equivalents/100 μl serum). The authors applied the method for the analysis of serum samples from residents of Chongming Island, China.

Wild et al. [105] observed a 10-fold higher level of AFB_1-albumin adducts when measured by ELISA than by HPLC fluorescence in human sera from Kenya, while the difference in sera from rats treated with AFB_1 was only four- to fivefold. One possible explanation for this difference was the presence of AFG_1-albumin adducts in addition to AFB_1-albumin adducts in human samples. Therefore, a method was developed to quantify AFG_1-albumin adducts. Sabbioni and Wild [109] identified and characterized a major AFG_1-albumin adduct in rats following exposure to AFG_1. The product isolated from a pronase digest of in vivo-modified albumin was identical as determined by chromatographic retention time to the

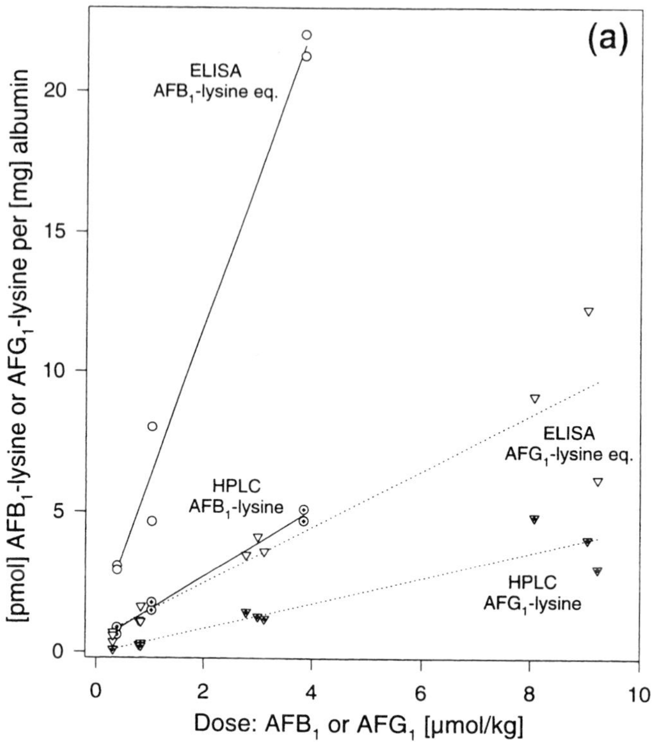

Figure 5 (a) Albumin from rats dosed with AFB_1 (AFG_1) were digested with pronase and analyzed by HPLC with fluorescence detection and by ELISA. The results of the HPLC analyses are expressed as AFB_1-lysine (AFG_1-lysine) (pmol)/mg albumin. The results of the ELISA analyses are expressed as AFB_1-lysine-equivalents (AFG_1-lysine-equivalents)/mg albumin [110]. (b) Comparison of AFB_1-lysine and AFG_1-lysine quantified by HPLC and by ELISA [110].

synthetic product obtained by the acylase-catalyzed deacetylation product of N_α-acetyl-L-lysine with 8,9-dihydro-8,9-dibromo-AFG_1. The in vitro product, AFG_1-lysine, was characterized by UV, fluorescence, ^{1}H- and ^{13}C-NMR spectroscopy, and fast atom bombardment mass spectrometry. A competitive ELISA for this adduct was established using polyclonal antibodies to AFB_1, and this was used together with an HPLC-fluorescence technique to quantitate the in vivo formation of AFG_1-albumin adducts in comparison to AFB_1. A linear dose-response relationship ($y = 0.45 \times -0.04$; $r = .95$) was observed in rats following single exposures to 0.304 to 9.137 µmol AFG_1/kg (Fig. 5a). Thus, in AFG_1-treated rats, binding to albumin was measured as 0.45 pmol of AFG_1-lysine/mg

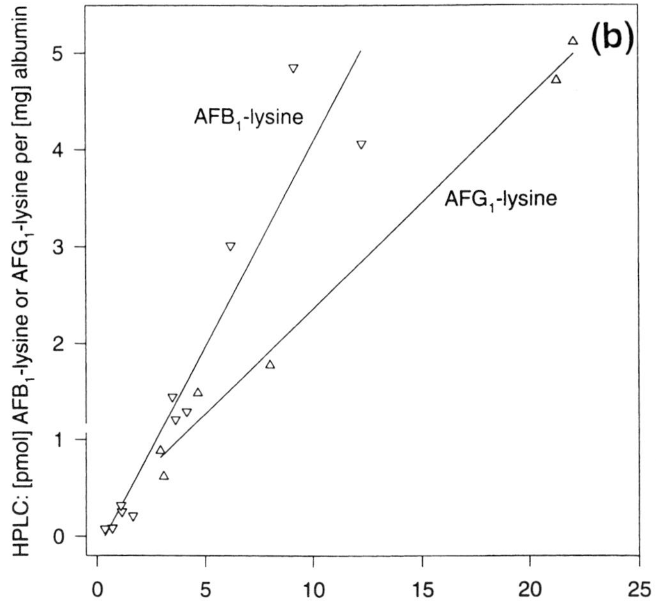

FIGURE 5 Continued.

albumin per μmol/kg dose of AFG$_1$. In the case of the ELISA AFG$_1$ binding to albumin is also linear, with the dose y = 1.0 × +0.50; r = .93. Thus, there was 1 pmol AFG$_1$-lysine equivalent/mg albumin/μmol/kg dose of AFG$_1$ (Fig. 5a). The levels of AFG$_1$-albumin adducts were determined to be 5.4- and 2.6-fold lower than with equivalent doses of AFB$_1$ as determined by immunoassay and HPLC fluorescence, respectively. The lower binding of AFG$_1$ and the lower levels in the human food supply compared to AFB$_1$ suggest that the newly identified adduct could be added as an internal standard for the analysis of AFB$_1$-lysine. However, the presence of AFG$_1$-lysine in human samples should be carefully investigated before using albumin of AFG$_1$-dosed rats as an internal standard. Furthermore, it appears that in contrast to AFB$_1$-lysine, AFG$_1$-lysine is not retained after regenerating the immunoaffinity columns. Therefore, for all analyses involving AFG$_1$-lysine analyses, the immunoaffinity columns might not be reusable. This is a major disadvantage of this adduct as an internal standard.

Wild et al. [105] described and compared three complementary approaches to the quantitation of AFB$_1$-albumin adducts: (1) ELISA performed directly on intact albumin (direct ELISA); (2) ELISA performed on an albumin hydrolyzate (hydrolysis ELISA); and (3) HPLC fluorescence detection of AFB$_1$-lysine adduct after albumin hydrolysis and immunoaffinity purification. These techniques have been validated by direct comparison with rat albumin samples modified to a known extent. In rat samples the recoveries of the AFB_1-albumin was determined. The recovery by direct ELISA is 64% ± 10%, by ELISA-hydrolysis 22.7% ± 2.2%, measuring against AFB_1-lysine as a standard. For the HPLC-fluorescence method the recovery of the adduct is 5.5% ± 1.3%. Detection limits of approximately 219, 11.0, and 11.0 fmol AFB_1-lysine equivalents/mg human albumin were determined for the three methods, respectively. Samples obtained from individuals from Thailand, Gambia, Kenya, and France have been used to validate the measurement of AFB_1-albumin adducts by these three methods. Levels of 15 to 740 fmol AFB_1-lysine equivalents/mg albumin were observed in the samples from Thailand and Gambia while no adduct were detected in the samples from France. The relative properties of the three assays, with special regard to their application in epidemiological studies, were considered. The authors concluded that a combination of the hydrolysis ELISA for large-scale screening following by confirmatory analyses in positive samples by HPLC-fluorescence is the most appropriate methodological approach.

3. Determination of Protein Adducts in Human Studies

China and Africa have the highest intake of aflatoxin and the highest incidences of liver cancer. The following studies were undertaken to explore the relationship between dietary intake of aflatoxins, the serum albumin adducts, the HBV-carrier status, and the excretion of the major aflatoxin-DNA adduct and other metabolites into the urine of chronically exposed people.

In a pilot study involving several research groups and different biomonitoring methods, blood and urine of 42 residents of Guangxi Province (China) were analyzed. The daily aflatoxin intake was compared with urine metabolites AFM_1 [50,111–113], AFP_1 and AFB_1-guanine [50,112], and serum albumin adducts determined either by RIA [111] or by HPLC fluorescence [101,106,107]. The following protocol was developed for this investigation. The diets of 30 males and 12 females (25 to 64 years old) were monitored for 1 week and aflatoxin intake levels determined each day. The average intake of AFB_1 by men was 154.9 nmol/day, giving a total mean exposure during the study period of 886.3 nmol. The average daily intake by women was 246.5 nmol/day, resulting in a total average exposure during the 7-day period of 1737.4 nmol AFB_1. Starting on the fourth day, total urine volumes were obtained in consecutive 12-h fractions for 3 or 4 days. HPLC and competitive RIA analyses were used to determine the relationships between excretion of total aflatoxin metabolites (AFB_1-guanine, AFM_1, AFP_1, and AFB_1) and AFB_1 intake values [77]. Initial efforts to characterize aflatoxin metabolites in urine samples were by competitive RIA. The analyses by linear regression of AFB_1 intake per day and total aflatoxin metabolite excretion per day showed a correlation coefficient of only 0.26. Therefore, the individual metabolites were analyzed by HPLC after immunoaffinity purification. The AFB_1-guanine excretion and AFB_1 intake from the previous day showed a correlation coefficient of 0.65. Similar analysis for AFM_1 resulted in a correlation coefficient of 0.55, whereas there was no positive statistical association between exposure from the diet and AFP_1 excretion, despite AFP_1 being quantitatively a major metabolite. However, plotting the total AFB_1-guanine excretion in the urine during the complete collection period against the total AFB_1 concentration in the diet for each of the individuals, the correlation coefficient increases to $r = .80$ (Fig. 6). Evidently, a summation of excretion and exposure status smooths day-to-day variations. Blood specimens were obtained during the same period that urine was collected and the diet was sampled [111]. Serum albumin was isolated from blood by affinity chromatography on Reactive Blue 2-Sepharose. After enzymatic proteolysis using pronase, immunoreactive products were purified by immunoaffinity chromatography and quantified by competitive RIA. A highly significant correlation ($r = .60$) of AFM_1 excretion with adduct level was observed. An equally

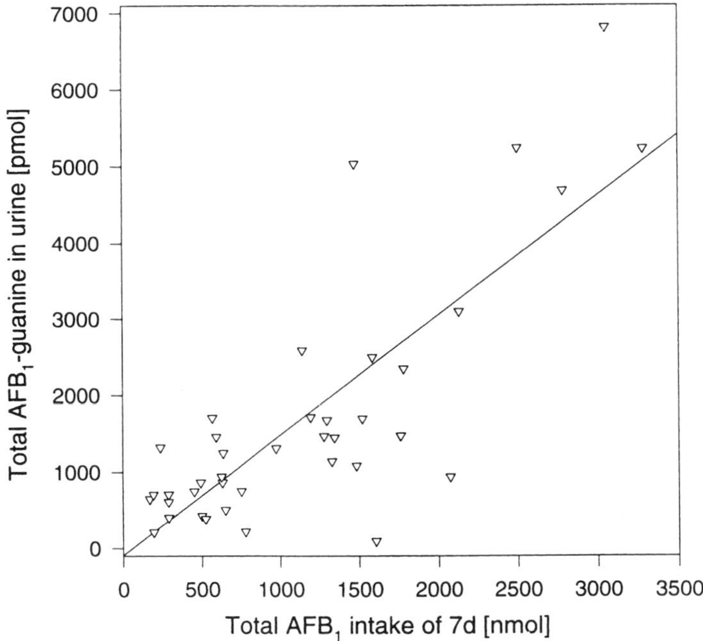

FIGURE 6 Analysis of samples of males and females from the Guangxi Province, People's Republic of China. Regression analysis of AFB_1 intake over 7 days with AFB_1-DNA (AFB_1-guanine) levels in 3-day urine [77]; $r = .80$.

significant correlation of AFB_1-intake ($r = .69$) with adduct level was observed. From the slope of the regression line for adduct level as a function of intake, it was calculated that 1.4% to 2.3% of ingested AFB_1 bound covalently to serum albumin. This value is similar to that observed when rats are administered AFB_1.

Gan et al. [114] reanalyzed the same samples with laser-induced fluorescence spectroscopy of the undigested albumin [111]. A highly significant correlation of AFB_1 serum albumin adduct level with daily AFB_1 intake was observed: $r = .75$ $P < 10^{-6}$. The slope of the regression line is 9.72 pmol adduct (= AFB_1 equivalent) g albumin/nmol AFB_1 intake (per day). About 4% of ingested AFB_1 was covalently bound to albumin. The results from the two methods, laser-induced fluorescence and RIA, correlated well (r= .72, $P < 10^{-6}$, slope = 1.2). The detection limit is limited by the fluorescence of the native albumin. Therefore, the lowest detection limit was 0.5 pmol AFB_1/mg albumin. This limit is far above the other methods. However, for this highly exposed population from China it was appropriate.

Sabbioni et al. analyzed the same serum samples by HPLC with fluorescence detection [101]. The authors did not find any AFB_1-lysine in albumin-depleted plasma proteins. Therefore, whole serum was digested with pronase, and the adducts were purified by monoclonal antibody IAC and quantified by HPLC. It should be noted that the recovery of AFB_1-lysine adduct is lower from digestion in whole plasma than in the presence of purified albumin (unpublished results). This was shown by spiking rat serum in to both a solution of human serum albumin and in to human serum. The recovery was 78% ± 10% in the presence of serum. The data depicted in Figures 6 to 10 compare the relationship among AFB_1 consumption per day, serum albumin adducts, and DNA adducts. Linear dose-response relationships were seen with correlation coefficients for males, females, and total population of 0.82 (Fig. 7), 0.67 (Fig. 8), and 0.59, respectively. These are highly significant relationships with P values of < .00001, .023, and .0001, respectively. Earlier analysis of the combined male and female data revealed a single outlying point which, if removed from the regression analysis, increased the overall correlation coefficient from .59 to .67. The slopes of the regression lines for the males, females, and total population are 518, 137, and 260 fmol AFB_1-lysine per nmol AFB_1 daily exposure, respectively. According to these data it appears that females bind ca. 4 times less AFB_1 to albumin than males. This fact should be further investigated and correlated with liver cancer incidents. The levels determined by RIA [111] are 3.6 times higher. The two totally independent determinations, RIA and HPLC-fluorescence, correlate with $r = .57$ ($P = .0002$), y (= HPLC) = 0.13 × (= RIA) − 1.55 [107]. According to this correlation factor these two methods do not compare as well as the values obtained with ELISA and HPLC, $r \leqslant .95$ (Figs. 5b, 11). In the former case the determinations were totally independent and 2 years apart. In the latter case the determinations were performed in the same time period and from the same purified albumin solutions.

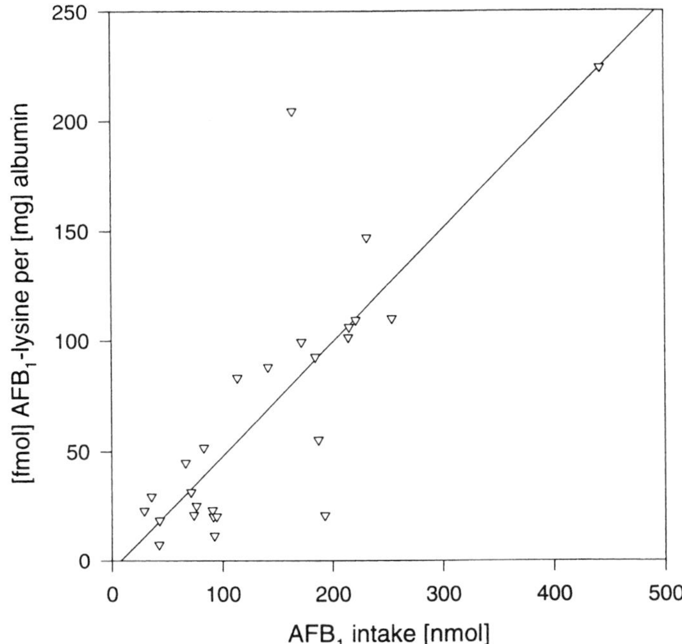

FIGURE 7 Analysis of samples of males from the Guangxi Province, People's Republic of China. Regression analysis of AFB_1 intake with serum albumin adduct (AFB_1-lysine) levels [102]; $y = 0.518 \times -3.8$, $r = .82$.

AFB_1-guanine in urine has been determined from the same individuals [78,112]. Total intake of AFB_1 (7 days) and total AFB_1-guanine in urine (3 days) correlate with $r = .80$ (Fig. 6). AFB_1-lysine correlates with the total AFB_1-guanine excreted in 3 days. For males the correlation is $r = .64$ ($P = .0004$) (Fig. 9), for females 0.76 ($P = .010$) (Fig. 10), and for both males and females $r = .56$ ($P = .0004$) (Sabbioni and Groopman, unpublished data). According to the respective regression lines, males (females) bind 25.5 fmol (12.2 fmol)/mg albumin/pmol total excreted AFB_1-guanine. These data suggest that females are at higher risk for genotoxic damage by AFB_1. However, this trend should be confirmed in larger cohorts.

A further comparison of aflatoxin exposure in different countries was performed by Wild et al. [115]. Albumin adducts were found in 12% to 100% of children and adults from various African countries with levels up to 766.9 fmol AFB_1-lysine equivalent/mg albumin. In Thailand, lower levels and prevalence of this adduct were observed, whereas no positive sera were detected from France or Poland.

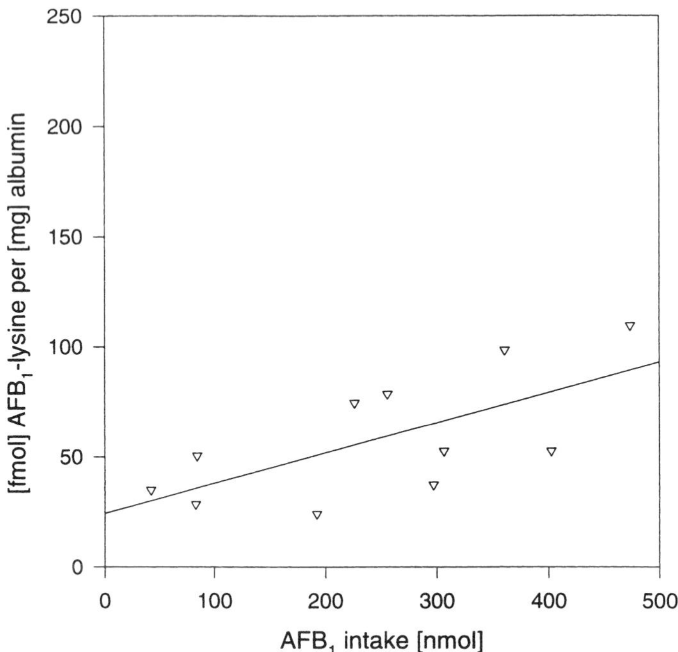

Figure 8 Analysis of samples of females from the Guangxi Province, People's Republic of China. Regression analysis of AFB_1 intake with serum albumin adduct (AFB_1-lysine) levels [102]; $y = 0.137\times + 24.4$; $r = .67$.

Wild et al. [116,117] found that in East and West African countries, the majority (75% to 100%) of individuals were positive (> 11 fmol AFB_1-lysine eq./mg albumin) for the AFB_1-albumin adduct, with levels reaching up to 1.58 pmol/mg. Levels of adduct to date have been age- and sex-independent, while marked seasonal variations were seen, for example, in Gambia. Exposure also occurs in utero; AFB_1 adducts have been found in umbilical cord blood.

In Denmark, AFB_1 is suspected as an etiological factor in the increased risk of primary liver cancer among workers in plants processing animal feed [118,119]. The exposure to AFB_1 in these animal feed processing plants was assessed using binding of AFB_1 to serum albumin. The albumin fraction was digested with pronase and then purified on a C18 Sep-pak column and an immunoaffinity column before quantification by ELISA. The limit of detection was 11 fmol/mg albumin. The workers served as their own controls, as blood samples were taken upon return from vacation and after 4 weeks of work. A total of seven to 45 samples were positive for AFB_1, with an estimated average daily intake of 204.9 pmol AFB_1/kg body weight. Three positive workers had been unloading a cargo

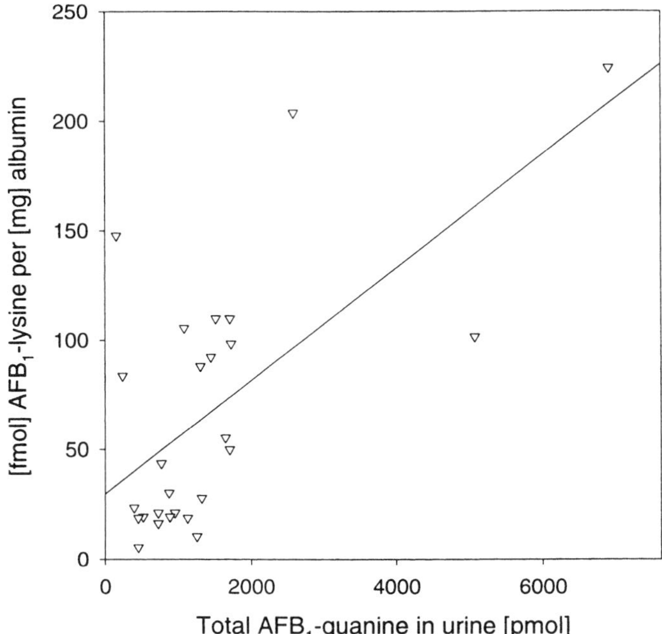

FIGURE 9 Analysis of samples of males from the Guangxi Province, People's Republic of China. Regression analysis of serum albumin adduct (AFB_1-lysine) levels with AFB_1-DNA (AFB_1-guanine) levels in 3-day urine (Sabbioni and Groopman, unpublished results); $y = 0.0255\times + 30.3$, $r = .64$.

with an AFB_1 level of 83.2 μmol/kg raw material. The exposed workers had been handling cargoes contaminated with AFB_1 or working at places where the dust contained detectable amounts of AFB_1. Dust samples collected at different sites showed considerable variation in the amount of AFB_1 (nondetectable to 25.6 μmol/kg dust). The sera from the exposed workers had a significantly higher titer against an AFB_1-epitope than a nonexposed Danish control group. The level of exposure could partly explain the increased incidence of liver cancer in workers in the animal feed processing industry.

4. Comparison of Albumin Adducts with Individual Susceptibilities and Cancer Prevention Measures

Dithiolethiones, including oltipraz and the unsubstituted molecule 1,2-dithiole-3-thione, are potent inhibitors of aflatoxin-induced hepatic tumorigenesis in rats [120–124]. Therefore, such compounds might be used for chemoprotection. The evaluation of dithiolethiones or other chemoprotective agents in human clinical

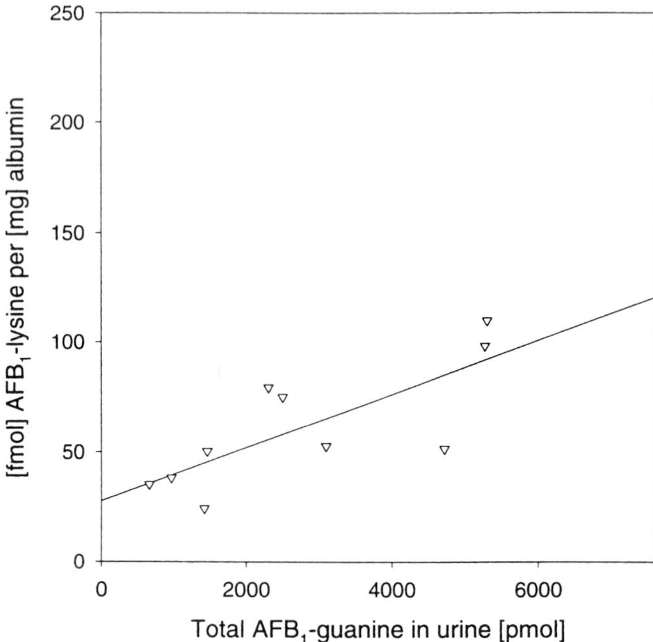

FIGURE 10 Analysis of samples of females from the Guangxi Province, People's Republic of China. Regression analysis of serum albumin adduct (AFB_1-lysine) levels with AFB_1-DNA (AFB_1-guanine) levels in 3-day urine (Sabbioni and Groopman, unpublished results); $y = 0.0122x + 27.8$, $r = .76$.

trials requires noninvasive biomarkers to assess the efficacy of these interventions [123]. Groopman et al. [123] performed animal experiments with male F344 rats to test the effect of the chemopreventive agents by measuring DNA and protein adducts. One-half of the rats were fed a diet supplemented with 0.03% 1,2-dithiole-3-thione to lower their risk of tumors, and the other half were fed an unsupplemented diet. Levels of hepatic aflatoxin-DNA adducts, serum aflatoxin-albumin adducts, and excreted AFB_1-guanine adducts in urine were determined following multiple administrations of 800 nmol AFB_1/kg body weight on days 0–4 and 7–11. In the rats fed 1,2-dithiole-3-thione, the overall reduction in the levels of hepatic DNA adducts, urinary AFB_1-guanine, and serum aflatoxin-albumin adducts over the 2-week exposure period were 76%, 62%, and 66%, respectively. This parallel reduction of urinary AFB_1-guanine and serum aflatoxin-albumin adducts levels relative to target organ DNA adduct burden implies that these biomarkers are noninvasive dosimeters which can be used to evaluate the efficacy of chemoprotective interventions.

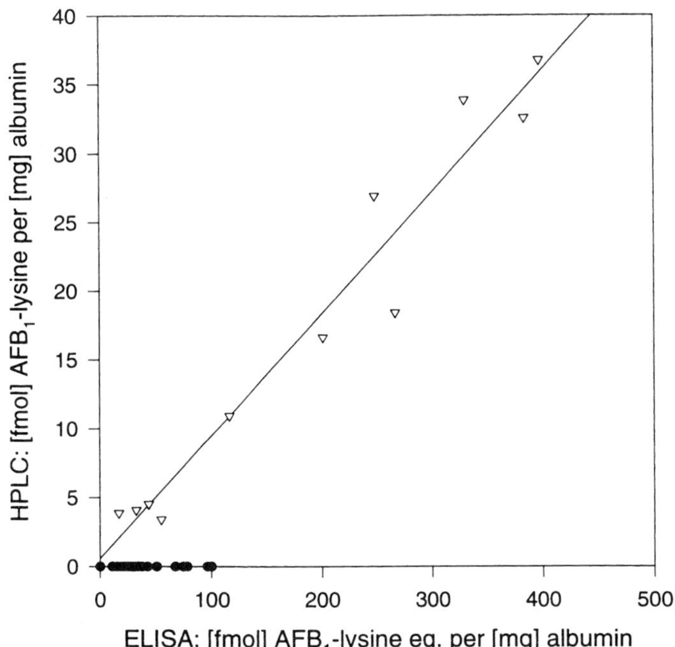

FIGURE 11 AFB_1 albumin adducts in 34 human sera from Gambia. Correlation of the 11 samples that were positive by both ELISA and HPLC fluorescence after digestion with pronase [109]; $y = 0.0887\times + 0.58$, ($r = .97$).

In studies with chronically exposed rats, Egner et al. [122] confirmed that long-term intervention with oltipraz produced a reduction in levels of the aflatoxin-albumin biomarker at all times throughout aflatoxin exposure. By contrast, in an experiment with a delayed transient intervention protocol, the decrease of biomarker levels was not seen until the ninth day of the intervention. Once achieved, the difference from the control group were significant for the rest of the aflatoxin exposure period. These changes in aflatoxin biomarker levels were consistent with the cancer outcomes seen in chemoprotective intervention studies in rats [120]. These results confirm the utility of measuring albumin adducts as a means of assessing the efficacy of chemoprotective interventions in individuals at high risk for aflatoxin exposure.

5. Comparison of Albumin Adducts with HBV Carrier Status

The relative roles of HBV and aflatoxin and their possible mechanism of interaction in the etiology of hepatocellular carcinoma (HCC) are not understood

[2,3,126,127]. To investigate this aspect, several authors have measured the hepatitis carrier state, the metabolites of AFB_1 in urine, and AFB_1 covalently bound to albumin in peripheral blood from humans living in countries with high liver cancer incidences [78,108]. One study was performed in Gambia with 20 residents of Keneba, West Kiang. Dietary intake of aflatoxin was measured at the individual level over a 7-day period and correlated with the level of aflatoxin bound to peripheral blood albumin at the beginning and end of the study. All subjects were exposed to aflatoxin originating from several food sources. The average daily intake was 4.48 μmol/day. The aflatoxin adducts were determined by ELISA and by HPLC. A significant correlation was observed between the dietary intake and the level of albumin-bound aflatoxin at the end of the study. In addition, a good agreement was obtained with the two analysis techniques. A comparison of matched chronic hepatitis B surface antigen carriers with noncarriers did not reveal any difference in adduct formation for a given dietary intake of aflatoxin. This result is in contrast to another study involving a larger number of people. Wild et al. [115,117] investigated 323 children (age 3 to 78 yr) from Gambia. The AFB_1-albumin adduct levels were examined with respect to HBV infection and ethnic group. Over 95% of all sera contained detectable adducts; children positive for HBV surface antigen (HBsAg) had significantly higher adduct levels than children with markers of past infection or who had never been infected (180, 124, and 126 fmol AFB_1-lysine equivalents/mg albumin, respectively). In addition, there were highly significant differences in adduct levels among the three major ethnic groups (Wollof, 180; Fulla, 126; Mandinka, 89 fmol AFB_1-lysine equivalents/mg albumin). Wollof children were also more likely to be HBsAg-positive than the other two groups. These data suggest that ethnic group and HBV infection can influence AFB_1 metabolism.

In another study in Guinea [127], sera were collected from 75 men. The sera were analyzed by immunoassay for aflatoxin covalently bound to serum albumin. More than 90% of the sera contained detectable adduct levels. The highest level was 843 fmol AFB_1-lysine equivalent per mg albumin. Eleven subjects (14.7%) were positive for HBsAg in the serum. These subjects had a tendency to have higher aflatoxin-albumin adduct levels than the other subjects (mean level 154 fmol/mg compared to 97 fmol/mg), but the difference was not statistically significant.

The mechanism of the synergetic effects of HBV and AFB_1 exposure is being studied by several groups [128,129]. Kirby et al. found that HBV alters the expression of phase I and phase II enzymes involved in the metabolism of AFB_1 [129]. Similar studies were performed in animals and humans infected with the liver fluke *Opisthorchis viverrini* [130,131]. The results indicate a high local expression of CYP isoenzymes in *Opisthorchis viverrini*-infested livers. This could be a contributing risk factor in the development of liver cancers associated with parasitic hepatitis.

6. Comparison of Albumin Adducts with Biological Effects

A recent study by Wild et al. [132] compared AFB_1-DNA adduct levels in liver with AFB_1-albumin adduct levels in different rodent species to determine whether the latter could serve as a marker of hepatic DNA adduct levels irrespective of species. Rats (three strains), hamster, guinea pig, and mice were treated PO with up to 14 daily doses of AFB_1. Animals were sacrificed 24 h after 1, 3, 7, or 14 days of treatment. A dose response for AFB_1-albumin and AFB_1-DNA adducts (AFB_1-FAPY) was seen for all species and strains with steady-state adduct levels at 14 days. The levels of both adducts was in the following order: rat > guinea pig > hamster > mouse. When the levels of the albumin and DNA adducts at 14 days were plotted against each other for all species and all strains, a correlation was observed ($r = .83$) suggesting a constant relationship between the level of binding of AFB_1 to serum albumin and liver DNA. The amounts of AFB_1-albumin adduct reflect the relative susceptibility of the different species to AFB_1 carcinogenesis; the rat is sensitive, and the hamster and mouse are resistant. The level of AFB_1-albumin adduct in rodents was compared to data from humans environmentally exposed to AFB_1. The adduct levels in rats were determined after a single dose. The levels in humans were extrapolated to a single dose assuming that with chronic intake the steady state level is ca. 30 times higher than after a single dose. In rats these levels were between 0.65 and 1.11 fmol AFB_1-lysine equivalent/mg albumin/3.2 nmol AFB_1/kg body weight and a value for the mouse of 0.054 fmol. The highest estimate for people from Gambia and southern China [101] was 3.14 fmol/mg albumin for the same exposure. These data suggest that humans exposed to AFB_1 form amounts of albumin adducts closer to those observed in AFB_1-sensitive and 1 to 2 orders of magnitude higher levels than AFB_1-resistant species.

Not only do serum albumin adducts correlate with the amount of DNA adducts in the target organ but also with biological effects such as micronuclei and chromosomal aberrations in the bone marrow. Anwar et al. investigated this in rats and mice [133]. Animals were treated with single doses at different concentrations of AFB_1. The level of AFB_1 bound covalently to albumin in the peripheral blood was compared to the frequency of chromosomal aberrations and micronuclei in the bone marrow. Compared to the control group, both chromosomal aberrations and micronuclei were significantly increased in treated rats at doses above 0.32 μmol/kg. In contrast, in mice, a slight increase in chromosome aberrations was seen in the highest dose group (3.2 μmol/kg). However, no increase in micronuclei was observed at any of the doses. The level of chromosomal aberrations was about 10 times lower in mice than in rats at the highest dose of AFB_1. AFB_1-albumin adducts did not show a strong dose-response increase after treatment in mice. However, in rats the levels increased linearly with dose of AFB_1. There were strong correlations at the individual rat level with both chromosomal aberrations ($r = .92$; $P < .0001$) and micronucleus frequency ($r = .86$; $P < .0001$). These results indicate that the albumin adduct levels may reflect the level of genetic alteration, which arises from the initial binding of AFB_1 to cellular DNA. Thus, albumin adducts of AFB_1 can provide information not only on exposure but also on the risk of genetic alterations resulting from that exposure. In an initial analysis of 35 human samples from Gambia, no correlation was found between albumin adducts and various cytogenetic alterations (chromosomal aberrations, micronuclei, sister chromatid exchange) [134].

The mutations in tumor suppressor genes are an important step in the etiology of human cancer. The p53 tumor suppressor gene is the most common mutated gene in human cancer. The number and type of mutations in this gene occur in specific hot spots that vary with tumor type [135]. Two independent research groups [136,137] found a high frequency of G to T transversions at the third base of codon 249 in hepatocellular carcinomas from populations exposed to AFB_1. In countries with little aflatoxin exposure, no mutations at the third base of codon 249 were found in hepatocellular carcinomas [138]. AFB_1 is known to cause G to T transversions in bacteria. Furthermore, Puisieux et al. [139] showed that in vitro AFB_1-epoxide can bind to codon 249 of p53 in a plasmid. These experiments provide additional indirect evidence for a putative role of AFB_1 exposure in p53 mutagenesis.

In addition to the HBV carrier status, individual susceptibilities were investigated in a preliminary study by McGlynn et al. [140]. Genetic variation in two AFB_1 detoxification genes, epoxide hydrolase and GSTM1, was contrasted with the presence of serum AFB_1-albumin adducts, the presence of hepatocellular carcinoma (HCC), and with p53 codon 249 mutations. These results indicate that individuals with mutant genotypes for epoxide hydrolase and GSTM1 may be at greater risk of developing AFB_1 adducts, p53 mutations, and HCC when exposed to AFB_1. Hepatitis B carriers with the high-risk genotypes may be at an even greater risk than carriers with low-risk genotypes. These findings support the existence of genetic susceptibility in humans to the environmental carcinogen AFB_1 and indicate that there is a synergistic increase in risk of HCC with the combination of HBV infection and susceptible genotypes.

7. Summary

The achievements of the scientific community with the studies on aflatoxin represent a key model for the development of biomonitoring methods and strategies for the other toxic compounds. These findings will help to improve risk assessment based on presumptive intake data. Especially, the rapidly excised AFB_1-guanine adduct in tissues and the more persistent AFB_1-albumin adduct are appropriate dosimeters for estimating exposure status and risk in individuals consuming mycotoxins. Most progress has been made in developing and applying methods to quantify biomarkers of internal dose and of biological effective dose. Single urine metabolite determinations are less laborious than AFB_1-albumin adduct determinations. However, to avoid an overestimation of the AFB_1 burden, urine metabolites have to be determined over a few days. In conclusion, single determinations of albumin adducts are more practical and more significant than urine analysis to monitor exposure to AFB_1. Furthermore, blood samples can be used to determine individual enzyme genotypes and other disease parameters. These individual markers of exposure may also permit the identification of individual susceptibilities and of other human diseases, that could be associated with AFB_1 exposure—for example, the immunosuppressive properties of AFB_1 [141]. Preliminary studies were performed to compare these biomarkers with biological effects and disease. Such comparison should take into account that urine metabolites reflect only very recent exposure to AFB_1 and albumin adducts are relevant to exposure over the previous 2 or 3 months. In the near future, more efforts should be put in to identify markers with longer lifetime, since most diseases have a long lag time—for example, adducts to collagen in tissue. Collagens are a group of stable, structural proteins that are common in the extracellular matrix. The immobility and low turnover rate of the collagens suggest that they may be useful as tissue-specific proteins for adduct biomonitoring. Recently, Skipper et al. identified adducts of styrene and methyltetrahydrophthalic anhydride to collagen [142,143].

Despite the excitement induced from the present success and for the promising future of chemoprevention, one must not lose sight of the importance to provide the resources necessary to protect food from aflatoxin contamination. Therefore, improvements in food storage and HBV vaccination should be used to drastically reduce the levels of liver cancer.

REFERENCES

1. C Garner, CN Martin. Fungal toxins, aflatoxins, and nucleic acids. In: PL Grover, ed. Chemical Carcinogens and DNA. West Palm Beach; CRC Press, Vol I, 1979, p 187.
2. WF Busby, GN Wogan. Aflatoxins. In: CE Searle, ed. Chemical Carcinogens. 2nd ed. Washington; American Chemical Society, 1984, p 945.
3. GN Wogan. Aflatoxins as risk factors for primary hepatocellular carcinoma in humans. In: GA Bray, DH Ryan, eds. Mycotoxins, Cancer and Health. Baton Rouge; Louisiana State University Press, 1991, p 18.
4. S Yeh, MC Yu, C-C Mo, S Luo, MJ Tong, BJ Henderson. Hepatitis B virus, aflatoxins, and hepatocellular carcinoma in southern Guangxi, China. Cancer Res. 49:2506, 1989.
5. K Ross, J-M Yuan, MC Yu, et al. Urinary aflatoxin biomarkers and risk of hepatocellular carcinoma. Lancet 339:943, 1992.
6. GS Qian, RK Ross, MC Yu, et al. A follow-up study of urinary markers of aflatoxin exposure and liver cancer risk in Shanghai, People's Republic of China. Cancer Epidemiol Biomarkers Prev 3:3, 1994.
7. SJ van Rensburg, P Cook-Mozaffari, DJ van Schalkwyk, JJ van der Watt, TJ

Vincent, IF Purchase. Hepatocellular carcinoma and dietary aflatoxin in Mozambique and Transkei. Br J Cancer 51:713, 1985.
8. National Research Council Committee on Biological Markers. Biological markers in environmental health research. Environ Health Perspect 74:3, 1987.
9. Commission of the European Communities. Indicators for Assessing Exposure and Biological Effects of Genotoxic Chemicals. Brussels; ECSC-EEC-EAEC, 1988.
10. JM Essigmann, RG Croy, AM Nadzan, et al. Structural identification of the major DNA adduct formed by aflatoxin B_1 in vitro. Proc Natl Acad Sci USA 74:1870, 1977.
11. RC Garner, EC Miller, JA Miller. Liver microsomal metabolism of aflatoxin B_1 to a reactive derivative toxic to *Salmonella typhimurium* TA1530. Cancer Res 32:2058, 1972.
12. DH Swenson, EC Miller, JA Miller. Aflatoxin B_1-2,3-oxide. Evidence for its formation in rat liver in vivo and by human liver microsomes in vitro. Biochem Biophys Res Commun 60:1036, 1974.
13. GE Neal, PJ Colley. The formation of 2,3-dihydro-2,3-dihydroxy AFB_1 by the metabolism of AFB_1 in vitro by rat liver microsomes. FEBS Lett 101:382, 1979.
14. BF Coles, AM Welch, PJ Hertzog, JRL Smith, RC Garner. Biological and chemical studies on 8,9-dihydroxy-8,9-dihydro-aflatoxin B_1 and some of its esters. Carcinogenesis 1:79, 1980.
15. SW Baertschi, KD Raney, MP Stone, TM Harris. Preparation of the 8,9-epoxide of the mycotoxin aflatoxin B_1: the ultimate carcinogenic species. J Am Chem Soc 110: 7929, 1988.
16. DH Swenson, JA Miller, EC Miller. Reactivity and carcinogenicity of aflatoxin B_1-2,3-dichloride, a model for the putative 2,3-oxide metabolite of aflatoxin B_1. Cancer Res 35:3811, 1975.
17. CP Gorst, PS Steyn, PL Wessels. Oxidation of the bisdihydrofuran moieties of aflatoxin B_1 and sterigmatocystgin; conformation of tetrahydrofurobenzofurans. J Chem Soc Perkin I:1360, 1977.
18. ML Wood, JRL Smith, RC Garner. Structural characterization of the major adducts obtained after reaction of an ultimate carcinogen aflatoxin B_1-dichloride with calf thymus DNA in vitro. Cancer Res 48:5391, 1988.
19. JM Essigmann, CL Green, RG Croy, KW Fowler, GH Büchi, GN Wogan. Interaction of aflatoxin B_1 and alkylating agents with DNA: structural and functional studies, Cold Spring Harbor Symp Quant Biol 47:327, 1983.
20. RG Croy, JM Essigmann, VN Reinhold, GN Wogan. Identification of the principal aflatoxin B_1-DNA adduct formed in vivo in rat liver. Proc Natl Acad Sci USA 75: 1745, 1978.
21. RG Croy, GN Wogan. Temporal patterns of covalent DNA adducts in rat liver after single and multiple doses of aflatoxin B_1. Cancer Res 41:197, 1981.
22. RA Bennett, JM Essigmann, GN Wogan. Excretion of an aflatoxin-guanine adduct in the urine of AFB_1-treated rats. Cancer Res 41:650, 1981.
23. PJ Hertzog, JRL Smith, RC Garner. A high pressure liquid chromatography study on the removal of DNA-bound aflatoxin B_1 in rat liver and in vitro. Carcinogenesis 1:787, 1980.

24. HG Raj, PD Lotlikar. Urinary excretion of thiol conjugates of AFB_1 in rats and hamsters. Cancer Lett 22:125, 1984.
25. HG Raj, K Santhanam, RP Gupta, TA Venkitasubramanian. Oxidative metabolism of aflatoxin B_1. Formation of epoxide-glutathione conjugate. Chem-Biol Interact 11:301, 1975.
26. GH Degen, H-G Neumann. The major metabolite of AFB_1 in the rat is a glutathione conjugate. Chem-Biol Interact 22:239, 1978.
27. EJ Moss, DJ Judah, M Przyblyski, GE Neal. Some mass spectral and NMR analytical studies of a glutathione conjugate of aflatoxin B_1. Biochem J 210:227, 1983.
28. DL Eaton, HS Ramsdell, GE Neal. Biotransformation of aflatoxins. In: DL Eaton, JD Groopman eds. The Toxicology of Aflatoxins. San Diego; Academic, 1994, p 45.
29. EJ Moss, GE Neal. The metabolism of AFB_1 by human liver. Biochem Pharmacol 34:3193, 1985.
30. KD Raney, B Coles, FP Guengerich, TM Harris. The *endo*-8,9-epoxide of aflatoxin B_1: a new metabolite. Chem Res Toxicol 5:333, 1992.
31. KD Raney, T Shimada, DH Kim, JD Groopman, TM Harris, FP Guengerich. Oxidation of aflatoxins and sterigmatocystin by human liver microsomes: significance of aflatoxin Q_1 as a detoxification product of aflatoxin B_1. Chem Res Toxicol 5:202, 1992.
32. RS Iyer, BF Coles, KD Raney, R Thier, FP Guengerich, TM Harris. DNA adduction by the potent carcinogen aflatoxin B_1: mechanistic studies. J Am Chem Soc 116:1603, 1994.
33. TM Harris, S Gopalakrishnan, SW Baertschi, KD Raney, S Byrd, MP Stone. Synthesis and characterization of aflatoxin B_1-epoxide and its adducts with oligodeoxynucleotides. In: GA Bray, DH Ryan, eds. Mycotoxins, Cancer and Health. Baton Rouge; Louisiana State University Press, 1991, p 142.
34. KD Raney, DJ Meyer, B Ketterer, TM Harris, FP Guengerich. Glutathione conjugation of aflatoxin B_1 *exo*- and *endo*-epoxides by rat and human glutathione S-transferases. Chem Res Toxicol 5:470, 1992.
35. YH Liu, J Taylor, P Linko, GW Lucier, CL Thompson. Glutathione S-transferase μ in human lymphocyte and liver: role in modulating formation of carcinogen-derived DNA adducts. Carcinogenesis 12:2269, 1991.
36. S Tsuchida, K Sato. Glutathione transferases and cancer. Crit Rev Biochem Molec Biol 27:337, 1992.
37. GM Kirby, CR Wolf, GE Neal, et al. In vitro metabolism of AFB_1 by normal and tumorous liver tissue from Thailand. Carcinogenesis 14:2613, 1993.
38. JT Heinonen, R Fisher, K Brendel, DL Eaton. Determination of aflatoxin B_1 biotransformation and binding to hepatic macromolecules in human precision liver slices. Toxicol Appl Pharmacol 136:1, 1996.
39. S Langouet, B Coles, F Morel, et al. Inhibition of CYP1A2 and CYP3A4 by Oltipraz results in reduction of aflatoxin B_1 metabolism in human hepatocytes in primary culture. Cancer Res 55:5574, 1995.
40. J Bujons, DPH Hsieh, NY Kado, A Messeguer. AFM_1 8,9-epoxide: preparation and mutagenic activity. Chem Res Toxicol 8:328, 1995.

41. TC Campbell, JP Caedo Jr, J Bulatao-Jayme, L Salamat, RW Engel. AFM_1 in human urine. Nature 227:403, 1970.
42. CEA Lovelace, H Njapau, LF Salter, AC Bayley. Screening method for the detection of aflatoxin and metabolites in human urine: aflatoxins B_1, G_1, M_1, B2a, G2a, aflatoxicols I and II. J Chromatogr 227:256, 1982.
43. MW Trucksess, GE Wood. Recent methods of analysis for aflatoxins in foods and feeds. In: DL Eaton, JD Groopman eds. The Toxicology of Aflatoxins. San Diego; Academic, 1994, p 451.
44. FS Chu. Development of antibodies against aflatoxins. In: DL Eaton, JD Groopman eds. The Toxicology of Aflatoxins. San Diego; Academic, 1994, p 451.
45. A Haugen, JD Groopman, I-C Hsu, GR Goodrich, GN Wogan, CC Harris. Monoclonal antibody to aflatoxin B_1-modified DNA detected by enzyme immunoassay. Proc Natl Acad Sci USA 78:4124, 1981.
46. P Sizaret, C Malaveille, R Montesano, C Frayssinet. Detection of aflatoxins and related metabolites by radioimmunoassay. J Natl Cancer Inst 69:1375, 1982.
47. RC Garner, R Ryder, R Montesano. Monitoring of aflatoxins in human body fluids and application to field studies. Cancer Res 45:922, 1985.
48. RC Garner, I Dvorackova, F Tursi. Immunoassay procedures to detect exposure to AFB_1 and benzo[a]pyrene in animals and man at the DNA level. Int Arch Occup Environ Health 60:145, 1988.
49. JD Groopman, LJ Trudel, PR Donahue, A Marshak-Rothstein, GN Wogan. High-affinity monoclonal antibodies for aflatoxins and their application to solid-phase immunoassays. Proc Natl Acad Sci USA 81:7728, 1984.
50. JD Groopman, JA Hasler, LJ Trudel, A Pikul, PR Donahue, GN Wogan. Molecular dosimetry in rat urine of aflatoxin-N7-guanine and other aflatoxin metabolites by multiple monoclonal antibody affinity chromatography and immunoaffinity/high performance liquid chromatography. Cancer Res 52:267, 1992.
51. S Wu, Z Sun, Y Wu, Y Wei, J Gu, Z Lu. Urinary excretion of aflatoxin M_1 (AFM_1) in Beijing and Qidong [China] inhabitants. Zhonghua Zhongliu Zazhi 6:163, 1984.
52. T Sun, S Wu, Y Wu, Y Chu. Measurement of individual aflatoxin exposure among people having different risk to primary hepatocellular carcinoma. Proc Int Symp Princess Takamatsu Cancer Res Fund. Volume Date 1985, 16th Diet, Nutr, Cancer 1986, p 225.
53. JQ Zhu, LS Zhang, X Hu, et al. Correlation of dietary AFB_1 levels with excretion of AFM_1 in human urine. Cancer Res 47:1848, 1987.
54. Z Liu, W Tu, D Li, et al. A new method for the quantitation of AFM_1 in urine by high performance liquid chromatography and its application to the etiologic study of hepatoma. Biomed Chromatogr 4:83, 1990.
55. CB Nyathi, CF Mutiro, JA Hasler, CJ Chetsanga. A survey of urinary aflatoxin in Zimbabwe. Int J Epidemiol 16:516, 1987.
56. TA Bean, DM Yourtee, B Akande, J Ogunlewe. Aflatoxin metabolites in the urine of Nigerians: comparison of chromatographic methods. J Toxicol Toxin Rev 8:43, 1989.
57. LO Dragsted, I Bull, H Autrup. Substances with affinity to a monoclonal AFB_1 antibody in Danish urine samples. Food Chem Toxicol 26:233, 1988.

58. JD Groopman, KF Donahue. Aflatoxin a human carcinogen: determination in foods and biological samples by monoclonal antibody affinity chromatography. J Assoc Off Anal Chem 71:861, 1988.
59. M Holcomb, HC Thompson Jr. Analysis of aflatoxins (B_1, B2, G_1, and G2) in rodent feed by HPLC using postcolumn derivatization and fluorescence detection. J Agric Food Chem 39:137, 1991.
60. A Kussak, B Andersson, K Andersson. Automated sample clean-up with solid-phase extraction for the determination of aflatoxins in urine by liquid chromatography. J Chromatogr Biomed Appl 616:235, 1993.
61. A Kussak, B Andersson, K Andersson. Determination of AFQ_1 in urine by automated immunoaffinity column clean-up and liquid chromatography. J Chromatogr B: Biomed Appl 656:329, 1994.
62. A Kussak, C-A Nilsson, B Andersson, J Langridge. Determination of aflatoxins in dust and urine by liquid chromatography/electrospray-ionization tandem mass spectrometry. Rapid Commun Mass Spectrom 9:1234, 1995.
63. P Degan, G Montagnoli, CP Wild. Time-resolved fluoroimmunoassay of aflatoxins. Clin Chem 35:2308, 1989.
64. JB Coulter, SM Lamplugh, GI Suliman, MI Omer, RG Hendrickse. Aflatoxins in human breast milk. Ann Trop Paediatr 4:61, 1984.
65. SM Lamplugh, RG Hendrickse, F Apeagyei, DD Mwanmut. Aflatoxins in breast milk, neonatal cord blood, and sera of pregnant women. Br Med J 296:968, 1988.
66. SM Maxwell, F Apeagyei, DD Mwanmut, RG Hendrickse. Aflatoxins in breast milk, neonatal cord blood, and sera of pregnant women. J Toxicol Toxin Rev 8:19, 1989.
67. CP Wild, FA Pionneau, R Montesano, CF Mutiro, CJ Chetsanga. Aflatoxin detected in human breast milk by immunoassay. Int J Cancer 40:328, 1987.
68. A Zarba, CP Wild, AJ Hall, R Montesano, GJ Hudson, JD Groopman. Aflatoxin M_1 in human breast milk from the Gambia, West Africa, quantified by combined monoclonal antibody immunoaffinity chromatography and HPLC. Carcinogenesis 13:891, 1992.
69. HS El-Nezami, G Nicoletti, GE Neal, DC Donohue, JT Ahokas. Aflatoxin M_1 in human breast milk samples from Victoria, Australia and Thailand. Food Chem Toxicol 33:173, 1995.
70. JD Groopman, LG Cain, TW Kensler. Aflatoxin exposure in human populations: measurements and relationship to cancer. CRC Crit Rev Toxicol 19:113, 1988.
71. IR McConnell, RC Garner. DNA adducts of aflatoxins, sterigmatocystin and other mycotoxins. IARC Sci Publ 125 DNA Adducts: Identification and Biological Significance 49, 1994.
72. CN Martin, RC Garner, F Tursi, et al. An enzyme-linked immunosorbent procedure for assaying AFB_1. IARC Sci Publ 59 Monit Hum Exposure Carcinog Mutagen Agents 313, 1984.
73. JD Groopman, PR Donahue, J Zhu, J Chen, GN Wogan. Aflatoxin metabolism in humans: detection of metabolites and nucleic acid adducts in urine by affinity chromatography. Proc Natl Acad Sci USA 82:6492, 1985.
74. H Autrup, KA Bradley, AKM Shamsuddin, J Wakhisi, A Wasunna. Detection of putative adduct with fluorescence characteristics identical to 2,3-dihydro-2-(7′-

guanyl)-3-hydroxy-AFB_1 in human urine collected in Murang'a district, Kenya. Carcinogenesis 4:1193, 1983.

75. H Autrup, J Wakhisi, K Vahakangas, A Wasunna, CC Harris. Detection of 8,9-dihydro-7′-guanyl-9-hydroxy-AFB_1 in human urine. Environ Health Perspect 62: 105, 1985.
76. H Autrup, T Seremet, J Wakhisi, A Wasunna. Aflatoxin exposure measured by urinary excretion of AFB_1-guanine adduct and hepatitis B virus infection in areas with different liver cancer incidence in Kenya. Cancer Res 47:3430, 1987.
77. JD Groopman, AJ Hall, H Whittle, et al. Molecular dosimetry of aflatoxin -N7-guanine in human urine obtained in the Gambia, West Africa. Cancer Epidemiol Biomarkers Prev 1:221, 1992.
78. JD Groopman, J Zhu, PR Donahue, et al. Molecular dosimetry of urinary aflatoxin-DNA adducts in people living in Guangxi Autonomous Region, People's Republic of China. Cancer Res 52:45, 1992.
79. VM Weaver, JD Groopman. Fluorescence quantification of aflatoxin N7-guanine adducts. Cancer Epidemiol Biomarkers Prev 3:669, 1994.
80. CC Harris, G LaVeck, JD Groopman, VL Wilson, D Mann. Measurement of aflatoxin B_1, its metabolites, and DNA adducts by synchronous fluorescence spectrophotometry. Cancer Res 46:3249, 1986.
81. JC Harrison, RC Garner. Immunological and HPLC detection of aflatoxin adducts in human tissues after an acute poisoning incident in SE Asia. Carcinogenesis 12:741, 1991.
82. JC Harrison, M Carvajal, RC Garner. Does aflatoxin exposure in the United Kingdom constitute a cancer risk? Environ Health Perspect 99:99, 1993.
83. JC Harrison, M Carvajal, NJ Froggatt, SH Leveson, RC Garner. AFB_1-DNA adducts present in human colorectal cancer. Recent Adv Toxicol Res [World Conf Anim Plant Microb Toxin], 10th, Volume 3, (P. Gopalakrishnakone, CK Tan, eds.) Singapore; Natl Univ Singapore, Venom Toxin Res Group, 1992, p 242.
84. RC Garner. Monitoring aflatoxin exposure at a macromolecular level in man with immunological methods. In: S Natori, K Hashimoto, Y Ueno, eds. Mycotoxins Phycotoxins '88. Amsterdam; Elsevier Science, 1989, p 29.
85. WK Lutz. Dose-reponse relationships and low dose extrapolation in chemical carcinogenesis. Carcinogenesis 11:1243, 1990.
86. LL Hsieh, S-W Hsu, D-S Chen, RM Santella. Immunological detection of aflatoxin B_1-DNA adducts formed in vivo. Cancer Res 48:6328, 1988.
87. P Scoppa, E Marafante. Interaction of AFB_1 with albumin spectrophotometric study using difference spectra. Boll Soc Ital Biol Sper 47:198, 1971.
88. HW Dirr, JC Schabort. Aflatoxin B_1 transport in rat blood plasma. Binding to albumin in vivo and in vitro and spectrofluorometric studies into the nature of the interaction. Biochim Biophys Acta 881:383, 1986.
89. O Bassir, EA Bababunmi. Binding of AFB_1 with serum albumin. Biochem Pharmacol 22:132, 1973.
90. F-J Lu, K-H Ling. Binding of AFB_1 to cellular components. 1. Binding of AFB_1 to albumin in vitro. Tai-wan I Hsueh Hui Tsa Chih 72:434, 1973.
91. ZA Wong, DPH Hsieh. Aflatoxicol: major AFB_1 metabolite in rat plasma. Science 200:325, 1978.

92. JI Dalezios. Aflatoxin P_1: A new metabolite of aflatoxin B_1. Its isolation and identification. PhD thesis, Massachusetts Institute of Technology, Cambridge, 1971.
93. JI Dalezios, GN Wogan. Metabolism of aflatoxin B_1 in rhesus monkeys. Cancer Res 32:2297, 1972.
94. AY Nassar, SE Megalla, HM Abd El-Fattah, AH Hafez, TS El-Deap. Binding of AFB_1, G_1 and M_1 to plasma albumin. Mycopathologia 79:35, 1982.
95. BS Appleton, MP Goetchius, TC Campbell. Linear dose-response curve for the hepatic macromolecular binding of aflatoxin B_1 in rats at very low exposures. Cancer Res 42:3659, 1982.
96. PL Skipper, DH Hutchins, RJ Turesky, G Sabbioni, SR Tannenbaum. Carcinogen binding to serum albumin. Proc Am Assoc Res 26:356, 1985.
97. G Sabbioni, PL Skipper, SR Tannenbaum. Aflatoxin B_1 binding to serum albumin. Proc Am Assoc Res 27:339, 1986.
98. G Sabbioni, PL Skipper, G Büchi, SR Tannenbaum. Isolation and characterization of the major serum albumin adduct formed by AFB_1 in vivo in rats. Carcinogenesis 8: 819, 1987.
99. CP Wild, RC Garner, R Montesano, F Tursi. AFB_1 binding to plasma albumin and liver DNA upon chronic administration to rats. Carcinogenesis 7:853, 1986.
100. G Sabbioni. Chemical and physical properties of the major serum albumin adduct of AFB_1 and their implications for the quantification in biological samples. Chem-Biol Interact 75:1, 1990.
101. G Sabbioni, S Ambs, GN Wogan, JD Groopman. The aflatoxin-lysine adduct quantified by high-performance liquid chromatography from human serum albumin samples. Carcinogenesis 11:2063, 1990.
102. ZSC Okoye, J Riley, DJ Judah, GE Neal. The in vivo site of formation of a carcinogen-serum albumin adduct. Biochem Pharmacol 40:2560, 1990.
103. IO Olubuyide, K Makarananda, DJ Judah, GE Neal. Investigation of the assay of AFB_1-albumin adducts using proteolysis products in ELISA. Int J Cancer 48:468, 1991.
104. B Chapot, CP Wild. ELISA for quantification of aflatoxin-albumin adducts and their application to human exposure assessment. Tech Diagn Pathol 2:135, 1991.
105. CP Wild, Y Jiang, G Sabbioni, B Chapot, R Montesano. Evaluation of methods for quantitation of aflatoxin-albumin adducts and their application to human exposure assessment. Cancer Res 50:245, 1990.
106. JD Groopman, G Sabbioni. Detection of aflatoxin and its metabolites in human biological fluids. In: GA Bray, DH Ryan, eds. Mycotoxins, Cancer and Health. Baton Rouge; Louisiana State University Press, 1991, p 18.
107. JD Groopman, G Sabbioni, CP Wild. Molecular dosimetry of human aflatoxin exposures. In: PL Skipper, JD Groopman eds. Molecular Dosimetry of Human Cancer; Epidemiological Analytical and Social Considerations. Boca Raton; CRC Press, 1991, p 263.
108. CP Wild, GJ Hudson, G Sabbioni, et al. Dietary intake of aflatoxins and the level of albumin-bound aflatoxin in peripheral blood in the Gambia, West Africa. Cancer Epidemiol Biomarkers Prev 1:229, 1992.
109. G Sabbioni, CP Wild. Identification of an AFG_1-serum albumin adduct and its relevance to the measurement of human exposure to aflatoxins. Carcinogenesis 12: 97, 1991.

110. FZ Sheabar, JD Groopman, GS Qian, GN Wogan. Quantitative analysis of aflatoxin-albumin adducts. Carcinogenesis 14:1203, 1993.
111. LS Gan, PL Skipper, X Peng, et al. Serum albumin adducts in the molecular epidemiology of aflatoxin carcinogenesis: correlation with AFB_1 intake and urinary excretion of AFM_1. Carcinogenesis 9:1323, 1988.
112. JD Groopman, CP Wild, J Hasler, C Junshi, GN Wogan, TW Kensler. Molecular epidemiology of aflatoxin exposures: validation of aflatoxin-N7-guanine levels in urine as a biomarker, in experimental rat models and humans. Environ Health Perspect 99:107, 1993.
113. JD Groopman. Molecular dosimetry methods for assessing human aflatoxin exposures. In: DL Eaton, JD Groopman eds. The Toxicology of Aflatoxins. San Diego; Academic, 1994, p 259.
114. LS Gan, MS Otteson, MM Doxtader, PL Skipper, RR Dasari, SR Tannenbaum. Quantitation of carcinogen bound protein adducts by fluorescence measurements. Spectrochim Acta Part A 45A:81, 1989.
115. CP Wild, YZ Jiang, SJ Allen, LAM Jansen, AJ Hall, R Montesano. Aflatoxin-albumin adducts in human sera from different regions of the world. Carcinogenesis 11:2271, 1990.
116. CP Wild, FN Rasheed, MFB Jawla, AJ Hall, LAM Jansen, R Montesano. In-utero exposure to aflatoxin in West Africa. Lancet 337:1602, 1991.
117. CP Wild, SM Shrestha, WA Anwar, R Montesano. Field studies of aflatoxin exposure, metabolism and induction of genetic alterations in relation to HBV infection and hepatocellular carcinoma in the Gambia and Thailand. Toxicol Lett 64–65:455, 1992.
118. JL Autrup, J Schmidt, T Seremet, H Autrup. Determination of exposure of aflatoxins among Danish workers in animal-feed production through the analysis of AFB_1 adducts to serum albumin. Scand J Work Environ Health 17:436, 1991.
119. JL Autrup, J Schmidt, H. Autrup. Exposure to AFB_1 in animal-feed production plant workers. Environ Health Perspect 99:195, 1993.
120. BD Roebuck, Y-L Liu, AE Rogers, JD Groopman, TW Kensler. Protection against aflatoxin B_1-induced hepatocarcinogenesis in F344 rats by 5-(2-pyrazinyl)-4-methyl-1,2-dithiol-3-thion (Oltipraz): predictive role for short term molecular dosimetry. Cancer Res 51:5501, 1991.
121. TW Kensler, EF Davies, MG Bolton. Strategies for chemoprotection against aflatoxin-induced liver cancer. In: DL Eaton, JD Groopman, eds. The Toxicology of Aflatoxin. San Diego, Academic, 1994, p 281.
122. PA Egner, SJ Gange, PM Dolan, JD Groopman, A Munoz, TW Kensler. Levels of aflatoxin-albumin biomarkers in rat plasma are modulated by both long-term and transient interventions with Oltipraz. Carcinogenesis 16:1769, 1995.
123. JD Groopman, P DeMatos, PA Egner, A Love-Hunt, TW Kensler. Molecular dosimetry of urinary aflatoxin-N7-guanine and serum aflatoxin-albumin adducts predicts chemoprotection by 1,2-dithiole-3-thione in rats. Carcinogenesis 13:101, 1992.
124. MG Bolton, N Munoz, LP Jacobson, JD Groopman, TW Kensler. Transient intervention with Oltipraz against aflatoxin induced tumorogenesis. Cancer Res 53:3499, 1993.

125. CP Wild, LAM Jansen, L Cova, R Montesano. Molecular dosimetry of aflatoxin exposure: contribution to understanding the multifactorial etiopathogenesis of primary hepatocellular carcinoma with particular reference to hepatitis B virus. Environ Health Perspect 99:99, 1993.
126. TC Campbell, J Chen, C Liu, J Li, B Parpias. Nonassociation of aflatoxin with primary liver cancer in a cross-sectional ecological survey in the People's Republic of China. Cancer Res 50:6882, 1990.
127. MS Diallo, A Sylla, K Sidibe, BS Sylla, CR Trepo, CP Wild. Prevalence of exposure to aflatoxin and hepatitis B and C viruses in Guinea, West Africa. Nat Toxins 3:6, 1995.
128. P Bannasch, N Imani Khoshkhou, HJ Hacker, et al. Synergistic hepatocarcinogenic effect of hepadneviral infection and dietary aflatoxin B_1 in woodchucks. Cancer Res 55:3318, 1995.
129. GM Kirby, I Chemin, R Montesano, FV Chisari, MA Lang, CP Wild. Induction of specific cytochrome P450s involved in aflatoxin B_1 metabolism in hepatitis B virus transgenic mice. Mol Carcinogen 11:74, 1994.
130. GM Kirby, P Pelkonen, V Vatanasapt, A-M Camus, CP Wild, MA Lang. Association of liver fluke (*Opisthorchis viverrini*) infestation with increased expression of cytochrome P450 and carcinogen metabolism in male hamster liver. Mol Carcinogen 11:81, 1994.
131. K Makarananda, CP Wild, JZ Jiang, GE Neal. IARC Sci Publ 105 Nr Relevance Hum Cancer N-Nitroso Compd, Tob Mycotoxins 96, 1991.
132. CP Wild, R Hasegawa, L Barraud, et al. Aflatoxin-albumin adducts: a basis for comparative carcinogenesis between animals and humans. Cancer Epidemiol Biomarkers Prev 5:179, 1996.
133. WA Anwar, MM Khalil, CP Wild. Micronuclei, chromosomal aberrations and aflatoxin-albumin adducts in experimental animals after exposure to AFB_1. Mutat Res 322:61, 1994.
134. M Miele, F Donato, AJ Hall, et al. Aflatoxin exposure and cytogenetic alterations in individuals from the Gambia, West Africa. Mutat Res 349:209, 1996.
135. CC Harris. The 1995 Walter Hubert Lecture—Molecular epidemiology of human cancer: insights from the mutational analysis of the p53 tumour-suppressor gene. Br J Cancer 73:261, 1996.
136. B Bressac, M Kew, J Wands, M Ozturk. Selective G to T mutations of p53 gene in hepatocellular carcinoma from southern Africa. Nature 350:429, 1991.
137. IC Hsu, RA Metcalf, T Sun, JA Welsh, NJ Wang, CC Harris. Mutational hotspot in the p53 gene in human hepatocellular carcinomas. Nature 350:427, 1991.
138. M Ozturk. p53 Mutation in hepatocellular carcinoma after aflatoxin exposure. Lancet 338:1356, 1991.
139. A Puisieux, S Lim, J Groopman, M Ozturk. Selective targeting of p53 gene mutational hotspots in human cancers by etiologically defined carcinogens. Cancer Res 51:6185, 1991.
140. KA McGlynn, EA Rosvold, ED Lustbader, et al. Susceptibility to hepatocellular carcinoma is associated with genetic variation in the enzymic detoxification of aflatoxin B_1. Proc Natl Acad Sci USA 92:2384, 1995.

141. GJ Jakab, PR Hmieleski, A Zarba, DR Hemenway, JD Groopman. Respiratory aflatoxicosis: suppression of pulmonary and systemic host defenses in rats and mice. Toxicol Appl Pharmacol 125:198, 1994.
142. L Skipper, X-C Peng, CK Soohoo, SR Tannenbaum. Protein adducts as biomarkers of human carcinogen exposure. Drug Metab Rev 26:111, 1994.
143. AG Jönsson, JS Wishnok, PL Skipper, WG Stillwell, SR Tannenbaum. Lysine adducts between methyltetrahydrophthalic anhydride and collagen in guinea pig lung. Toxicol Appl Pharmacol 135:156, 1995.
144. AJ Hall, CP Wild. Aflatoxin biomarkers [letter; comment]. Lancet 339:1413, 1992.

7
Mechanistic Interactions of Mycotoxins: Theoretical Considerations

Ronald T. Riley
United States Department of Agriculture, Agricultural Research Service, Athens, Georgia

I. INTRODUCTION

Mechanisms of mycotoxicity were recently reviewed by Riley and Norred [1]. This review covered many of the mycotoxins that commonly occur on foods and feeds. In addition, it described many of the pitfalls commonly encountered when attempting to interpret the results of in vitro studies and extrapolate those results to in vivo studies and the natural diseases associated with consumption of mycotoxins. Briefly, the factors that tend to confound in vitro mechanistic studies include: (1) failure to differentiate secondary effects from the primary biochemical lesion; (2) failure to relate the effective intracellular concentration in vitro to the tissue concentration of toxin which causes disease in vivo; (3) choice of an in vitro model which is either deficient in the biochemical target or unresponsive due to other inadequacies of the model system; and (4) failure to adequately model the complexity of the in vivo exposure with regard to potential interactions with other toxins, drugs, environmental, and/or nutritional factors. This chapter will attempt to expand on the earlier review and provide additional insights into the potential mechanistic interactions of mycotoxins.

The study of mechanism of action of toxic compounds is founded in the belief that cells are molecular machines. To understand how mycotoxins interfere with the cellular machinery, it is only necessary to understand how toxic compounds alter the behavior of the molecules of life. Examples of mycotoxins

Table 1 Probable Primary Biochemical Lesions and the Early Cellular Events in the Cascade of Cellular Events Leading to Toxic Cell Injury or Cellular Deregulation of Selected Mycotoxins[a]

Mycotoxin	Initial lesion → cascade of events	References
Aflatoxin	Metabolic activation → DNA modification → cell deregulation → cell death/transformation (metabolic activation → disruption of macromolecular synthesis → cell deregulation → cell death (apoptotic))	2, 3
Adenophostins	ER IP_3 receptor → Ca^{2+} release → cell deregulation → ?	4
Anthraquinones	Mitochondrial uncoupler → loss of respiratory control → cell death (apoptotic?)	5
Beauvericin	K^+ ionophore → K^+ loss → cell deregulation → cell death/apoptosis (inhibition of cholesterol acytransferase → disruption of cholesterol metabolism → ?)	6, 7
Citrinin	Loss of selective membrane permeability → cell disruption → cell death (apoptotic?) (disruption of macromolecular synthesis → ?)	8
Cyclopiazonic	ER & SR Ca^{2+}-ATPase → disruption of Ca^{2+} homeostasis → cell deregulation → cell death	9
Cytochalasins	Cytoskeleton → disruption of endocytosis → cell deregulation → cell death	10
Deoxynivalenol	Inhibition of protein synthesis → disruption of cytokine regulation → altered cell proliferation → cell death/apoptosis?	11
Fumonisins	Sphinganine N-acyltransferase → disrupted lipid metabolism → cell deregulation → cell death/apoptosis (disrupted delta-6-desaturase activity → disrupted fatty acid and arachidonic acid metabolism → cell death)	12–14
Gliotoxin	Calcium homeostasis → zinc homeostasis → endonuclease activation → apoptosis (radical mediated damage → oxidative stress → cell death) (inhibition of protein synthesis → apoptosis?)	15
Griseofulvin	Cytoskeleton → deregulation of cytoskeletal control → cell death	16
Luteoskyrin	Radical mediated damage → oxidative stess → lipid peroxidation → cell death (apoptotic?)	17

Moniliformin	Pyruvate and α-ketoglutarate decarboxylation → loss of respiratory control → cell death	18
Ochratoxin	Disruption of phenylalanine metabolism → reduced PEPCK → reduced glyconeogenesis → cell death (metabolic activation → inhibition of protein/DNA synthesis → apoptosis?) (altered membrane permeability → disrupt calcium homeostasis → cell deregulation → cell death)	19–21
Patulin	Nonprotein sulfhydryl depletion → altered ion permeability and/or altered intercellular communication → oxidative stress → cell death (inhibition of macromolecular biosynthesis → cell death)	22–24
Sphingofungin or ISP-1	Serine palmitoyltransferase → decreased sphingolipids → cell deregulation → cell death/apoptosis	25, 26
Sporidesmin	Radical mediated damage → oxidative stress → thiol depletion → cell death (disrupt calcium homeostasis → zinc homeostasis → endonuclease activation → apoptosis)	27
Swainsonine	Manosidase II → disrupted glycoprotein processing → cell deregulation → cell death	28
Tremorgens	GABA receptors → altered Cl^+ permeability → disrupted neuromuscular control → ?	29
T-2 toxin	Inhibition of protein synthesis → ? → cell death (apoptotic?) (transient Ca^{2+} elevation → endonuclease activation → apoptosis)	30–32
Wortmannin	Inhibition of phosphatidylinositol 3-kinase → responsiveness to insulin/growth factors/apoptosis (inhibition of myosin light chain kinase → inhibition of IP3 signaling pathway → ?) (inhibition of phospholiapse A2 → ?)	33–35
Zearalenone	Cytosolic estrogen receptor → estrogenic response → disruption of hormonal control → ?)	36

[a]Additional references in Riley and Norred [1]. Primary cellular biochemical lesion = the molecular interaction between the parent toxin and target biomolecules which occur within the cell and from which the cascade of events (secondary effects) leading to celluar deregulation and/or toxic cell injury is initiated. It should be recognized that there is often more than one mechanism of action with different dose dependencies and time courses and that these mechanisms may run parallel paths both leading to toxic cell injury (Fig. 1). Where there is more than one proposed mechanism of action, the alternative mechanism is presented in parentheses. Question marks indicate uncertainty, or events not documented in vivo, or repeated in more than one study or model system. From Ref. 1.

interacting directly with the molecules of life or preventing their biosynthesis are easily found in the literature (Table 1) [1–36].

The pursuit of mechanism of action is an important area of mycotoxin research for several reasons, for example: (1) to reveal the initial biochemical lesion leading to the onset and progression of the natural diseases; (2) to be able to differentiate between those biological effects that can only occur at high dosages from those more likely to occur at environmentally relevant dosages; (3) to be able to predict downstream biochemical effects which will develop as a consequence of the initial biochemical lesion; (4) to be able to predict potential chronic toxicity and the potential for interaction with other mycotoxins or bioactive agents (toxins or drugs); and (5) to understand the mechanism of action in animal cells and provide insight into strategies for reducing the phytotoxicity of mycotoxins that are virulence factors. Finally, mechanism of action studies can provide insight into why fungi produce chemicals that appear to functionally mimic bioactive animal and plant molecules [37]. An understanding of the mechanism of action can provide the basis for testable hypotheses concerning the role of mycotoxins in pathogenesis [13,38].

II. DIVERSITY OF MECHANISMS AND INTERACTIONS

Unfortunately, mycotoxins as a group cannot be classified based upon their mechanism of action. This is not surprising when one considers the diversity of chemical structures that are encompassed by the group of secondary fungal metabolites called mycotoxins. Secondary fungal metabolites are metabolites that are not essential for growth of the organism but are biosynthesized from primary metabolites and secreted. For the purposes of this review, mycotoxins are defined as secondary metabolites which interact with cells and reduce their ability to survive, grow, or otherwise carry out their normal function. In order to grasp the potential diversity of mechanisms of action of mycotoxins it is necessary to understand the potential number of mycotoxins in nature.

The discovery of new mycotoxins is occurring at a high rate, and while there are few human diseases definitively caused by mycotoxins [39], there is considerable evidence to support the association of mycotoxins as agents of animal disease. The birth of modern mycotoxicology began in 1960 with the report by Blount [40] of a new turkey disease (turkey "X" disease) problem in England characterized by heavy mortality. The cause of turkey X disease [41] was attributed to aflatoxin, which was soon thereafter found to be carcinogenic to rats [42]. Since the discovery of aflatoxins as agents of animal disease, many new mycotoxins have been discovered and implicated in clinical cases of farm animal disease. Most recently, the fumonisins have been identified as the cause of many cases of equine and porcine mortality in the United States and elsewhere [43]. There is little doubt that mycotoxins kill farm animals; however, perhaps of greater importance is the fact that consumption of mold-contaminated feeds by farm animals has been associated with clinical cases involving reproductive effects, feed refusal, emesis, decline in productivity, lameness, bruising, decreased resistance to infectious agents, and other nonlethal effects [44]. Thus, mycotoxins are potentially contributing factors in chronic disease and may confound therapeutic and/or other interventions.

The precise number of secondary metabolites of mycotoxins is uncertain. Knowing the actual occurrence and distribution of toxins in foods and feeds is important because exposure to unknown bioactive agents could be an important confounding factor in the attempt to explain the etiology of chronic disease in animals and man. It is possible to roughly estimate the number of fungal metabolites and potential mycotoxins. In 1971 Turner [45] cataloged approximately 1200 secondary fungal metabolites produced by approximately 500 species of fungi. In 1983 Turner and Alderidge [46] cataloged 2000 more metabolites produced by approximately 1100 species. Thus, as an estimate, there were approximately two unique secondary metabolites/fungal species. In 1991, Hawksworth [47] estimated that there were 69,000 known fungal species which represented approximately 5% of the total species in the world—estimated at 1,500,000; other conservative estimates are on the order of 100,000 total species [48]. Based on the work of Hawksworth [47] and the assumption of two unique secondary metabolites/fungal species, there may be as many as 3 million unique secondary fungal metabolites; however, a conservative estimate would be 200,000. Between 1971 and 1983 the total number of known secondary fungal metabolites increased from 1200 [45] to 3200 [46]. Assuming that the rate of discovery (accessible in the published literature) remained at a similar rate, there would be approximately 5400 described fungal secondary metabolites by 1996. This is $< 0.2\%$ of the 3 million postulated metabolites, or only 3% of the 200,000 postulated by the more conservative estimate.

No matter which estimate one is comfortable with, the number of secondary metabolites that are undiscovered is quite large. Cole and Cox [49] listed approximately 300 secondary fungal metabolites as mycotoxins. Approximately 10% of the secondary fungal metabolites described by Turner [45] and Turner and Alderidge [46] were classified as mycotoxins in Cole and Cox [49]. Thus, there are potentially between 20,000 and 300,000 unique mycotoxins. Given the potentially large number of mycotoxins and the diversity of the mechanisms of action, the potential for extremely complex toxin interactions is also great. Figure 1 illustrates how exposure to multiple toxins can lead to extremely complicated biological responses within a simple system like a single cell. Mycotoxins with similar modes of action would be expected to have at least additive effects. Conversely,

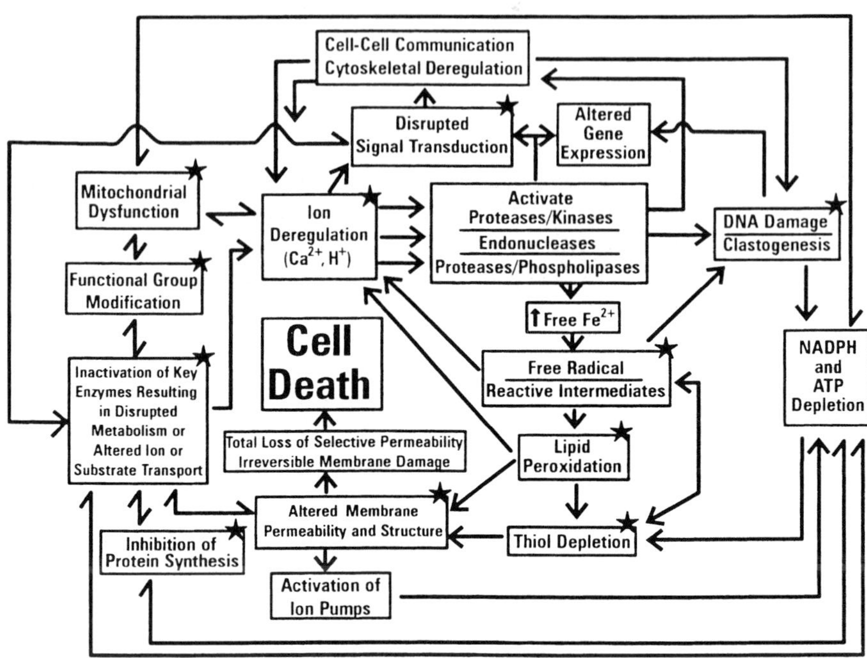

FIGURE 1 A flow chart showing the interrelationship between various toxin-initiated biochemical changes which can lead to cell death. Stars indicate interactions which could serve as initiating events. (From Ref. 1.)

some interactions could have subtractive effects; for example, the ability of cyclopiazonic acid to prevent the lipid peroxidation induced by patulin [24] or the ability of fungal serine palmitoyltransferase inhibitors such as sphingofungins to prevent the fumonisin-induced accumulation of free sphingoid bases [13].

Typically, mycotoxins occur naturally in feeds and foods in low concentrations, and several different mycotoxins are present. For example, *Fusarium proliferatum* on corn might produce several fumonisins (B_1, B2, B3, B4, etc.), fusaric acid, beauvericin, fusarins, moniliformin, and fusaproliferin [50,51]. In addition, co-occurrence of other *Fusarium* species and *Aspergillus* species can result in cocontamination of corn with fumonisins, T-2 toxin, deoxynivalenol, nivalenol, aflatoxin B_1, cyclopiazonic acid, zearalenone, and nitrosamines [52–56]. Thus, in a diverse human diet, exposure will be to multiple mycotoxins, intermittently and at low concentrations, over a long period of time. Acute toxicity studies cannot reveal the risks associated with this type of exposure, and chronic toxicity studies of all mycotoxins and combinations of mycotoxins are logistically impossible. An understanding of the mechanism of action in simple in vitro systems can provide a rational basis for predicting possible toxin, drug, and/or nutritional interactions.

The remainder of this chapter will be devoted to examples of how possible interactions might be predicted based on the known mechanisms of action. Coupling of this approach with the known co-occurrence of mycotoxins will maximize the possibility for discovery of those interactions that pose a health risk from consumption of contaminated foods or feeds. For example, fumonisins and aflatoxins are hepatotoxic, co-occur on corn, and have quite different mechanisms of action (Table 1). Similarly, deoxynivalenol, fumonisins, and ochratoxins are all nephrotoxic, are common contaminates of human foods, and have quite dissimilar mechanisms of action. Because the mechanisms are different, is there a possibility for synergy or additive risk for multiple mycotoxin exposure? Are there potential adverse interactions with infectious agents, drugs, pesticides, stress, or nutritional deficiencies that might be aggravated by exposure to specific mycotoxins or combinations of mycotoxins?

III. SELECTED EXAMPLES OF POTENTIAL MECHANISTIC INTERACTIONS

Many food grains, fruits, vegetables, and nuts can be contaminated with low levels of mycotoxins. Therefore, it is probable that consumption of small quantities of mycotoxins occurs at every meal or feeding. In addition, fungal spores, sclerotia, and mycelia also can contain mycotoxins which can be respired (for example, fumonisins and AAL-toxin [57] and aflatoxins [58]). For each of us who consume a varied diet that might include wheat, corn, and nut products, we also probably consume small amounts of aflatoxins, ochratoxins, fumonisins, and trichothecenes. How might these combinations of mycotoxins affect cells? Are there other risk factors which might predispose individuals to the adverse effects of specific mycotoxins, or are there interactions which could be beneficial? The following interactions where chosen only to provide examples of how the interactions might be viewed from the perspective of mechanism of action. This of course assumes that the mechanisms presented are in fact correct—an assumption that may be debatable.

A. Zearalenone and Phytoestrogens

Zearalenone reacts specifically with the cytosolic estrogen receptor (Fig. 2). Weak estrogen agonists such as zearalenone and phytoestrogens such as coumestrol are quite common [36]. Environmental estrogens are estrogen agonists found in pesticides and synthetic chemical mixtures such as DDT and polychlorinated biphenyls (PCBs). Environmental estrogens have been blamed for increased

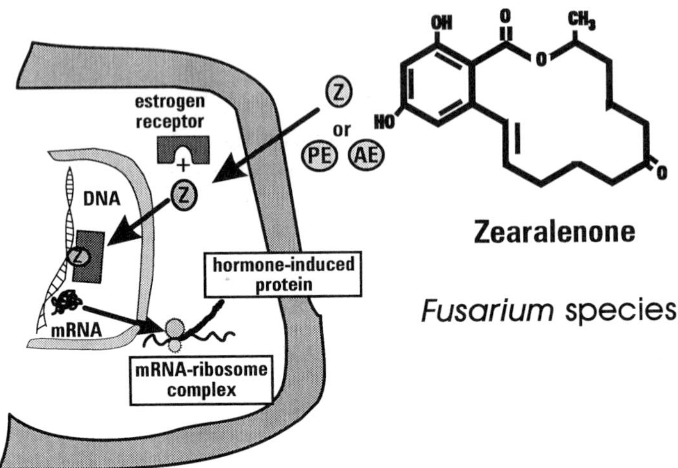

FIGURE 2 Interaction of zearalenone (Z), phytoestrogens (PE), and antiestrogens (AE) with the cytosolic estrogen receptor illustrating hormonal mimicry [1]. Zearalenone (Z)-like phytoestrogens (PE) and environmental estrogens passively cross the cell membrane and bind to the cytosolic estrogen receptor. The receptor-Z complex (or PE complex) is rapidly transferred into the nucleus, where it binds to specific nuclear receptors and generates estrogenic responses via gene activation, resulting in the production of mRNAs that code for proteins that are normally expressed by receptor-estrogen complex binding. Dietary phytoantiestrogens (APE) are also known to exist and, like the weak estrogen agonists, bind the estrogen receptor. However, in the case of antiestrogens, the effect is to block the response to estrogen or the weak agonists such as zearalenone.

incidence of breast cancer [59] and male reproductive problems [60]. Conversely, weak estrogens such as zearalenone have been hypothesized to act as antagonists to endogenous estrogens and thus reduce the risk of estrogen-induced breast cancer [61]. It has been hypothesized that the lifetime risk for breast cancer and some male reproductive disorders is proportional to the lifetime exposure to bioavailable estrogens [59]. Thus, the animal and human risk from zearalenone-contaminated foods and feeds must be viewed from the perspective of exposure to all dietary chemicals that mimic estrogen and the possibility that low levels may exert some beneficial effects, for example, in postmenopausal women [61]. Since many animal feeds are known to contain phytoestrogens (for example, soybeans, alfalfa, clover), antiestrogens, and possibly environmental estrogens (for example, pesticides), the effects or lack of effects from consumption of zearalenone-contaminated feeds or foods can only be understood if the total estrogenic potential of the food or feed is known.

B. Aflatoxin B_1, Patulin, Cyclopiazonic Acid, and Fumonisin B_1

Aflatoxin B_1 and cyclopiazonic acid are produced by the fungus *Aspergillus flavus*, and both mycotoxins are commonly found on corn and peanuts [62]. Patulin can be produced by *A. flavus* and several other genera of fungi and has been found in animal feeds and cereal grains but not corn [63]. However, patulin is a frequent contaminant of apple and grapefruit juice [64]. Fumonisins produced by *Fusarium moniliforme* and other *Fusarium* species can co-occur with aflatoxin B_1 on corn [52,53]. Thus, the consumption of low levels of all four mycotoxins by animals and humans is probable.

How might these four mycotoxins interact? Aflatoxin B_1, fumonisin B_1, and cyclopiazonic are hepatoxic in rat but have quite different mechanisms of action (Figs. 3–5). In rat liver, aflatoxin B_1 is a good initiator and a proven complete carcinogen [2]. Fumonisin B_1 is a tumor promotor and has been reported to be a complete carcinogen [14]. Cyclopiazonic acid has a mechanism of action that is quite similar to the known tumor promotor, thapsigargin [9]. In cultured cells, patulin rapidly depletes intracellular glutathione [23], which is an essential component in the detoxification of aflatoxins [2]. Thus, the co-occurrence of these four mycotoxins poses the potential of an interesting interaction with regard to rat liver tumor promotion and initiation.

Aflatoxin B_1 acts by covalent binding of the aflatoxin B_1-8,9-epoxide to cellular DNA and proteins (Fig. 3) [2]. The molecular targets in AFB_1 carcinogenesis include mutations in codon 12 of c-*ras* oncogenes, increased expression of c-*myc*, overexpression of protein kinase C, and mutations at codon 249 in the p53 tumor suppressor gene. It is believed that the codon 249 mutation results in the inactivation of the p53 tumor suppressor gene. The p53 gene is believed to be responsible for monitoring so that damaged DNA can be repaired or cells can be eliminated via the normal apoptotic pathway [66]. In addition, aflatoxin B_1 is cytotoxic to liver cells and thus increases the rate of compensatory cell proliferation as a consequence of tissue repair. Increased compensatory cell proliferation in target tissues is a suspected cancer risk factor [66].

Like aflatoxin B_1, fumonisin B_1 and cyclopiazonic acid are also hepatoxic. However, fumonisin B_1 and cyclopiazonic acid do not appear to be mutagenic or genotoxic. Both fumonisin B_1 and cyclopiazonic acid act by mechanisms which can induce alterations in normal cell cycle progression through disruption of intracellular calcium homeostasis and effects on key kinases, phosphatases, and phopholipases involved in cellular regulation [9,12,13]. While compensatory cell proliferation is a normal response to tissue injury, disruption of signaling pathways can lead to alterations in cell proliferation and to inappropriate entry or exit from the cell cycle [67].

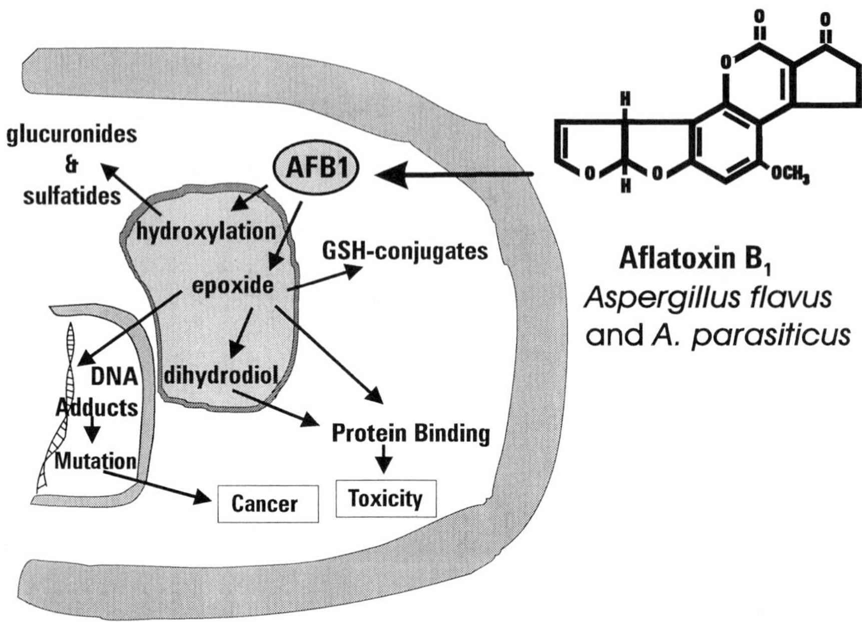

FIGURE 3 The pathways of aflatoxin B_1 (AFB1) metabolism leading to cancer and cytotoxicity [2]. Briefly, aflatoxin B_1 enters the cell and is either metabolized via monooxygenases in the endoplasmic reticulum to hydroxylated metabolites which are further metabolized to glucuronide and sulfate conjugates or is oxidized to the reactive epoxide which undergoes spontaneous hydrolysis to the AFB_1-8,9-dihydrodiol which can bind to proteins resulting in cytotoxicity. The epoxide can react with DNA or protein or be detoxified by an inducible glutathione S-transferase to the glutathione (GSH) conjugate.

Fumonisins are specific inhibitors of de novo sphingolipid biosyntheses via inhibition of the enzyme sphinganine N-acyltransferase (Fig. 4) [68,69]. Fumonisin-induced elevation of free sphinganine can inhibit the phosphorylation of protein kinases by protein kinase C [70], and consequently agonist-induced responses mediated via protein kinase C can be altered. Increased concentration of intracellular sphinganine 1-phosphate (sphinganine degradation product) can trigger calcium release from intracellular calcium stores [71] and thus induce the conversion of inactive cytosolic protein kinase C to membrane-associated protein kinase C. Disruption of calcium homeostasis by fumonisins and cyclopiazonic acid can alter calcium-dependent kinases, phosphatases, phospholipases, endonucleases, and proteases, which can alter normal cell cycle progression, induce apoptosis, and lead to increased cytotoxicity. The net result of sphinganine inhibition of protein kinase C (of which there are many isoforms) and the calcium-mediated transition of inactive cytosolic to active membrane-associated protein kinase C and activation of calcium-dependent proteins could be very difficult to predict. In addition, fumonisin-induced depletion of ceramide and sphingomyelin and increases in free sphingoid bases can have diverse effects on many other key enzymes involved in cell cycle progression, proliferation, and apoptosis [12,13].

In cultured cells, patulin rapidly depletes intracellular nonprotein thiols [24] and, in particular, reduced glutathione [22,23] (Fig. 6). The amount of the aflatoxin B_1 epoxide that binds to DNA is decreased by conjugation with glutathione [2]. The rate of activation to the epoxide relative to the rate of inactivation is a critical determinant in species susceptibility to aflatoxin carcinogenesis [2]. Whether patulin can deplete glutathione in vivo without inducing cell death in the glutathione depleted cells is not known.

It is difficult to predict the outcome of the combined cellular effects of all these mycotoxins acting in concert. Patulin depletion of reduced glutathione could increase the proportion of aflatoxin B_1 protein and DNA adducts, leading to increased liver cancer risk. The increased apoptosis induced by fumonisin might reduce the population of cells that have aflatoxin-induced mutations of the p53 gene. This would result in a reduced cancer risk. If the increased cell death due to fumonisin-induced apoptosis randomly selects cells, then the relative population of p53-deficient cells will not decrease, while the demand for tissue repair will increase resulting in an additional increase in cancer risk. In addition, calcium release mediated by cyclopiazonic acid or fumonisin-induced sphinganine-1-phosphate would be expected to increase the proportion of membrane-associated protein kinase C. However, free sphinganine [70] and fumonisin B_1 [72] have both been shown to reduce the activity of protein kinase C. Because protein kinase C has been shown to be overexpressed in aflatoxin-transformed cells [73], the interaction of cyclopiazonic acid, aflatoxin B_1, and fumonisin B_1 would be quite complex.

C. Adenophostin A and Cyclopiazonic Acid

Adenophostin A and cyclopiazonic acid are both fungal metabolites which have marked effects on calcium homeostasis and thus affect the processes associated with signal transduction pathways within which regulated modulation of intracellular calcium is critical (Fig. 7). Signal transduction pathways couple receptor binding of extracellular first messengers (hormone or agonist) with production of intracellular second messengers [74]. Signal transduction involves first messenger binding, transduction of the signal from the extracellular membrane surface to the intracellular membrane surface (transmembrane proteins and associated G-protein), amplification of the first message (e.g., activation of adenylate cyclase, phospholipase C), production of an intracellular second messenger (e.g., Ca^{2+}, cAMP, phosphoinositides, ceramide), activation or inactivation of internal effectors (e.g., protein kinases, transcription factors), and cellular responses (e.g., effects on ion channels, gene expression). Mycotoxins which interfere with agonist-induced signal transduction systems can be expected to have diverse effects on cellular metabolism, growth, proliferation, and differentiation.

Adenophostin A is produced by a strain of *Penicillium brevicompactum* isolated from soil [4]. Nothing is known about the natural occurrence of adenophostin A; however, the reported mechanism of action makes it a very intriguing molecule. Adenophostin A is a potent agonist for the inositol-1,4,5-trisphosphate receptor on the endoplasmic reticulum (Fig. 7). This receptor mediates the rapid release of calcium from the endoplasmic reticulum calcium store. The inositol-1,4,5-trisphosphate calcium pool plays an extremely critical role in the regulation of numerous cell functions, many of which are hormone-initiated. Cyclopiazonic

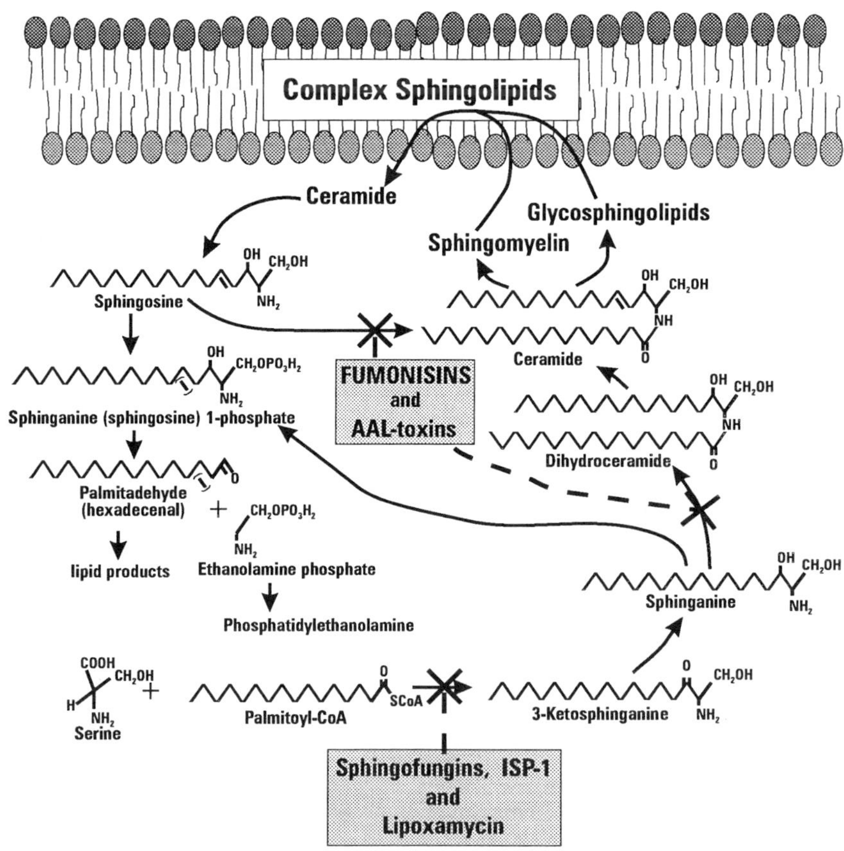

acid, a specific inhibitor of P-type calcium ATPases [9], is known to cause the release of the calcium from the IP_3-sensitive calcium pool [e.g., 75,76]. Thus, cyclopiazonic acid will indirectly interfere with agonist-induced responses which are mediated via phospholipase C activated signal transduction pathways (Fig. 5). The combination of cyclopiazonic acid inhibition of the endoplasmic reticulum calcium pump (Fig. 5) and the potentiation of release of calcium from the inositol-1,4,5-trisphosphate calcium pool (Fig. 7) would effectively disrupt all cellular events mediated by intracellular calcium. There are several high-risk groups that might be quite sensitive to these mycotoxins—for example, animals susceptible to stress-induced malignant hyperthermia or humans taking drugs such as calcium antagonists designed to carefully control calcium homeostasis.

D. Wortmannin, Beauvericin, and Fumonisin B_1

There are many mycotoxins that can induce apoptosis. These include T-2 toxin, deoxynivalenol, luteoskyrin, ochratoxin A, citrinin, aflatoxin B_1, sporodesmin, gliotoxin, wortmannin, beauvericin, and fumonisin B_1 [3,6,15,27,31,77]. Many of these mycotoxins have quite different initial biochemical targets (Table 1). For example, wortmannin disrupts signaling pathways mediated by activation of

---

FIGURE 4 Inhibitors of de novo sphingolipid biosynthesis. Structures of some long-chain bases and more complex sphingolipids and pathway of sphingolipid metabolism and site(s) of inhibition by fumonisins (*Fusarium* spp.), AAL-toxins (*Alternaria alternata* f.sp. *lycopercici*), sphingofungins (*Aspergillus fumigatus* and others), ISP-1 (*Isaria sinclairii* and others), and lipoxamycin (*Streptomyces*). Fumonisin B_1, B_2, B_3, hydrolyzed B_1, and AAL-toxin inhibit ceramide synthase (sphingosine and sphinganine N-acyltransferase). Sphingofungins and ISP-1 are specific inhibitors of serine palmitoyltransferase, which is a pyridoxal phosphate-dependent enzyme. Inhibition of sphinganine (sphingosine) N-acyltransferase results in blockage of complex sphingolipid biosynthesis, accumulation of sphinganine (and sometimes sphingosine), and diversion of sphinganine to sphinganine 1-phosphate, which is further broken down to a fatty aldehyde and ethanolamine phosphate. The ethanolamine phosphate is directed into phophatidylethanolamine, and the fatty aldehyde to other lipids. Fumonisins also block the reacylation of dietary sphingosine and the sphingosine that is released by the turnover of more complex sphingolipids. As a consequence of disruption of sphingolipid metabolism, many bioactive lipid intermediates are created and some are lost. Some examples of the biological activity of these intermediates are increases in intracellular free sphingoid bases, which can inhibit protein kinase C and other kinases; increase in sphinganine 1-phosphate, which triggers endoplasmic reticulum calcium release; increase in free sphinganine, which can affect cell proliferation, cell cycle progression, and initiate apoptosis or necrotic cell death; and depletion of complex sphingolipids, which are integral components of pathways that regulate cell growth and cell cycle progression [12,13].

phosphatidylinositol 3-kinase (PI 3-kinase) (Fig. 8). This enzyme catalyzes the phosphorylation of phosphatidylinositol-4,5-bisphosphate (PIP_2) to phosphatidylinositol-3,4,5-trisphosphate (PIP_3). While the function of PIP_3 is not established, it is known that inhibition of PI-3 kinase induces apoptosis [35,78]. Wortmannin inhibition of PI-3 kinase mimics the apoptosis that is induced in cells by growth factor deprivation.

Beauvericin is a cyclic depsipeptide produced by *Fusarium* species (79,80). Beauvericin can occur on corn that is concurrently contaminated with fumonisin B_1 (80). Beauvericin has been shown to be a specific potassium ionophore (Fig. 9) which behaves like the well-known potassium carrier ionophore, valinomycin [6]. Beauvericin-induced disruption of intracellular potassium homeostasis has been shown to lead to disruption of intracellular calcium homeostasis and consequent activation of calcium-dependent endonucleases, leading ultimately to increased apoptosis [6]. Whether beauvericin is active in vivo in mammals or can be absorbed from the diet is unknown. This is a common problem with many studies which are done exclusively using in vitro models.

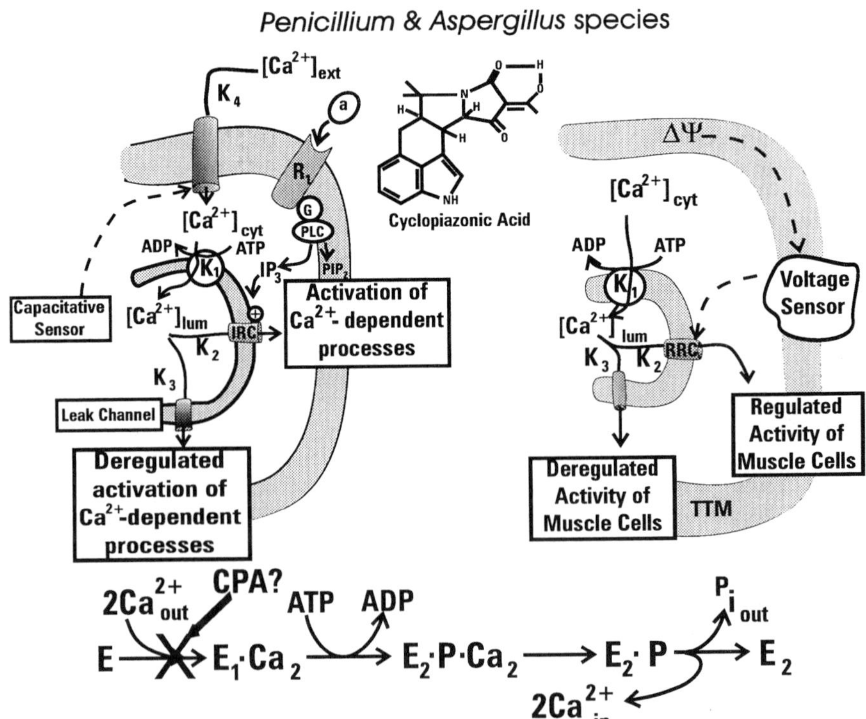

Fumonisin B_1 induces increased apoptosis both in vitro and in vivo [81,82]. The mechanism for this effect has not been clearly demonstrated; however, studies with human keratinocytes have shown that the fumonisin-induced accumulation of free sphinganine is correlated with increased apoptotic cell death in both a time- and dose-dependent manner [83]. Both exogenous addition of sphinganine and C2-ceramide induced apoptosis [83]. Ceramide, a product of sphingomyelin turnover, has been shown to be a mediator of apoptosis, cell cycle arrest, and programmed cell death [84]. Both inhibition of sphingolipid biosynthesis by

---

FIGURE 5 Chemical structures of cyclopiazonic acid and a schematic showing inhibition of sarcoplasmic and endoplasmic reticulum calcium-dependent ATPases. (Left) Agonist-induced Ca^{2+} release from the endoplasmic and sarcoplasmic reticulum and capacitative Ca^{2+} entry across the plasma membrane (PM) of a nonskeletal muscle or smooth muscle cell. (Right) Model for depolarization-induced CA^{2+} release from the sarcoplasmic reticulum (SR) ryanodine receptor channel (RRC) of skeletal or cardiac muscle cells. Agonist-induced Ca^{2+} release (left) from the endoplasmic and sarcoplasmic reticulum and capacitative Ca^{2+} entry across the plasma membrane (PM) of a nonskeletal muscle or smooth muscle cell is initiated by Ca^{2+} release via the inositol-1,4,5-triphosphate (IP_3)-regulated channel (IRC) which is initiated by agonist (a) binding to the PM receptor (R) which activates, via a GTP binding protein (G), phospholipase C (PLC) hydrolysis of phosphatidylinositol 4,5-bisphosphate (PIP_2) to IP_3 and diacylglycerol (not shown). The soluble IP_3 binds to its receptor on the IRC, triggering the release of the endoplasmic reticulum Ca^{2+} pool ($[Ca^{2+}]_{lum}$). The state of filling of the endoplasmic reticulum Ca^{2+} pool is believed to signal the opening of PM associated Ca^{2+} channels and influx of extracellular Ca^{2+} ($[Ca^{2+}]_{ext}$; capacitative entry). K_1 is the rate of Ca^{2+} entry into the endoplasmic reticulum via the cyclopiazonic acid-sensitive Ca^{2+} transporter; K_2 is the rate of efflux via the IRC which is initiated by IP_3; K_3 is the rate of Ca^{2+} efflux through the leak channel; and K_4 is the rate of extracellular Ca^{2+} influx via PM Ca^{2+} channels. Normally, K_1 is much greater than K_3; however, in the presence of cyclopiazonic acid, K_3 is greater than K_1, allowing the accumulated Ca^{2+} in the endoplasmic reticulum to be depleted. Depletion of the endoplasmic reticulum Ca^{2+} pool will also reduce K_2 initiated by IP_3 and can induce Ca^{2+} entry via the PM channels. In sarcoplasmic reticulum (right) the initiating event for calcium release is the electrical depolarization ($\Delta\Psi$) of the T-tubule membrane (TTM). The cytosolic Ca^{2+} concentration ($[Ca^{2+}]_{cyt}$) is approximately 0.1 μM, whereas the Ca^{2+} concentration in the lumen of the sarcoplasmic reticulum ($[Ca^{2+}]_{lum}$) is approximately 65 to 70 mM. K_1 is the rate of Ca^{2+} entry via the cyclopiazonic acid-sensitive Ca^{2+} transporter; K_2 is the rate of efflux via the RRC which is initiated by depolarization of the TTM; and K_3 is the rate of Ca^{2+} efflux through the leak channel. Normally, K_1 is much greater than K_3; however, in the presence of cyclopiazonic acid, K_3 is greater than K_1, causing the accumulated Ca^{2+} in the sarcoplasmic reticulum to be slowly depleted. Depletion of the sarcoplasmic reticulum Ca^{2+} pool will also reduce the rapid efflux via the RRC (K_2) initiated by depolarization of the TTM. (From Refs. 99–101.)

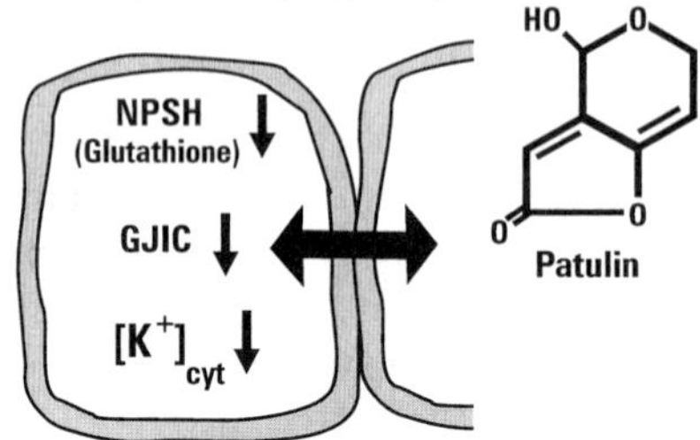

FIGURE 6 A schematic showing the cellular consequences of exposure to patulin in the absence of antioxidants [summarizing the data from 22–24]. Patulin-induced depletion of nonprotein sulfhydrals (NPSH), primarily reduced glutathione, results in simultaneous suppression of gap junction-mediated intercellular communication (GJIC) in clone 9 rat liver cells and simultaneous potassium efflux in LLC-PK_1 pig kidney epithelial cells.

inhibitors such as ISP-1 (Fig. 4) and high doses of sphingoid bases can induce apoptosis [85]. Beta-chloroalanine (which inhibits fumonisin-induced sphinganine accumulation) partially protected keratinocytes from fumonisin B_1-mediated apoptosis [83]. Thus, elevated free sphinganine contributed to fumonisin B_1-induced apoptotic cell death in cultured keratinocytes (83).

These three mycotoxins, wortmannin, beauvericin, and fumonisin B_1, each initiate increased apoptosis by distinctly different biochemical mechanisms. At what point downstream of the initial biochemical lesion the pathways might intersect is unknown. Whether or not the three mechanisms acting in concert will result in a more than additive increase in apoptosis could prove interesting. The specificities of wortmannin for PI 3-kinase, fumonisin for ceramide synthase, and beauvericin as a potassium ionophore have all been questioned (Table 1) [7,34,86].

E. Fumonisins, Sphingofungins, and ISP1

Fumonisin B_1 was the first discovered naturally occurring inhibitor of de novo sphingolipid biosynthesis [69]. Fungal metabolites which act by disruption of sphingolipid biosynthesis are now known to be produced by nine separate genera of fungi [12,13]. In vitro studies with cultured neuronal cells, renal cells, fibroblasts, and keratinocytes have shown that either reduction of free sphinganine with serine palmitoyltransferase inhibitors (Fig. 4) or supplementation of growth medium with ceramide completely or partially reverse the effects of fumonisin B_1 on cell growth and/or cell death [83,87–89]. For example, a recent study in LLC-PK_1 cells [89] found that coaddition of fumonisin B_1 and β-chloroalanine (a serine

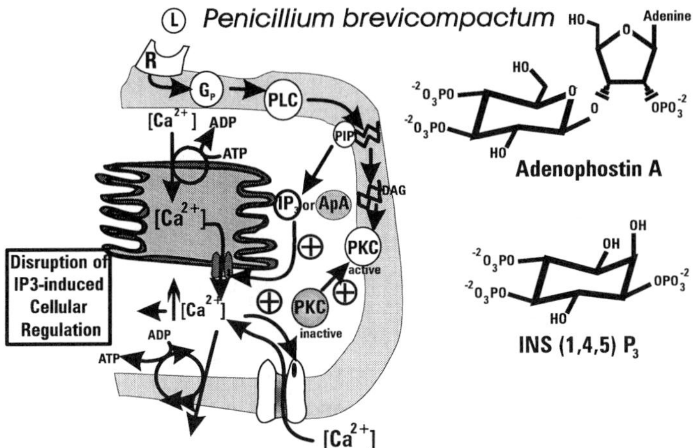

FIGURE 7 A schematic demonstrating how adenophostin A (ApA) could interfere with an agonist-induced cellular responses mediated by phospholipase C (PLC) generated second messengers (IP_3, DAG, Ca^{2+}) and protein kinase C (PKC). Other abbreviations used are L = ligand, G_p = G-protein, PIP = phosphoinositol 4,5-bisphosphate, DAG = diacylglycerol, IP_3 = inositol-1,4,5-trisphosphate. Also shown are the endoplasmic reticulum calcium-dependent ATPase, the IP_3-sensitive endoplasmic reticulum release channel, the plasma membrane calcium-activated calcium channel, and the plasma membrane calcium extrusion pump (calmodulin-dependent calcium ATPase).

palmitoyltransferase inhibitor) reduced the cytotoxic effects of fumonisin B_1 by approximately 50% to 60%. The fumonisin B_1 increase in free sphingoid bases was reduced by approximately 90%, but depletion of more complex sphingolipids was greater than when either agent was added alone. Whether or not coadministration of a serine palmitoyltransferase inhibitor such as ISP1 or sphingofungin [25,26] can reduce the fumonisin-induced increase in free sphingoid base concentration in vivo and reduce the in vivo hepatotoxicity or nephrotoxicity of fumonisins has not been determined. If fungal inhibitors of serine palmitoyltransferase prevent the fumonisin-induced increase in free sphingoid bases in vivo and do not reduce the fumonisin nephrotoxicity or hepatotoxicity, then elevation in free sphinganine could be discounted as a direct cause of the toxicity. Likewise, if serine palmitoyltransferase inhibitors deplete more complex sphingolipids without inducing nephrotoxicity or hepatoxicity, then the role of more complex sphingolipid depletion in these fumonisin-induced toxicities would be disproved. Thus, the interaction of fungal serine palmitoyltransferase inhibitors and ceramide synthase inhibitors in vivo could prove interesting.

Fusarium species & Talaromyces wortmannii

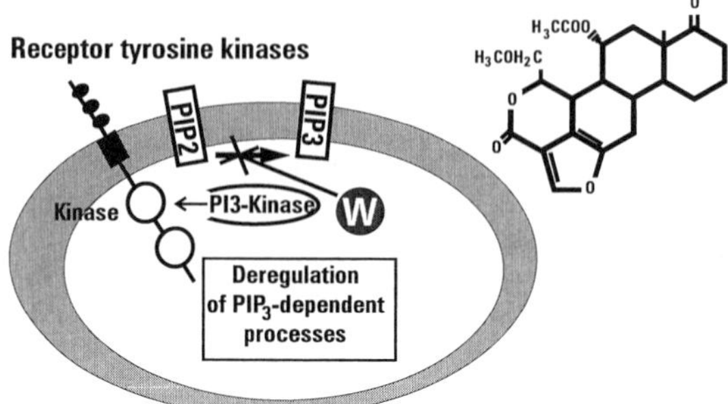

FIGURE 8 Wortmannin disruption of receptor tyrosine kinase activated PI3-kinase-mediated processes. Wortmannin inhibits the enzymatic processes which catalyze the conversion of the membrane lipid, phosphatidylinositol 4,5-bisphosphate (PIP2) to phosphatidylinositol-3,4,5-trisphosphate (PIP3). This signaling pathway has been shown to affect many aspects of cellular regulation including the ability of growth factors to prevent apoptosis and activation of mitogen-activated protein kinase (MAP kinase) [33].

F. Interaction of Mycotoxins with Infectious Agents

In addition to the possibility of mycotoxins mechanistically interacting with other mycotoxins or xenobiotics, the interaction with infectious agents is also possible. For example, weak estrogen agonists might ameliorate paracoccidioidomycosis, which is a common disease in Latin American men but not women. The causative agent for this disease is *Paracoccidioides brasiliensis*, which possesses a cytosolic estrogen receptor [38]. In women, the presence of estrogen prevents the formation of the asexual conidia that are believed to be responsible for the disease [38]. Thus, the relative lack of estrogen in men predisposes them to the disease. In this case, males consuming weak estrogen agonists such as zearalenone might be protected from the disease.

As with zearalenone, there are some interesting implications for interactions between aflatoxin and infectious agents. For example, in humans, relative liver cancer risk is greatly increased with aflatoxin exposure and hepatitis B virus infection [2]. The production of codon 249 mutations by aflatoxin B_1 may be enhanced by hepatitis virus infection [2]. Wortmannin inhibition of PI 3-kinase has been reported to block p53 induction following DNA damage [90]. This could exacerbate the carcinogenicity associated with aflatoxin B_1/hepatitis virus infection, since one of the known targets of aflatoxin B_1 mutation is the p53 gene [2].

Fumonisins and cyclopiazonic acid may also alter viral infectivity. Inhibition of sphingolipid biosynthesis has been shown to inhibit the cytopathic effects,

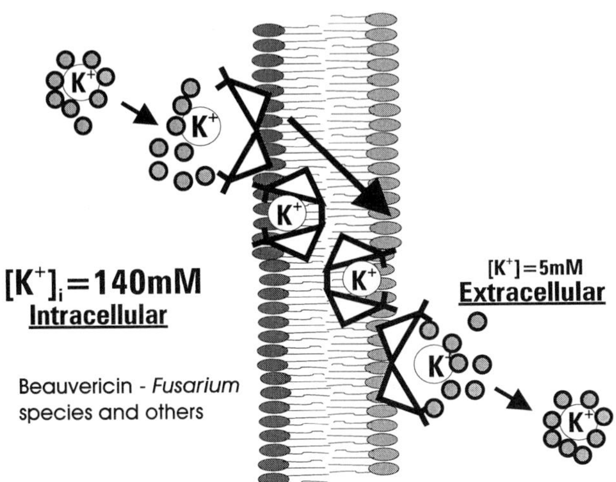

FIGURE 9 A schematic representation of the mode of action of a potassium-specific carrier ionophore such as valinomycin. Presumably, like valinomycin, beauvericin dissipates the transmembrane potassium concentration difference by carrying potassium ions down the concentration gradient in the same manner as the well-studied potassium ionophore valinomycin. The potassium atom is shown surrounded by water molecules. This charged sphere of hydration must be removed so that the carrier-ion complex can diffuse through the hydrophobic core of the phospholipid bilayer.

replication, and infectivity of HIV-1 [91]. Conversely, cyclopiazonic acid inhibition of the endoplasmic reticulum calcium-dependent ATPase in T-lymphocyte cells stimulates HIV expression [92]. Recently, cyclopiazonic acid was shown to dramatically increase IL-2 and IL-5 levels and stimulate cell proliferation in a T-cell model [93]. The ability of cyclopiazonic acid or fumonisins to alter immune response in vivo and the effect of fumonisins and cyclopiazonic acid on hepatitis B infectivity and pathogenicity have not been studied.

The fungal serine palmitoyltransferase inhibitor ISP-1 has been shown to be a potent immunosuppressant [26]. The ability of fumonisins to act as immunosuppressants is unknown; however, like ISP-1 [85], fumonisins disrupt sphingolipid metabolism and induce apoptosis in many cell types [77]. In addition, sphingolipids are known to be receptors for many microbial toxins and viruses [94,95]. Thus, mycotoxins which disrupt sphingolipid metabolism could alter the ability of cells to respond to infectious agents or modify the cell surface recognition sites for microbial toxins, bacteria, viruses, or other infectious agents which require specific surface receptors in order to enter cells. For example, fumonisin B_1 has been shown to inhibit the association, intracellular sorting, and translocation of Shiga toxin in A431 cells [96].

The ability of deoxynivalenol (and other trichothecenes) to suppress an animal's immune response to pathogens is well documented [11]. This prevalent mycotoxin has been shown to superinduce cytokine gene expression and secretion and inhibit cell proliferation both in vitro and in vivo [97,99]. Superinduction of cytokines is associated with many allergic and autoimmune diseases including IgA nephropathy [97]. Certain high-risk groups with preexisting autoimmune diseases, many of which are associated with previous bacterial infections, could be highly sensitive to deoxynivalenol in the diet. It is easy to image combinations of mycotoxins and infectious agents that could interact so as to modulate the immune response in a manner resulting in specific tissue injury (Fig. 10). For example, fumonisin B_1 (disruption of cytokine signaling systems), deoxynivalenol (increased cytokine response), ochratoxin A (nephrotoxic), and Shiga toxin produced by *Escherichia coli* O157:H7 (hemolytic uremic syndrome) are all known to cause renal dysfunction in animal models. All of these agents could co-occur. The consequences of such an interaction would be difficult to predict; however, an increased risk of renal dysfunction is one possibility.

IV. CONCLUSIONS

There is an extremely large number of potential mycotoxin interactions that can be studied. Unfortunately, the resources available to conduct these studies are limited. Understanding the mechanism of action can help to maximize the probability of choosing for study those interactions which are relevant to human, animal, and plant safety. Because there are so many mycotoxins yet to be discovered, the actual combined health risk from exposure to mycotoxins is unknown. For this reason McLachlan [36] proposed that chemicals should be classified by their function in vitro as a guide for toxicological studies. This same concept which has been used extensively in drug discovery efforts is also appropriate for the study of feeds and foods naturally contaminated with fungi and suspected of being involved in human or animal disease outbreaks. Functional "food toxicology" would screen suspected feeds and foods for mechanisms known to be underlying causes of chronic disease. For example, cereal grains, milling fractions, or fungal culture material could be screened for the ability to inhibit sphinganine N-acyltransferase (fumonisin-like activity), inhibit IP_3 binding to the IP_3 receptor (adenophostin A-like activity), produce $G \rightarrow T$ transversion mutations at codon 249 (aflatoxin-like activity), bind to the SERCA-ATPase (cyclopiazonic acid-like activity), or bind to the cytosolic estrogen receptor (zearalenone-like activity). The number of possible biochemical targets is great and unraveling the source of the biological activity would be difficult in a contaminated food or feed. The

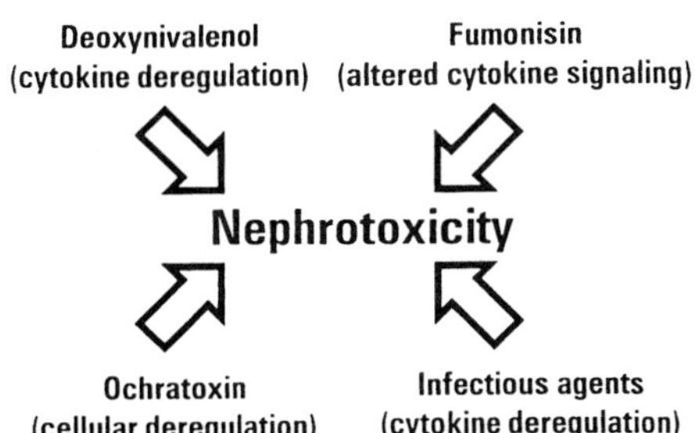

FIGURE 10 A theoretical example of mycotoxins and an infectious agent acting in concert to initiate or exacerbate renal dysfunction. Fumonisins acting through disruption of the cytokine signaling pathway [102,103], deoxynivalenol acting through induction of an increased cytokine response [97], ochratoxin A acting as a known nephrotoxin [19,21], and Shiga toxin produced by *Escherichia coli* O157:H7, internalized via a sphingolipid receptor and initiating renal dysfunction via disruption of endothelial cell function acting through a cytokine-mediated pathway [104].

screening of fungal isolates would help to differentiate biological activities of fungal origin from those originating from other components of the food.

REFERENCES

1. RT Riley, WP Norred. Mechanisms of mycotoxicity. In: DH Howard, JD Miller, eds. The Mycota Vol. VI. Berlin; Springer, 1996, p 194.
2. DL Eaton, EP Gallager. Mechanisms of aflatoxin carcinogenesis. Annu Rev Pharmacol Toxicol 34:135, 1994.
3. Y Ueno, K Umemori, E Niimi, et al. Induction of apoptosis by T-2 toxin and other natural toxins in HL-60 human promyelotic leukemia cells. Nat Toxins 3:129, 1995.
4. M Takahashi, K Tanzawa, S Takahashi. Adenophostins, newly discovered metabolites of *Penicillium brevicompactum*, act as potent agonists of the inositol 1,4,5-trisphosphate receptor. J Biol Chem 269:369, 1994.
5. K Kawai, J Kitamura, T Hamasaki, Y Nozawa. The mode of action of anthraquinone mycotoxins on ATP synthesis in mitochondria. In: AE Pohland, VR Dowell Jr, JL Richard, eds. Microbial Toxins in Foods and Feeds: Cellular and Molecular Modes of Action. New York; Plenum Press, 1990, p 369.
6. DM Ojcius, A Zychlinsky, LM Zheng, JD Young. Ionophore-induced apoptosis: role of DNA fragmentation and calcium fluxes. Exp Cell Res 197:43, 1991.
7. H Tomada, XH Huang, J Cao, et al. Inhibition of acyl-CoA: cholesterol acyltransferase activity by cyclodepsipeptide antibiotics. J Antibiot Tokyo 45:1626, 1992.
8. RA Ansari, RS Thakran, WO Berndt. The effects of potassium chromate and citrinin on rat renal membrane transport. Fund Appl Toxicol 16:701, 1991.

9. RT Riley, DE Goeger, WP Norred. Disruption of calcium homeostasis: the cellular mechanism of cyclopiazonic acid toxicity in laboratory animals. In: M Eklund, JL Richard, K Mise, eds. Molecular Approaches to Food Safety—Issues Involving Toxic Microorganisms. Fort Collins; Alaken, Inc., 1995, p 461.
10. MF Carlier, P Criquet, D Pantaloni, ED Korn. Interaction of cytochalasin D with actin filaments in the presence of ADP and ATP. J Biol Chem 261:2041, 1986.
11. BA Rotter, DB Prelusky, JJ Pestka. Toxicology of deoxynivalenol (vomitoxin). J Toxicol Environ Health 48:101, 1996.
12. AH Merrill Jr, DC Liotta, RT Riley. Fumonisins: naturally occurring inhibitors of ceramide synthesis. Trends Cell Biol 6:218, 1996.
13. RT Riley, E Wang, JJ Schroeder, et al. Evidence for disruption of sphingolipid metabolism as a contributing factor in the toxicity and carcinogenicity of fumonisins. Nat Toxins 4:3, 1996.
14. WCA Gelderblom, SD Snyman, S Abel, et al. Hepatocarcinogenicity and carcinogenicity of the fumonisins in rats: a review regarding mechanistic implications for establishing risk in humans. In: L Jackson, JW DeVries, LB Bullerman, eds. Fumonisins in Food. New York, Plenum Publishing Corp., 1996, p 279.
15. P Waring. DNA fragmentation induced in macrophages by gliotoxin does not require protein synthesis and is preceded by raised inositol triphosphate levels. J Biol Chem 265:14476, 1990.
16. M Cadrin, SW French, PT Wong. Alteration in molecular structure of cytoskeleton proteins in griseofulvin-treated mouse liver: a pressure tuning infrared spectroscopy study. Exp Mol Pathol 55:170, 1991.
17. T Masuda, N Miyasaka, T Kato, Y Ueno. Formation of the 8-hydroxydeoxyguanosine moiety in hepatic DNA of mice orally administered with luteoskyrin, a bisanthraquinoid mycotoxin. Toxicol Lett 58:287, 1991.
18. PG Thiel. A molecular mechanism for the toxic action of moniliformin, a mycotoxin produced by *Fusarium moniliforme*. Biochem Pharmacol 27:483, 1978.
19. EE Creppy. Ochratoxin A in food: molecular basis of its chronic effects and detoxification. In: M Eklund, JL Richard, K Mise, eds. Molecular Approaches to Food Safety—Issues Involving Toxic Microorganisms. Fort Collins; Alaken, Inc., 1995, p 445.
20. J Fink-Gremmels, A Jahn, MJ Blom. Toxicity and metabolism of ochratoxin A. Nat Toxins 3:214, 1995.
21. RR Marquardt, AA Frolich. A review of recent advances in understanding ochratoxicosis. J Anim Sci 70:3968, 1992.
22. R Barhoumi, RC Burghardt. Kinetic analysis of the chronology of patulin- and gossypol-induced cytotoxicity in vitro. Fund Appl Toxicol 30:290, 1996.
23. RC Burghardt, R Barhoumi, EH Lewis, et al. Patulin-induced cellular toxicity: a vital fluorescence study. Toxicol Appl Pharmacol 112:235, 1992.
24. RT Riley, JL Showker. The mechanism of patulin's cytotoxicity and the antioxidant activity of indole tetramic acids. Toxicol Appl Pharmacol 109:108, 1991.
25. MM Zweerink, AM Edison, GB Wells, W Pinto, RL Lester. Characterization of a novel, potent, and specific inhibitor of serine palmitoyltransferase. J Biol Chem 267:25032, 1992.

26. Y Miyake, Y Kozutsumi, S Nakamura, T Fugita, T Kawasaki. Serine palmitoyltransferase is the primary target of a sphingosine-like immunosuppressant, ISP-1/myriocin. Biochem Biophys Res Commun 211:396, 1995.
27. P Waring, M Egan, A Braithwaite, A Mullbacher, A Sjaada. Apoptosis induced in macrophages and T blasts by the mycotoxin sporidesmin and protection by Zn^{2+} salts. Int J Immunopharmacol 12:445, 1990.
28. M Hino, O Nakayama, Y Tsurumi, et al. Studies of an immunomodulator, swainsonine I. Enhancement of immune response by swainsonine in vitro. J Antibio 7:926, 1985.
29. DB Gant, RJ Cole, JJ Valdes, ME Eldefrawi, AT Eldefrawi. Action of tremorgenic mycotoxins on $GABA_a$ receptor. Life Sci 41:2207, 1987.
30. Y Ueno. Biochemical mode of action of mycotoxins. In: JE Smith, RS Henderson, eds. Mycotoxins and Animal Foods. Boca Raton; CRC Press, 1991, p 437.
31. N Yoshino, M Takizawa, H Akiba, et al. Transient elevation of intracellular calcium ion levels as an early event in T-2 toxin-induced apoptosis in human promyelotic cell line HL-60. Nat Toxins 4:234, 1996.
32. DL Bunner, ER Morris. Alteration of multiple cell membrane functions in L-6 myoblasts by T-2 toxin: an important mechanism of action. Toxicol Appl Pharmacol 92:113, 1988.
33. GI Welsh, EJ Foulstone, SW Young, JM Tavare, CG Proud. Wortmannin inhibits the effects of insulin and serum on the activities of glycogen synthase kinase-3 and mitogen-activated protein kinase. Biochem J 303:15, 1994.
34. MJ Cross, A Stewart, MN Hodgkin, DJ Kerr, MJ Wakelam. Wortmannin and its structural analogue demethoxyviridin inhibit stimulated phospholipase A2 activity in Swiss 3T3 cells. J Biol Chem 270:25352, 1995.
35. R Yao, GM Cooper. Requirement for phophatidylinositol-3 kinase in the prevention of apoptosis by nerve growth factor. Science 267:2003, 1995.
36. JA McLachlan. Functional toxicology: a new approach to detect biologically active xenobiotics. Environ Health Perspect 101:386, 1993.
37. RT Riley, DE Goeger. Cyclopiazonic acid: speculation on its function in fungi. In: D Bhatnager, EB Lillehoj, DK Arora, eds. Handbook of Applied Mycology: Mycotoxins in Ecological Systems. Vol 5. New York, Marcel Dekker, 1992, p 385.
38. JE Cutler, Y Han. Fungal factors implicated in pathogenesis. In: DH Howard, JD Miller, eds. The Mycota. Vol. VI. Berlin; Springer, 1996, p 3.
39. CP Wild, AJ Hall. Epidemiology of mycotoxin-related disease. In: DH Howard, JD Miller, eds. The Mycota. Vol. VI. Berlin; Springer, 1996, p 213.
40. WP Blount. A new turkey disease problem in England characterized by heavy mortality. Q Poul Bull 27:1, 1960.
41. WP Blount. Turkey "X" disease. Turkeys 9:52, 1961.
42. JM Barnes, WH Butler. Carcinogenic activity of aflatoxin to rats. Nature 202:1016, 1964.
43. WP Norred. Fumonisins—mycotoxins produced by *Fusarium moniliforme*. J Toxicol Environ Health 38:309, 1993.
44. PB Hamilton. Mycotoxins and farm animals. Refush Vet 39:17, 1982.
45. WB Turner. Fungal Metabolites. London; Academic Press, 1978, p 446.

46. WB Turner, DC Alderidge. Fungal Metabolites II. London, Academic Press, 1983, p 631.
47. DL Hawksworth. The fungal dimension of biodiversity: magnitude, significance, and conservation. Mycol Res 95:641, 1991.
48. K Esser, PA Lemke. Series preface. In: K Esser, PA Lemke, eds. The Mycota. Vol. VI. Berlin; Springer, 1996.
49. RJ Cole, RH Cox. Handbook of Toxic Fungal Metabolites. New York; Academic Press, 1981, p 937.
50. PE Nelson, AE Desjardins, RD Plattner. Fumonisins, mycotoxins produced by *Fusarium* species: biology, chemistry, and significance. Annu Rev Phytopathol 31: 233, 1993.
51. A Ritieni, V Fogliano, G Randazzo, et al. Isolation and characterization of fusaproliferin, a new toxic metabolite from *Fusarium proliferatum*. Nat Toxins 3:17, 1995.
52. WJ Chamberlain, CW Bacon, WP Norred, KA Voss. Levels of fumonisin B1 in corn naturally contaminated with aflatoxins. Feed Chem Toxicol 31:995, 1993.
53. FS Chu, GY Li. Simultaneous occurrence of fumonisin B_1 and other mycotoxins in moldy corn collected from the People's Republic of China in regions with high incidences of esophageal cancer. Appl Environ Microbiol 60:847, 1994.
54. JP Rheeder, EW Sydenham, WFO Marasas, et al. Ear-rot fungi and mycotoxins in South African corn of the 1989 crop exported to Taiwan. Mycopathologia 127:35, 1994.
55. DS Wang, YX Liang, NT Chau, LD Dien, T Tanaka, Y Ueno. Natural co-occurrence of *Fusarium* toxins and aflatoxin B1 in corn for feed in North Vietnam. Nat Toxins 3:445, 1995.
56. R Widiastuti, R Maryam, BJ Blaney, Salfina, DR Stoltz. Cyclopiazonic acid in combination with aflatoxin, zearalenone and ochratoxin A in Indonesian corn. Mycopathologia 104:153, 1988.
57. HK Abbas, RT Riley. The presence and phytotoxicity of fumonisins and AAL-toxin in *Alternaria alternata*. Toxicon 34:133, 1996.
58. DT Wicklow, RJ Cole. Tremorgenic indole metabolites and aflatoxins in sclerotia of *Aspergillus flavus*: an evolutionary perspective. Can J Bot 60:525, 1982.
59. DL Davis, HL Bradlow, M Wolff, T Woodruff, DG Hoel, H Anton-Culver. Medical hypothesis: xenoestrogens as preventable causes of breast cancer. Environ Health Perspect 101:372, 1993.
60. RM Sharpe, NE Skakkeback. Are estrogens involved in falling sperm counts and disorders of male reproductive tract? Lancet 341:1392, 1993.
61. WG Helferich. Food phytoestrogens: an abundance of weak estrogen agonists. Fund Appl Toxicol (suppl) 30:87, 1996.
62. T Urano, M Trucksess, R Beaver, D Wilson, J Dorner, F Dowell. Co-occurrence of cyclopiazonic acid and aflatoxins in corn and peanuts. J AOAC Int 75:838, 1992.
63. MF Dutton, K Westlake. Occurrence of mycotoxins in cereals and animal feedstuffs in Natal, South Africa. J Assoc Off Anal Chem 68:839, 1985.
64. G Blunden, OG Roch, DJ Rogers, RD Coker, N Bradburn, AE John. Mycotoxins in food. Med Lab Sci 48:271, 1991.

65. ML Hooper. The role of *p53* and *Rb-1* genes in cancer, development, and apoptosis. J Cell Sci (suppl) 18:13, 1994.
66. SM Cohen, LB Ellwein. Cell proliferation and carcinogenesis. Science 249:1007, 1990.
67. SE Egan, RA Weinberg. The pathway to signal achievement. Nature 365:781, 1993.
68. H-S Yoo, WP Norred, E Wang, AH Merrill Jr, RT Riley. Fumonisin inhibition of *de novo* sphingolipid biosynthesis and cytotoxicity are correlated in LLC-PK_1 cells. Toxicol Appl Pharmacol 114:9, 1992.
69. E Wang, WP Norred, CW Bacon, RT Riley, AH Merrill Jr. Inhibition of sphingolipid biosynthesis by fumonisins: implications for diseases associated with *Fusarium moniliforme*. J Biol Chem 266:14486, 1991.
70. VL Stevens, S Nimkar, WC Jamison, DC Liotta, AH Merrill Jr. Characteristics of the growth inhibition and cytotoxicity of long-chain (sphingoid) bases for Chinese hamster ovary cells: evidence for an involvement of protein kinase C. Biochim Biophys Acta 1051:37, 1990.
71. S Spiegel, A Olivera, RO Carlson. The role of sphingosine in cell growth regulation and transmembrane signaling. Adv Lipid Res 25:105, 1993.
72. C Huang, M Dickman, G Henderson, C Jones. Repression of protein kinase C and stimulation of cyclic AMP response elements by fumonisin, a fungal encoded toxin which is a carcinogen. Cancer Res 55:1655, 1995.
73. JA Dunn, MB Faletto, SJ Kasper, HL Gurtoo. Aflatoxin-transformed C3H/10T1/2 cells overexpress protein kinase-C and have an altered response to phorbol ester treatment. Cancer Res 52:990, 1992.
74. Y Nishizuka, ed. Signal transduction: crosstalk. TIBS 17:367, 1992.
75. N Demaurex, DP Lew, KH Krause. Cyclopiazonic acid depletes intracellular Ca^{2+} stores and activates an influx pathway for divalent cations in HL-60 cells. J Biol Chem 267:2318, 1992.
76. WP Schilling, OA Cabello, L Rajan. Depletion of the inositol 1,4,5-trisphosphate-sensitive intracellular Ca^{2+} store in vascular endothelial cells activates the agonist-sensitive Ca^{2+} influx pathway. Biochem J 284:521, 1992.
77. WH Tolleson, WB Melchior J, SM Morris, et al. Apoptosis and antiproliferative effects of fumonisin B_1 in human keratinocytes, fibroblasts, esophageal epithelial cells, and hepatoma cells. Carcinogenesis 17:239, 1996.
78. MP Scheid, RW Lauener, V Duronio. Role of phosphoinositol 3-OH-kinase activity in the inhibition of apoptosis in haemopoietic cells: phosphoinositol 3-OH-kinase inhibitors reveal a difference in signalling between interleukin-3 and granulocyte-macrophage colony stimulating factor. Biochem J 312:159, 1995.
79. S Gupta, SB Krasnoff, NL Underwood, JA Renwick, DW Roberts. Isolation of beauvericin as an insect toxin from *Fusarium semitectum* and *Fusarium moniliforme* var *subglutinans*. Mycopathologia 115:185, 1991.
80. A Bottalico, A Logrieco, A Ritieni, A Moretti, G Randazzo, P Corda. Beauvericin and fumonisin B1 in preharvest *Fusarium moniliforme* maize ear rot in Sardinia. Food Addit Contam 12:599, 1995.
81. CW Lim, HM Parker, RF Vesonder, WM Haschek. Intravenous fumonisin B_1 induces cell proliferation and apoptosis in the rat. Nat Toxins 4:34, 1996.

82. WH Tolleson, WG Sheldon, JD Thurman, TJ Bucci, PC Howard. The mycotoxin fumonisin induces apoptosis in cultured human cells and in liver and kidney of rats. In: L Jackson, JW DeVries, LB Bullerman, eds. Fumonisins in Foods. New York; Plenum Publishing Corp., 1996, p 237.
83. WH Tolleson, WB Melchior Jr, SM Morris, et al. The effects of fumonisin B_1 on human keratinocytes are similar to those caused by sphinganine and ceramide. Abstracts of the Annual Meeting of South Central Flow Cytometry Association, 1995.
84. S Jayadev, B Liu, AE Bielawska, et al. Role for ceramide in cell cycle arrest. J Biol Chem 270:2047, 1995.
85. S Nakamura, Y Kozutsumi, Y Sun, Y Miyake, T Fujita, T Kawasaki. Dual roles of sphingolipids in signaling of the escape from and onset of apoptosis in a mouse cytotoxic T-cell line, CTLL-2. J Biol Chem 271:1255, 1996.
86. WCA Gelderblom, SD Snyman, L van der Westhuizen, WFO Marasas. Mitoinhibitory effect of fumonisin B_1 on rat hepatocytes in primary culture. Carcinogenesis 16: 625, 1995.
87. JJ Schroeder, HM Crane, J Xia, DC Liotta, AH Merrill Jr. Disruption of sphingolipid metabolism and stimulation of DNA synthesis by fumonisin B_1. J Biol Chem 269: 3475, 1994.
88. R Harel, AH Futerman. Inhibition of sphingolipid synthesis affects axonal outgrowth in cultured hippocampal neurons. J Biol Chem 268:14476, 1993.
89. H-S Yoo, WP Norred, JL Showker, RT Riley. Elevated sphingoid bases and complex sphingolipid depletion as contributing factors in fumonisin-induced cytotoxicity. Toxicol Appl Pharmacol 138:211, 1996.
90. BD Price, MB Youmell. The phosphatidylinositol 3-kinase inhibitor wortmannin sensitizes murine fibroblasts and human tumor cells to radiation and blocks induction of *p53* following DNA damage. Cancer Res 56:246, 1996.
91. Y Mizrachi, M Lev, Z Harish, SK Sundaram, A Rubinstein. L-cycloserine, an inhibitor of sphingolipid biosynthesis, inhibits HIV-1 cytopathic effects, replication, and infectivity. J AIDS Human Retrovirol 11:137, 1996.
92. B Papp, RA Byrn. Stimulation of HIV expression by intracellular calcium pump inhibition. J Biol Chem 270:10278, 1995.
93. ML Marin, J Murtha, W Dong, JJ Pestka. Effects of mycotoxins on cytokine production and proliferation in EL-4 thymoma cells. J Toxicol Environ Health 48: 101, 1996.
94. PH Fishman, T Pacuszka, PA Orlandi. Gangliosides as receptors for bacterial enterotoxins. Adv Lipid Res 25:165, 1993.
95. CA Lingwood. Verotoxins and their glycolipid receptors. Adv Lipid Res 25:189, 1993.
96. K Sandvig, O Garred, A van Helvoort, G Van Meer, B Van Deurs. Importance of glycosphingolipid synthesis for butryic acid-induced sensitization to Shiga toxin and intracellular sorting of toxin in A431 cells. Mol Biol Cell 7:1391, 1996.
97. JI Azcona-Olivera, YL Ouyang, J Murtha, FS Chu, JJ Pestka. Induction of cytokine mRNAs in mice after oral exposure to the trichothecene vomitoxin (deoxynivalenol): relationship to toxin distribution and protein synthesis. Toxicol Appl Pharmacol 133:109, 1995.

98. YL Ouyang, JI Azcona-Olivera, J Murtha, J Pestka. Vomitoxin-mediated IL-2, IL-4, and IL-5 superinduction in murine CD4$^+$ T cells stimulated with phorbol ester and calcium ionophore: relation to kinetics of proliferation. Toxicol Appl Pharmacol 138:324, 1996.
99. DL Gill, TK Ghosh, J Bian, AD Short, RT Waldron, SL Rybak. Function and organization of the inositol 1,4,5-trisphosphate-sensitive calcium pool. In: JW Putney, ed. Advances in Second Messenger and Phosphoprotein Research. New York; Raven Press, 1992, p 265.
100. JW Putney. Inositol phosphate and calcium entry. In: JW Putney Jr, ed. Advances in Second Messenger and Phosphoprotein Research. New York, Raven Press, 1992, p 143.
101. AP Somlyo, AV Somlyo. Signal transduction and regulation in smooth muscle. Nature 372:231, 1994.
102. LR Ballou, SLF Laulederkind, EF Rosloniec, R Raghow. Ceramide signalling and the immune response. Biochim Biophys Acta 1301:273, 1996.
103. AH Merrill Jr, E-M Schmelz, DL Dillehay, et al. Sphingolipids—the enigmatic lipid class: biochemistry, physiology, and pathophysiology. Toxicol Appl Pharmacol 142: 208, 1997.
104. VL Tesh, B Ramegowda, JE Sammuel. Purified Shiga-like toxins induce expression of pro-inflammatory cytokines from murine peritoneal macrophages. Infect Immun 62:5085, 1994.

8
Mycotoxin Formation and Environmental Factors

David Abramson
Agriculture and Agri-Food Canada, Winnipeg, Manitoba, Canada

I. INTRODUCTION

A. General Considerations

Several aspects of mycotoxin research have been discussed in reviews since 1990. Various topics have been summarized including fungal production [1–3], chemical properties [4], detection and occurrence in agricultural products [5], biological effects [6], and storage ecology [7,8]. Ongoing coverage of mycotoxin assay methods and commodity surveys is provided by the Association of Official Analytical Chemists International, which publishes an annual summary in the January issue of its journal [9].

In this chapter, the term "mycotoxin" is used to describe fungal secondary metabolites which occur naturally as contaminants of agricultural products, and which show toxicity via a natural route of administration, primarily the oral route. This definition covers many of the *Aspergillus* and *Penicillium* toxicants, most of the naturally occurring *Fusarium* trichothecenes, and one of the *Alternaria* toxins—tenuazonic acid. This definition excludes secondary metabolites which to date have been produced solely in laboratory fermentation cultures, such as the rubratoxins, and metabolites without proven oral toxicity, such as erythroskyrin, rugulosin [10], alternariols, and altenuenes [11,12]. Although zearalenone and its derivatives have widespread occurrence and proven estrogenic activity, they have negligible oral toxicity [13] and will be termed "fungal estrogens" rather than mycotoxins.

B. Taxonomy of Toxigenic Species

Mycotoxins discussed in this chapter are often produced by a single genus, but some toxins produced by *Penicillium* species, e.g. cyclopiazonic acid and ochratoxin A, are produced by *Aspergillus* species as well [10]. Classification of toxin-producing *Penicillium* species has recently undergone revision on the basis of several new taxonomic criteria. The scheme currently in widest use is based on assays of homogeneous *Penicillium* cultures for known secondary metabolites by liquid chromatography and diode-array spectrometry [14,15]. Although there is some controversy about correct identification [16] of some of the secondary metabolites, this scheme is gaining acceptance, and the current taxonomic classification of *Penicillium* species [17] will be used in this chapter. The *Fusarium* species discussed in this review follow the taxonomic system of Nelson et al. [18].

II. PREHARVEST PRODUCTION

A. *Fusarium* Species

Wheat, barley, and corn together account for about two-thirds of the world production of cereals, and unfortunately are the crops most susceptible to *Fusarium* fungal disease and trichothecene toxin contamination. *Fusarium* fungi and toxins are also a problem in rye, oats, and triticale [19–21], but the incidence and levels of contamination are generally much lower.

Some species of *Fusarium* downgrade cereal crops first by reducing quality and yield, and second by producing mycotoxin contamination. *F. graminearum* and *F. culmorum* develop during midsummer to produce fusarium head blight (known as "scab" in the U.S.) in wheat and barley, along with trichothecene mycotoxins. An epidemic affecting less than a quarter of the heads in a field may still produce a crop that exceeds accepted tolerances for mycotoxins in human food or animal feed. The main toxin associated with fusarium head blight is deoxynivalenol, although acetyl-deoxynivalenols may also be present. Fortunately, deoxynivalenol is one of the least toxic substances produced by *Fusarium* fungi (Table 1). Recent regulations in North America [22] specify limits of 1 ppm deoxynivalenol for wheat (U.S.), 2 ppm deoxynivalenol in uncleaned soft white winter wheat for adult food (Canada), and 1.2 ppm in uncleaned soft white winter wheat intended for baby food (Canada). Current guidelines recommend that feed intended for swine contain no more than 1 ppm deoxynivalenol in order to avoid feed refusal problems, but other livestock such as poultry and cattle appear to be far more tolerant [23].

In corn crops, *F. moniliforme* and the related fungi *F. subglutinans* and *F. proliferatum* cause fusarium kernel rot and contamination by moniliformin and by a recently characterized group of toxins known as the fumonisins [24]. Acceptable

TABLE 1 Oral Toxicity of Some Compounds Isolated from *Fusarium* Fungi

Compound	Test animal	LD_{50} (mg/kg)	Reference
Type-A trichothecenes			
T-2 toxin	Rat	5.2	127
	Chick	5.0	128
HT-2 toxin	Chick	7.2	128
Diacetoxyscirpenol	Rat	7.3	127
	Chick	3.8	128
Type-B trichothecenes			
Deoxynivalenol (DON)	Mouse	46.0	127
3-Acetyl-DON	Mouse	34.0	
15-Acetyl-DON	Mouse	34.0	129
Nivalenol	Rat	19.5	130
Fusarenone-X	Mouse	4.5	127
	Chick	33.8	128
Nontrichothecenes			
Moniliformin	Cockerel[a]	4.0	131
Fumonisin B_1	Horse	[b]	132
Estrogens			
Zearalenone	Mouse	>5000	13

[a]One day old.
[b]Leukoencephalomalacia arising from 1 to 4 mg/kg fumonisin B_1.

levels of moniliformin and fumonisins in human and animal diets have not been determined. In corn, *F. graminearum* and other *Fusarium* fungi also cause *Gibberella* ear rot, and contamination by deoxynivalenol and zearalenone, and by T-2 toxin and diacetoxyscirpenol if *F. sporotrichioides* and *F. poae* are involved.

1. Toxigenic Species

Wheat and Barley. Fusarium head blight (known as "scab" in the U.S.) is a disease of wheat and other small grains caused by several *Fusarium* species. *F. graminearum*, the species found most often in affected wheat and barley, is not always the prime cause of *Fusarium* head blight in all regions, and there is usually some year-to-year variability in the various species isolated. The most commonly found species, *F. graminearum* and *F. culmorum*, are the most pathogenic.

F. graminearum (sexual stage *Gibberella zeae*) predominates in blighted wheat from Canada, the U.S., and China [2,25]. It is highly pathogenic to both wheat and corn. North American isolates of this species produce zearalenone [26] and deoxynivalenol along with the related 8-ketotrichothecenes 15-acetyl-deoxynivalenol and nivalenol. Isolates from Australia, New Zealand, China, and northern Europe tend to produce 3-acetyl-deoxynivalenol instead of 15-acetyl-deoxynivalenol [26]. *F. graminearum* isolates from western Canadian grains can also produce significant levels of fusarenone-X (4-acetyl-nivalenol) in culture [27]. The visual symptoms arising from infection by *F. graminearum* in wheat are generally quite apparent. For example, *Fusarium* head blight in western Canadian hard spring wheat produces characteristic "tombstone" kernels which are typically smaller than normal, shriveled in appearance, and white or pale pink in color [28]. In corn, North American isolates of *F. graminearum* can produce high levels of deoxynivalenol, 15-acetyl-deoxynivalenol and zearalenone; Abbas et al. [29] found 20, 16, and 5 ppm, respectively, in a sample of corn involved in feed refusal by swine.

F. culmorum (no sexual stage identified) is the predominant species in cooler wheat-growing areas of northern Europe, eastern Europe, and Scandinavia [19,30–32]. Its pathogenicity toward wheat and corn ranks high to moderate [2]. This species produces mainly deoxynivalenol and 3-acetyl-deoxynivalenol. As with *F. graminearum*, visual symptoms arising from infection are generally conspicuous.

F. crookwellense (no sexual stage identified) occurs occasionally in wheat from Canada [28] and Poland [19], but is the main source of *Fusarium* head blight on irrigated land in Orange Free State, South Africa [33]. Its pathogenicity toward wheat is rated low to moderate, and toward corn is rated moderate. An isolate of this species from corn in Manitoba produced fusarenone-X, deoxynivalenol, and 15-acetyl-deoxynivalenol in culture [27].

F. avenaceum (sexual stage *Gibberella avenacea*) is also common in wheat from all regions studied. Its pathogenicity toward wheat and corn ranks low [2]. Compared to *F. graminearum* and *F. culmorum*, this species is a relatively weak producer of deoxynivalenol in culture [27]. Recent evidence indicates that *F. avenaceum* in durum wheat may cause serious moniliformin contamination [34]. Isolates (17/20) of this fungus from durum wheat in Austria produced 32 to 800 ppm moniliformin on corn grits in culture.

F. sporotrichioides and *F. poae* are isolated from wheat and corn in low to moderate frequencies, often under cooler conditions of a late harvest. Both species are weakly pathogenic to cereals, but are capable of producing the comparatively toxic type-A trichothecenes (Table 1). It has been suggested that these fungal species cannot produce type-B trichothecenes [35], but current evidence indicates that *F. poae* isolates are quite capable of making the type-B trichothecenes fusarenone-X [27] and nivalenol [36].

Corn. In corn, *Gibberella zeae* (sexual stage of *F. graminearum*) and *F. culmorum* cause *Gibberella* ear rot, also called pink ear rot. This disease is prevalent in northern temperate climates, especially in wet years. In outbreaks of this disease, *F. graminearum* is predominant in Canada and northern U.S., while *F. culmorum* tends to be the dominant species in north-central and eastern Europe [37,38].

A second disease of corn, *Fusarium* kernel rot, occurs during warm dry weather and is facilitated by insect damage. It is caused by *F. moniliforme, F. subglutinans* (sexual stage *G. subglutinans*), and *F. proliferatum. F. moniliforme* and *F. subglutinans* are very common in corn kernels in Canada, the U.S., and Europe. In southern U.S. and tropical locales, *F. moniliforme* is the main cause of *Fusarium* kernel rot [39,40], with insect damage and warm dry weather as facilitating factors. Insect damage appears to promote *F. moniliforme* occurrence regardless of moisture conditions.

F. moniliforme, F. subglutinans, F. proliferatum, and other species produce the mycotoxin moniliformin. This toxin was found originally in naturally infected corn from South Africa [41,42] and Germany [43]. In Canada, Scott and Lawrence [44] later found moniliformin in two corn samples. Canadian isolates of *F. moniliforme* and *F. subglutinans* produced moniliformin in culture, and the former species also produced this toxin in field-inoculated corn [45]. Lew et al. [46] found that *F. subglutinans* produced large amounts of this toxin in Austrian corn. Sharman et al. [47] have found moniliformin in samples of corn from the U.S., Great Britain, Netherlands, France, Italy, several African countries, and Thailand. They also found this toxin was produced by *F. subglutinans* in corn from Poland, and by *F. avenaceum* in wheat, rye, and oats. Several other *Fusarium* species also produce moniliformin [35].

F. moniliforme [48] and *F. subglutinans* [49] also produce the fumonisin group of mycotoxins, although this group is also produced by other *Fusarium* species. Fumonisins B_1 and B_2 are reported to cause leukoencephalomalacia in horses, pulmonary edema in pigs, and hepatic cancer in rats [49–51]. Fumonisins occur naturally in corn in the U.S. [52,53], Poland [46], South Africa [24,54], and other countries [20,55]. Fumonisin-producing isolates of *F. moniliforme* are also widely distributed among various crops in Taiwan [56]. High fumonisin levels in corn associated with human esophageal cancer [54] have prompted a suggestion of 1 ppm as a preliminary tolerance level for foods in Switzerland [57].

In addition to the *Gibberella* ear rot, arising from *F. graminearum* and *F. culmorum*, and *Fusarium* kernel rot, arising from *F. moniliforme*, there appears to be a third type of *Fusarium* disease in corn called red ear rot or red fusariosis [38]. Far less common than the gibberella ear rot described above, this disease is associated with several species including *F. tricinctum, F. equiseti, F. sporotrichioides, F. crookwellense*, and *F. avenaceum*, and trichothecenes of the A type, including T-2 toxin [58].

2. Bioclimatic Conditions: Inoculum, Temperature, and Moisture

Wheat and Barley. Variations in temperature and moisture generally interact to produce differential growth of fungal species, including field fungi such as fusaria [59]. Consequently, *Fusarium* diseases of cereals are associated with different species under various climatic conditions. For the highly pathogenic species, *F. graminearum* predominates on cereals grown under warmer conditions, while *F. culmorum* is less prevalent. The less pathogenic species *F. sporotrichioides, F. poae*, and *F. nivale* generally occur under cooler conditions. Regional and year-to-year predominance of pathogenic species are dependent on temperature; from coolest to warmest conditions, the most abundant species are *F. culmorum, F. crookwellense, F. avenaceum*, and *F. graminearum* [2].

Although copious rainfall promotes *Fusarium* head blight, the incidence is evidently most affected by moisture at anthesis [60,61]. This implies that the disease results from the timing of humidity and precipitation rather than the amount. Ascospores can be released from debris to initiate infection (Fig. 1),

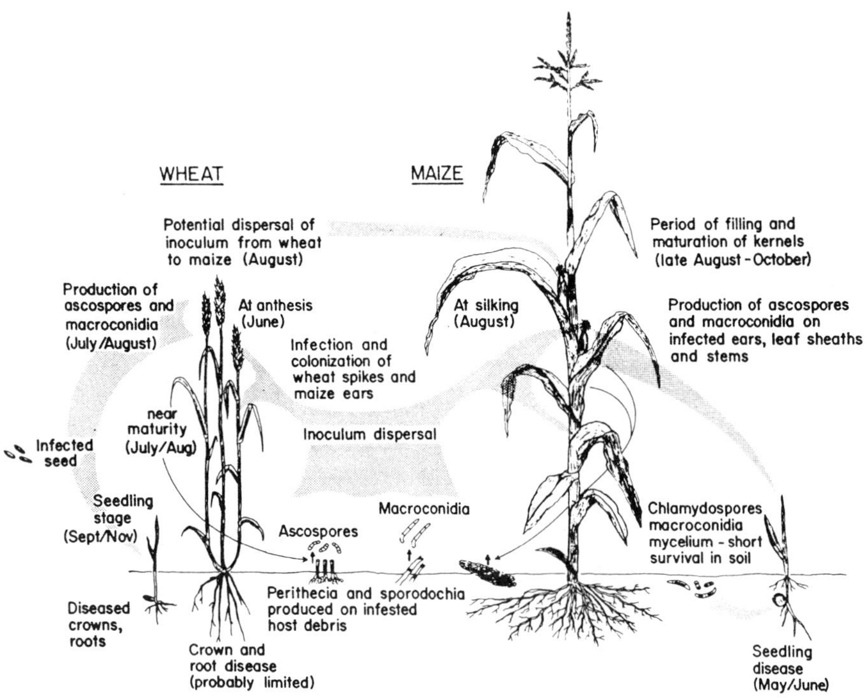

FIGURE 1 Disease cycle of *Fusarium graminearum* in wheat and corn. (From Ref. 60.)

which is followed by the production of macroconidia under favorable conditions. Dispersal of ascospores has been found to be related to high relative humidities. Although wheat cultivars having resistance to *Fusarium* head blight are most susceptible at anthesis [62], susceptible cultivars continue to be receptive to infection up to head maturity [63]. Epidemics where an appreciable percentage of the heads in a field are infected appear to require moisture for some time after anthesis, allowing the fungus to sporulate on the heads and infect adjacent plants and late-flowering tillers. This explanation is consistent with high measured concentrations of airborne conidia over wheat fields. In most years, the most important factor in determining severity of infection remains moisture during the susceptible period at anthesis [64]. In addition, insect damage apparently facilitates entry of *F. graminearum* and other species that cause head blight [60].

In new "hull-less" varieties of barley, removal of the hull from the caryopsis at harvest exposes visual evidence of *Fusarium* damage similar to that seen in wheat. Symptoms of infection in regular barley, however, are quite different from those observed in wheat. This arises from the retention of the lemma and palea on the harvested seed, which covers any damage occurring to the caryopsis. Official grading tolerances have been established [65] in Canada for *Fusarium* damage in barley—specifically, orange sporodochia and black perithecia on the seed.

Unlike wheat, where the proportion of visibly *Fusarium*-damaged kernels is loosely related to the level of *Fusarium* mycotoxins in the sample, in barley it is not. The presence or absence of visible *Fusarium* damage in a barley sample is consequently a poor criterion for estimating the risk of trichothecene contamination. Barley samples containing seeds infected by *F. graminearum* often display a dark smudge at the basal end of the seed. This discoloration is, however, not specific for *Fusarium* infection, since it is also associated with infection of seeds by the field fungi *Cochliobolus sativus* and *Alternaria alternata*. Without a reliable visual criterion for *Fusarium* damage in barley, postharvest assessment of the disease level is restricted to indirect measurements such as deoxynivalenol content or fungal assessment via plating techniques [66].

Corn. The development of *F. graminearum* and *Gibberella* ear rot in corn is dependent on moisture at silk emergence, and on damp weather and moderate temperatures later in the season. Fortuitously, in the northern temperate zone, conditions of continuous rain are usually accompanied by temperatures too low for the rapid growth of *F. graminearum* [60]. For rapid growth of *F. graminearum*, a period of approximately 15 growing degree-days with average temperature >5%C appears necessary from field trials with experimentally infected ears [67].

Corn is susceptible to *F. graminearum* infection from airborne spores for a period of about 7 to 10 days after silking begins (Fig. 1). Inoculum sources are conidia or ascospores from crop debris on the soil surface, or spores released from growth on weeds or grasses near the cornfield. If weather earlier in the season is especially conducive to ascospore production on crop debris, the inoculum abundance will increase. Insects such as picnic beetles and the European corn borer commonly act as spore vectors, and may facilitate inoculation by damaging the plant. Sufficient moisture and adequate temperature are needed to establish fungal colonization. Adequate moisture for plant growth is adequate for fungal growth [67]. For severe epidemics, *Gibberella* ear rot requires high rainfall and normal to above normal temperatures to allow reinfection from rain-splashed, airborne, and/or insect-vectored propagules (or propagules that land on the husks or silks and become dormant).

Although the presence of species such as *F. graminearum* and *F. culmorum* in corn is normally influenced by temperature and moisture, the ubiquity of *F. moniliforme* is not. Since *F. moniliforme* appears to be endemic in kernels, current research suggests that kernel damage by insects and birds and predisposing weather are sufficient to promote *Fusarium* kernel rot disease [24]. Plant injury by insects such as the European corn borer was found to increase *Fusarium* kernel rot and fumonisin formation [46]. Schaafsma et al. [68] suggested that plant damage resulting from *F. graminearum* infection might also be conducive to *F. moniliforme* development and fumonisin formation. Generally, moderate rain at silk emergence, alternating with periods of warm dry weather, appears to be required for *Fusarium* kernel rot in corn, and is facilitated by insect damage.

There appears to be a third type of *Fusarium* disease in corn, called red ear rot or red fusariosis, mentioned previously. *F. sporotrichioides* is associated with this disease, and growth of the fungus occurs over extended periods of wet and cold weather [38]. This species produces fewer symptoms than more pathogenic species such as *F. graminearum*, but can produce low concentrations of T-2 toxin [69].

3. Other Factors

Both host and nonhost plants can serve as sources of *F. graminearum* inoculum. The main nonhost plants include soybeans, sorghum, and wild oats and other kinds of weeds [70,71]. Crop rotation has been recognized as an important factor in the spread of the inoculum. In field trials, a corn/soybean rotation yielded a less extensive disease outbreak than corn/corn [72]. The phenomenon of severe *F. graminearum* infection in both wheat and corn has led to the recommendation of not planting wheat after corn. Wheat, planted in the same field following a corn crop, was reported to undergo more *Fusarium* head blight than nonrotated wheat. This may arise from longer persistence of the fusaria on infected corn debris fragments, which are typically larger than those from infected wheat [73].

Foliar and seed-treatment fungicides have not proven to be worthwhile against *Fusarium* head blight. In a trial involving foliar fungicides on wheat, symptoms were reduced compared with controls, but deoxynivalenol content was unaffected [70]. Foliar fungicides have been shown to have little effect on *Fu-*

sarium head blight incidence [74]. In laboratory tests, certain concentrations of the fungicide tridemorph actually increased production of T-2 toxin by *F. sporotrichioides* [75]. In field trials, barley inoculated with *F. sporotrichioides* and treated with tridemorph showed greater toxin concentrations than untreated control barley after 21 days [76]. Seed treatment fungicides have had little effect on *Fusarium* head blight, suggesting that crop debris in soil, as discussed above, is a more important source of inoculum [77,78]. As an alternative to fungicides, breeding new varieties using germ plasm from genotypes showing resistance to *Fusarium* head blight offers some promise. Wong et al. [79] recently tested 11 wheat cultivars against seven *Fusarium* species, and found high levels of resistance, expressed as low disease severity combined with low trichothecene production, in several Chinese cultivars.

B. *Alternaria* Species

Although *Alternaria* secondary metabolites, e.g., the alternariol and altenuene types, have been found to occur naturally in American sorghum [11,80] and in Polish wheat and rye [81], most of these metabolites do not demonstrate any oral toxicity [11,12]. One notable exception is tenuazonic acid, which is produced by *Alternaria alternata*. In young chickens, the single-dose LD_{50} is 37.5 mg/kg [82], while in rats the single-dose LD_{50} for the sodium salt is 174 mg/kg [83]. Culture material containing crude tenuazonic acid showed oral toxicity when added to the diets of young chickens and rats [11].

Tenuazonic acid can be found in fruits, vegetables, and grains, notably in mandarin oranges at 174 ppm [84], in tomatoes at 0.4 to 70 ppm [85], in olives at 0.26 ppm [86], and in sorghum and ragi up to 6 ppm [87]. Tenuazonic acid has also been found in diseased rice plants [88].

Alternaria species are widely distributed in crops and soil. These fungi require a moisture content of about 28% to 34% for growth, with the result that most *Alternaria* infection of cereals occurs preharvest, when the moisture content is sufficiently high. *Alternaria* secondary metabolites usually accumulate only when drying in the field is delayed by rain or humid weather, resulting in extensive infection. In fruits and vegetables, postharvest *Alternaria* proliferation, even at refrigeration temperatures, is facilitated by high moisture content, overripening, and surface damage [89]. The disease cycles and pathogenicity of different *Alternaria* species have recently been reviewed in depth [90]. The taxonomy of *Alternaria alternata* and related taxa is still under revision [91].

C. *Aspergillus* Species

Owing to their health threat and global distribution, the aflatoxins, produced by *Aspergillus* species, are the most studied mycotoxins in agricultural crops. The aflatoxins are formed by three closely related species: *A. flavus, A. parasiticus*, and

A. nomius [92]. Aflatoxins are very widespread, since they are produced in all tropical and subtropical environments, preharvest and postharvest, on a variety of crops [93]. In terms of acute toxicity, aflatoxin B_1 has an oral LD_{50} in rats of 7.2 mg/kg [10]; it is also carcinogenic, immunosuppressive, and teratogenic. The other aflatoxins—B_2, G_1, and G_2—are also toxic, but less so.

Although the aspergilli are considered to be mainly storage molds, aflatoxins can form in the field, complicating control measures. Aflatoxins are primarily a problem in corn, and not in other cereals. In the U.S., where large quantities of corn are grown, and where preharvest contamination by aflatoxins is a major concern [94], *A. flavus* rather than *A. parasiticus* is the main causative species. In American corn-growing areas, *A. flavus* infection begins in two basic ways: (1) kernels damaged by insects and birds become colonized with the fungus; and (2) conidia from air or insects contaminate the silk and grow into the ear. Plant stress through drought, heat, or nutrient imbalance is conducive to *A. flavus* colonization [95].

Insect activity appears to be the major cause of *A. flavus* contamination of corn in the U.S. [94,96]. The insects not only damage the kernels, but also act as inoculum vectors, and *A. flavus* conidia are common on insects late in the growing season [95]. Application of insecticides to corn silk reduces the eventual kernel contamination by *A. flavus* [97]. In the U.S., corn debris after harvest and *A. flavus* sclerotia act as a source of inoculum for the next crop, since the sclerotia can germinate and produce conidia at the time of silking in the corn [98].

III. POSTHARVEST PRODUCTION

A. General Considerations

The growth of storage fungi, and their production of secondary metabolites, presents a major problem to agriculture, feed, and food industries worldwide. Fungal invasion of stored agricultural products leads to quality loss in many ways: unacceptable physicochemical changes in the products such as deterioration of color, texture, and taste [99,100]; development of fungal odors [7,101]; reduction of seed germination [102]; energy and nutritional loss [102]; production of allergens affecting livestock [103]; and formation of toxic metabolic products such as mycotoxins. Although fungal cultures are used for production of antibiotics and other therapeutic substances [104], products from fungi usually present a toxic hazard when present in food or feed. Fungal growth and mycotoxin production are frequent and obvious signs of storage problems.

The subject of mycotoxins in stored products, especially cereals and other feed ingredients, has been reviewed several times since 1990 [3,5,6,105]. The aflatoxins, ochratoxin A, ergot alkaloids, and *Fusarium* trichothecenes are often considered the most significant toxins for human and animal health in terms of their direct effects, but several other toxins may also cause human and animal disease. There is increasing concern about the synergism among mycotoxins [106], and there is the possibility that several mycotoxins may compromise animal and human immune systems, leading to subsequent bacterial infection [107].

In the past, storage fungi have been considered generalists, capable of growing on all kinds of food and feed substrates. Today, modern species identification techniques [17] and storage simulation studies point toward an improved understanding of the spoilage process. Evidence indicates that, in defined bioclimatic locales under typical conditions, each agricultural commodity contains only certain active saprophytic mycoflora. Furthermore, studies on fungal species indicate some specificity in the production of secondary metabolites on artificial media [108]. Whether stored agricultural commodities have any specific profiles of potential mycotoxins determined by ecological factors remains to be definitively established.

B. Mycotoxins and Fungi in Stored Crops

Mycotoxins that have proven oral toxicity and that have been found occurring naturally in stored agricultural commodities are indicated in Table 2. Aflatoxins may also be produced under field conditions, mainly in corn (see above). Although Table 2 deals mainly with postharvest cereals and oilseeds, other agricultural commodities such as stored cheeses, fruits, and meat products are also suitable substrates for toxigenic fungi such as *P. commune* and *P. expansum*.

Crops entering storage, especially cereal grains, are commonly contaminated by propagules of *Aspergillus* and *Penicillium* species. Dormant conidia originating from soil particles, plant debris, and residues from harvesting equipment often occur on seed surfaces [109]. The aspergilli are widespread, and grow on many agricultural products including cereals in storage environments. The main toxigenic species are *A. flavus* and *A. parasiticus* for aflatoxins, and *A. ochraceus* for ochratoxins. Penicillia and aspergilli are ubiquitous fungi, with species well able to grow at the moisture contents encountered in storage environments. Once fungal growth begins in storage, moisture is released through metabolism, leading to growth of other species and to heating. At these increased water activity levels, approximately 0.86 and above for penicillia, production of mycotoxins such as ochratoxin A begins [110].

Although toxin production in laboratory cultures appears to be affected mainly by moisture, temperature, and time, a large number of environmental variables affect mycotoxin contamination of cereals in storage. This has created problems in establishing cause-and-effect relationships among the various factors, and predictive problems arise in typical multispecies storage situations. Traditional laboratory studies in cultures and controlled environments can provide preliminary information to account for mycotoxin contamination of cereals, but

Table 2 Mycotoxins Occurring in Stored Agricultural Products

Mycotoxin	Producers[a]	Food or feed contamination	Oral LD_{50}, mg/kg	Reference
Aflatoxin B_1	*A. flavus*	Corn	7.2, rat	10
	A. parasiticus	Nuts		
	A. nomius	Cottonseed		
		Copra		
Cyclopiazonic acid	*A. flavus*	Corn	36, rat	133
	A. tamarii	Peanuts		
	P. griseofulvum	Cheese		
	P. commune			
	P. camemberti			
Ochratoxin A	*P. verrucosum*	Wheat	22, rat	10
	A. ochraceus	Barley		
	A. ostianus	Rice		
Citrinin	*P. verrucosum*	Wheat	56, turkey	134
	P. citrinum			
	A. terreus			
Patulin	*P. expansum*	Apples	35, mouse	135
	P. griseofulvum			
Penicillic acid	*P. aurantiogriseum*	Corn	90, chicken	136
	P. aurantiovirens	Beans	600, mouse	
	P. cyclopium			
	P. freii			
	P. viridicatum			
Penitrem A	*P. crustosum*	Wheat bread	10, mouse	137
	P. melanoconidium	Nuts		
		Cheese		
Secalonic acid D	*P. oxalicum*	Corn (dust)	25, rat	138
Viomellein	*P. freii*	Barley	[b]	10
	P. cyclopium			
	P. viridicatum			
Xanthomegnin	*P. freii*	Corn	[c]	10
	P. cyclopium	Cereal grains		
	P. viridicatum			

[a]According to Ref. 17.
[b]Jaundice in male mice at 456 mg viomellein/kg feed.
[c]Jaundice in male mice at 448 mg xanthomegnin/kg feed.

the predictive ability of the results has been questioned in both preharvest and storage situations [111]. The simple extrapolation from controlled-environment studies to an actual agricultural context rarely applies, and this has prompted an increasing emphasis on multiple-variable studies.

Studies of mycotoxin development in storage have dealt mainly with investigating the natural development of mycotoxins and associated fungal growth. In some early work, Trenk and Hartman [112] stored naturally infected corn at varying temperature and moisture content (MC) values, and found aflatoxin B_1 production by indigenous strains of *A. flavus* at MC of 17.5% and above, and at temperatures of 24°C and above. Other *Aspergillus* and *Penicillium* species were also present. Seitz et al. [113] studied fungal growth as measured by ergosterol content of grain, loss of dry matter, and levels of aflatoxins during the storage of high-moisture corn in bins aerated from the bottom. Fungal invasion of kernels was found to be highest at the top, and lowest at the bottom of the bins. Abramson et al. [114–117] monitored quality changes and mycotoxin development in moist bin-stored Canadian cereals. When six-row barley, hard red spring wheat, and oats were stored at 20.5% MC for 20 weeks, ochratoxin A production occurred in the barley and to a lesser extent in the wheat, and was associated with *Penicillium* species. No mycotoxin production was observed in oats [114]. When six-row barley was stored at 16% and 20% MC for 66 weeks, ochratoxin A and sterigmatocystin were produced at 20% MC starting at 20 weeks. Ochratoxin A production was associated with *Penicillium*, and sterigmatocystin with *A. versicolor* [115]. Indigenous isolates of *A. versicolor* from stored products were verified as producers of sterigmatocystin in culture [118]. When corn was stored at 16% and 21% MC for 66 weeks, ochratoxin A was produced at the latter MC starting at 4 weeks. Ochratoxin A production was associated with *Penicillium*, and reached a maximum of 3.6 ppm by 8 weeks [116]. To assess substrate effect in a granary situation, a typical two-row barley, used for malting, and a six-row barley, used for feed, were stored at 19% MC for 60 weeks [119]. The six-row barley produced significantly more ochratoxin A than the two-row, and *P. verrucosum* was the associated fungus. The effect of substrate was again apparent with amber durum wheat stored at 15% and 19% MC for 60 weeks [117]. Ochratoxin A and citrinin reached maximum levels of 8 ppm and 85 ppm, respectively, by 48 weeks, attaining some of the highest levels yet encountered in stored grain.

C. Factors Affecting Postharvest Mycotoxin Formation

The formation of mycotoxins at harvest time is affected by environmental factors (e.g., temperature, MC, rapidity of drying, rewetting, ambient humidity), degree of exposure of the crop to mechanical injury, and extent of inoculation. These factors, along with favorable postharvest conditions, can facilitate further toxin development, but mycotoxins are often absent at harvest [5]. Contamination usually arises from predisposing storage conditions.

Fungal growth, crop spoilage, and mycotoxin formation result from the interaction of numerous factors illustrated in Figure 2. At the beginning of the deteriorative process, the agricultural crop is contaminated by fungal spores from the air, other grain, or dockage. Mechanical damage from harvesting equipment, and from rodents, birds, or insects, can break the outer seed coat in cereals and further facilitate infection. Insects, through their movement, also aid in spreading the inoculum. Fungal growth depends on available oxygen, time, heat, and moisture. The last two factors are augmented by insect activity and by the metabolism of the fungi themselves, as illustrated. Moisture to initiate the spoilage process can also come from rain seepage, the melting of snow driven into the bin during winter, or moisture migration and condensation on the storage bin walls. The eventual development of mycotoxins depends on the growth and predominance of toxigenic fungal strains, heat for further fungal development, a substrate suitable for the fungus, and time for fungal metabolism to enter into its secondary phase [7]. Chemical byproducts of fungal growth, such as free fatty acids, CO_2, ergosterol, and the specific fatty alcohols and fatty ketones that constitute fungal odors, may be useful as signs of impending mycotoxin formation.

Several of the above factors connected with mycotoxin production have been investigated in various stored cereals, including wheat, barley, oats, and corn [114–117]. The following effects, in addition to increasing mycotoxin levels and the succession of fungal species, are generally apparent: increases in MC, grain temperature, fungal propagule count, CO_2, and free fatty acids; decreases in O_2 and seed germination; and the appearance of fungal odors. Fungal volatiles such as 3-methyl-1-butanol, 1-octanol, 3-methyl-furan, and 3-octanone have been characterized in culture [101,120,121], and some have been detected in granary storage experiments [114,115].

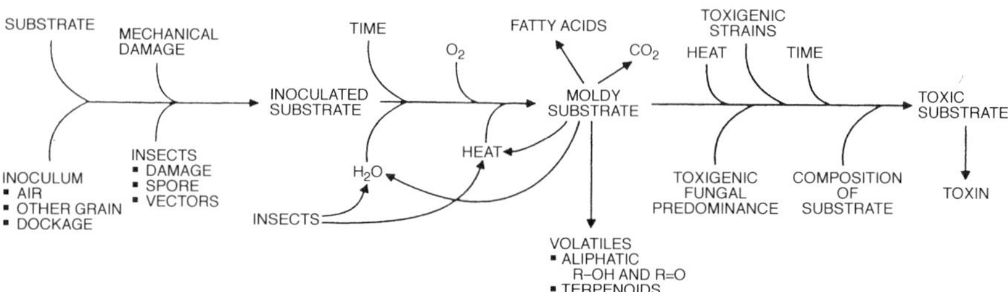

FIGURE 2 Some factors affecting mycotoxin formation in stored cereal crops.

To numerically assess the significance of some of these factors as variables within a stored-grain ecosystem, a statistical procedure based on principal component analysis (PCA) has been used to group together interrelated factors. Generally an optimum linear combination of variables is produced that better accounts for variance in the data than do other statistical treatments. The first principal component, C1, is generally the best summary of linear interrelations evident from the data; the second principal component, C2, is the second-best linear combination of variables, with C2 being uncorrelated to C1, and accounting for the largest proportion of the residual variance after the effects of C1 are removed from the data set [122].

In a granary study, wheat of a new high-yield medium-protein class was stored at 15% and 19% initial MC for 60 weeks [123]. Ochratoxin A began to appear during the fourth week of storage, along with the characteristic changes in temperature, O_2, CO_2, MC, germination, fatty acids, and fungal species populations. The data for the 60-week period were analyzed using PCA, and using factors with eigenvalues >0.30, a C1 was produced, characterized by increasing MC, ochratoxin A, *Penicillium* incidence, and total fungal propagule count, and decreasing *Alternaria* incidence and seed germination.

In a similar study in Germany, feed-grade corn was stored at 15% and 19% initial MC for 20 weeks [124]. Ochratoxin A began to appear during the 12th week of storage, along with the characteristic changes in temperature, CO_2, germination, and fungal species populations. Data for the 20-week period were analyzed using PCA, and using factors with eigenvalues >0.30, a C1 was produced, characterized by increasing CO_2, ochratoxin A, *Aspergillus* and *Penicillium* incidence, total fungal propagule count, and decreasing seed germination.

Because of the interaction of multiple factors such as moisture, temperature, time, substrate, O_2, fungal species, etc., it is necessary to study mycotoxin formation in actual storage environments. When specific storage situations and the particular processes associated with mycotoxin formation can be characterized, early-warning indicators and intelligent intervention procedures can be proposed. As an early warning indicator, CO_2 is sensitive and easy to measure, but is also produced by insects. Fungal volatiles are far more specific [125,126]. For these odor volatiles, the development of a simple and rapid test, which can be performed in the field by untrained personnel, would be valuable and timely.

ACKNOWLEDGMENTS

The author thanks Randy Clear, Reg Sims, Diane Smith, Mike Malyk, Gaye Miller, Dr. Andy Tekauz, and Dr. John Mills for their help in the preparation of the manuscript.

REFERENCES

1. JI Pitt, L Leistner. Toxigenic *Penicillium* species. In: JE Smith, RS Henderson, eds. Mycotoxins and Animal Foods. Boca Raton: CRC Press, 1991, p 81.
2. JD Miller. Epidemiology of *Fusarium* ear diseases of cereals. In: JD Miller, HL Trenholm, eds. Mycotoxins in Grain: Compounds Other Than Aflatoxin. St. Paul: Eagan Press, 1994, p 19.
3. JC Frisvad. Mycotoxins and mycotoxigenic fungi in storage. In: D Jayas, NDG White, WE Muir, eds. Stored-Grain Ecosystems. New York: Marcel Dekker, 1995, p 251.
4. PM Scott. *Penicillium* and *Aspergillus* toxins. In: JD Miller, HL Trenholm, eds. Mycotoxins in Grain: Compounds Other Than Aflatoxin. St. Paul: Eagan Press, 1994, p 261.
5. DM Wilson, D Abramson. Mycotoxins. In: DB Sauer, ed. Storage of Cereal Grains and Their Products. 4th ed. St. Paul: American Association of Cereal Chemists, 1992, p 341.
6. JE Smith, RS Henderson, eds. Mycotoxins and Animal Foods. Boca Raton: CRC Press, 1991.
7. D Abramson. Development of molds, mycotoxins and odors in moist cereals during storage. In: J Chelkowski, ed. Cereal Grain: Mycotoxins, Fungi and Quality in Drying and Storage. Amsterdam: Elsevier, 1991, p 119.
8. KH Ominski, RR Marquardt, RN Sinha, D Abramson. Ecological aspects of growth and mycotoxin production by storage fungi. In: JD Miller, HL Trenholm, eds. Mycotoxins in Grain: Compounds Other Than Aflatoxin. St. Paul: Eagan Press, 1994, p 287.
9. MW Trucksess. Mycotoxins. J AOAC Int 78:135, 1995.
10. RJ Cole, RH Cox. Handbook of Toxic Fungal Metabolites. New York: Academic Press, 1981.
11. DB Sauer, LM Seitz, R Burroughs, et al. Toxicity of *Alternaria* metabolites found in weathered sorghum grain at harvest. J Agric Food Chem 26:1380, 1978.
12. GF Griffin, FS Chu. Toxicity of the *Alternaria* metabolites alternariol, alternariol methyl ether, altenuene, and tenuazonic acid in the chicken embryo assay. Appl Environ Microbiol 46:1420, 1983.
13. PH Hidy, RS Baldwin, RL Greasham, JR McMullen. Zearalenone and some derivatives: production and biological activities. Adv Appl Microbiol 22:59, 1977.
14. JC Frisvad, U Thrane. Standardized high-performance liquid chromatography of 182 mycotoxins and other fungal metabolites based on alkylphenone retention indices and UV-VIS spectra (diode array detection). J Chromatogr 404:195, 1987.
15. JC Frisvad. The use of high-performance liquid chromatography and diode array detection in fungal chemotaxonomy based on profiles of secondary metabolites. Bot J Linn Soc 99:81, 1989.
16. RRM Paterson, C Kemmelmeier. Natural alkaline and difference ultraviolet spectra of secondary metabolites from *Penicillium* and other fungi and comparisons to published maxima from gradient high-performance liquid chromatography with diode-array detection. J Chromatogr 511:195, 1990.
17. JC Frisvad, U Thrane. Mycotoxin production by food-borne fungi. In: RA Samson,

ES Hoekstra, JC Frisvad, U Thrane, eds. Introduction to Food-Borne Fungi. 4th ed. Baarn: Centraalbureau voor Schimmelcultures, 1995, p 251.
18. PE Nelson, TA Toussoun, WFO Marasas. *Fusarium* Species: An Illustrated Manual for Identification. University Park: Pennsylvania State University Press, 1983.
19. J Chelkowski. Formation of mycotoxins produced by *Fusarium* in heads of wheat, triticale and rye. In: J Chelkowski, ed. *Fusarium* Mycotoxins, Taxonomy and Pathogenicity. Amsterdam: Elsevier, 1989, p 63.
20. PM Scott. Fumonisins. Int J Food Microbiol 18:257, 1993.
21. J Perkowski, T Meidaner, HH Geiger, H-M Müller, J Chelkowski. Occurrence of deoxynivalenol (DON), 3-acetyl-DON, zearalenone and ergosterol in winter rye inoculated with *Fusarium culmorum*. Cereal Chem 72:205, 1995.
22. HP Van Egmond. Current situation on regulations for mycotoxins. Overview of tolerances and status of standard methods of sampling and analysis. Food Addit Contam 6:139, 1989.
23. HL Trenholm, RMG Hamilton, DW Friend, BK Thompson, KE Hartin. Feeding trials with vomitoxin (deoxynivalenol)-contaminated wheat: effects in swine, poultry, and dairy cattle. J Am Vet Med Assoc 185:527, 1984.
24. PG Thiel, WFO Marasas, EW Sydenham, GS Shephard, WCA Gelderblom, JJ Nieuwenhuis. Survey of fumonisin production by *Fusarium* species. Appl Environ Microbiol 57:1089, 1991.
25. YZ Wang, JD Miller. Screening techniques and sources of resistance to *Fusarium* head blight. In: AR Khlatt, ed. Wheat Production: Constraints in Tropical Environments. Mexico City: Centro Internacional de Mejoramiento de Maiz y Trigo, 1988, p 239.
26. CJ Mirocha, HK Abbas, CE Windels, W Xie. Variation in deoxynivalenol, 15-acetyldeoxynivalenol, 3-acetyldeoxynivalenol and zearalenone production by *Fusarium graminearum* isolates. Appl Environ Microbiol 55:1315, 1989.
27. D Abramson, RM Clear, DM Smith. Trichothecene production by *Fusarium* spp. isolated from Manitoba grain. Can J Plant Pathol 15:147, 1993.
28. D Abramson, RM Clear, TW Nowicki. *Fusarium* species and trichothecene mycotoxins in suspect samples of 1985 Manitoba wheat. Can J Plant Sci 67:611, 1987.
29. HK Abbas, CJ Mirocha, J Tuite. Natural occurrence of deoxynivalenol, 15-acetyldeoxynivalenol, and zearalenone in refusal factor corn stored since 1972. Appl Environ Microbiol 51:841, 1986.
30. A Ylimaki. The mycoflora of cereal seeds and some feedstuffs. Ann Agric Fenn 20:74, 1981.
31. CHA Snijders, J Perkowski. Effects of head blight caused by *Fusarium culmorum* on toxin production and weight of wheat kernels. Phytopathology 80:566, 1990.
32. L Saur. Recherche de géniteurs désistance à la fusariose de l'épi causée par *Fusarium culmorum* chez le blé et les espèces voisines. Agronomie 11:535, 1991.
33. DB Scott, EJH de Jager, PS van Wyk. Head blight of irrigated wheat in South Africa. Phytophylactica 20:317, 1988.
34. A Adler, H Lew, W Brodacz, W Edinger, M Oberforster. Occurrence of moniliformin, deoxynivalenol, and zearalenone in durum wheat (*Triticum durum* Desf.). Mycotoxin Res 11:9, 1995.

35. U Thrane. *Fusarium* species and their specific production of secondary metabolites. In: J Chelkowski, ed. *Fusarium* Mycotoxins, Taxonomy and Pathogenicity. Amsterdam: Elsevier, 1989, p 199.
36. H Pettersson, R Hedman, B Engström, K Elwinger, O Fossum. Nivalenol in Swedish cereals—occurrence, production and toxicity towards chickens. Food Addit Contam 12:373, 1995.
37. HK Abbas, CJ Mirocha, JD Pokorney, SL Gould, T Kommedahl. Mycotoxins of *Fusarium* spp. associated with infected ears of corn in Minnesota. Appl Environ Microbiol 54:1039, 1988.
38. J Chelkowski. Mycotoxins associated with corn cob fusariosis. In: J Chelkowski, ed. *Fusarium* Mycotoxins, Taxonomy and Pathogenicity. Amsterdam: Elsevier, 1989, p 53.
39. TE Ochor, LE Trevathan, SB King. Relationship of harvest date and host genotype to infection of maize kernels by *Fusarium moniliforme*. Plant Dis 71:311, 1987.
40. C de Leon, S Pandey. Improvement of resistance to ear and stalk rots and agronomic traits in tropical maize gene pools. Crop Sci 29:12, 1989.
41. PG Thiel, JC Meyer, WFO Marasas. Natural occurrence of moniliformin together with deoxynivalenol and zearalenone in Transkeian corn. J Agric Food Chem 30:308, 1982.
42. PG Thiel, WCA Gelderblom, WFO Marasas, PE Nelson, TM Wilson. Natural occurrence of moniliformin and fusarin C in corn screenings known to be hepatocarcinogenic in rats. J Agric Food Chem 34:773, 1986.
43. A Thalmann, S Metzenauer, S Gruber-Schley. Untersuchungen uber das Vorkommen von Fusarientoxinen in Getreide. Ber Landwirtsch 63:257, 1985.
44. PM Scott, GA Lawrence. Liquid chromatographic determination and stability of the *Fusarium* mycotoxin moniliformin in cereal grain. J Assoc Off Anal Chem 70:850, 1987.
45. JM Farber, GW Sanders, GA Lawrence, PM Scott. Production of moniliformin by Canadian isolates of *Fusarium*. Mycopathologia 101:187, 1988.
46. H Lew, A Adler, W Edinger. Moniliformin and the European corn borer (*Ostrinia nubilalis*). Mycotoxin Res 7:71, 1991.
47. M Sharman, J Gilbert, J Chelkowski. A survey of the occurrence of the mycotoxin moniliformin in cereal samples from sources worldwide. Food Addit Contam 8:459, 1991.
48. WCA Gelderblom, K Jasiewicz, WFO Marasas, et al. Fumonisins—novel mycotoxins with cancer-promoting activity produced by *Fusarium moniliforme*. Appl Environ Microbiol 54:1806, 1988.
49. PE Nelson, RD Plattner, DD Shackleford, AE Desjardins. Fumonisin B_1 production by *Fusarium* species other than *F. moniliforme* in section Liseola and by some related species. Appl Environ Microbiol 58:984, 1992.
50. PF Ross, LG Rice, GD Osweiler, PE Nelson, JL Richard, TM Wilson. A review and update of animal toxicoses associated with fumonisin contaminated foods and production of fumonisins by *Fusarium* isolates. Mycopathologia 117:109, 1992.
51. WCA Gelderblom, ME Cawood, SD Snyman, WFO Marasas. Fumonisin B_1 dosimetry in relation to cancer initiation in rat liver. Carcinogenesis 15:209, 1994.

52. PF Ross, LG Rice, RD Plattner et al. Concentrations of fumonisin B_1 in feeds associated with animal health problems. Mycopathologia 114:129, 1990.
53. RD Plattner, WP Neffed, CW Bacon, et al. A method of detection of fumonisins in corn samples associated with field cases of equine leukoencephalomalacia. Mycologia 82:698, 1990.
54. EW Sydenham, PG Thiel, WFO Marasas, et al. Natural occurrence of some *Fusarium* mycotoxins in corn from low and high esophageal cancer prevalence areas of the Transkei. southern Africa. J Agric Food Chem 38:1900, 1990.
55. LB Bullerman, W-Y Tsai. Incidence and levels of *Fusarium moniliforme, Fusarium proliferatum* and fumonisins in corn and corn-based foods and feeds. J Food Prot 57:541, 1994.
56. T-C Tseng, K-L Lee, T-S Deng, C-Y Liu, J-W Huang. Production of fumonisins by *Fusarium* species of Taiwan. Mycopathologia 130:117, 1995.
57. O Zoller, F Sager, B Zimmerli. Occurrence of fumonisins in foods. Mitt Geb Lebensm Hyg 85:81, 1994.
58. J Chelkowski, P Zajkowski, H Kwasna, A Visconti, A Bottalico. *Fusarium sporotrichioides* Sherb. and trichothecenes associated with fusarium ear rot of corn before harvest. Mycotoxin Res 3:111, 1987.
59. N Magan, J Lacey. Water relations of some *Fusarium* species from infected wheat ears and grain. Trans Br Mycol Soc 83:281, 1984.
60. JC Sutton. Epidemiology of wheat head blight and maize ear rot caused by *Fusarium graminearum*. Can J Plant Pathol 4:195, 1982.
61. JA Duthie, R Hall, AV Asselin. *Fusarium* species from seed of winter wheat in eastern Canada. Can J Plant Pathol 8:282, 1986.
62. CHA Snijders. Breeding for resistance to *Fusarium* in wheat and maize. In: JD Miller, HL Trenholm, eds. Mycotoxins in Grain: Compounds Other Than Aflatoxin. St. Paul: Eagan Press, 1993, p 37.
63. RD Wilcoxson, RH Busch, EA Ozman. *Fusarium* head blight resistance in spring wheat cultivars. Plant Dis 76:658, 1992.
64. RM Clear, SK Patrick. *Fusarium* species isolated from wheat samples containing tombstone (scab) kernels from Ontario, Manitoba and Saskatchewan. Can J Plant Sci 70:1057, 1990.
65. Official Grain Grading Guide. Winnipeg: Canadian Grain Commission, 1995.
66. RM Clear, SK Patrick, RG Platford, M Desjardins. Occurrence and distribution of *Fusarium* species in barley and oat seed from Manitoba in 1993 and 1994. Manitoba Agriforum '95, Winnipeg, Manitoba, 1995, pp 70–79.
67. JD Miller, JC Young, HL Trenholm. *Fusarium* toxins in field corn. I. Time course of fungal growth and production of deoxynivalenol and other mycotoxins. Can J Bot 61:3080, 1983.
68. AW Schaafsma, JD Miller, ME Savard, RJ Ewing. Ear rot development and mycotoxin production in corn in relation to inoculation method, corn hybrid, and species of *Fusarium*. Can J Plant Pathol 15:185, 1993.
69. PM Scott. The natural occurrence of trichothecenes. In: VR Beasley, ed. Trichothecene Toxicosis: Pathophysiological Effects. Vol 1. Boca Raton: CRC Press, 1989, p 2.

70. RA Martin, HW Johnston. Effects and control of *Fusarium* disease of cereal grains in the Atlantic provinces. Can J Plant Pathol 4:210, 1982.
71. MR Fernandez. Recovery of *Cochliobolus sativus* and *Fusarium graminearum* from living and dead wheat and nongramineous winter crops in southern Brazil. Can J Bot 69:1900, 1991.
72. PE Lipps, IW Deep. Influence of tillage and crop rotation on yield, stalk rot and recovery of *Fusarium* and *Trichoderma* spp. from corn. Plant Dis 75:828, 1991.
73. AH Teich. Epidemiology of wheat (*Triticum aestivum* L.) scab caused by *Fusarium* spp. In: J Chelkowski, ed. *Fusarium* Mycotoxins, Taxonomy and Pathogenicity. Amsterdam: Elsevier, 1989, p 269.
74. RA Martin, JA MacLeod, C Caldwell. Influences of production inputs on incidence of infection by *Fusarium* species on cereal seed. Plant Dis 75:784, 1991.
75. MO Moss, JM Frank. Influence of the fungicide tridemorph on T-2 toxin production by *Fusarium sporotrichioides*. Trans Br Mycol Soc 84:585, 1985.
76. MO Moss, JM Frank. Production of trichothecenes in winter barley inoculated with *Fusarium sporotrichioides*. Proc 2nd World Congr Foodborne Infections and Intoxicants, Berlin, 1986, Vol 2, pp 903–906.
77. AH Teich, JR Hamilton. Effect of cultural practices, soil phosphorus, potassium and pH on the incidence of *Fusarium* head blight and deoxynivalenol levels in wheat. Appl Environ Microbiol 49:1429, 1985.
78. L Mihuta-Grimm, RL Forster. Scab of wheat and barley in southern Idaho and evaluation of seed treatments for eradication of *Fusarium* spp. Plant Dis 73:769, 1989.
79. LSL Wong, D Abramson, A Tekauz, D Leisle, RIH McKenzie. Pathogenicity and mycotoxin production of *Fusarium* species causing head blight in wheat cultivars varying in resistance. Can J Plant Sci 75:261, 1995.
80. LM Seitz, DB Sauer, HE Mohr, R Burroughs. Weathered grain sorghum: natural occurrence of alternariols and storability of the grain. Phytopathology 65:1259, 1975.
81. J Grabarkiewicz-Szczesna, J Chelkowski, P Zajkowski. Natural occurrence of *Alternaria* mycotoxins in the grain and chaff of cereals. Mycotoxin Res 5:77, 1989.
82. JJ Giambrone, ND Davis, UL Diener. Effect of tenuazonic acid on young chickens. Poult Sci 57:1554, 1978.
83. ER Smith, TN Frederickson, Z Hadidian. Toxic effects of the sodium and the *N,N′*-dibenzylethylenediamine salts of tenuazonic acid (NSC-525816 and NSC-82260). Cancer Chemother Rep 52:579, 1968.
84. A Logrieco, A Visconti, A Bottalico. Mandarin fruit rot caused by *Alternaria alternata* and associated mycotoxins. Plant Dis 74:415, 1990.
85. ME Stack, PB Mislivec, JAG Roach, AE Pohland. Liquid chromatographic determination of tenuazonic acid and alternariol methyl ether in tomatoes and tomato products. J Assoc Off Anal Chem 68:640, 1985.
86. A Visconti, A Logrieco, A Bottalico. Natural occurrence of *Alternaria* mycotoxins in olives—their production and possible transfer into the oil. Food Addit Contam 3:323, 1986.
87. AA Ansari, AK Shrivastava. Natural occurrence of *Alternaria* mycotoxins in sorghum and ragi from North Bihar, India. Food Addit Contam 7:815, 1990.

88. N Umetsu, J Kaji, K Tamari. Isolation of tenuazonic acid from blast-diseased rice plants. Agric Biol Chem 37:451, 1973.
89. A Visconti, A Sibilia. *Alternaria* toxins. In: JD Miller, HL Trenholm, eds. Mycotoxins in Grain: Compounds Other Than Aflatoxin. St. Paul: Eagan Press, 1994, p 315.
90. J Chelkowski, A Visconti, eds. *Alternaria*—Biology, Plant Diseases and Metabolites. Amsterdam: Elsevier, 1992.
91. EG Simmons. *Alternaria* taxonomy: current status, viewpoint, challenge. In: J Chelkowski, A Visconti, eds. *Alternaria*—Biology, Plant Diseases and Metabolites. Amsterdam: Elsevier, 1992, p 1.
92. JC Frisvad. The connection between the penicillia and aspergilli and mycotoxins with emphasis on misidentified isolates. Arch Environ Contam Toxicol 18:452, 1989.
93. RV Bhat. Aflatoxins: successes and failures of three decades of research. In: BR Champ, E Highley, AD Hocking, JI Pitt, eds. Fungi and Mycotoxins in Stored Products. Canberra: Austral Center Intl Agric Research, 1991, p 80.
94. GA Payne. Aflatoxin in maize. Curr Rev Plant Sci 10:423, 1992.
95. UL Diener, RJ Cole, RA Hill. Epidemiology of aflatoxin formation by *Aspergillus flavus*. Annu Rev Phytopathol 25:249, 1987.
96. DT Wicklow. The mycology of stored grain: an ecological perspective. In: D Jayas, NDG White, WE Muir, eds. Stored-Grain Ecosystems. New York: Marcel Dekker, 1994, p 197.
97. WW McMillian. Maize plant resistance to insect damage and associated aflatoxin development. In: MS Zuber, EB Lillehoj, BL Renfro, eds. Aflatoxin in Maize. Mexico City: Centro Internacional de Mejoramiento de Maiz y Trigo, 1987, p 250.
98. DT Wicklow, DM Wilson. Germination of *Aspergillus flavus* sclerotia in a Georgia maize field. Trans Br Mycol Soc 87:651, 1986.
99. JT Mills. Spoilage and Heating of Stored Agricultural Products: Prevention, Detection and Control. Ottawa: Agriculture Canada Publication 1823, Canadian Govt. Pub. Center, 1989.
100. DB Brooker, FW Bakker-Arkema, CW Hall. Drying and Storage of Grains and Oilseeds. New York: AVI Van Nostrand Reinhold, 1992.
101. E Kaminski, E Wasowicz. The usage of volatile compounds produced by molds as indicators of grain deterioration. In: J Chelkowski, ed. Cereal Grain: Mycotoxins, Fungi and Quality in Drying and Storage. Amsterdam: Elsevier, 1991, p 229.
102. DB Sauer. Effects of fungal deterioration on grain: nutritional value, toxicity, germination. Int J Food Microbiol 7:267, 1988.
103. B Flannigan, EM McCabe, F McGarry. Allergenic and toxigenic micro-organisms in houses. J Appl Bacteriol 70:61S, 1991.
104. PS Masurekar. Therapeutic metabolites. In: DB Finkelstein, C Ball, eds. Biotechnology of Filamentous Fungi: Technology and Products. Boston: Butterworth-Heineman, 1991, p 241.
105. J Chelkowski, ed. Cereal Grain: Mycotoxins, Fungi and Quality in Drying and Storage. Amsterdam: Elsevier, 1991.
106. BJ Blaney. *Fusarium* and *Alternaria* toxins. In: BR Champ, E Highley, AD Hocking, JI Pitt, eds. Fungi and Mycotoxins in Stored Products. Canberra: Austral Center Intl Agric Research, 1991, p 86.

107. JJ Petska, GS Bondy. Alteration of immune function following dietary mycotoxin exposure. Can J Physiol Pharmacol 68:1009, 1990.
108. JC Frisvad. Chemometrics and chemotaxonomy: a comparison of multivariate statistical methods for the evaluation of binary fungal secondary metabolite data. Chemometrics Intell Lab Syst 12:253, 1992.
109. B Flannigan. Primary contamination of barley and wheat grain by storage fungi. Trans Br Mycol Soc 71:37, 1978.
110. J Lacey. Prevention of mould growth and mycotoxin production through control of environmental factors. In: S Natori, K Hashimoto, Y Ueno, eds. Mycotoxins and Phycotoxins '88. Amsterdam: Elsevier, 1989, p 161.
111. EB Lillehoj, F Elling. Environmental conditions that facilitate ochratoxin contamination of agricultural commodities. Acta Agric Scand 33:113, 1983.
112. HL Trenk, PA Hartman. Effects of moisture content and temperature on aflatoxin production in corn. Appl Microbiol 19:781, 1970.
113. LM Seitz, DB Sauer, HE Mohr, DF Aldis. Fungal growth and dry matter loss during bin storage of high-moisture corn. Cereal Chem 59:9, 1982.
114. D Abramson, RN Sinha, JT Mills. Mycotoxin and odor formation in moist cereal grain during granary storage. Cereal Chem 57:346, 1980.
115. D Abramson, RN Sinha, JT Mills. Mycotoxin and odor formation in barley stored at 16 and 20% moisture in Manitoba. Cereal Chem 60:350, 1983.
116. D Abramson, RN Sinha, JT Mills. Mycotoxin formation and quality changes in granary-stored corn at 16 and 21% moisture content. Sci Aliment 5:653, 1985.
117. D Abramson, JT Mills, RN Sinha. Mycotoxin production in amber durum wheat stored at 15 and 19% moisture content. Food Addit Contam 7:617, 1990.
118. JT Mills, D Abramson. Production of sterigmatocystin by isolates of *Aspergillus versicolor* from western Canadian stored barley and rapeseed/canola. Can J Plant Pathol 8:151, 1986.
119. D Abramson, RN Sinha, JT Mills. Mycotoxin formation in moist 2-row and 6-row barley during granary storage. Mycopathologia 97:179, 1987.
120. T Börjesson, U Stöllman, P Adamek, A Kaspersson. Analysis of volatile compounds for detection of molds in stored cereals. Cereal Chem 66:300, 1989.
121. T Börjesson, U Stöllman, J Schnürer. Volatile metabolites produced by six fungal species compared with other indicators of fungal growth on cereal grains. Appl Environ Microbiol 58:2599, 1992.
122. JO Kim. Factor analysis. In: HH Nie, CH Hull, JG Jenkins, K Steinbrenner, D Brent, eds. Statistical Packages for the Social Sciences. New York: McGraw-Hill, 1975, p 468.
123. D Abramson, RN Sinha, JT Mills. Mycotoxin formation in HY-320 wheat during granary storage at 15 and 19% moisture content. Mycopathologia 111:181, 1990.
124. D Abramson, W Richter, J Rintelen, RN Sinha, M Schuster. Ochratoxin A production in Bavarian cereal grains stored at 15 and 19% moisture content. Arch Environ Contam Toxicol 23:259, 1992.
125. CE Ericksson, E Kaminski, P Adamek, T Börjesson. Volatile compounds and off-flavor produced by microorganisms in cereals. In: G Charalambous, ed. Off-Flavors in Foods and Beverages. Amsterdam: Elsevier, 1992, p 37.

126. P Adamek, B Bergstrom, T Börjesson, U Stöllman. Determination of volatile compounds for the detection of molds. In: RA Samson, AD Hocking, JI Pitt, AD King, eds. Modern Methods in Food Mycology: Second International Workshop on Standardization of Methods for Mycological Examination of Foods. Amsterdam: Elsevier, 1992, p 327.
127. Y Ueno, K Ishii. Chemical and biological properties of trichothecenes from *Fusarium sporotrichioides*. In: J Lacey, ed. Trichothecenes and Other Mycotoxins. New York: John Wiley, 1985, p 123.
128. Y Ueno. General toxicology. In: Y Ueno, ed. Trichothecenes: Chemical, Biological and Toxicological Aspects. Amsterdam: Elsevier, 1983, p 135.
129. JH Forsell, R Jensen, J-H Tai, M Witt, WS Lin, JJ Petska. Comparison of the acute toxicities of deoxynivalenol (vomitoxin) and 15-acetyldeoxynivalenol in the B6C3F1 mouse. Food Chem Toxicol 25:155, 1987.
130. Y Kawasaki, O Uchida, K Sekita, et al. Single and repeated oral administration toxicity studies of nivalenol in F344 rats. Shokuhin Eiseigaku Zasshi 31:144, 1990.
131. RJ Cole, JW Kirksey, HG Cutler, BL Doupnik, JC Peckham. Toxins from *Fusarium moniliforme*: effects on plants and animals. Science 179:1324, 1973.
132. TS Kellerman, WFO Marasas, PG Thiel, WCA Gelderblom, M Cawood, JAW Coetzer. Leukoencephalomalacia in two horses induced by oral dosing of fumonisin B_1. Onderstepoort J Vet Res 57:269, 1990.
133. IFH Purchase. The acute toxicity of the mycotoxin cyclopiazonic acid to rats. Toxicol Appl Pharmacol 18:114, 1971.
134. C Hanika, WW Carlton. Toxicology and pathology of citrinin. In: GC Llewellyn, WV Dashek, CE O'Rear, eds. Biodeterioration Research 4. New York: Plenum Press, 1994, p 41.
135. G Engel, M Teuber. Patulin and other small lactones. In: V Betina, ed. Mycotoxins: Production, Isolation, Separation, and Purification. Amsterdam: Elsevier, 1984, p 291.
136. MF Murnaghan. The pharmacology of penicillic acid. J Pharmacol Ep Ther 88:119, 1946.
137. Y Ueno, I Ueno. Toxicology and biochemistry of mycotoxins. In: K Uraguchi, M Yamazaki, eds. Toxicology, Biochemistry and Pathology of Mycotoxins. New York: John Wiley, 1978, p 107.
138. CS Reddy, AW Hayes, WL Williams, A Ciegler. Toxicity of secalonic acid D. J Toxicol Environ Health 5:1159, 1979.

9

Process of Contamination by Aflatoxin-Producing Fungi and Their Impact on Crops

Gary A. Payne
North Carolina State University, Raleigh, North Carolina

I. INTRODUCTION

The three major groups of fungi associated with mycotoxin contamination are the *Aspergillus* species, the *Fusarium* species, and the *Penicillium* species. There is a large body of information on the biology and epidemiology of these fungi. The epidemiology of each of these genera, excluding *A. flavus* and *A. parasiticus*, has been reviewed recently [1]. Because of the extensive information on the biology and epidemiology of *A. flavus* and *A. parasiticus*, this review will focus solely on these two fungi.

Aflatoxins are produced by three species of *Aspergillus, A. flavus* Link ex. Fries, *A. parasiticus* Speare, and *A. nomius* Kurtzman, Horn, and Hesseltine. Only the first two species appear to be important in the colonization of agricultural commodities, and thus this review will focus on these two species. Both *A. flavus* and *A. parasiticus* produce a family of aflatoxins. The dominant aflatoxins produced by *A. flavus* are B1 and B2, whereas *A. parasiticus* produces two additional aflatoxins, G1 and G2. Although these fungi can produce aflatoxins on any commodity, they are of major concern on corn, peanuts, cottonseed, and nuts. *A. flavus* and *A. parasiticus* are frequent storage molds, and their growth on stored commodities with the subsequent production of aflatoxin is a major concern throughout the world. In certain geographical regions, these fungi also colonize and invade developing seeds in the field. The preharvest contamination of crops by these two fungi has been the major focus of current control strategies, and also will be the subject of this review. In order to devise effective preharvest control strategies, it is important to understand the biology of the aflatoxin-producing fungi and to understand the factors important in the epidemiology of the disease.

The objective of this review is to summarize what is known about the preharvest infection process by these fungi and the environmental factors that lead to their successful colonization and invasions of seeds. I will attempt to identify common factors conducive to disease development in the four major crops (corn, peanuts, cotton, and tree nuts) affected by aflatoxin contamination. At the onset, however, it is important to note that these two fungi cause disease on very diverse plant species (from peanuts to pistachios), and they affect seeds produced above ground and below ground. For these reasons, each host-parasite interaction has unique features. However, there appear to be environmental factors common to all of the host-parasite interactions that contribute to aflatoxin contamination. Because more is known about the host–parasite interaction between corn and *A. flavus* [2–5], general conclusions will be drawn from this interaction.

II. BIOLOGY OF AFLATOXIGENIC FUNGI

Viewed from an agronomic perspective, *A. flavus* and *A. parasiticus* are plant pathogens. However, from an ecological perspective, living tissue is only a minor substrate for these soilborne filamentous fungi. *A. flavus* and *A. parasiticus* are saprophytic during most of their life cycle, and grow on a wide variety of substrates including decaying plant and animal debris. Thus, the populations of these organisms in the soil are dependent upon how well these fungi compete with the existing soil microflora. The two major factors that influence soil populations of these two fungi are soil temperature and soil moisture. *A. flavus* and *A. parasiticus* can grow at temperatures from 12 to 48°C and at water potentials as low as −35 MPa [6]. The optimum temperature for growth is 25 to 42°C. Thus these organisms are semithermophilic and semixerophytic. Under conditions of high temperature and low water activity, conditions associated with drought in temperate agricultural crops, these two fungi become very competitive and may become dominant fungal species in the soil. These two factors, more than any other, contribute to the epidemiology of these two fungi. These environmental factors are discussed in the sections below.

As just stated, *A. flavus* and *A. parasiticus* are considered saprophytes. However, it is now clear that these two fungi have limited parasitic abilities. Not only do they colonize the injured seeds, but in some cases these fungi directly invade seeds. Few factors involved in pathogenicity of the two fungi have been characterized. However, *A. flavus* appears to be more aggressive than *A. parasiticus* [7] and is the more dominant species on all commodities. Cleveland and Cotty [8] compared isolates of *A. flavus* that differed in their aggressiveness on cotton bolls. They found that weakly aggressive strains could not degrade the intercarpellary membrane in cotton bolls and their growth was restricted to individual locules. Although both strains produced three pectinase activities, the highly aggressive strain produced an additional enopolygalacturonase designated P2c [9]. The gene (*pecA*) coding for this pectinase has been cloned and characterized [10]. Proof of involvement of this gene awaits gene disruption experiments. *A. flavus* also has been shown to produce cutinase. Guo et al. [11] found that the addition of cutinase to corn kernels increased colonization and aflatoxin accumulation, whereas the addition of a cutinase inhibitor reduced both growth and aflatoxin accumulation. Wax and cutin layers in some maize genotypes have been associated with resistance to aflatoxin accumulation [12].

Because these fungi have limited parasitic abilities, several environmental conditions must be met for infection and aflatoxin contamination to occur. The two most important environmental factors are drought stress and temperature. Under favorable environmental conditions, these fungi compete well with other soil organisms and produce abundant inoculum. It is becoming evident that in many cases inoculum levels of these fungi are very important and may be limiting in certain years and certain locations. Under conditions of drought stress and high temperatures, these pathogens outcompete other microflora on seeds of corn, peanut, cotton, and nuts. Not only do these conditions favor the aflatoxin-producing fungi, but they favor insect activity and often compromise the defense system of the host plant. Several factors must interplay before extensive colonization and aflatoxin contamination occur.

III. FACTORS AFFECTING INOCULUM, INITIAL COLONIZATION, AND INFECTION

In this review I will use the term colonization to denote the growth of the fungus on plant surfaces, and the term infection to denote a more intimate relationship with the host. In this context, colonization does not imply that the fungus is obtaining its nutrition from living cells, whereas infection does. *A. flavus* both colonizes and infects living plant tissue.

A. Corn

1. Colonization

The host-parasite relationship of *A. flavus* with corn is better understood than for any other host of the fungus. Several field, greenhouse, and controlled-environment studies have examined the host-parasite interactions between these two. Many of these studies have been done in the southern U.S., where *A. flavus* is commonly

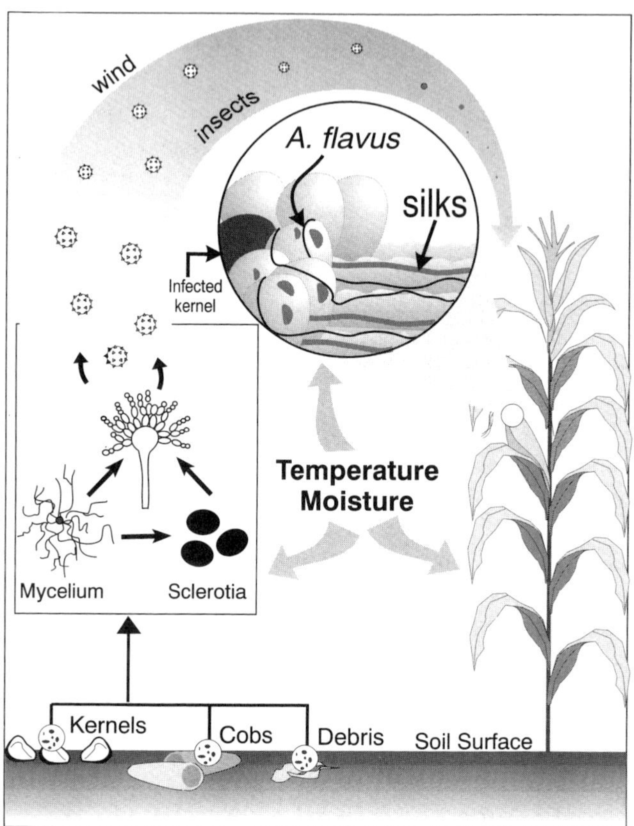

FIGURE 1 Schematic diagram showing possible sources of inoculum of *Aspergillus flavus* and the routes of colonization of the fungus on corn.

found on corn ears during the growing season and aflatoxin contamination is a chronic problem. Figure 1 is a schematic showing a model for the factors involved in this host-parasite interaction.

Corn kernels become colonized with *A. flavus* early after silking [3, 13–20]. Spores of the fungus can be brought to the kernel surfaces by insects, as will be discussed later. Alternatively, the fungus can colonize silk tissue and grow down into the ear [13,17,21]. Silks yellow-brown in color but still turgid support the best growth of *A. flavus* [14,17,21]. In contrast, unpollinated silks appear resistant to colonization by *A. flavus*. Spores of *A. flavus* applied to silks can germinate within 4 to 8 h and produce extensive hyphal growth with lateral branching [17]. The growth of the fungus is most extensive around pollen grains. Although much of the hyphal growth appears on the surface of the silks, the fungus can penetrate the silks directly or through cracks and intercellular gaps. Once inside the silk, the hyphae grow within the parenchymatous tissues down the silk axis [17]. The fungus can sporulate on silk tissue within 24 h after inoculation and grow down the silks to the base of the ear within 4 to 13 days [14].

Colonization of silks and kernel surfaces of naturally contaminated ears in the field was extensively studied in North Carolina [14]. *A. flavus* was readily isolated after pollination from silks external to the husk and from the ear within the husk throughout the season. The fungus was present throughout the ear but was more prevalent at the tip than at the base. *A. flavus* was isolated from the silks most frequently at the milk stage of kernel development, whereas the percentage of ears with colonized kernels increased to dough stage of kernel development. Even though 47% of the ears were colonized by dough stage, the average percentage of kernels infected in these ears at harvest was only 3.3%.

In a companion study, Marsh and Payne [14] also inoculated corn ears with a color mutant or wild-type strain of *A. flavus* and followed colonization. The pattern of tissue colonization was similar to that found for naturally infected ears. Growth proceeded from the ear tip toward the base; the fungus colonized the silks first, then the glumes (by the milk stage), the kernel surfaces, and, rarely, the cob pith. Even though external colonization of the kernels was extensive in some of the ears, none of the kernels were internally infected prior to the early dent stage of kernel development. Twelve days later, when the ears were harvested, 28% of the kernels were infected.

These findings are similar to those of Hill et al. [15]. They sampled Georgia corn fields every 15 days from silk to harvest for the presence of *A. flavus* and *A. parasiticus*. They found the incidence of the two fungi to increase during the growing season. At maturity, 75% of the kernels were colonized with these two fungi; *A. flavus* was the dominant species. Zummo and Scott [20] also monitored the colonization of kernels in inoculated and noninoculated ears in Mississippi. Profuse fungal growth and sporulation of *A. flavus* was found on silks and kernel surfaces of inoculated ears. They also found that 83% of the cobs from inoculated and noninoculated ears were colonized. Even though over 25% of the surface of some ears were colonized with *A. flavus*, only 4% of the kernels were invaded.

The prevailing data argue that *A. flavus* can colonize silk tissue, grow down the silks, and colonize kernel surfaces. Colonization of the silks and kernel surfaces occurs soon after silking and may continue throughout the season. Later the fungus moves to the glumes of the kernels and the fungus infects the kernel as it matures. Colonization of kernel surfaces by *A. flavus* may be extensive, but internal infection is usually low. All of the above processes have been shown to occur without insect injury [13,14,21].

2. Mode of Entry

The precise mode of entry of *A. flavus* into corn seeds is still in question. Jones et al. [22] proposed that the fungus enters through the stylar canal, as has been proposed for other fungi [23]. This would be consistent with the observations of Anderson et al. [24], who reported the presence of blue-greenish yellow fluorescence in the crown area near the silk scar of kernels infected with *A. flavus*. However, *A. flavus* also has been found in the pedicel region of the seed [17,18,20], and thus the entry could be similar to that proposed for other seed-infecting fungi [25–29]. Also, Tsuruta et al. [28] found this to be the route of infection in detached kernels inoculated with *A. flavus*.

The most detailed study on infection was done by Smart et al. [18]. They characterized the route of infection by *A. flavus* in a controlled environment chamber, and found, as others have [14,20], that *A. flavus* readily colonizes kernel surfaces and the glume tissue surrounding the kernel. They proposed two routes of invasion into seeds. In one route the fungus grows superficially over the surface of the rachis and spikelet and invades at the junction of the bracts with their rachillas, where the cells are large, thin-walled, and highly vacuolate. In some cases they found the crushed epidermal cells at this junction to be torn, creating continuity between the aerenchyma cells of the rachilla and the exterior environment within the ear. These tears provided a direct path of entry with little physical obstruction.

Alternatively, the fungus was found to grow up through the cob into the spikelet through the continuous air space in these tissues. Because this tissue may be as much as 50% airspace, the fungus has no physical barrier to overcome. Once inside this tissue, the fungus moves up the aerenchyma cells within the rachillae and into the innermost layer of the pericarp. The fungus then enters the seed through breaks in the seed coat cuticle. In this study the fungus was found only over the embryo and in the center of the dorsal face of the seed. Hyphae were found commonly in all embryonic tissues, but particularly in the scutellum. Interestingly, in this tissue the fungal mycelium was densely cytoplasmic and the hyphae ramified inter- and intracellularly. The appearance of the cells suggested that the fungus was dissolving the cell walls of the scutellum in its path as the integrity of the scutellum cell cytoplasm was lost ahead of the hyphae. The fungus was rarely found in the endosperm. Smart et al. [18] also noted that the morphology of the fungal mycelium inside the kernel was different from that of mycelium outside the kernel. Whereas the mycelium outside the seed coat appeared vacuolated and sometimes collapsed, the mycelium ramifying inter- and intracellularly within the seed appeared to be broad and densely cytoplasmic.

The localization of the fungus in seeds grown in a controlled environment agrees with observations made in field studies. Fennell et al. [30] examined undamaged and slightly cracked kernels and observed *A. flavus* most often in the germ, next in the tip cap, and least in the endosperm. Jones et al. [13] in field studies also found extensive growth of *A. flavus* over the germ, with growth in the endosperm present only in severely infected kernels. Jones et al. [13], like Smart et al. [18], observed sporulation of the fungus in airspaces within the seed.

The data of Marsh and Payne [14] and Smart et al. [18] would argue that *A. flavus* and *Fusarium moniliforme* may have a common site of entry. Studies by Zummo and Scott [20], however, argue that the route of infection by these two fungi may not be the same. In a 2-year field study, they sectioned undamaged kernels from inoculated ears and examined the sections for growth of the two fungi. They found *F. moniliforme* more frequently in the pedicel than apical portions of transversely cut kernels. In contrast, *A. flavus* was detected more frequently in the apical sections, and was isolated more from the middle sections than the apical sections, arguing that if *A. flavus* penetrates at the tip or at the silk scar, it does so infrequently. From their data they argue that infection occurs over the surface of the kernels at cracks in the pericarp. They found, as did Smart et al. [18], *A. flavus* in the cob tissue. *A. flavus* colonized 83% of the cobs from inoculated and noninoculated ears. *F. moniliforme* was isolated less frequently from the cobs.

Thus the route of entry for *A. flavus* has not been clearly resolved. It is likely that the fungus has multiple sites of entry depending on the environmental conditions and corn genotype. Even if *F. moniliforme* and *A. flavus* do not share the same infection site, *F. moniliforme* does appear to compete with *A. flavus* on the corn ear. Wicklow et al. [31] showed that *F. moniliforme* could interfere with infection and aflatoxin accumulation in developing maize seeds. Hill et al. [15] also showed a negative correlation between the presence of *A. flavus* and *F. moniliforme*. Further, Zummo and Scott [32] found that inoculation of kernels with *A. flavus* did not affect colonization by *F. moniliforme*, but kernel infection by *A. flavus* was significantly reduced when ears were inoculated with both fungi at the same time. These data argue that the two fungi may either colonize the same site or compete for other resources within the ear.

Even though corn kernels are colonized soon after silking, infection of the kernels occurs late in their development [14,15,19,21]. Figure 2 shows the percentage of kernels infected on corn grown in a controlled environment and inoculated with a color mutant of *A. flavus*. Note that little invasion of kernels occurred before the seed reached physiological maturity (32%) [33]. This is consistent with field studies [19] showing a rapid increase in the rate of kernels infected as kernel moisture fell below 28%. There may be several reasons for this timing of infection. The main reason is likely the reduced physiological activity of the kernel and the lack of its ability to express active defense against the fungus. However, the increase could also be due to increased competition of *A. flavus* with other microflora on the ear. As shown in Figure 3, an increase in colonization of kernels with *A. flavus* parallels a decline in colonization by *F. moniliforme*.

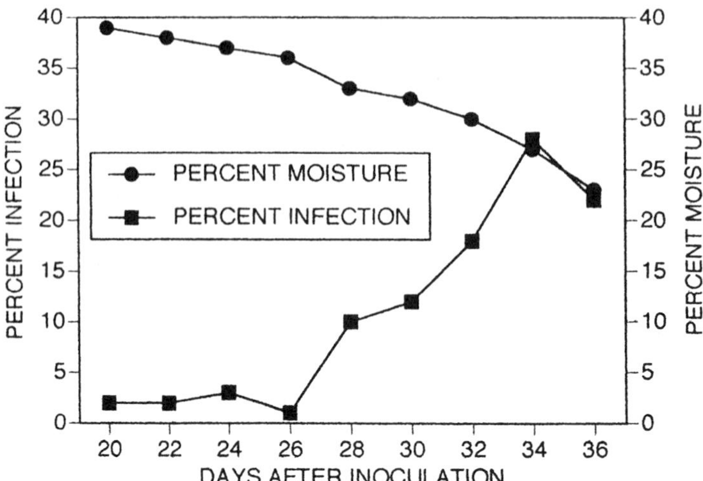

FIGURE 2 Percent kernels infected by *Aspergillus flavus* and percent moisture of kernels from plants grown in a controlled environment with a 15-h day at 34°C and 9-h night at 30°C and harvested every 2 days for 36 days after inoculation.

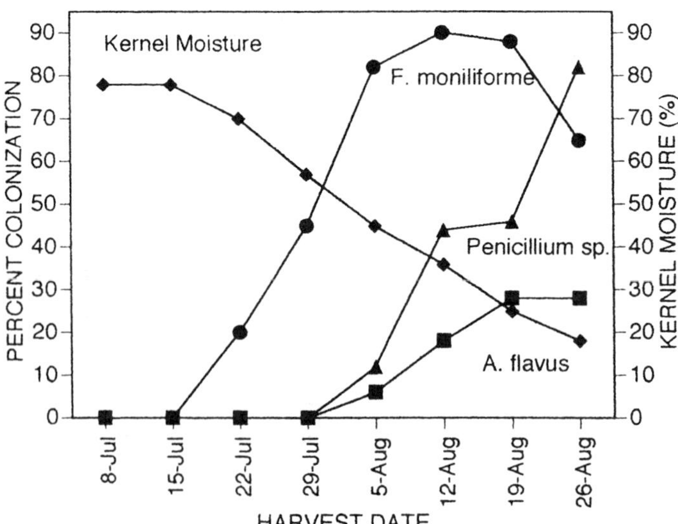

FIGURE 3 Percent kernels colonized by *Fusarium moniliforme* (solid circles), *Penicillium* species (triangles), and *Aspergillus flavus* (squares) and percent moisture (diamonds) of kernels grown in the field in Clayton, NC, and harvested each week for 8 wk.

3. Insects

Thus, as shown in Figure 1, *A. flavus* colonizes corn silks, grows down the silks, and colonizes the kernel surfaces. Insects are not required for this initial colonization, but they play several important roles in kernel colonization. The role of insects in the infection process has been covered by several excellent reviews [3,34–37]. Insects may contribute to the infection of kernels in four ways. They may (1) transport primary inoculum to the ears; (2) move inoculum from the silks into the ear; (3) disseminate inoculum within the ear; and (4) facilitate colonization and infection of the kernels by injuring the kernels. There is little doubt that insects can be involved in all of the above processes. Insects have been shown to carry conidia of *A. flavus* internally and externally [37], and the application of conidia to maize weevils has been shown to increase the number of infected kernels and aflatoxin contamination [24]. Thus, it is clear that insects are good vectors of fungal spores, and there is no doubt that insects are involved in bringing *A. flavus* to the ears and moving the fungus within the ear. Conidia of *A. flavus*, however, have been shown to be airborne [38–42], consistent with the morphology of the conidia and the asexual reproductive structures. Thus, the relative contribution of insects to the initial colonization of the ears is not clear.

Marsh and Payne [13] mapped the distribution of two naturally infected corn ears, one apparently free of insect damage (Fig. 4A) and one with insect damage (Fig. 4B). As can be seen in Figure 4, colonized kernels were scattered across the ear. The distribution of infected kernels from the tip to the base of the ear is consistent with findings of Jones et al. [39] in an earlier field study. However, colonized kernels appeared to be randomly distributed across the ear and not banded in the middle on the ear, as found by Jones et al. [12] in a controlled-environment study. It is evident from Figure 4 that colonization of maize kernels is not always associated with insect injury. In some cases there was colonization without insect injury (Fig. 4A, B), and conversely, there was insect injury without colonization (Fig. 4B). It is not clear what affects the distribution of colonized kernels, but it could be a function of the environmental conditions present at the time of colonization, or to the distribution of competing fungi.

The relative contribution of insects to any one of the four processes listed above likely differs depending on environmental conditions and insect populations. Under favorable environmental conditions, *A. flavus* can become an aggressive pathogen and grow down silks, colonize the surface of kernels, and infect developing kernels. When this occurs, insects are not required to bring the inoculum to the ear, distribute it within the ear, or provide a site for entry of the fungus. For example, over 100% of the kernels on an ear have been shown to be colonized by *A. flavus* under favorable environmental conditions [14]. In contrast, when the environmental conditions are less favorable for *A. flavus*, only a few kernels may be colonized in the absence of insect injury. The interaction between

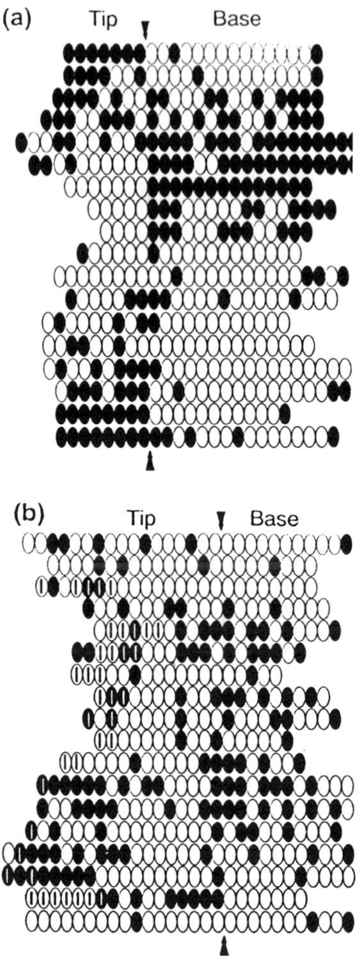

Figure 4 Schematic diagram of naturally infected ears of corn from a field in North Carolina without (a) or with (b) visible insect injury. Closed circles represent kernels colonized by *Aspergillus flavus*; the letter I represents kernels with insect damage. Arrow denotes the arbitrary separation of the tip from the base of the ear.

insects and *A. flavus* is complex. This likely explains why there is often a lack of correlation between insect injury and aflatoxin contamination [13,43–50].

4. Inoculum

Neither *A. flavus* nor *A. parasiticus* has a known sexual stage, and conidia are assumed to be the primary inoculum. Regardless of the source, inoculum likely arrives at the silks and kernels by two routes, wind and insects (Fig. 1). The role of insects as vectors of spores is discussed above. Even though insects may be involved, it is clear that airborne conidia of *A. flavus* are present each year. The airborne populations, however, differ greatly across years. Holtmeyer and Wallin [38] trapped airborne conidia in Missouri and detected *A. flavus* in the air over fields at least 17 days in each year. Jones et al. [39] monitored airborne conidia in irrigated and nonirrigated plots and found more airborne conidia in nonirrigated plots. They also found that hybrids pollinating during the weeks with higher levels of airborne conidia had a greater percentage of ears with visible growth of *A. flavus*. Zummo and Scott [32] sampled the air for 30 sec every 30 min at 1 m above the soil in a Mississippi cornfield. Daily populations ranged from 2 to 1208 conidia and averaged 229 in July and 315 in August. McGee et al. [40] also found airborne conidia of *A. flavus* in all plots of an Iowa study.

The source of inoculum for *A. flavus* is likely the soil. Shearer et al. [39] found extensive colonization of crop residues and soils in Iowa fields in which the corn was highly contaminated with aflatoxin. McGee et al. [40] sampled soil populations and airborne conidia in Iowa in 1991 and 1993. *A. flavus* was found in soils in tillage and rotation plots in both years. They found significantly higher populations in July than in June, August, or September. *A. flavus* was recovered more frequently from corn crop residue in continuous corn culture than in a corn-soybean rotation. Airborne conidia were found in all plots in each year, and *A. flavus* was found on leaves of growing corn plants in all treatments in each year.

The survival structure for *A. flavus* in the soil is not known. It is clear that it exists in debris as mycelium and possibly in soil as conidia. Boller and Schroeder [52] reported that moistened conidia lose viability in 21 to 60 days. In contrast, Angle et al. [53] found that *A. flavus* can overwinter in soil as mycelia and as conidia. Both *A. flavus* and *A. parasiticus* produce sclerotia [54], and Wicklow and Donahue [55] showed that sclerotia of both fungi undergo sporogenic germination. Further, sclerotia have been found [56] in naturally infected corn kernels at harvest in Georgia, and sclerotia have been shown capable of germinating on the soil surface in a Georgia cornfield 8 days before silking [57]. Thus it appears that sclerotia may be an important survival structure for these fungi. Wicklow et al. [56], however, suggest that there must be other fungal propagules in the soil in addition to sclerotia, as sclerotia could not account for all of the colonies that developed on the soil dilution plates in his study. Zummo and Scott [41] also found sclerotia of *A. flavus*, but not of *A. parasiticus*, in the pith tissues of uninoculated corn cobs in Mississippi. Cob tissue may serve as an important reservoir for *A. flavus*, as this tissue persists in Mississippi whereas other corn debris does not. It is interesting that sclerotia have not been found in the Midwest.

Thus the data indicate that *A. flavus* survives in the soil as conidia, mycelia, and sclerotia (Fig. 1). The presence of preharvest colonized corn debris such as kernels, cobs, and leaf residue appear to increase populations of *A. flavus* in the soil.

B. Peanuts

1. Colonization

Colonization and infection of peanuts and corn differ, as may be expected due to the way the peanut fruit is formed and its development underground. Because colonization and the mode of entry in peanuts have not been well characterized as separate events in peanuts, these two processes will be discussed together.

The peanut is unusual in that the flower is fertilized above ground, but the fruit develops in the soil from a peg that forms after fertilization and grows into the soil. Both *A. flavus* and *A. parasiticus* are soilborne pathogens, and it thus seems logical that infection of the peanut fruit may occur in the soil. In fact, attempts to obtain high levels of infection of flowers and above-ground parts has been difficult. Griffin and Garren [58] recovered *A. flavus* from approximately 7% of washed peanut flowers in both 1974 and 1975. Recovery from other flower parts was much lower. Only 1.5% of washed pegs and 0.3% of surface-sterilized pegs contained *A. flavus*.

Cole et al. [59] reexamined the colonization of peanut flowers by inoculating plants with a color mutant of *A. parasiticus*. Plants were inoculated with a conidial suspension of the mutant and subjected either to adequate moisture or to drought stress. Infection of flowers and aerial pegs was 0.9%, whereas the number of kernels infected at harvest was 10.3%. Although the data do not conclusively rule out aerial infection as significant in the infection process, the higher number of infected kernels compared with flowers argues that soil invasion may be more important.

Invasion of peanut plant parts by fungi in the soil is consistent with the findings of Pitt et al. [60]. In a greenhouse maintained at day/night temperatures of 32 and 23°C, respectively, they found that peanuts grown in unsterilized soil were invaded by *A. parasiticus* and *A. flavus* as soon as they emerged from soil. The fungi spread throughout the plant, with *A. flavus* being more invasive than *A. parasiticus*. Levels of seed invasion ranged from 1% to 4%, and all isolates taken from seeds were *A. flavus*. Young plants were invaded and invasion was more frequent in plant parts under or immediately adjacent to the soil (i.e., roots, stems) rather than distal parts (petiole, leaves). More than 80% of stem and root pieces were invaded in treatments with *A. flavus* in the soil, with or without *A. parasiticus* being present. Levels were lower when *A. parasiticus* alone was present. In plant pieces, >90% of the invasion was by *A. flavus*. Fungal hyphae were observed both on surfaces of stem pieces and within them, in interstitial spaces between cells, and in vascular bundles. Direct penetration of cell walls was rarely observed.

Kisyombe et al. [61] also observed colonization of fresh peanut tissue. They grew peanuts in field microplots and inoculated the plots with cracked corn colonized with a brown color mutant of *A. parasiticus* (ATTC 24690). *A. parasiticus* was observed growing from fresh pods, pegs, and tap and fibrous roots at all sampling dates. Recovery was the highest from tap roots and pod pieces. The fungus was isolated less frequently from other tissue such as pegs and fibrous roots. The degree of colonization was dependent on the environmental conditions. In 1983 and 1984, respectively, 81% and 48% of pegs, pods, and root tissues sampled were infected with *A. parasiticus*. In contrast to 1984, which was a near-ideal year for peanut production, 1983 was hot and dry. In 1983 the range of infected kernels for 14 peanut lines was 4% to 28%. From this study and others [62], it has been shown that peanuts without obvious damage can be invaded by *A. flavus* and contaminated with aflatoxin in the field before digging.

2. Insects

Although damage to peanut pods is not required for colonization of peanut tissue and the subsequent invasion of the kernels [63], insects play a major role in aflatoxin contamination of peanuts. Damaged kernels have much higher concentrations of aflatoxin than sound, mature kernels. Both mites [64] and the lesser corn stalk borer larvae (*Elasmopalpus lingosellus*) [65] are known vectors of *A. flavus* and *A. parasiticus*. It is the lesser corn stalk borer, however, that is responsible for most of the damage to peanuts, especially during drought stress [66]. Both internal and external pod damage (scarification of the shell) by the lesser corn stalk borer have been shown to increase the percentage of kernels infected by *A. flavus* [65]. Even microscopic damage to the peanut pods increases infection by *A. flavus* [67]; thus, insect damage need not be severe to facilitate infection by *A. flavus*.

3. Inoculum

Soilborne propagules of *A. flavus* and *A. parasiticus* are assumed to be the source of inoculum for the colonization of peanuts. Populations of these two fungi in the soil vary across years and locations. Criffen and Garren [68] sampled the soil around peanut pods in Virginia and isolated *A. flavus* at a frequency of 0.8–12.8 and 0.5–57.3 propagules per gram of soil in 1971 and 1972, respectively. Horne et al. [69] examined the effect of corn and peanut rotations in Georgia on soil populations of *A. flavus* and *A. parasiticus*. At the beginning of the study the two fungi were present at a similar proportion in all fields. Initial soil populations ranged from 30 to 3600 colony-forming units (cfu) per gram of soil. In 2 of the 3 years of the study, the levels of *A. flavus* and *A. parasiticus* remained constant for a 12-month period, but levels of *A. flavus* were higher in fields previously planted to peanut. In a third year of the study, the levels of the two fungi also were similar in fields planted to either corn or peanuts the year before, but the populations of *A. flavus* increased from 200 to 6400 cfu following the harvest of drought-stressed corn contaminated with *A. flavus* and aflatoxin.

In an attempt to identify the propagules of the aflatoxigenic fungi most important in infection, Horne et al. [7] inoculated field plots, whose moisture and temperature could be regulated, with the *A. parasiticus* brown spore mutant ATTC 24690. Conidia, sclerotia, and mycelium grown on peanut seeds were incorporated into the soil prior to planting. They found that soil populations of the mutant changed little in plots inoculated with sclerotia. Sclerotia of the fungus remained viable during the study, but they showed little evidence of sporogenic germination when collected from the soil at monthly intervals. Sclerotia of wild-type *A. parasiticus* were found on 0.6% of seeds from damaged pods. In contrast, populations of the mutant strain increased dramatically, from 12 to 8.8×10^4 cfu/g, in plots inoculated with mycelial fragments. Both the mutant strain and wild-type *A. flavus* and *A. parasiticus* populations increased with the onset of drought. Populations of the wild-type aflatoxigenic fungi increased from 440 cfu/g soil in August to 1.1×10^4 cfu/g in October. Horne et al. [7] also found a positive correlation between the soil populations of the mutant and percent infection of kernels in intact pods. Data from their study argue that soil populations of the fungi are important for infection of peanuts, and that soil propagules other than sclerotia are the primary inoculum for peanut seed invasion.

Even though the ratio of *A. flavus* to *A. parasiticus* shifted from 7:3 to 1:1 during drought stress, *A. parasiticus* accounted for only 10% to 12% of the two species in peanut seeds. Thus it appears that peanuts, like corn, are more susceptible to infection by *A. flavus* than *A. parasiticus*. Calvert et al. [70] have shown that *A. flavus* can outcompete *A. parasiticus* when double-inoculated onto corn.

C. Cottonseed

1. Colonization

A. flavus and aflatoxin contamination are associated with early-formed bolls in cotton [71]. However, the mode of entry of *A. flavus* into cottonseeds has not been well established. There is some controversy as to whether the fungus can directly infect the cottonseed through nectaries and leaf scars, or if injury or insects are required for the infection process. To determine if *A. flavus* could directly enter cotton bolls, Klich et al. [72] dusted *A. flavus* spores onto the involucral nectaries of cotton flowers. They found the fungus present in 20% to 58% of the inoculated immature bolls harvested 25 to 35 days after anthesis. The surface of the developing seed was the most frequent focus of fungal growth in developing bolls followed by the fiber. The fungus was observed least in placental tissue and the intercarpellary membrane. *A. flavus* was virtually the only fungus present within the green bolls. The percentage of seed within these bolls invaded by *A. flavus* was not determined.

2. Mode of Entry

Studies by Klich et al. [73,74] suggest that *A. flavus* can colonize fresh cotyledonary leaf scars and nectaries. In a field study, Klich et al. [73] inoculated cotylendonary leaf scars of seedlings, unopened flower buds, and involucral nectaries or sterigmata of newly opened flowers. They found that seeds from plants inoculated at cotyledonary nodes, involucral nectaries, and flower buds had significantly higher *A. flavus* infection rates than uninoculated controls. The highest level of infection resulted from inoculation of the involucral nectaries. They argued from their data that the fungus enters the plant in some manner at these natural openings and moves into the seeds. Klich and Chmielewski [74] found maximum seed infection (28%) to occur when involucral nectaries were inoculated 7 days after anthesis. There was a sharp decline in the percentage of infected seeds if nectaries were inoculated after 25 days. They argue that the fungus moves in the vascular system to the seed, and the decline after 25 days may be related to the degeneration of the funiculus attaching the developing seed to the placenta. Degeneration of the funiculus has been shown to occur 30 days after anthesis.

Huizar et al. [75] attempted to determine the path of colonization by *A. flavus* in seeds collected from cotton fields in Arizona. Using light and transmission electron microscopy, they found hyphae characteristic of *A. flavus* located throughout the cotyledons and in nonlignified layers of the seed coat. There was no evidence that *A. flavus* invaded the lignified colorless and palisade layers of the seed coat. Hyphae were found within cotton fibers external to the seed coat and also at the point of attachment to the epidermis. They postulated that *A. flavus* may enter through the chalazal cap, as the palisade layer terminates there. Lee et al. [76] found evidence of fungal penetration of the chalazal cap of seed harvested after 30 days and air-dried. Hyphae were found as well in the vessel elements within the seed coat, supporting, but not proving, the hypothesis that *A. flavus* may enter seeds through the vascular tissue.

3. Insects

Even though there is evidence for direct infection of cotton bolls by *A. flavus*, insects appear to be required for high levels of aflatoxin contamination in the field. The major insect is the pink bollworm larva *Pectinophora gossypiella* Saunders. Data suggest that it is the exit holes and not the entrance holes that provide entry courts for *A. flavus* [77]. Ashworth et al. [77] contaminated eggs of the pink bollworm with conidia of *A. flavus* and showed that the first instar does not carry conidia to the interior of the boll, nor do the entrance holes provide an entry for the fungus. In contrast, when exit holes made by the insect were inoculated with conidia of *A. flavus*, 16% of the locules in these bolls were contaminated with the fungus [77]. Beetles of the Nitidulidae family had no significant effect on the transmission of *A. flavus* into the interior of the bolls even though they were found to occur commonly in the exit holes. The role of insects in aflatoxin contamination has also been confirmed by treatment of bolls with insecticides. In two separate studies [78,79], aflatoxin contamination in cotton was shown to be significantly reduced with frequent chemical applications for control of pink bollworm. Even though these chemical applications greatly reduced aflatoxin contamination, they did not always reduce the level of contamination below the FDA guidelines.

The timing of insect injury also contributes to the levels of aflatoxin found in the cotton at harvest. Lee et al. [76] wounded cotton bolls to simulate insect injury and inoculated the wound site with conidia of *A. flavus*. They found that inoculation 33 days after flowering resulted in the highest levels of aflatoxin. At this stage of development the bolls are still green but close to maturity. Cotty [80] also found high levels of aflatoxin in bolls wounded and inoculated with *A. flavus* 28 to 32 days after flowering. In contrast to the findings of Lee et al. [76], Cotty found low concentrations of aflatoxin in adjacent nonwounded locules.

4. Inoculum

A. flavus inoculum in cotton fields probably comes from windborne conidia released by the fungus in the soil. Lee et al. [81] have shown the populations of *A. flavus* in the soil to increase during the growing season in Arizona and to reach a maximum in August. Over the six locations examined, the maximum propagules of *A. flavus* per gram of soil ranged from 42 to 632. Airborne populations increased dramatically in July, and at one location reached 750 propagules per cubic foot of air on August 5. Similarly, Ashworth et al. [82] found a four- to eightfold increase in inoculum potential between April and December in cotton fields of southern California. They found a maximum of 6×10^3 colonies per gram soil.

D. Tree Nuts

The incidence of aflatoxin contamination in tree nuts is low, indicating that *A. flavus* does not colonize tree nuts as effectively as the other commodities discussed in this review. However, high levels of aflatoxin can develop in a small percentage of nuts [83]. Infection of tree nuts by *A. flavus* is almost always associated with insect injury, and in pistachios, aflatoxin contamination is almost always associated with "early splits."

1. Colonization

In general, intact nuts are resistant to infection by either *A. parasiticus* or *A. flavus*. Pistachio nuts with intact cuticles have been shown to be resistant to colonization by *A. flavus* [84]. Damage of the cuticular layer, however, results in rapid colonization of the hull by *A. flavus* [85]. Despite the rapid colonization of the hulls by *A. flavus*, Mahoney and Rodriguez [85] were unable to detect any aflatoxin accumulation in the hulls. Intact kernels of pistachios with the seed coat also support little colonization and production of aflatoxin. Resistance to colonization in the seed may also be conferred in part by a cuticle layer.

In almond, the seed coat also appears to be a barrier to infection by *A. flavus*. Gradziel and Wang [86] found that the seed coat of all almond varieties examined had a high level of resistance to colonization by *A. flavus*. There also appeared to be some level of cotyledonary resistance in cultivars Ne Plus, Ruby, and Carrion.

Whatever the mechanism of resistance, it is clear that the hull plays a major role in resistance to *A. flavus* and aflatoxin contamination. Thomson and Mehdy [87] found that pistachios sampled prior to split were relatively free of *A. flavus*, but at harvest the fungus was present in 11.8% of the nuts. In Turkey, Heperkan et al. [88] found *A. flavus* on 39% of the shells at harvest, but infection in the endosperm was only 13%. In pistachios, the hull normally remains intact until harvest. In some cases the hull splits early, causing a condition known as "early splits." Doster and Michailides [89] have shown that the early splits are the most important source of preharvest contamination of pistachios in California. They further showed [90] that the earlier the split, the more aflatoxin that accumulates. Earlier splits allow more time for colonization of the seed and aflatoxin contamination.

2. Insects

Insects play a dominant role in preharvest aflatoxin contamination of tree nuts [91]. The main insect involved in aflatoxin contamination of pistachios and almond is the navel orange worm larvae. Even though *A. flavus* can colonize pistachios with split hulls, the degree of colonization and the amount of aflatoxin produced are much greater in the presence of insects. Doster and Michailides [89] found that the early splits were colonized to a greater degree by many fungi, including *A. flavus*. They found that 84% of the aflatoxin occurred in kernels with navel orange worm damage. Thomson and Mehdy [87] found *A. flavus* in 11% of nuts not damaged by insects, compared to 28% of insect-damaged nuts. Most of the damage was caused by navel orange worm larvae.

3. Inoculum

The dominant fungal flora on pistachios appears to be *Aspergillus*, with *A. niger* being the dominant species [88,89]. Doster and Michailides [90] found *A. niger* in over 30% of the early split nuts. In contrast, *A. flavus* and *A. parasiticus*, respectively, were found in only 0.7% and 0.1% of the kernels. *A. niger* also appears to be the dominant *Aspergillus* species on almond [92,93]. *A. flavus* was found on

60% of the nondisinfested kernels but was found on only 0.4% of the disinfested kernels [93].

The source of inoculum for contamination of tree nuts is assumed to be the orchard floor. *Aspergillus* species were found in debris, especially in the male inflorescences, and soil [90]. Again, as is the case with pistachio nuts, incidence of *A. flavus* appears to be rare [89]. Further, many of the isolates of *A. flavus* found did not produce aflatoxin. Doster and Michailides [94] found that only 43% of the *A. flavus* isolates collected from California orchards produced aflatoxin. Heperkan et al. [88] found that 35% of the *A. flavus* isolates collected from orchards in Turkey produced aflatoxin. In California [89] and Iran [95], *A. flavus* was found more commonly than *A. parasiticus*. Doster and Michailides [89] examined several cultural practices including the presence or absence of cover crops or type of irrigation for their effect on the occurrence of *Aspergillus* species in pistachio kernels but found no effect.

IV. CONDITIONS AFFECTING INVASION, GROWTH, AND AFLATOXIN ACCUMULATION

A. Corn

The two major factors associated with aflatoxin contamination are increased temperature and drought stress. Years in which aflatoxin contamination is a serious problem are characterized as having above-average temperatures and below-average rainfall. In Iowa in 1983, a year noted for high levels of aflatoxin contamination, temperatures in July and August were 2 to 3°C higher than normal [96]. Also, the areas in the state with the highest contamination of aflatoxin received 15 cm less rainfall than normal.

McMillian et al. [97] monitored several environmental factors in a 5-year study in Georgia. They found temperature and net evaporation rates to be correlated with aflatoxin contamination. Further studies [43] indicated that high maximum and high minimum daily temperatures, especially during periods of high net evaporation, are more important for the development of aflatoxin than humidity or average precipitation during the same period. High temperatures have been shown to favor kernel infection [21] in controlled-environment studies. Payne et al. [21] examined the effect of temperature regimes with a 9-h day and a 15-h night on kernel infection. The day/night temperatures and respective thermal units (TU) were: 26/22 (13.5); 34/22 (16.5); 26/30 (18.5); and 34/30 (21.5). The maximum infection rate of 28% was found at the highest temperature regimen tested, but rates as high as 7% were obtained at 16.5 TU. It is likely to expect a temperature regimen generating at least 16.5 thermal units to occur in the field in years conducive for aflatoxin contamination. Such infection rates have been reported for field studies [13,14,39,98,99].

In field studies it has been difficult to separate the effect of high temperatures from that of drought, the two of which often occur together. However, in most years irrigation to reduce water stress reduces aflatoxin contamination. Jones et al. [39] found irrigation to reduce aflatoxin levels in a 2-year study in North Carolina. They found the effect of drought to be the most pronounced when drought occurred during the silking to late dough stage of grain development. In a 4-year study, Payne et al. [99] found that irrigation or subsoiling to reduce water stress reduced aflatoxin contamination. Drought stress likely contributes to both greater infection of kernels and increased growth of the fungus.

Nitrogen stress also contributes to aflatoxin accumulation. Payne et al. [100] found aflatoxin levels to be negatively correlated with corn yield, silk leaf nitrogen, and grain nitrogen in a 2-year study in North Carolina. Silk- or wound-inoculated ears from plants receiving no added nitrogen contained an average of 28% more aflatoxin than ears from plants that received optimum nitrogen fertilization. In the U.S., where adequate nitrogen is usually applied, nitrogen stress does not likely contribute to aflatoxin contamination.

B. Peanuts

The two major factors affecting aflatoxin contamination of peanuts are drought stress and temperature, and these two factors are interrelated. Neither high temperature nor drought stress alone will lead to increased concentrations of aflatoxin. The influence of temperature and drought stress on fungal infection and aflatoxin contamination has been extensively studied in Georgia [62,101–105]. These studies were done in field plots that allowed the manipulation of soil temperature and moisture. Results from these studies showed that with adequate moisture there is no contamination of undamaged peanuts with aflatoxin. Similarly, no significant aflatoxin contamination of kernels occurred in intact peanuts with prolonged drought stress before harvest if the mean geocarposphere temperature between drought stress was <25 or $>32°C$. Colonization by *A. flavus* and aflatoxin contamination was maximum at 30.5°C. At this temperature 97.5% of the kernels were colonized and aflatoxin concentration reached 500 ng/ml. The most critical time for drought and temperature stress appears to be during the last 3 to 6 weeks of the growing season [105].

C. Cottonseed

Contamination of cotton with aflatoxin can be a serious problem in Texas, Arizona, and southern California. Interestingly, although it occurs in the southern U.S., it is rarely a serious problem [71]. Ashworth et al. [71] argue that the disease is generally confined to the regions of the country having high relative humidity and temperature, and which are free of a significant amount of precipitation during the midsummer. The Texas-New Mexico region is humidified by moist air from the Gulf of Mexico, and the southern California-western Arizona regions are humidified from July to September by moist air from the Gulf of California and the Pacific Ocean.

The effects these environmental conditions have on the infection and aflatoxin accumulation in cottonseeds is not well characterized. Ashworth et al. [71] have argued that the higher humidity causes the bolls to remain moist longer once they open. Russell [106], however, has argued that temperature, not relative humidity, is the key factor responsible for greater aflatoxin contamination of Western-grown cotton as compared to the Southeast. He argued that high day temperatures are common to all cotton-growing areas, but high night temperatures of 25 to 34°C are unique to the lower elevations of Arizona and the Imperial Valley of California. In Arizona, aflatoxin levels are much higher in seed harvested at the hotter, lower elevations than at elevations above 550 m.

The effect of temperature on growth and aflatoxin production may be complex. Ashworth et al. [107] compared three temperature regimens for their effect on growth and aflatoxin production. The fungus grew rapidly at 20 to 30°C, with maximum seed infection by *A. flavus* occurring between 30 and 35°C. At 15 to 20°C, competing fungi dominated the seeds. Even though growth was maximum at 30 to 35°C, aflatoxin was maximal at 25 to 30°C. They argued that the slower rate of drying at 25 to 30°C gave more time for fungal development than the faster rate of drying. Their data support this, but previous studies have shown that the fungus also produces more aflatoxin between 25 and 30°C [108].

Moisture appears to affect both the early and late phases of aflatoxin contamination. Klich [109] has shown that drought stress can increase the percentage of seed infected. She inoculated the involucral nectaries of cotton with conidia of *A. flavus* and determined the effect of xylem water potential on the percentage of infected kernels. Plants with water potentials in range of 1.6 to 1.9 MPa had the greatest percentage of infected seed.

After the initial infection of seed by *A. flavus*, high humidity or boll rewetting increases the level of aflatoxin contamination [110]. Cotty [110] found aflatoxin concentrations in cottonseed to increase over a broad range of temperatures if the relative humidity was 93% or greater. Further, Russell et al. [111] showed that aflatoxin levels could be decreased by reductions in late-season irrigation. For these reasons, early harvest is recommended for reduction of aflatoxin contamination in Arizona [111].

D. Tree Nuts

Most of the almonds in the U.S. are grown in the central valley of California. Purcell et al. [90] found that the incidence of aflatoxin contamination differed within this region. *A. flavus* was found most frequently in the southern regions of the central valley, and least frequently in the northern areas. They found a significant correlation between the average temperature and incidence of *A. flavus*. *A. flavus* was more common in orchards from an area with high average day and night temperatures, and on almonds taken from sunny sites within the orchard. Irrigation has been shown to reduce aflatoxin contamination in pistachios by reducing the number of early splits [112].

V. SUMMARY

Although *Aspergillus flavus* and *A. parasiticus* are considered "weak" parasites, under favorable environmental conditions they can colonize and infect living plant tissue, and contaminate seeds with aflatoxin. Even in cases of serious aflatoxin contamination the percentage of seeds infected is often low. Because high levels of aflatoxin can be produced in an individual seed, and the tolerance for aflatoxin contamination is so low, even a small number of infected seeds can be economically important. Only a few contaminated seeds can result in the rejection of the entire lot of a commodity.

As discussed in this chapter, the interactions of *A. flavus* and *A. parasiticus* with their hosts differ. There are, however, several common elements. The primary source of inoculum for these fungi appears to be the soil. The survival structure of the fungi has not been determined, but existing data suggest that fungal mycelium in debris is likely the primary soil propagule. The presence of sclerotia in infected tissue and in the soil in the southern U.S. argues that these structures also may play an important role in the survival of *A. flavus*. In the case of peanuts, populations of the fungi in the soil are important in the epidemiology of the disease, as the pods appear to be infected once they enter the soil. In the other hosts discussed, airborne populations of inoculum appear to be important.

The two overriding environmental factors affecting aflatoxin contamination are temperature and moisture. These two conditions, more than any other, dictate the outcome of the host-parasite interaction. Temperature and moisture have a dual effect on the interaction because they directly affect both the host plant and the fungus. Under conditions optimum for these fungi, i.e., high temperature and low moisture, they thrive and outcompete other soil and plant microflora. Under such conditions, the fungi are able to produce abundant inoculum, and the inoculum is easily dispersed in the air as conidia or windblown debris. High temperatures, and the lower water activity of the kernels as they mature, also provide a competitive advantage for *A. flavus* and *A. parasiticus*. These conditions allow the fungi to outcompete other microflora on the kernel surfaces, placing them in an ideal position to colonize seeds that are injured or have a compromised defense system.

By contrast, the host plant is compromised under conditions favorable for the fungi. Many of the physiological defense systems are compromised due to high temperatures and drought stress. Further, these conditions often lead to cracks in the seed, allowing the fungi to breach the seed's morphological barriers.

Injury, especially that caused by insects, is very important in the epidemiology of the disease. Sporulation of *A. flavus* and *A. parasiticus* is often observed on seeds damaged by insects. Injury not only provides an easy entry site for the fungi, but it also causes the dehydration of the kernels and thus a more favorable environment for the fungi. In many cases the damage caused by insects can be severe. In peanuts, however, it has been shown that only minor injury is needed to increase aflatoxin contamination.

The association of high concentrations of aflatoxin with seed injury argues that morphological barriers are very important in resistance. Recent data, however, argue that additional resistance factors are involved [113–117]. Successful management of aflatoxin contamination will require the employment of host resistance combined with management strategies.

REFERENCES

1. JD Miller, HL Trenholm. Mycotoxins in Grain: Compounds Other Than Mycotoxins. St. Paul, MN: Eagan Press, 1994.
2. UL Diener, RJ Cole, TH Sanders, GA Payne, LS Lee, MA Klich. Epidemiology of aflatoxin formation by *Aspergillus flavus*. Annu Rev Phytopathol 25:249, 1987.
3. GA Payne. Aflatoxin in maize. Crit Rev Plant Sci 423:423, 1992.
4. NW Widstrom. The aflatoxin problem with corn grain. In: DL Sparks, ed. Advances in Agronomy. New York: Academic Press, 1996, p 220.
5. DM Wilson, GA Payne. Factors affecting *Aspergillus flavus* group infection and aflatoxin contamination of crops. In: DL Eaton, JD Groopman, eds. The Toxicology of Aflatoxins. San Diego: Academic Press, 1994, p 309.
6. MA Klich, LH Tiffany, G Knaphus. Ecology of the aspergilli of soils and litter. In: JW Bennett, MA Klich, eds. *Aspergillus* Biology and Industrial Applications. Boston, MA: Butterworth-Heineman, 1994, p 309.
7. BW Horn, JW Dorner, RL Greene, PD Blankenship, RJ Cole. Effect of *Aspergillus parasiticus* soil inoculum on invasion of peanut seeds. Mycopathologia 125:179, 1994.
8. TE Cleveland, PJ Cotty. Invasiveness of *Aspergillus flavus* isolates in wounded cotton bolls is associated with production of a specific fungal polygalacturonase. Phytopathology 81:155, 1991.
9. TE Cleveland, SP McCormick. Identification of pectinases produced in cotton bolls infected with *Aspergillus flavus*. Phytopathology 77:1498, 1987.
10. MP Whitehead, MT Shieh, TE Cleveland, JW Cary, RA Dean. Isolation and characterization of polygalacturonase genes (*pecA* and *pecB*) from *Aspergillus flavus*. Appl Environ Microbiol 61:3316, 1995.

11. BZ Guo, JS Rusin, TE Cleveland, RL Brown, KE Damann. Evidence for cutinase production by *Aspergillus flavus* and its possible role in infection of corn kernels. Phytopathology 86:824, 1996.
12. BZ Guo, JS Russin, TE Cleveland, RL Brown, NW Widstrom. Wax and cutin layers in maize kernels associated with resistance to aflatoxin production by *Aspergillus flavus*. J Food Protect 58:296, 1995.
13. RK Jones, HE Duncan, GA Payne, KJ Leonard. Factors influencing infection by *Aspergillus flavus* in silk-inoculated corn. Plant Dis 64:859, 1980.
14. SF Marsh, GA Payne. Preharvest infection of corn silks and kernels by *Aspergillus flavus*. Phytopathology 74:1284, 1984.
15. RA Hill, DM Wilson, WW McMillian, et al. Ecology of the *Aspergillus flavus* group and aflatoxin formation in maize and groundnut. In: J Lacey, ed. Tricothecenes and Other Mycotoxins. New York: John Wiley & Sons, 1985, pp 8, 79, 95.
16. LL Ilag. *Aspergillus flavus* infection of pre-harvest corn, drying corn and stored corn in the Philippines. Philippine Phytopathol 9:37, 1975.
17. SF Marsh, GA Payne. Scanning EM studies on the colonization of dent corn by *Aspergillus flavus*. Phytopathology 74:557, 1984.
18. MG Smart, DT Wicklow, RW Caldwell. Pathogenesis in *Aspergillus* ear rot of maize: light microscopy of fungal spread from wounds. Phytopathology 80:1287, 1990.
19. GA Payne, WM Hagler Jr, CR Adkins. Aflatoxin accumulation in inoculated ears of field-grown maize. Plant Dis 72:422, 1988.
20. N Zummo, GE Scott. Cob and kernel infection by *Aspergillus flavus* and *Fusarium moniliforme* in inoculated, field-grown maize ears. Plant Dis 74:627, 1990.
21. GA Payne, DL Thompson, EB Lillehoj, MS Zuber, CR Adkins. Effect of temperature on the preharvest infection of maize kernels by *Aspergillus flavus*. Phytopathology 78:1376, 1988.
22. RK Jones. The epidemiology and management of aflatoxins and other mycotoxins. In: JG Horsfall, EB Cowling, eds. Plant Disease, An Advanced Treatise. Vol. 4. How Pathogens Induce Disease. New York: Academic Press, 1979, p 381.
23. MJ Wolf, CL Buzan, MM MacMasters, CE Rist. Structure of the mature corn kernel. II. Microscopic structure of pericarp, seed coat, and hilar layer of dent corn. Cereal Chem 29:234, 1952.
24. HW Anderson, EW Nehring, WR Wichser. Aflatoxin contamination of corn in the field. J Agric Food Chem 23:775, 1975.
25. B Koehler. Natural mode of entrance of fungi into corn ears and some symptoms that indicate infection. J Agric Res 64:421, 1942.
26. H Johann. Histology of the caryopsis of yellow dent corn, with reference to resistance and susceptibility to kernel rots. J Agric Res 51:855, 1935.
27. AM Salama, AG Mishricky. Seed transmission of maize wilt fungi with special reference to *Fusarium moniliforme* Sheld. Phytopathol Zeitschrift 77:356, 1973.
28. O Tsuruta, S Gohara, M Saito. Scanning electron microscopic observations of a fungal invasion of corn kernels. Trans Mycol Soc (Japan) 22:121, 1981.
29. TF Manns, JF Adams. Parasitic fungi internal of seed corn. J Agric Res 23:495, 1923.
30. DI Fennell, RJ Bothast, EB Lillehoj, RE Peterson. Bright greenish-yellow fluores-

cence and associated fungi in white corn naturally contaminated with aflatoxin. Cereal Chem 50:404, 1973.
31. DT Wicklow, BW Horn, OL Shotwell, CW Hesseltine, RW Caldwell. Fungal interference with *Aspergillus flavus* infection and aflatoxin contamination of maize grown in a controlled environment. Phytopathology 78:68, 1988.
32. N Zummo, GE Scott. Interaction of *Fusarium moniliforme* and *Aspergillus flavus* on kernel infection and aflatoxin contamination in maize ears. Plant Dis 76:771, 1992.
33. SR Aldrich, WO Scott, ER Long. Modern Corn Production. 2nd ed. Champaign, IL: A & L Publications, 1975.
34. NW Widstrom. The role of insects and other plant pests in aflatoxin contamination of corn, cotton, and peanuts—a review. J Environ Qual 8:5, 1979.
35. D Barry. Insects of maize and their association with aflatoxin contamination, in Aflatoxin in maize. In: MS Zuber, EB Lillehoj, BL Renfro, eds. Proceedings of the Workshop. CIMMYT, Mexico City, 1987, p 201.
36. WW McMillian. Role of anthropods in field contamination. In: UL Diener, RA Asquith, JW Dickens, eds. In: Southern Cooperative Service Bulletin 279. Auburn: Alabama Agricultural Experiment Station, 1983, p 20.
37. WW McMillian. Relation of insects to aflatoxin contamination in maize grown in the southeastern USA, in aflatoxin in maize. In: MS Zuber, EB Lillehoj, BL Renfro, eds. Proceedings of the Workshop. CIMMYT, Mexico City, 1987, p 194.
38. MG Holtmeyer, JR Wallin. Identification of aflatoxin-producing atmospheric isolates of *Aspergillus flavus*. Phytopathology 70:325, 1980.
39. RK Jones, HE Duncan, PB Hamilton. Planting date, harvest date, and irrigation effects on infection and aflatoxin production by *Aspergillus flavus* in field corn. Phytopathology 71:810, 1981.
40. DC McGee, OM Olanya, GM Hoyos. Populations of *Aspergillus flavus* in the Iowa cornfield ecosystem in years not favorable for aflatoxin contamination of corn grain. Plant Dis 80:742, 1996.
41. N Zummo, GE Scott. Relative aggressiveness of *Aspergillus flavus* and *A. parasiticus* on maize in Mississippi. Plant Dis 74:978, 1990.
42. RJ Bothast, LR Beuchat, BS Emswiller, MG Johnson, MD Pierson. Incidence of airborne *Aspergillus flavus* spores in corn fields of five states. Appl Environ Microbiol 35:627, 1978.
43. NW Widstrom, WW McMillian, RW Beaver, DM Wilson. Weather-associated changes in aflatoxin contamination of preharvest maize. J Prod Agric 3:196, 1990.
44. DI Fennell, EB Lillehoj, WF Kwolek. *Aspergillus flavus* and other fungi associated with insect-damaged field corn. Cereal Chem 52:314, 1975.
45. EB Lillehoj, WF Kwolek, RE Peterson, OL Shotwell, CW Hesseltine. Aflatoxin contamination, fluorescence and insect damage in corn infected with *Aspergillus flavus* before harvest. Cereal Chem 53:505, 1976.
46. DM Wilson, NW Widstrom, LR Marti, BD Davis. *Aspergillus flavus* group, aflatoxin, and bright greenish-yellow fluorescence in insect-damaged corn in Georgia. Cereal Chem 58:40, 1981.
47. EB Lillehoj, WF Kwolek, MS Zuber, et al. Aflatoxin in corn before harvest: interactions of hybrids and locations. Crop Sci 20:731, 1980.

48. WW McMillian, DM Wilson, NW Widstrom. Insect damage, *Aspergillus flavus* ear mold, and aflatoxin contamination in south Georgia corn fields in 1977. J Environ Qual 7:564, 1978.
49. WW McMillian, DM Wilson, NW Widstrom, RC Gueldner. Incidence and level of aflatoxin in preharvest corn (*Zea mays*) in south Georgia, USA in 1978. Cereal Chem 57:83, 1980.
50. NW Widstrom, AN Sparks, EB Lillehoj, WF Kwolek. Aflatoxin production and lepidopteran insect injury on corn in Georgia. J Econ Entomol 68:855, 1975.
51. JF Shearer, LE Sweets, NK Baker, LH Tiffany. A study of *A. flavus/parasiticus* group in Iowa crop fields. Plant Dis 76:19, 1992.
52. RA Boller, HW Schroeder. Production of aflatoxin by cultures derived from conidia stored in the laboratory. Mycologia 66:61, 1974.
53. JS Angle, RL Lindgren, D Gilbert-Effiong. Survival of *Aspergillus flavus* conidia in soil. Biodeteriorat Res 2:245, 1989.
54. DT Wicklow. Taxonomic features and ecological significance of sclerotia in aflatoxins. In: UL Diener, RA Asquith, JW Dickens, eds. Aflatoxin and *Aspergillus flavus* in Corn. Southern Cooperative Series Bulletin 279. Auburn: Alabama Agricultural Experiment Station, 1983, pp 16, 19.
55. DT Wicklow, JE Donahue. Sporogenic germination of sclerotia in *Aspergillus flavus* and *A. parasiticus*. Trans Br Mycol Soc 82:621, 1984.
56. DT Wicklow, BW Horn, WR Burg, RJ Cole. Sclerotium dispersal of *Aspergillus flavus* and *Eupenicillium ochrosalmoneum* from maize during harvest. Trans Br Mycol Soc 83:299, 1984.
57. DM Wilson, NW Widstrom, WW McMillian, RW Beaver. Aflatoxins in corn. Proceedings of the 44th Annual Corn and Sorghum Research Conference, American Seed Association, Washington, DC, 1989.
58. GJ Griffin, KH Garren. Colonization of aerial peanut pegs by *Aspergillus flavus* and *A. niger*-group fungi under field conditions. Phytopathology 66:1161, 1974.
59. RJ Cole, RA Hill, PD Blankenship, TH Sanders. Color mutants of *Aspergillus flavus* and *Aspergillus parasiticus* in a study of preharvest invasion of peanuts. Appl Environ Microbiol 52:1128, 1986.
60. JI Pitt, SK Dyer, S McCammon. Systemic invasion of developing peanuts by *Aspergillus flavus*. Lett Appl Micro 13:16, 1991.
61. CT Kisyombe, MK Beute, GA Payne. Field evaluation of peanut genotypes for resistance to infection by *Aspergillus parasiticus*. Peanut Sci 12:12, 1985.
62. TH Sanders, RA Hill, RJ Cole, PD Blankenship. Effect of drought on occurrence of *Aspergillus flavus* in maturing peanuts. J Am Oil Chem Soc 58:966A, 1981.
63. UL Diener. Preharvest aflatoxin contamination of peanuts, corn and cottonseed: a review. In: GC Llewellyn, CE O'Rear, eds. Biodeterioration Research 2. New York: Plenum Press, 1989, p 217.
64. JL Aucamp. The role of mite vectors in the development of aflatoxin in groundnuts. J Stored Prod Res 5:245, 1969.
65. RE Lynch, DM Wilson. Enhanced infection of peanut, *Arachis hypogaea* L., seeds with *Aspergillus flavus* group due to external scarification of peanut pods by the lesser cornstalk borer, *Elasmopalpus lignosellus* (Zeller). Peanut Sci 18:110, 1991.

66. PD Blankenship, RJ Cole, TH Sanders, RA Hill. Effect of geocarposphere temperature on pre-harvest colonization of drought stressed peanuts by *Aspergillus flavus* and subsequent aflatoxin contamination. Mycopathologia 85:69, 1984.
67. DM Porter, FS Wright, JL Steele. Relationship of microscopic shell damage to colonization of peanut by *Aspergillus flavus*. Oleagineux 41:23, 1986.
68. GJ Griffin, KH Garren. Population levels of *Aspergillus flavus* and the *A. niger* group in Virginia peanut fields and soils. Phytopathology 64:322, 1974.
69. BW Horn, RL Greene, JW Dorner. Effect of corn and peanut cultivation on soil populations of *Aspergillus flavus* and *A. parasiticus* in southwestern Georgia. Appl Environ Microbiol 61:2472, 1995.
70. OH Calvert, EB Lillehoj, WF Kwolek, MS Huber. Aflatoxin B_1 and G_1 production in developing *Zea mays* kernels from mixed inocula of *Aspergillus flavus* and *A. parasiticus*. Phytopathology 68:501, 1978.
71. LJ Ashworth Jr, JL McMeans, CM Brown. Infection of cotton by *Aspergillus flavus*: epidemiology of the disease. J Stored Prod Res 5:193, 1969.
72. MA Klich. Presence of *Aspergillus flavus* in developing cotton bolls and its relation to contamination of mature seeds. Appl Environ Microbiol 52:963, 1986.
73. MA Klich, SH Thomas, JE Melton. Field studies on the mode of entry of *Aspergillus flavus* into cotton seeds. Mycologia 76:665, 1984.
74. MA Klich, MA Chmielewski. Nectaries as entry sites for *Aspergillus flavus* in developing cotton bolls. Appl Environ Microbiol 50:602, 1985.
75. HE Huizar, CC Bertke, MA Klich, JM Aronson. Cytochemical localization and ultrastructure of *Aspergillus flavus* in cottonseed. Mycopathologia 110:43, 1990.
76. LS Lee, PE Lacey, WR Goynes. Aflatoxin in Arizona cottonseed: a model study of insect-vectored entry of cotton bolls by *Aspergillus flavus*. Plant Dis 71:997, 1987.
77. LJ Ashworth Jr, RE Rice, JL McMeans, CM Brown. The relationship of insects to infection of cotton bolls by *Aspergillus flavus*. Phytopathology 61:488, 1971.
78. TJ Henneberry, LA Bariola, T Russell. Pink bollworm: chemical control in Arizona and relationship to infestations, lint yield, seed damage, and aflatoxin in cottonseed. J Econ Entomol 71:440, 1978.
79. JL McMeans, CM Brown, LL Parker, RL McDonald. Aflatoxins in cottonseed: effects of pink bollworm control. Crop Sci 16:259, 1976.
80. PJ Cotty. Effects of cultivar and boll age on aflatoxin in cottonseed after inoculation with *Aspergillus flavus* at simulated exit holes of the pink bollworm. Plant Dis 73: 489, 1989.
81. LS Lee, LV Lee Jr, TE Russell. Aflatoxin in Arizona cottonseed: field inoculation of bolls by *Aspergillus flavus* spores in wind-driven soil. J Am Oil Chem Soc 63:530, 1986.
82. LJ Ashworth Jr, JL McMeans, CM Brown. Infection of cotton by *Aspergillus flavus*: epidemiology of the disease. J Stored Prod Res 5:193, 1969.
83. JE Schade, AD King Jr. Fluorescence and aflatoxin content of individual almond kernels naturally contaminated with aflatoxin. J Food Sci 49:493, 1984.
84. SB Rodriguez, NE Mahoney, DW Irving, AD King Jr. Aflatoxin production in pistachios. Aflatoxin Elimination Workshop. U.S. Department of Agriculture, Fresno, CA, 1996.

85. NE Mahoney, SB Rodriguez. Aflatoxin variability in pistachios. Appl Environ Microbiol 62:1197, 1996.
86. TM Gradziel, D Wang. Susceptibility of California almond cultivars to aflatoxigenic *Aspergillus flavus*. Hort Science 29:33, 1994.
87. SV Thomson, MC Mehdy. Occurrence of *Aspergillus flavus* in pistachio nuts prior to harvest. Phytopathology 68:1112, 1978.
88. D Heperkan, N Aran, M Ayfer. Mycoflora and aflatoxin contamination in shelled pistachio nuts. J Sci Food Agric 66:273, 1994.
89. MA Doster, TJ Michailides. *Aspergillus* molds and aflatoxins in pistachio nuts in California. Phytopathology 84:583, 1994.
90. MA Doster, TJ Michailides. The relationship between date of hull splitting and decay of pistachio nuts by *Aspergillus* species. Plant Dis 79:766, 1995.
91. NF Sommer, JR Buchanan, RJ Fortlage. Relation of early splitting and tattering of pistachio nuts to aflatoxin in the orchard. Phytopathology 76:692, 1986.
92. SL Purcell, DJ Phillips, BE Mackey. Distribution of *Aspergillus flavus* and other fungi in several almond-growing areas of California. Phytopathology 70:926, 1980.
93. DJ Phillips, B Mackey, WR Ellis, TN Hansen. Occurrence and interaction of *Aspergillus flavus* with other fungi on almonds. Phytopathology 69:829, 1979.
94. MA Doster, TJ Michailides. Development of *Aspergillus* molds in litter from pistachio trees. Plant Dis 78:393, 1994.
95. H Mojtahedi, CJ Rabie, A Lubben, M Steyn, D Danesh. Toxic aspergilli from pistachio nuts. Mycopathologia 67:123, 1979.
96. SG Schmitt, CR Hurburgh Jr. Distribution and measurement of aflatoxin in 1983 Iowa corn. Cereal Chem 66:165, 1989.
97. WW McMillian, DW Wilson, NW Widstrom. Aflatoxin contamination of preharvest corn in Georgia: a six-year study of insect damage and visible *Aspergillus flavus*. J Environ Qual 14:200, 1985.
98. CW Hesseltine, OL Shotwell, WF Kwolek, EB Lillehoj, WK Jackson, RJ Bothast. Aflatoxin occurrence in 1973 corn at harvest. II. Mycological studies. Mycologia 69: 328, 1977.
99. GA Payne, DK Cassel, CR Adkins. Reduction of aflatoxin contamination in corn due to irrigation and tillage. Phytopathology 76:679, 1986.
100. GA Payne, EJ Kamprath, CR Adkins. Increased aflatoxin contamination in nitrogen-stressed corn. Plant Dis 73:556, 1989.
101. RJ Cole, RA Hill, PD Blankenship, TH Sanders, KH Garren. Influence of irrigation and drought stress on invasion by *Aspergillus flavus* of corn kernels and peanut pods. Dev Ind Microbiol 23:229, 1982.
102. RA Hill, PD Blankenship, RJ Cole, TH Sanders. Effects of soil moisture and temperature on preharvest invasion of peanuts by the *Aspergillus flavus* group and subsequent aflatoxin development. Appl Environ Microbiol 45:628, 1983.
103. RJ Cole, TH Sanders, RA Hill, PD Blankenship. Mean geocarposphere temperatures that induce preharvest aflatoxin contamination of peanuts under drought stress. Mycopathologia 91:41, 1985.
104. RJ Cole, PD Blankenship, RA Hill, TH Sanders. Effect of geocarposphere temperature on preharvest colonization of drought stressed peanuts by *Aspergillus flavus* and

subsequent aflatoxin contamination. In: H Kurata, Y Ueno, eds. Toxigenic Fungi: Their Toxins and Health Hazard. Tokyo: Kodansha Publishing, 1984.
105. RJ Cole, JW Dorner, CC Holbrook. Advances in mycotoxin elimination and resistance. In: HE Pattee, HT Stalker, eds. Advances in Peanut Science. Stillwater, OK: American Peanut Research and Education Society, 1995, p 456.
106. TE Russell. Aflatoxin contamination of cottonseed climate, insects, cultural practices. Proceeding Conference, Special Session, Aflatoxin. St. Louis, MO, 1980.
107. LJ Ashworth Jr, JL McMeans, CM Brown. Infection of cotton by *Aspergillus flavus*: the influences of temperature and aeration. Phytopathology 59:669, 1969.
108. KK Maggon, SK Gupta, TA Venkitasubramanian. Biosynthesis of aflatoxins. Bacteriol Rev 41:822, 1977.
109. MA Klich. Relation of plant water potential at flowering to subsequent cottonseed infection by *Aspergillus flavus*. Phytopathology 77:739, 1987.
110. PJ Cotty. Effect of harvest date on aflatoxin contamination of cottonseed. Plant Dis 75:312, 1991.
111. TE Russell, TF Watson, GF Ryan. Field accumulation of aflatoxin in cottonseed as influenced by irrigation termination dates and pink bollworm infestation. Appl Environ Microbiol 31:711, 1976.
112. TJ Michailides. Altering agronomic practices to improve management of aflatoxin contamination. Aflatoxin Elimination Workshop, Fresno, CA, 1996, p 43.
113. Zhengyu Huang, DG White, GA Payne. Corn seed proteins inhibitory to *Aspergillus flavus* and aflatoxin biosynthesis. Phytopathology 87:622, 1997.
114. RL Brown, TE Cleveland, GA Payne, CP Woloshuk, KW Campbell, DG White. Determination of resistance to aflatoxin production in maize kernels and detection of fungal colonization using an *Aspergillus flavus* transformant expressing *Escherichia coli* β-glucuronidase. Phytopathology 85:983, 1995.
115. KW Campbell, DG White. Inheritance of resistance to *Aspergillus* ear rot and aflatoxin in corn genotypes. Phytopathology 85:886, 1995.
116. JN Neucere, RL Brown, TE Cleveland. Correlations of antifungal properties and β-1, 3-glucanases in aqueous extracts of kernels from several varieties of corn. J Agric Food Chem 43:275, 1995.
117. RA Norton. Effect of selected corn metabolites on growth and aflatoxin production by *Aspergillus flavus*. In 8th Aflatoxin Elimination Workshop. Atlanta: U.S. Dept Agric Res Serv, 1995.

10

Involvement of Arthropods in the Establishment of Mycotoxigenic Fungi Under Field Conditions

Patrick F. Dowd
National Center for Agricultural Utilization Research, United States Department of Agriculture, Agricultural Research Service, Peoria, Illinois

I. INTRODUCTION

Arthropods are invertebrate organisms with jointed legs, consisting primarily of the insects. The purpose of this review is to explore the involvement of insects and other arthropods in the establishment of fungi that produce mycotoxins in different commodities. Because initial problems appear to occur through contamination in the field, preharvest will be the main focus in this review. However, additional information on storage interactions will also be discussed.

Although there is ample evidence for the involvement of insects and other arthropods in either vectoring or otherwise facilitating the establishment of mycotoxigenic fungi, the interactions are often not straightforward, and involve interactions with many other factors in the environment. The intent here is to enable the reader to determine how insects and other arthropods have been demonstrated to facilitate the establishment of mycotoxigenic fungi so that it will be possible to determine the importance in situations for which there is limited information. As various aspects of this topic have been reviewed in some form in the past [1–21], the reader is directed to these earlier reviews for additional information and sometimes different perspectives.

The present discussion will describe different types of insect facilitation of mycotoxigenic fungi, review examples from different commodities, offer general-

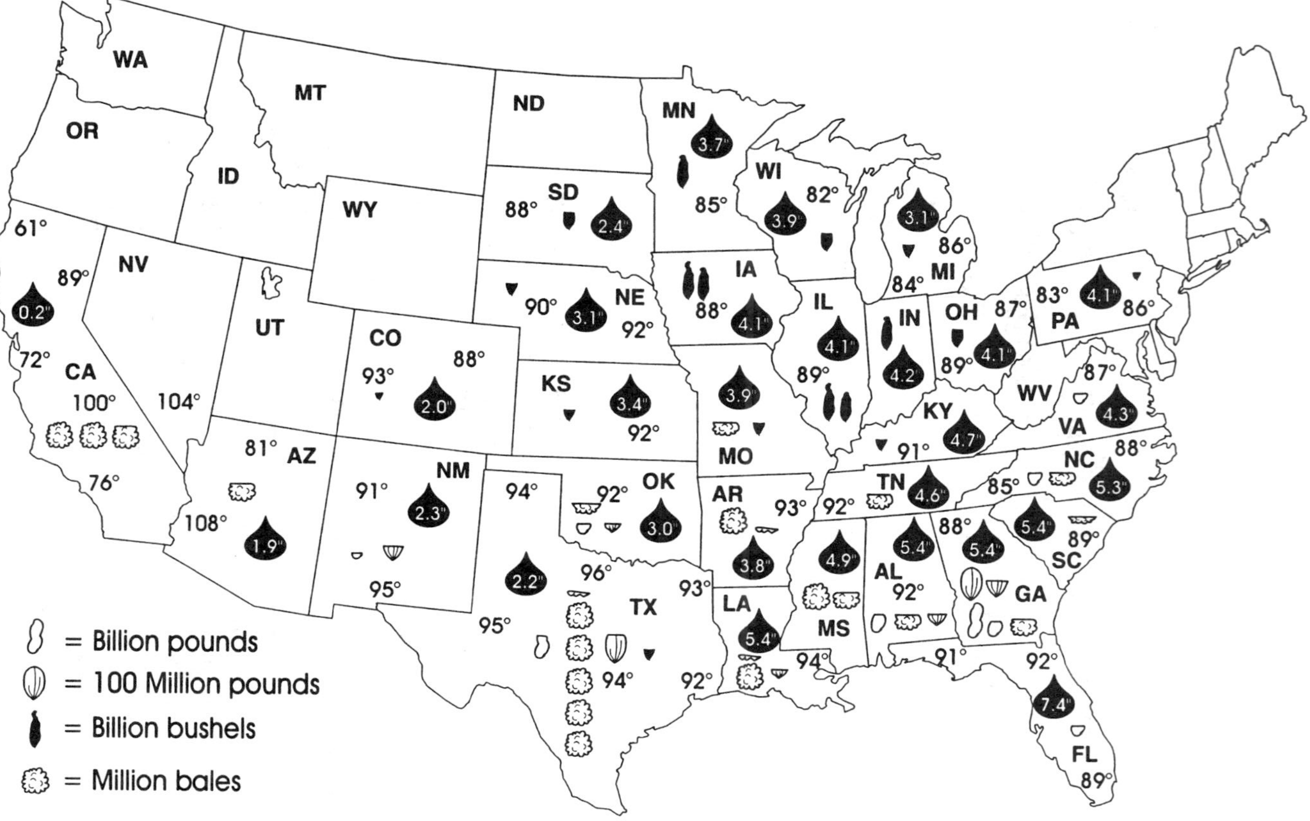

FIGURE 1 Distribution of crops reported to have insect associations with mycotoxigenic fungi (peanuts, pecans, corn, and cottonseed), along with temperature and rainfall indicators, in the U.S. Additional crop values not shown (all in California) are almonds, 490 million pounds; figs, 51,000 tons; and English walnuts, 260,000 tons. Temperature given is average maximum for July and is positioned at the approximate location of the reporting site; rainfall given is the average for July and is reported on a state basis. (From Refs. 207–211.)

izations, and describe arthropod-related management techniques. Additional information on peripheral areas may be found in other chapters in this book. Because most of the information covers different crops and areas, with environmental influences, in the 50 different states of the U.S. (unless otherwise indicated), the reader should refer to Figure 1 if unfamiliar with crop areas and climate.

II. INSECT VECTORS

A. What Constitutes an Insect Vector

Vectors are carriers of (for example) fungal inoculum. The quantity carried is sufficient and compatible such that the fungus can become established. Although many definitions for satisfying the criteria of a "vector" are available, the criteria defined by Leach [22] are generally accepted and utilized, although the source of the criteria is seldom acknowledged. These criteria are:

1. A close, although not necessarily a constant, association of the insect with diseased plants must be demonstrated;
2. It must be demonstrated that the insect also regularly visits healthy plants under conditions suitable for the transmission of the disease;
3. The presence of the pathogen in or on the insect in nature or following visitation to a diseased plant must be demonstrated;
4. The disease must be produced experimentally by insect visitation under controlled conditions with adequate checks.

In all cases Koch's postulates should be fulfilled [22]. Prior work on insects associated with the occurrence of mycotoxigenic fungi on crops will be examined to determine if this work fulfills the vectoring criteria, although all examples of interest will be considered.

B. Characteristics of a Successful Vector

A successful vector must naturally encounter the fungal agent. It must carry the fungal agent to the crop/site in a viable condition. This transport may be directly to the plant or in close enough proximity for the fungus to arrive at the plant site by other means, such as other insects, water, or air. The vector may have morphological adaptations that enhance its ability to carry fungal inoculum of a particular type. For example, pocket-like pouches in the head of bark beetles (Coleoptera: Scolytidae) are thought to help them collect and carry yeasts/fungi to host trees [23]. The hairy legs and bodies of many species of sap beetles (Coleoptera: Nitidulidae) can promote carrying of dry conidia. This adaptation also allows the beetles to be important in pollinating different plants, such as *Opuntia* spp., a cactus [24], and custard apple and related fruit (*Annona* spp.) [25,26]. The scales of moths and hairs of maize weevils (Coleoptera: Curculionidae) may also promote collection of dry conidia (see later discussion). Internal adaptations may also be involved.

Appropriate adaptations may include an absence of the enzymes that digest (or low enough levels that some material survives) fungal material, such that the material is also internally carried. Evaginations of the insect gut may serve as areas where propagules accumulate and are released slowly, as in sap beetles (Fig. 2). Adaptations on the part of the fungi may include the production of powdery materials that can coat the hairs (as described) or sticky matrix that clings to the insects—e.g., ergot and flies [22], *Ceratocystis*, and sap beetles [27,28]. Production of structures or materials attractive to the insect in sufficient quantity (in addition to, or mimicking, those of the host) will increase the probability the insect would encounter the fungus.

Temporal associations are also important. The insect needs to be inhabiting material colonized by the fungus and then move from this material to the plant host when an appropriate inoculum stage is present and when the plant is in a stage that is susceptible to the fungus. As indicated earlier, vectoring may not necessarily be directly to the plant part ultimately infected by the fungus. For example, sap beetles feeding on molded corn ears that have overwintered are first attracted

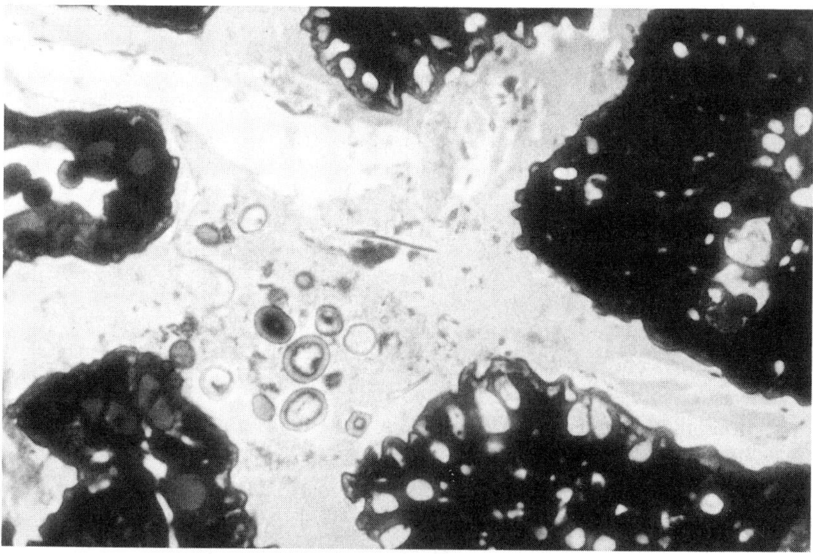

FIGURE 2 Dusky sap beetle gut evaginations. (Photo by S. K. Shen.)

Figure 3 Dusky sap beetles feeding on molded pollen and anthers in corn leaf axils. (Photo by P. F. Dowd.)

to pollen that is accumulating in leaf axils of corn [29]. It is then possible for the mold to become established on the highly nutritious pollen (Fig. 3). This situation is commonly observed in the Midwest corn belt of the U.S., with *Fusarium* spp., in some cases approaching 90% of several hundred leaf axils sampled in a particular year (Dowd, personal observation).

The inoculum can then build up in the axils. As sap beetles move to ears, or as corn borers (Lepidoptera: Pyralidae) enter ear shanks from the axils or the ear-leaf ear gap (and ultimately ears) (Fig. 4), the fungus can be carried into the ear. Once in the ear, it can be further dispersed within the ear or to other ears when suitable inoculum stages are present. Fungal colonization of silks and kernel surfaces (Fig. 5) can provide an inoculum source ready to enter kernels when damaged by insects [12].

In cases where toxic compounds are produced by the fungus, if the material is fed on by insects, the effect on the insect will determine its ability to vector. Obviously if the insect is harmed by the toxins, it is less likely to be able to survive and carry inoculum. Compounds considered mycotoxins are generally toxic to nonadapted insects at naturally occurring levels [4,30]. The resistance to selected *Aspergillus*, *Penicillium*, and *Fusarium* mycotoxins is at least partially due to the ability of sap beetle larvae to enzymatically detoxify the compounds [5].

FIGURE 4 Corn ear with husk removed showing mold and entry hole on husk (which was adjacent to ear leaf) and kernel damage by European corn borer. Sap beetle frass (dark specks) is also present. (Photo by P. F. Dowd.)

FIGURE 5 Silk molded by *Fusarium*, with adjacent kernels damaged by dusky sap beetle larvae also showing mold. (Photo by P. F. Dowd.)

C. Why Vectoring Is Important to the Fungi

Although viable propagules are theoretically infective under benign and optimal nutritional conditions, in reality the establishment of the fungus is likely to be difficult under natural conditions, due to the defenses of the plant host, unfavorable environmental conditions (moisture, nutrients), and competing organisms. Thus, a particular inoculum density within a time period is necessary for establishment (see chapters in this book by Abramson and Payne). The quantity necessary for establishment may not occur with quantities that are air- or waterborne (based on viability of propagules—under conditions of desiccation by the time germination occurs). Because of the high mobility and the ability to carry large quantities of inoculum (in a protected state if carried internally), insect vectors can help the fungi to become established. Damage caused by insects may provide further locations for fungal introduction. Sap beetle vectors of mycotoxigenic fungi have been considered to have a symbiotic relationship with *Ceratocystis fagacearum* [27], *Graphium rigidum* and *Ophiostoma pluriannulata* [28], and a trend for this type of relationship with mycotoxigenic fungi [3].

In other cases it may be necessary for the vector to carry the fungal inoculum past the physical barriers of the host plant. For example, before reaching the highly nutritious corn kernel germ, the physical barriers of the husk, the silks, and then the pericarp of the kernel itself must be bypassed. Similar barriers in other materials are the shell and testa (skin) of peanuts [1], the carpel and shell of almonds [31], and the carpel and shell of pistachio nuts [32]. Stored-products insects, through their physiological processes, may generate sufficient moisture in stored grain for fungi to grow [33].

III. NON-VECTOR–FACILITATED ESTABLISHMENT

Even if insects do not carry the fungi to a site, they may be responsible for providing conditions whereby fungal propagules may enter. Entry and emergence holes made by caterpillar in fruit, nuts, cotton bolls, and corn ears are obvious means whereby spores may enter and become established. Again, the establishment of the fungi will depend on a sufficient quantity of the fungal propagule reaching the opening. Through insect feeding and elimination of waste, a "processing" of plant material (that may otherwise be chemically and/or biochemically resistant) can occur, whereby the fungus may become more readily established. For example, insect frass is often colonized by ear molds, and can be a source of inoculum buildup [34].

Determination of the role of insects in facilitating colonization by mycotoxigenic fungi is complicated by the interaction of environmental conditions that favor the fungus (such as *A. flavus*), the availability of fungal inoculum, and insect damage [35]. Fungus that has colonized silks may be unable to enter intact corn kernels until they are damaged by insects. However, aflatoxin in the 100 ppb range can occur with only silk inoculation of some varieties under appropriate environmental conditions in the absence of insects [36]. Without insect damage, *A. flavus* does not penetrate (as opposed to surface infect) kernels until denting occurs [37].

Interpretation is complicated by relative resistance of different tissues of different varieties of the same plant. For example, in one variety not subject to "stress cracking" or other kernel-induced rupture of the pericarp, i.e., silk-cut or popping [38], insect damage may be highly associated with degree of mold colonization and hence toxin levels. A stress crack-susceptible variety may have no insect damage yet be highly colonized by fungus and have high levels of toxin, because the fungus is able to enter through the cracks in the pericarp (seed coat) [16]. These ruptures may only be manifested and obvious under particular weather conditions [39,40]. It now appears that visible moldiness does not necessarily correlate with toxin levels [41]; apparently there are different mechanisms of resistance to fungal growth and toxin products, which may not co-occur. Thus, where insect damage occurs and visible mold is present, toxin levels may not be related to either.

This relationship may be due to the ability of some components of the damaged material to not provide optimum conditions for toxin elaboration, or to inhibit toxin production, and even degrade the toxin after formation. As mentioned earlier, the presence of biocompetitive organisms that are not readily noticeable, and thus not considered, can also confound interpretation. Another confounding factor is the apparent seed colonization by strains of *Fusarium moniliforme* that can subsequently be found in other parts of the plant, including the ears [42,43]. There are also reports of *A. flavus* systemic colonization of corn [44]. These types of variation further complicate the determination of insect association with molds and mycotoxins.

IV. INSECT VECTORS AND FACILITATORS

A. Introduction

Information on demonstrated and suspected insect vectors, or insect facilitation of mycotoxigenic fungi, including both negative and positive examples, will be discussed in this section. It is intended to allow the reader to determine for themselves when a vector situation may result. It is more likely that this will be prompted by noticing the presence of a particular insect type on a crop at the appropriate time (when the infection process is initiated, and not just at harvest). Knowing that a particular insect can vector or promote the establishment of mycotoxigenic fungi in a particular crop should suggest that such an insect will have the capability of assuming this role in several crops. This will be readily demonstrated in the following discussion by considering the involvement of caterpillars, sap beetles, and weevils in vectoring and promoting mycotoxigenic fungi in several different and unrelated crops. Because introduction often occurs in the field, emphasis will be placed on field examples. As aspects of field will apply to storage, comparable insects will also be involved and will be mentioned in some cases. As many crops have a complex of insect pests, many insects have the potential to be involved in facilitating the growth of mycotoxigenic fungi in crops (see, for example, the list of corn insects in the U.S. [45]).

B. Caterpillars

Caterpillars are the larval form of butterflies and moths (Lepidoptera). Most caterpillars that are pests on crops are the larvae of moths. Many are extremely destructive, and are often the major pest of crops. Because of their importance, there is much information on the use of insecticides for indirect control of mycotoxins through control of insects, and use of insect-resistant varieties for indirect control of mycotoxins through control of insects. Because many species of caterpillars are important on a relatively limited number of crops, they will be considered on a crop-by-crop basis.

1. Corn

Corn earworms (*Helicoverpa zea*) are very destructive and primarily damage the kernels at the tip of the ear. Early researchers in the U.S. reported an association of caterpillars and ear mold. The relative incidence of ear molds other than *Diplodia zea* in corn in Kentucky was attributed to the relative incidence of *H. zea* [46]. Years when ear molds were more severe were attributed to more severe damage by *H. zea*[46]. The overall severity of corn ear molds throughout the U.S. also was attributed to the severity of damage by *H. zea* [47]. Higher rates of livestock deaths in certain years were attributed to a higher incidence of unspecified ear molds [47]. Larvae of *H. zea* collected from young tassels and ears of corn in Texas were contaminated with *A. flavus*, *A. niger*, *Penicillium*, and *Fusarium* fungi, and again the incidence of ear mold was related to the incidence of *H. zea* damage [48]. Ears in Illinois with corn earworm damage had more than twice the level of *F. moniliforme* compared to undamaged ones when husk coverage was complete [49]. Caterpillar frass appears to be highly colonized by various mycotoxigenic fungi and thus can provide a location for fungal biomass buildup [34].

Following these earlier reports, research became more oriented toward quantitative associations of ear mold fungi and *H. zea* in the U.S. Earworm-damaged corn at harvest from Missouri and Illinois was contaminated by 7.2% of *A. flavus* group and 6.5% by *Fusarium* spp. [50]. Of the earworms collected from corn in Missouri, 37% were contaminated by *A. flavus* group fungi, and 63% by

Fusarium spp. [50]. Insect-damaged kernels were 82% colonized by *Fusarium* but only 0.03% by *A. flavus* group fungi in Indiana in 1971 [51]. Over 80% of ears with detectable aflatoxin were presumably damaged by corn earworms [52]. In South Carolina, *A. flavus* was rare in soil, not present in soil debris, not present in air samples, and not recovered from debris or soil of corn that had been inoculated with *A. flavus* [53]. In general, *A. flavus* was not common on silks or kernels until mid-August [53]. Subsequent studies indicated that corn earworms contaminated with spores of *A. flavus* could be transmitted by over 90% of individuals from generation to generation, and that these caterpillars contaminated three times the number of ears with *A. flavus* than occurred with controls, and ca. 6 times the number of kernels [54].

In a study involving collection of corn earworms, fall armyworms and European corn borers from 10 states, including the Midwest, the Southeast, and Texas, caterpillars were externally infected by 7% to 0% and internally by 4% to 0%; only Indiana did not have contaminated caterpillars (but sample size was very small) [55]. Despite relatively uniform incidence on the insects, detectable aflatoxin occurred at the highest percentage from corn in Florida, Georgia, Kansas, and South Carolina, indicating environmental influences other than insects [55].

Subsequent studies concentrated mainly on adding insects, or simulating insect damage, and then determining effects on *A. flavus* or other fungal colonization and toxin production. Corn earworms added to silks inoculated with *A. flavus* increased aflatoxin by up to threefold in Missouri and nearly 20-fold in Georgia [56]. Corn earworm simulated damage of kernels using a razor in Missouri had over 30 times the level of aflatoxin B_1 compared to kernels that were not damaged, when inoculum was applied [57]. Plant feeding insects (*H. zea*, the European corn borer, and lady beetles) were contaminated with *A. flavus* group fungi at a fairly constant rate (10% to 20%) from silking to 70 days postsilking in Iowa and Illinois in 1978 [58]. However, increasing incidences of *A. flavus* group fungi (increasing from 3% to 54%) were noted on these insects as the season progressed in Georgia [58]. Ears with simulated corn earworm damage (slit pericarps) had significantly higher levels of aflatoxin than undamaged ears over 15 hybrids in South Carolina [59].

Over a 6-year period in Georgia, greater levels of *A. flavus*/aflatoxin were generally associated with greater insect (predominantly corn earworm) damage, but insect damage did not explain all of the aflatoxin present [60]. Compared to silk inoculation, wound-inoculated kernels often had 10 times the level of aflatoxin when examined in South Carolina [61], suggesting that insect damage can greatly increase the levels of aflatoxin compared to when colonization occurs in the absence of insects. There is also some evidence that these insects, as moths, may be able to vector the inoculum to corn. Contamination of moths during the seedling stage was 30%, compared to 70% at plant maturity, but incidence can vary from year to year in Georgia [10,62]. However, the validity of these data has been questioned because other insects entering traps may have contaminated moths [63].

No relationship was seen between insect damage and aflatoxin in corn among 14 hybrids planted in Texas [64]. In many cases, damage by fall armyworms (*Spodoptera frugiperda*) and pink scavenger caterpillar (*Sathrobrota rileyi*) will be intermixed with that of the corn earworm in the southeastern U.S. [18], so damage attributed to the corn earworm may involve these insects as well.

Varieties that are resistant to the corn earworm have been used to indicate the association between corn earworms and mycotoxigenic fungi; however, corresponding resistance of the variety itself to the fungi is often not considered. Taubenhaus [48] noted that the varieties Improved Indian Squaw and Surecropper had less ear mold in Texas, which he attributed to their resistance to *H. zea*. In an examination of three resistant and two susceptible hybrids over 12 states throughout the U.S., the corn earworm-resistant material generally had less aflatoxin than the corn earworm-susceptible material [65]. Overall insect damage in Florida, Georgia, and Tennessee of 12 hybrids was generally associated with aflatoxin levels [66]. In Georgia, the degree of insect damage could not be associated with levels of aflatoxin when three different varieties were used over 2 years [67]. In other words, varietal resistance to the fungus in this case appeared to be obscuring any insect-associated interactions. A corn earworm-resistant variety tested in Georgia, Iowa, and Missouri showed lower aflatoxin levels than a susceptible variety in Georgia and Missouri when *A. flavus* and corn earworms were added; only sporadic aflatoxin occurred in the Iowa material [68]. In an examination of 15 different hybrids primarily damaged by corn earworms and fall armyworms in South Carolina, insect damage was significantly correlated with aflatoxin levels overall [59]. Two varieties that had previously demonstrated low aflatoxin were examined along with one with high aflatoxin for association with insect damage in Georgia, Louisiana, and Missouri [69]. The two aflatoxin-resistant varieties had significantly lower levels of insect damage, suggesting that insect damage was influencing aflatoxin levels, although location effects were also present [69]. There was no direct correlation, perhaps because the susceptible hybrid, B73 × Mo17, is also one that is subject to internal "splitting" [70] and thus invasion would be less associated with insect damage.

Tight husk coverage has been shown to reduce levels of aflatoxin in corn. These types of material are original southern-adapted material, which also promote resistance to maize weevils, unlike higher-yielding corn belt-adapted material with looser husks [71]. These loose husked varieties apparently have increased aflatoxin problems because of increased insect damage [10]. A southern-adapted variety with a tight husk sustained up to threefold less damage by corn earworms compared to a loose husked, corn belt-adapted variety, and up to 60-fold less aflatoxin depending on the year [72,73]. Compared to loose-husked varieties, tight-husked varieties had five to 10 times lower aflatoxin when infested with both corn earworms and European corn borers and *A. flavus*, which was most apparent in Georgia (vs. Iowa or Missouri) [74]. Although the lower levels of aflatoxin in these tight-husked varieties has been attributed to reducing insect damage, the tight husk also appears to prevent stress on kernels due to heat compared to looser-husked varieties, which also reduces aflatoxin levels [64]. Unfortunately, these tight-husked varieties are inappropriate for growth in the corn belt and further north due to increased problems with *Fusarium* caused by slow drying [75].

Because insecticide treatments have been most often used with regard to controlling the corn earworm in corn (although other ear-feeding insects may be controlled at the same time), the issue of insecticide control for controlling mycotoxigenic fungi in corn will primarily be addressed here. In general, it appears that insecticide treatments, if properly and frequently applied, can reduce insect damage and, as a result, colonization by mycotoxigenic fungi and mycotoxin levels as well. However, in some cases it is difficult to interpret results across studies. Different hybrids can vary in their susceptibility to colonization of mycotoxigenic fungi, as well as have resistance to insects that is plant-based. In addition, with overhead (such as center pivot) irrigation often involved in cultural practices in many areas, the issue of wash-off of insect sprays, and consequently the efficacy of treatments, is also a concern.

Studies that do not determine the efficacy of insect control do not offer as much insight into this strategy for controlling mycotoxigenic fungi indirectly through insect control. In a trial involving 20 different varieties of corn, a single application of 5% DDT dust reduced damage by the corn earworm and reduced *Fusarium* ear rot from 26.5% to only 7% [76]. Treatments with stirofos on four different hybrids, every other day from silking to silk browning, reduced caterpillar damage by over 40-fold [77]. The percentage of kernels with bright greenish-yellow fluorescence (BGYF—often correlated with aflatoxin levels) was reduced by 20-fold in inoculated ears of an opaque-2 line [77]. Similar results were obtained with a commercial hybrid (PAG-653) under the same conditions, although the incidence of BGYF in kernels was only reduced by about sevenfold [77]. Treatments of carbaryl every other day for 6 weeks after silking substantially reduced aflatoxin in South Carolina and Florida for some hybrids but not others; hybrids with high levels of aflatoxin were helped the most [78]. Treatments with stirofos three times a week for 6 weeks significantly reduced incidence of corn earworms in Georgia, but not aflatoxin [79]. However, aflatoxin in control plots was only 1.4 ppb, and was half that in treated material [79], which may not be high enough (at the limits of detection) to have separated out variation. The percentage of plots with detectable aflatoxin was twice as high for untreated as treated ones [79]. Six applications of chlorpyrifos at weekly intervals reduced corn earworm damage by about two- to fourfold (depending on planting date) and aflatoxin by over 60-fold in nonirrigated areas in Louisiana [80]. Single treatment with adherent corn flour-encapsulated malathion can reduce the occurrence of visible *Fu-*

sarium in corn by threefold in both *Fusarium moniliforme*-susceptible and -resistant varieties, which is correlated with reduction of early sap beetle levels in axils and European corn borer incidence in milk stage ears [81,82].

The European corn borer (*Ostrinia nubilalis*) is a primary pest of corn in the corn belt (midwestern U.S.). It can cause significant damage in other corn-growing areas of the U.S. as well, but is not as important a corn pest as the corn earworm in the Southeast. The closely related southwestern corn borer (*Diatrea grandiosella*) is similar in habits and will also be considered here. Unlike the corn earworm, the European corn borer can be found in many different parts of the ear. It may be found in the base of the ear, having entered through the attachment of the ear to the stalk. It may enter the top of the ear and feed primarily in the ear tip. It may enter through the junction of the ear leaf and the ear, and cause damage to the central portion of the ear (Fig. 4). In many cases the extent of damage to the kernels is not obvious because the caterpillars will tunnel under the surface of the kernels and eat primarily the germ. The same statements about the association of mycotoxigenic fungi with avenues of entry, frass, and exit holes described for the corn earworm also apply for the European corn borer.

In surveys conducted in the U.S., *Aspergillus* spp., *Fusarium* spp., and *Gibberella zea* were isolated from larvae in corn stalks, with *Fusarium* spp. occurring in 35% of live larvae from corn in Minnesota [34]. Larvae were thought to be important in dissemination within the plant itself, and infested plants were usually more severely rotted than undamaged plants [34]. In Iowa, *A. parasiticus* was isolated from European corn borers, and it caused 31% mortality of larvae exposed to spores, suggesting that the insects were a reservoir of the fungus through pathogenic interactions with the insect [83]. Several species of *Penicillium* including *P. cyclopium*, were also isolated from the European corn borer [83]. Corn damaged by borers (European and southwestern considered together) from Missouri and Illinois at harvest was infected by 5% with *A. flavus* and 10% with *Fusarium* spp. [50]. Larvae collected from Missouri corn were 15% contaminated with *A. flavus* and 42% contaminated with *Fusarium* spp. [50]. Iowa corn that was infested with European corn borers had up to 23% kernels infected with *A. flavus* (vs. 0% for undamaged); insect-damaged material always had a higher percentage of kernels infected with *A. flavus* [84]. The amount of colonization by *F. moniliforme* of these ears was also generally greater in insect-damaged material, in some cases up to 14% (vs. 0% for undamaged kernels) [84].

European corn borers increase aflatoxin in *A. flavus* silk-inoculated ears by up to 48-fold in Missouri and Georgia, depending on location and planting date [56]. In a study run in Iowa, Missouri, and Georgia, larvae of European corn borers did not appear to be very important in introducing applied spore inoculum into ears, although inoculum applied to silks was more often introduced into ears as opposed to that applied to stalks (in most cases not exceeding 10%) [85]. The resistant variety did not appear to reduce the presence of the *A. flavus* in ear-colonizing larvae compared to the susceptible one [85]. It should be considered, however, that resistance is defined in terms of stalk resistance (for second generation) and not ear resistance.

However, in another Georgia study, a wider range of germplasm that was resistant to second-generation European corn borers did have substantially (ca. 10-fold) less aflatoxin than the susceptible hybrids when inoculated with *A. flavus* (NRRL 3357) and infested with larvae [86]. Damage caused by *O. nubilalis* increased infection by *F. moniliforme* by three- to ninefold over those with simple mechanical damage when the ears were sprayed with a spore suspension in Iowa [87]. When both larvae of the southwestern corn borer and *A. flavus* were applied to silks of hybrids made from resistant inbreds, high levels of aflatoxin were noted [88]. However, when only *A. flavus* was applied, aflatoxin levels were low [88]. This study suggests that resistance to both *A. flavus* and insects will be necessary to reduce aflatoxin levels in corn to acceptable levels. Compared to the corn earworm and fall armyworm, European corn borer-damaged ears had significantly more aflatoxin [89].

2. Peanuts

General insect damage was not responsible for high concentrations of aflatoxin in Spanish peanuts; instead, those with growth cracks had the highest levels [90]. However, there is good evidence that damage by the lesser cornstalk borer (*Elasmopalpus lignosellus*) does promote aflatoxin in peanuts. An association between damage by the larvae and *A. flavus* was noted in Spanish peanuts in Texas [91] and peanuts in North Carolina [92]. Insect-damaged material had over four times the incidence of *A. flavus* compared to other materials, greater levels of aflatoxin than mechanically damaged or distorted pods [92]. Of the four different varieties examined, there was a greater correlation with insect damage and presence of *A. flavus* than insect damage and aflatoxin [92], suggesting a situation similar to that of corn, where presence of the fungus and aflatoxin production appear to be governed by different resistance mechanisms.

Drier areas typically had more insect damage, more fungus, and more aflatoxin [92]. Laboratory studies indicated effective transfer of *A. parasiticus* inoculum to pods, with younger pods more severely affected [93]. Field-collected larvae were 50% contaminated with *A. flavus* group fungi, and pods injured by *E. lignosellus* had up to seven times the level of aflatoxin as compared to uninjured pods [93]. A high percentage of larvae of *E. lignosellus* were contaminated with *A. flavus*, and internal contamination also occurred [94]. Presence of *A. flavus* and levels of aflatoxin were correlated with damage by *E. lignosellus* in four different trials [94]. Treatments of granular chlorpyrifos, but not fonophos, often reduced damage by *E. lignosellus* and colonization by *A. flavus* [94]. However, timing of the treatment was important, and volatilization in hot weather posed a potential problem, so multiple treatments may be necessary [94]. High temperature and drought appear to favor this insect, which can promote damage. However, undamaged peanuts can also have high levels of aflatoxin under drought and temperature stress [95]. These temperature and drought factors appear to reduce the production of a phytoalexin that normally inhibits growth of *A. flavus* [96].

3. Cotton

The corn earworm is referred to as the bollworm when it feeds on cotton. Bollworms inoculated with *F. moniliforme* infected 55% of bolls, while uninoculated field-collected bollworms infected 72% of bolls in Mississippi [97]. Moths of the cabbage looper (*Trichoplusia ni*) treated in the same way and caged with flowers infected ca. 50% of bolls in all cases [97].

Damage by the pink bollworm, *Pectinophora gossypiella*, can be considered analogous to damage of corn by the caterpillars already considered. A small hole is made going in, and a large one (ca. 1.5 mm) coming out. Conclusions based on the corn situations should theoretically apply to cotton also, in terms of vectoring vs. promoting colonization. The degree of damage by pink bollworms was correlated with the number of BGYF spots per pound of lint in Texas [98]. Bolls with simulated pink bollworm exit holes increased colonization by *A. flavus* from 0% to 1.5% in California [99]. Insecticide was applied at an economic threshold for insect control and for "complete" control (15 applications on a 7-day schedule, vs. 17 applications on a 5-day schedule, respectively) in California, and aflatoxin levels were determined in seed [100]. The success of insect control was associated with relative levels of aflatoxin in seed, with the "complete" control reducing aflatoxin up to 500 times that of the economic control, and reducing the level below 20 ppb in 1 year [100]. Levels of aflatoxin in cottonseed from Arizona were associated with levels of pink bollworm damage [101]. Aflatoxin levels were reduced by up to sixfold using damage threshold-based applications of methyl-azinphos or methomyl [101]. However, 95% of bolls needed to be free of insects in order to reduce contamination of cotton seed below 20 ppb, requiring an application frequency that was below the economic threshold for cotton fiber yield reduction [101]. Nine to 15 insecticide applications on cotton in Arizona were needed to significantly reduce both pink bollworm damage and aflatoxin levels in seed below 20 ppb, whereas only five applications reduced aflatoxin levels by 2000-fold (but not below 20 ppb) [102].

In a study in California, over 50% of the bolls with exit holes had fibers with BGY fluorescence, especially just below the entry holes, but no fungi were isolated from callused-over entry sites [103]. This suggests that the fungus is not being carried in by the caterpillars, that it instead enters through the emergence holes. A study in Arizona designed to determine appropriate timing for insecticide application simulated exit holes at different times during boll development, indicating that exit holes formed 33 days after flowering produced contamination similar to that observed naturally, although seed drying and rupture of internal membranes appear to allow the fungus to ultimately enter the embryo [104]. When cottonseed was examined in Arizona, most of the aflatoxin-contaminated cottonseed came from pink bollworm-damaged bolls, with average levels over 1000 times higher in the seed from damaged bolls versus undamaged bolls [105].

4. Tree Nuts and Fruit

The navel orangeworm (*Amyelois transitella*) damages several different nut crops in California. Insect damage by this caterpillar was associated with almost threefold greater levels of *A. flavus* in pistachio nuts in some cases [106]. From two- to fourfold higher levels of occurrence of aflatoxin in pistachio were associated with insect damage by the navel orangeworm in early split nuts [107]. Overall, nuts damaged by the navel orangeworm had 84% of all aflatoxin detected in various grades of samples [108,109]. Spontaneous splitting of hulls increases infestation by *A. flavus* and other fungi and promotes navel orangeworm infestation, both of which are worse the earlier the splitting occurs [32,109,110]. This situation is analogous to drought-shortened corn husks and silk cut of corn. Cultural practices can help reduce the splitting [111] and thereby reduce the presence of inoculum [112,113].

The staining of the shells, and rough hulls, that are often associated with infestation by the navel orangeworm, allow for sorting removal of these nuts [110, 114]. The navel orangeworm also appears to be associated with the presence of aflatoxin in almonds [115]. Damage to the seedcoat by these insects breaches the barrier and allows for the possibility of carrying in spores as well [31,116]. Varieties resistant to the navel orangeworm had less infection by *A. flavus* [117]. Walnuts damaged by the navel orangeworm had 45-fold higher levels of *A. niger* and nearly eightfold higher levels of other fungi compared to nuts not damaged by insects [118]. Similar to the situation with pistachio nuts, fungal colonization could occur without insect damage due to splitting, and the earlier the splitting occurred, the more severe the fungal and insect damage observed [118]. However, in figs that were predominantly damaged by navel orangeworm, there was no significant difference in colonization by *Aspergillus* spp. compared to undamaged ones in California over 2 different years [110]. The lack of insect involvement was believed to be due to lack of a barrier against the fungus, as compared to other crops such as pistachio nuts, corn, or cotton, where fungal entry is restricted by physical barriers associated with the seed [119]. Levels of *A. niger* in Calimyrna figs is correlated (r^2 of .947) with soil dust on the trees, suggesting the soil blown up as part of the cultural practices (disking vs. noncultivation) serves to move the fungus to the trees as opposed to insects [120].

5. Caterpillar Generalizations

Caterpillar generalizations across the different commodities are possible. It appears that the caterpillars can introduce inoculum primarily by providing an entry for the fungus, as opposed to long-distance vectoring, although transmission of the inoculum to the plant by moths is possible. Varieties that are resistant to insects at the place of concern (which excludes European corn borer resistance, which has been defined in terms of stalk and not ear) can reduce insect damage and appear to indirectly reduce colonization and mycotoxin production by mycotoxigenic fungi. Insecticide treatments can also reduce mycotoxigenic fungi and mycotoxins indirectly by reducing insect damage. Several factors complicate the understanding of caterpillar associations with mycotoxigenic fungi. Different varieties may have resistance to colonization and inhibit toxin production by the fungi. Different strains of fungi used in studies may have differing abilities to produce mycotoxins.

Environmental variation at different locations may affect resistance levels in plant varieties, and also affect fungal survivability and rates of colonization. For example, data from studies in Honduras indicate different varieties may have equivalent levels of insect damage, but one may have an average of seven times more fungal damage than the other [121]. This suggests that in one variety, little spread from the site of insect damage occurs, while for the other one, either spread from insect damage is more extensive or infections occur in the absence of insect damage (Fig. 6). Differences in the degree of spread from simulated pink bollworm damage in cotton in different varieties has also been described [122]. Fungal strain aggressiveness may also account for the ability of *A. flavus* to spread from sites in cotton [104,122]. Fungal strains may also vary in their susceptibility to inhibition of growth or toxin production by plant defensive compounds [123].

It is suggested that if further studies are desired, a consistent strain of fungus (such as NRRL 3357 for *A. flavus*) and varieties with generally recognized susceptibility/resistance to insects and fungi (such as B73 x Mo17 as a susceptible corn hybrid for *A. flavus*) should be used under equivalent environmental conditions. Competitive micro-organisms can inhibit the growth of mycotoxigenic fungi, and in fact appear to have potential in management strategies for peanuts [124], cotton [125], corn [126,127], and almond [128]. Although it is likely to be difficult to control for this variable in the field, nevertheless it should be considered as yet another factor in complicating interpretation of insect vectoring and damage with the presence of mycotoxigenic fungi and mycotoxins themselves. Similar caveats will apply for other insect associations, as will be seen in the following discussions.

C. Beetles (Coleoptera)

1. Sap Beetles

Sap beetles (Coleoptera: Nitidulidae) are typically small beetles (< 5 mm) with clubbed antennae and, from above, part of the abdomen protruding beyond the wing covers. They are capable of vectoring many different disease causing organisms to a wide variety of crops [6]. This relationship has been recognized for

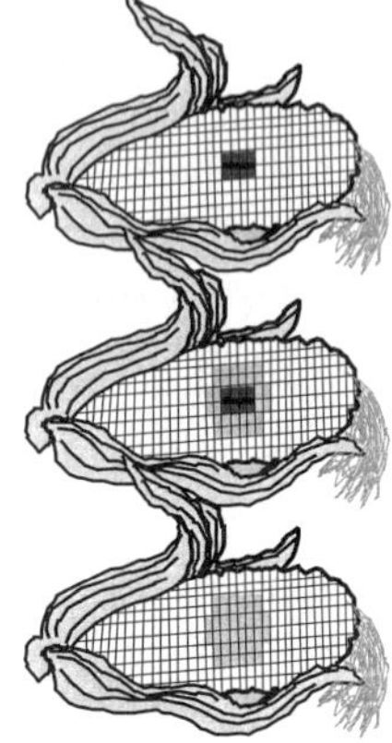

FIGURE 6 Example of how significant levels of aflatoxin can be present in an ear of the same variety under different environmental conditions. Top: invasion by *A. flavus* at insect damage sites (only eight kernels are assumed intact enough to survive harvest conditions and end up in end-use material). Center: invasion by *A. flavus* at insect damage sites and subsequent spread to adjacent kernels. Bottom: invasion by *A. flavus* not facilitated by insect damage, as in cases where kernel splitting or cracking occurs, such as high-temperature conditions.

the longest period for figs, where species were reported to carry disease organisms (primarily to figs with a hole, or ostiole, at the base that remains open). Sap beetle vectoring of "fig smut" (*A. niger*) was reported as early as 1925 [129]. Driedfruit beetles (*Carpophilus hemipterus*) were often collected from figs with smut, and were found in smut colonized decaying fruits that the beetles were overwintering in [129]. When spores were applied onto figs of all stages to the extent that they were visibly black, no figs were subsequently found to be colonized with *A. niger*, indicating that the spores had to be carried into the interior, such as by driedfruit beetles [129]. However, thrips and certain mites are also capable of introducing the organism into the figs [130].

Because not all figs contain this ostiole, sap beetle invasion does not occur consistently in all varieties. In spite of the overwhelming evidence of sap beetle vectoring of many organisms to figs, it has been concluded that they are not important in spreading *A. flavus* to ostiole figs [131]. This was based on data that indicated when mature, rip figs were dusted with *A. flavus* spores, the level of aflatoxin was approximately the same whether sap beetles dusted with spores were added or not [131]. There was no treatment where only dusted sap beetles were added. When insect-simulated damage was used on yellow (ripening) figs, aflatoxin levels were increased by about fivefold, but no increase in incidence was noted for mature figs when they were wounded [132]. This information suggests an explanation for the observation that insect damage on ripe, mature figs does not increase aflatoxin levels [132].

As discussed earlier, when disking is part of the cultural practice, which produces high levels of airborne dust, the incidence of *A. niger* is associated with the level of dust on the leaves [120]. However, frequent presence of *A. niger* spores on sap beetles in date gardens suggested that they may be carrying the spores to ripening, uninfected dates [133]. It appears that some sap beetle species are more likely to act as vectors than others, which depends on their preference for fresh vs. rotting fruit. When several sap beetle species were tested with brown rot, *Monilia fructicola*, *Carpophilus mutilatus*, a species that had no definite preference for molded vs. fresh peaches, was the most important vector [134]. This criterion would also apply to other fruits (such as figs) and, as will be seen later, for *Carpophilus lugubris* in field corn as well.

A clear association with sap beetles and mycotoxigenic fungi in corn was first reported in 1947. In that study, drought-stressed corn that was used for silage was highly contaminated with *Fusarium moniliforme* as well as with both adults and larvae of *Carpophilus lugubris* [135]. The *C. lugubris* species is a primary pest of sweet corn, often an equal or worse problem to corn earworms in causing ear problems [29]. It is more common in the corn belt regions and similar climatological regions across the United States. However, to the north of 34° latitude, up into Ontario, *Glischrochilus* spp. such as *G. quadrisignatus* appears to be more common [136]. Although *C. lugubris* is reported through Central and well into South America (the type specimen is from Venezuela [29]), drier regions (or perhaps lack of hardwood forest) appear to limit its presence.

Unfortunately presence of sap beetles is more often only recorded when evaluating corn at harvest. By harvest time, *C. lugubris* has left, but *C. freemani*, *C. mutilatus*, and a *C. dimidiatus*, which appear more tolerant of dry conditions [29], remain in the U.S. in corn from Texas (Dowd, unpublished data), and additional tropical sap beetle species remain in dry corn in Mexico (Rodriguez et al., unpublished data). However, larvae of *C. freemani* and *C. mutilatus* have been recovered from milk to soft-dough stage corn from Mississippi (Dowd, unpublished data). Based on doses required for mortality, the corn inhabiting sap beetles appears to be much more tolerant of mycotoxins, including aflatoxin [4], than other ear-feeding insects such as caterpillars [4,137].

A dependent relationship appears to exist, with a population increase (approximate doubling) just prior to corn pollen shed [138]. The adults of *C. lugubris* feed on pollen in anthers, and like larvae of *O. nubilalis*, also feed on pollen that falls into leaf axils [29]. Depending on canopy, leaf axils below the ear (as well as the ear leaf/ear juncture) remain relatively moist, even in dry sandy areas. These moist areas are often colonized by fungi, and it is not uncommon to find *Fusarium* sp. growing and sporulating in these locations; in some cases > 90% of axils sampled (Dowd, unpublished data) have visually sporulating mold. These insects are often reported feeding on molded ears and kernels on the ground that remain from the prior year's harvest, especially since the advent of no-till and minimum-tillage cultural methods. Prior to the adoption of these methods for control of soil erosion, deep tillage buried ears such that they would be unavailable for feeding by *Carpophilus* spp. [29].

It is easy to understand how the beetles can be responsible for carrying the mold to axils, and then on into ears. Both *Aspergillus flavus* and *Fusarium* spp. occur in sap beetles collected in spring (March and April) in the Midwest (Dowd, unpublished data). It appears that the fungi can overwinter with the insect in a manner similar to that described for the overwintering of oak wilt with sap beetles [28]. In addition, a weak pathogen association has been reported for these fungi and other insects, suggesting continual reservoirs (ca. 30% to 40%) [83]. This situation may occur with the sap beetles or may be a function of their feeding on substrates contaminated by these molds. Although yeast can be readily digested by some species [139], fungal spores from *Ceratocystis* spp. [28,140] and *Aspergillus flavus* [16] can retain high viability after passing through the digestive system of various sap beetles.

Studies with *C. fimbriata* have indicated that *C. freemani* is able to pass on spores 8 days after final contact [140]. The gut crypts (Fig. 2) may serve as "reservoirs" to restrict flow of spores over long periods of time, while retaining them in a protected environment (as opposed to externally). Sap beetles are capable of dispersing long distances very rapidly. Some species (probably *C. lugubris*) dispersed oak wilt spores one mile in 24 h [141], and marked driedfruit beetles were recaptured over 2 miles away from the release site after 4 days in another study [142].

Sap beetles have consistently been reported as vectors of mycotoxigenic fungi to corn, although their importance relative to other insects has probably been underestimated until more recently. The emergence holes that the caterpillars make when exiting the ear (which is also true for other ear-feeding caterpillars such as the European corn borer and fall armyworm) also provide entry sites for sap beetles, and their damage is highly attractive to these beetles compared to simple mechanical damage [143]. During studies in Illinois in the 1930s, it was observed that where ear tips were exposed, or the ear was slightly damaged by corn earworms, more than twice as many ears were infected with *F. moniliforme*; only kernels away from earworm damage were considered [49].

Based on the descriptions of sap beetle biology, and known presence in Illinois, this suggests that sap beetles were a common factor in both types of exposure. Of the *Carpophilus* beetles collected from corn (unspecified stage) in Arizona, 11% carried *A. flavus* [144]. Low levels of sap beetles (unspecified species) were found in < 5% of corn at harvest from Illinois and Missouri [50]. Of these insects examined from Missouri (the only state so examined), 7% were contaminated with *A. flavus*, and 60% with *Fusarium* spp. [50]. Although caterpillar damage was common in corn from Georgia, other sorts of insect damage were nearly as common in a loose-husked opaque-2 line [77]. Figure 3 of the Anderson et al. 1975 article [77] shows discolored, shriveled kernels at the base of the ear. This figure, coupled with the description that "other insects followed this (corn earworm) access and many times did additional damage to other parts of the ear ..." [77], suggests damage by sap beetles, although maize weevil damage cannot be ruled out. Corn sap beetles were the most common species observed on corn at silking compared to maize weevils, European corn borers, and corn earworms in Georgia, South Carolina, Texas, Missouri, Illinois, and Iowa in the U.S.; with 3.1% of larvae and 2.8% of adults contaminated with *A. flavus* (although contamination at silking vs. harvest was not discriminated)—higher levels than any other type of insect, including weevils and caterpillars [145].

When *G. quadrisignatus* were obtained from buried corn ears in Minnesota infested with *F. graminearum* and *F. moniliforme*, 44% were contaminated externally, and 15% internally [146]. When obtained from standing corn, 71% were contaminated by these fungi externally, and 22% internally [146]. "Soil-feeding" insects (sap beetles, ground beetles, and leaf beetles) were generally more highly contaminated with *A. flavus* group fungi (up to 69%) from corn silking through 70 days after silking compared to "plant-feeding" caterpillars and other insects (maximum of 20%) in Illinois and Iowa [59].

The *G. quadrisignatus* collected from fields in Ontario were also externally and internally contaminated with *F. graminearum* and *F. moniliforme* [147]. Adults of *G. quadrisignatus* transmitted spores of *F. graminearum* to corn and were more effective at silking than at later stages of development [148]. When beetles were present, the incidence of mold was increased by threefold [148]. Taken together, these studies satisfy Leach's [22] criteria of isolation from sap beetles, and ability to vector by contaminated beetles. In Kentucky, adults of *G. quadrisignatus* contaminated with *A. flavus* caused lower aflatoxin contamination when added to simulated corn earworm injured corn than when no insects were added 15 days after silking, but significantly increased aflatoxin compared to corn without arthropods after 21 days (but beetles were not surface-sterilized) [149]. It is possible that the initial organisms contaminating the beetles inhibited *A. flavus*, but subsequently *A. flavus* grew well into areas damaged by the beetles. Sap beetles can carry micro-organisms that can inhibit growth of mycotoxigenic fungi [127,150].

In Georgia, *A. flavus* was isolated from 80% of *C. lugubris* and 66% of *C. freemani* obtained from partly buried ears from the preceding season in a peanut field next to a cornfield that was silking in June [63]. The *C. lugubris* also carried a color mutant of *A. flavus* to damaged corn (*C. freemani* was not tested) [63]. No *A. flavus* was isolated from *H. zea* found in experimental ears [63]. Interceptive trapping for sap beetles (which did not affect caterpillars) could reduce populations in corn by 50% and subsequently mold to a corresponding level [6]. Both *Aspergillus* and *Fusarium* are associated with sap beetle damage [6], and these molds are more highly correlated with sap beetle than caterpillar damage [4,82]. Sap beetles can apparently carry *A. flavus* from corn spillage at corn bins to standing corn in Iowa [151–153]. Ears that were insecticide-treated and thus free from sap beetles showed little mold compared to those where sap beetles (but not caterpillars) were present (G. Munkvold, personal communication). Sap beetles carried a blue dye from an autoinoculator baited with pheromone and attractants for *C. lugubris* to a trap over 700 ft away [154]. From a single autoinoculator (in 1995), sap beetles could eventually spread the biocompetitive bacteria *Bacillus subtilis* to such an extent that it could be isolated from over 90% of traps and up to 70% beetles around a 40 acre (44 ha) area ca. 8 weeks after the single autoinoculator was put out [127].

Evidence for sap beetle vectoring of mycotoxigenic fungi to corn is compelling. However, sap beetle presence in corn fields may vary greatly over a single county area (Dowd, unpublished data), so, as for other insects, importance is related to occurrence. Although capable of entering on their own, depending on husk coverage (which may be drought-shortened), silk channel, and pericarp thickness, they are opportunistic. Because of their association with caterpillar damage, sap beetle presence may not be recognized when evaluations are done at harvest unless very carefully examined for characteristic feeding and presence of frass. Larvae may enter kernel undersides and hollow them out so that only a transparent pericarp is left, and the frass has a characteristic shape [3]. Thus, as for other insects, their role depends on the numbers present, variety being grown, weather factors that change husk coverage, and so on. Nevertheless, it is important to consider sap beetles in managing mycotoxigenic fungi in corn.

Some reports of sap beetles and cotton also exist. Three species of sap beetles, *C. hemipterus*, *C. dimidiatus*, and *C.* (*Urophorus*) *humeralis*, were found infesting pink bollworm-damaged bolls in California [155]. Beetles collected from soil and melon were found to be contaminated with *A. flavus* (ca. 10% overall, present in all samples) and *Fusarium* spp. (sporadic in samples but up to 100% contaminated in some cases) [155]. Contaminated beetles increased the amount of infection by *A. flavus* threefold when caged with bollworm inoculated bolls, including those bolls that were simply sprayed with spores; but uncovered bollworm damaged bolls had the same level of infection as that seen when contaminated beetles were added in the second study [155]. This information led the authors to conclude beetles were not involved in fungal contamination [155]. As bolls in the open field study were not examined for sap beetle presence, the validity of this conclusion is uncertain. Although *Carpophilus* spp. beetles were seldom encountered from bollworm-damaged bolls in Arizona cotton (compared to neighboring corn), 10% of those collected from cotton carried *A. flavus* [144]. In another California study, the 81% of bolls with pink bollworm exit holes contained "scavenger beetles" which were mostly sap beetles [103]. Of the sap beetles tested, 42% were contaminated by *A. flavus* [103].

Involvement of sap beetles in vectoring and promoting mycotoxigenic fungi appears strongest for corn, likely for figs for some fungi under certain conditions, and less common for cotton. Regardless, involvement will vary according to the location and suitability as a host. When these insects are common in an area, their involvement is likely to be great. Monitoring (with traps) or scouting the crop at the susceptible stage (axil samples in corn, low axils, ca. 10 days after pollination) is currently the only way to predict most accurately whether sap beetles will be important in a particular area. Adaptations and preferences, capability for long-distance dispersal, and dissemination of viable inoculum several days after exposure, are all potent capabilities suggesting importance both as a long-distance vector and in facilitating entry of fungi through mechanical damage.

2. Weevils (Curculionidae)

Maize weevils (*Sitophilus zeamais*) attack corn both in the field and in storage; other species of *Sitophilus* can cause similar problems. They are relatively uncommon in the field in the corn belt, but very important in the southeastern U.S. There appeared to be some association between rice weevil (*S. oryzae*) damage and *A. flavus* from corn in South Carolina [156]. Corn thought to be damaged by rice weevils in South Carolina contained up to 91% of the total aflatoxin in corn samples [157]. Rice weevils collected from corn at harvest in South Carolina were 92% contaminated with *A. flavus*; *Penicillium* was isolated from 72%, and *Fusarium* from 27% of insects (mostly rice weevils) collected (although selective media used for isolation were more favorable for *A. flavus* growth) [158]. The percentage of kernels with insect damage was correlated with the percentage of *A. flavus* and aflatoxin [158].

Depending on the hybrid, aflatoxin production was correlated with the degree of rice weevil damage in some cases, but not others, in South Carolina [159]. Maize weevils contaminated with *A. flavus* and released onto ears with simulated corn earworm damage about doubled the level of aflatoxin B_1 found [149]. Maize weevils increased the levels of aflatoxin in corn by two- to fivefold when dusted with spores and caged with ears in Missouri, Georgia, and Tennessee [160]. In a comparative study, *A. flavus* contaminated maize weevil-infested corn ears had aflatoxin levels ca. four- to sixfold higher than in those ears infested with fall armyworms, corn earworms, or European corn borers [73]. Maize weevil-infested ears had ca. 10-fold more aflatoxin than ears without insects [73]. The tighter-husked variety reduced aflatoxin contamination by up to 25-fold for maize weevils, and sometimes to a greater degree for other insect species [73].

Maize weevil-infested corn can also be a carryover problem into storage.

Weevils exposed to spores of *A. flavus* about doubled the overall infection by *A. flavus* of four different hybrids, although southern-adapted varieties received less insect damage [161]. Activity of the weevils can cause increased moisture levels in grain, provided conditions are more conducive to fungal growth [33]. Kernels damaged by weevils can also provide entry sites for infection, as weevils increased aflatoxin levels by sixfold when added in conjunction with *A. flavus* spores [33].

Boll weevils (*Anthonomus grandis grandis*) can be serious pests of the cotton boll. Boll weevils inoculated with *F. moniliforme* caused 100% infection of bolls they were caged with in Mississippi, and uninoculated field-collected weevils caused boll rot on 95% of the bolls they were caged with [97]. Boll weevils collected from cotton were up to 50% contaminated with *Aspergillus* spp. and *Fusarium* spp. in Georgia [162]. The inoculum appears to be introduced when the female weevil oviposits into the boll, although *A. flavus* is not carried internally [162]. The conidia of *Fusarium* remained viable after passage through the boll weevil; nearly all conidia were expelled by 24 h [163]. Approximately 60% of field-collected weevils were contaminated with *Fusarium* in Louisiana [164]. In south Texas, where insect damage to cotton is primarily due to boll weevils, 61% of the seed from insect-damaged locks of the boll had > 20 ppb of aflatoxin, vs. no aflatoxin in locks not damaged by insects in areas where mature cotton did not receive excessive rain [165]. Although rain over a 4-week period on mature cotton could cause a few hundred ppb of aflatoxin in undamaged locks, insect-damaged locks still had from three to nine times more aflatoxin [165]. Reduction of *A. flavus* and aflatoxin in cottonseed may now be occurring where the boll weevil has been eradicated.

Pecan weevils (*Curculio caryae*) can be serious pests of pecan. Culled pecans (which are broken, have poorly developed kernels, or are shriveled, which can be a sign of weevil damage) contained over twofold more nut meats molded with *A. flavus* or *A. parasiticus* [166]. Although *A. flavus* and *A. parasiticus* strains that produce aflatoxin (at least 80% of the *A. flavus* strains) have been isolated from marketed pecans originating in Georgia, Alabama, and Oklahoma, initially there were no reports of insect associations [167]. However, toxin-producing strains of *Aspergillus*, *Fusarium*, and *Penicillium* were later isolated from weevil-damaged pecans from Georgia [168,169].

3. Other Beetles

The corn rootworm complex of beetles (*Diabrotica*) are very damaging pests of corn as larvae, although adults can damage silks and kernels when accessible. Damaged silk, and of course kernels, can provide an avenue of infection for mycotoxigenic fungi. They also vector mosaic (virus) diseases and bacterial wilts to cucumbers and related plants [170]. Southern corn rootworm (*Diabrotica undecimpunctata*) adults inoculated with *F. moniliforme* caused no infection when caged with cotton bolls in Mississippi, but there was no indication if feeding damage occurred [97]. The *D. undecimpunctata* collected from cotton fields in Arizona were 29% contaminated with *A. flavus*, but they were relatively uncommon [144]. However, larvae of *D. undecimpunctata* increased colonization of peanuts by *Fusarium* spp. by 13-fold, but had no significant effect on colonization by *Aspergillus* spp. in Virginia [171]. The larvae of the northern corn rootworm, *D. longicornis*, in Minnesota, through root damage, increased the incidence of *Fusarium* (including *F. graminearum* and *F. moniliforme*) by up to eightfold in corn roots compared to when insecticide controlled the rootworms; all stages of the insect were contaminated with the fungi [172]. Larvae of *D. undecimpunctata howardi* commonly bore into peanut pods in Virginia and the Carolinas [173], so presumably they would be responsible for promoting *A. flavus* and *A. parasiticus* in a manner similar to that of the lesser cornstalk borer.

Scarab beetles were encountered feeding on axils below ears to a limited extent in some cornfields in Illinois at one site in 1994, and were relatively common (in some samples nearly 50% of axils) in 1995 at another site in Illinois [81,82]. These insects were also found in some caterpillar-damaged ears at milk stage [81,82]. A very similar scarab was also found in damaged corn from Mexico at harvest state (Dowd, unpublished data). Because these habits are very similar to those noted with sap beetles, these insects appear to have the ability to vector mycotoxigenic fungi to corn, at least from axil-colonized material to ears.

Nine-spotted lady beetles (*Coccinella novemnotata*) inoculated with spores of *F. moniliforme* caused no infection of cotton bolls in Mississippi [97]. Another predatory beetle, *Collops vittatis*, was somewhat common in cotton in Arizona, and was infected with *A. flavus* at a rate of 88% to 29% [144]. The degree of infection depended on the location from which the beetles were collected [144]. The beetle *Cantharus melanura* was first reported as a visitor of ergot-infected grasses exuding "honey dew" in 1847 [174]. Rolfs observed in 1901 that carabid beetles, which are also mainly predatory, can vector *Claviceps paspali* and *Claviceps rolfsii* spores [175]. These beetles first visit sporulating stroma formed from sclerotia in the ground, and then move up the grass stem and over the flowers, which become infected [175].

Flea beetles, such as the corn flea beetle (*Chaetocnema pulicaria*), are small (typically < 3 mm) hopping beetles that are pests on a variety of plants. The corn flea beetle is a well-known vector of the bacterial disease Stewart's wilt of corn [176]. The flea beetle *Systena blanda* was encountered fairly frequently at one site in cotton in Arizona, and 46% were contaminated with *A. flavus* [144].

D. Other Chewing Arthropods

Early work indicated an association between termites (*Odontotermes badius* and *O. latericus*) damage and the degree of *A. flavus* and level of aflatoxin in peanuts in Africa [177]. In Africa, termite damage of peanut pods has been associated with higher levels of *A. flavus* and aflatoxin contamination, with the types of damage similar to those caused by the lesser cornstalk borer [178]. Termite damage is also important in determining aflatoxin levels in peanuts in West Africa, and appears associated with increased dryness during the growing season [178].

Mites (which are not insects) have both positive and negative ability to transmit mycotoxigenic fungi. The blister mite (*Eriophyes fici*) was not thought to transmit *A. niger* to figs because there was no correlation with mite presence and *A. niger* contamination [129]. However, it was later shown that 54.2% of predaceous mites (*Sejus pomi* and *Cheyletus* sp. and two unidentified species) were contaminated with unspecified molds and smuts and were thought to be capable of introducing the organisms (although thrips were probably more effective by comparison) [130]. In a later study, mites appeared to be more important than thrips, based on numbers present [179]. Mites of the genera *Caloglyphus* and *Tyrophagus* from peanut pods in the soil in South Africa were highly contaminated (internally and externally) with spores of *A. flavus* and could enter pods and spread the fungus when natural pod openings occur [180]. Both *A. flavus* and *Fusarium* spp. were isolated from mites at low levels ($< 10\%$) from Illinois and Missouri corn at harvest [51].

Generally, the mites *Caloglyphus rodriguezi* and *Tyrophagus putrescentiae* did not increase the levels of aflatoxin when contaminated with *A. flavus* and added to corn with simulated corn earworm damage, although in one instance the presence of *C. rodriguezi* increased the level of aflatoxin by twofold [149]. The wheat curl mite was not an effective vector of *A. flavus* spores to corn, apparently due to its inability to damage the pericarp [160]. The small size and limited damage that may be produced by mites in some commodities (corn) suggests limited importance of mites in promoting the establishment of mycotoxigenic fungi, but in crops where tissue is less durable (figs), they may play a greater role.

The bronze psocid, *Ectopsocopsis cryptomeriae*, can be common on corn near harvest in the midwestern U.S. It has occasionally been observed damaging kernels and apparently spreading molds along the ear (Fig. 7) (Dowd, personal observation). Interestingly, the fig wasp, *Blastophaga psenes*, which is essential to the pollination of some figs, is also responsible for introducing the fig endosepsis organism, primarily *Fusarium moniliforme* [181]. Surface sterilized fig wasps did not infect figs with *F. moniliforme* [182]. Treatment for this problem now involves application of fungicide to the winter crop to obtain a fungus free spring crop [183]. The capability for wide dispersal appears limited, as wasp flights are typically < 100 m [183].

E. Sucking and Sponging Insects

Sucking insects such as aphids and leafhoppers are well known for vectoring many plant diseases, such as mosaics, streaks, and stunts [170]. In many cases,

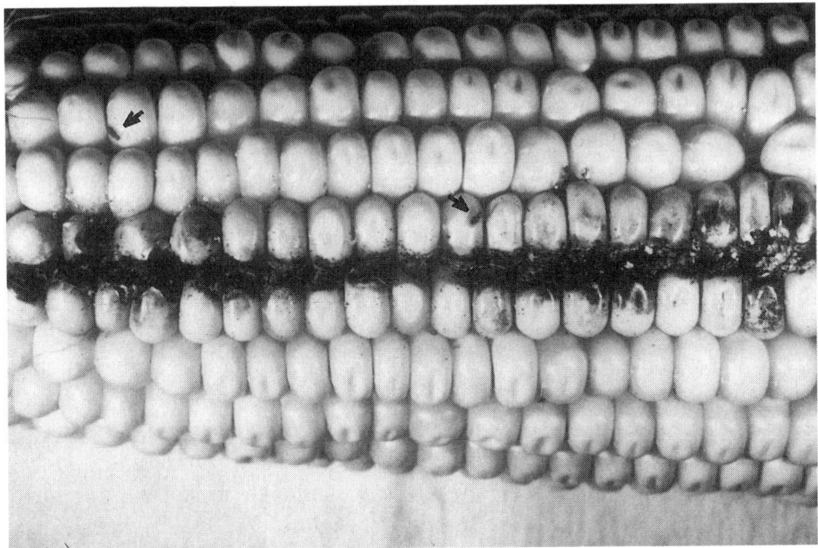

FIGURE 7 Bronze psocids (indicated by arrows) and associated moldy damage on corn at harvest. (Photo by P. F. Dowd.)

however, these micro-organisms can multiply within the insect, while other micro-organisms can be mechanically vectored [176]. Cross sections of the mouthparts of the green stink bug (*Nezara viridula*) indicated that this insect was capable of internally vectoring *Penicillium* fungi [184]. Insect honeydew (liquid waste products of sucking insects) appears to stimulate the growth of *A. flavus* and other fungi on cotton plant tissue [103].

When tarnished plant bugs (*Lygus lineolaris*) were inoculated with *F. moniliforme* and caged with cotton bolls in Mississippi, 65% of bolls were contaminated with the fungus [97]. When treated in a similar manner, uninoculated, field-collected tarnished plant bugs infected 40% of bolls with *F. moniliforme* [97]. However, three-cornered alfalfa hoppers caused no infection under the same conditions [97]. Several sucking insects that feed on cotton were contaminated with *A. flavus* in Arizona. Internal contamination by *A. flavus* of a lygus bug (*Lygus hesperus*) was 33%; for a stink bug (*Chlorochroa sayi*) it was 37% [144]. The degree of surface contamination of these two bugs was about twofold higher [144]. These insects occurred during boll development and apparently brought in initial inoculum, which later can spread into bolls through pink bollworm damage holes [144]. The fungus was also isolated at low levels from predatory sucking insects, such as assassin bugs (*Zelus* spp.) and lacewing larvae (*Chrysopa* spp.), but they were relatively uncommon on the plants [144]. Kernels at harvest from

Illinois and Missouri damaged by stink bugs had relatively low incidence of either *A. flavus* or *Fusarium* spp. (< 10%) [51]. The stink bug *Euchistus servus* appears relatively common at one site in Illinois in some years, and feeding damage is sometimes associated with *Fusarium* contamination of kernels (Fig. 8) (Dowd, unpublished data). Damage by southern green stink bugs (*Nezara viridula*) was associated with several toxic species of *Penicillium* in pecans from Georgia [185]. Thinner-shelled varieties (which are more susceptible to insect feeding) had significantly higher levels of infection [185]. Individuals of the stink bug *Thyanata pallidovirens* infested with spores of *A. flavus* eventually died, but caused aflatoxin in the pistachio nuts they fed on, compared to no aflatoxin where these stink bugs were not contaminated with *A. flavus* [186].

Thrips are typically small (< 3 mm) insects with rasping-sucking mouthparts, and are often associated with flowers. Western flower thrips (*Franklinella occidentalis*) control using one or two applications of acephate or carbaryl in corn in California was associated with reduced levels of *F. moniliforme* in corn when obvious control of thrips resulted [187]. In some cases treatment with acephate reduced ear rot by *F. moniliforme* by six- to eightfold [188]. Tighter-husked varieties appeared to limit thrip invasion and had less ear rot than looser-husked varieties (which allowed thrips to enter the ear) [187,189]. Sixty percent of thrips (*Franklinielli tritici*, *Thrips bremneri*, *Liothrips ilex*, and *Heliothrips fasciatus*,

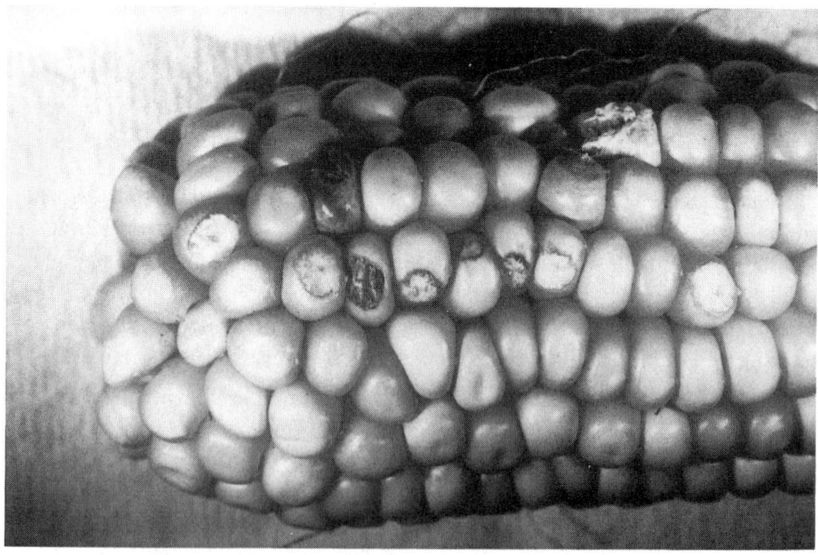

Figure 8 Stink bug damage on corn, associated with *Fusarium* mold. (Photo by P. F. Dowd.)

along with a predaceous species) were contaminated with smuts and molds, and were thought to be effective carriers of *A niger* to figs [130].

Many flies have lapping or sponging mouthparts. The pomace fly *Drosophila ampelophaga* was thought to be relatively unimportant compared to the driedfruit beetle in carrying *A. niger* to figs because of a lack of association with infected material early in the season [129]. Different species of flies appear to be important in vectoring the ergot *Claviceps purpurea*. An extensive discussion on the process is provided by Leach [22]. Conidia are liberated in a sticky liquid exuded from the flower head, that has a carrion-like odor especially attractive to flies [22]. Although 43 species of insects were listed as vectors of ergot fungi, based on reports in the literature at the time, *Rhagonycha fulva* and *Melanostoma mellina* were the most important, and mentioned in Germany in 1874 [174]. Another early example demonstrated the involvement of a species fungus gnat (*Sciara thomae*) in Europe, which carried spores both externally and internally [190]. Several other species of flies also have this capability [174]. When pomace flies (*Drosophila melanogaster*) inoculated with *F. moniliforme* were caged with cotton flowers, ca. 50% of the bolls were infected with *F. moniliforme* [97].

F. Vertebrate Associations

Mold has also been associated with vertebrate damage, such as that by birds and raccoons in corn. Exposed corn kernels at ear tips are often damaged by birds [38], which can be sites of invasion by *Fusarium* molds [191]. Ear tips with complete husk coverage may also be damaged by birds and raccoons while they are searching for and feeding on caterpillars at the ear tip (Dowd, personal observation). Raccoon-damaged ears are highly attractive to sap beetles, and often large numbers are found in these ears, along with large numbers of kernels visibly molded by *Fusarium* spp. (Dowd, unpublished data). Bird damage can also increase the levels of *Aspergillus* species in pistachio nut, but is typically less important than insect damage [108].

V. GENERALIZATIONS

Based on all of the above information, it should be clear that many different insects have the capability of promoting infection of various crops with mycotoxigenic fungi. Under equivalent conditions, ability to promote establishment of these fungi will depend on how common the insect is on the commodity and whether it can damage a particular plant material. For example, mite vectoring appears to be dependent on prior damage of some plant barrier (in corn and peanut) while for most other arthropods, they are capable of doing the damage themselves, depending on the condition of the material. There are relatively few instances where the mycotoxigenic fungus has been isolated from potential vectors prior to the occurrence of the fungus in the commodity, and thus a complete vector relationship conclusively proven. However, this lack of association may be more a function of not having done these studies than negative information. With the dispersal capability of sap beetles known to be 1 mile in 24 h [141], and with the known dispersal of corn earworm and fall armyworm moths from the south northward known to be hundreds of miles [e.g., 192], there is great potential for long-range dispersal of mycotoxigenic fungi to areas remote from the original source. In cases where the ability of insects to damage has been compared, European corn borers appear to have more potential than corn earworms or fall armyworms [89] while the maize weevil has more potential than these three caterpillars [73]. Similarly, in cotton, the boll weevil has more potential than the bollworm [97]. The ability of many different insects to be associated with mycotoxigenic fungi indicates that it is important to examine the material at the time the infection can occur for relative importance; scouting and trapping will be critical in cases where it is desirable to apply some sort of "rescue" control measure, such as insecticides.

VI. INSECT-ORIENTED STRATEGIES FOR MYCOTOXIN MANAGEMENT

Management of the insects themselves may provide an indirect control of mycotoxins in a particular commodity. Removal of insect-damaged pistachios can significantly reduce the total levels of aflatoxin in the pistachio crop [110,114]. With major variation in varietal susceptibility to fungal colonization of undamaged and damaged materials, possibly unrelated or related toxin production, and weather influences, predictability for insect control in corn becomes more difficult. However, assuming that the conditions and varieties are such that intact kernels are predictable (vs. splitting, popping, or silk cut), insect control should correlate well with control of mycotoxins. Adequate insect control has resulted in much lower levels of mold and/or toxins in corn and other crops treated in several different areas in the U.S., but treatment frequency with formulations and insecticides currently available is not considered economically feasible as economic thresholds for insect damage. If the ultimate cost of monitoring for mycotoxins, and losses caused by contaminated materials that are not detected, are considered, then the economic threshold for insecticide treatment of commodities where mycotoxin contamination is a problem needs to be readjusted. However, in most cases there is little direct economic incentive for the grower to produce materials with lower than mandatory threshold levels of mycotoxins, so it is unlikely at this point that increased use of conventional formulations of insecticides will be undertaken by growers to reduce mycotoxin levels. However, new formulations with lower levels of active ingredient, but longer residuality, can provide single-treatment control of insects and ear molds often comparable to conventional multiple-treatment regimens [81,82,126,193–195], and may prove economically viable if adopted.

Biological control or plant resistance can be an alterative to insecticides. As indicated earlier, varieties with long, tight husks can reduce insect damage and occurrence of aflatoxin in areas where it is endemic (such as the southeastern U.S.). Unfortunately, existing varieties with these characteristics do not yield as well as corn belt-adapted material and so are planted less by growers in the southeastern U.S. Because of the relatively small market for this material, there is less incentive for seed companies to develop varieties with long, tight husks that are higher-yielding. As indicated earlier, varieties with these characteristics are less desirable in the corn belt due to the slower drydown, and increased problems with *Fusarium* spp. as the ear mold. In many areas it is the trend to plant the longest-season variety possible to obtain the greatest yield per acre (with yield correlated to the length of time the plant is in a prereproductive state); thus, rapid drydown may allow a longer-season hybrid to be planted in more northerly areas than would otherwise be possible, with frost always a factor.

Crops with other forms of insect resistance also have the potential to indirectly reduce mycotoxin problems through insect control. However, the interaction of environmental factors with the mycotoxin problem does not make indirect control of insects as a means of controlling mycotoxins a predictable enough strategy for seed companies to expend resources in this area. It is more likely that mycotoxin control through insect control will be the result of efforts to obtain better control of insects as an end result of itself. Plant breeding continues to be a viable method for introducing insect resistance, provided this resistance can be directly associated with increases in yield or, in some cases, quality. However, greater levels of insect control will be needed than presently available in commercially bred varieties to make this a viable method for indirectly controlling mycotoxins.

Genetic engineering has potential for overcoming some of the barriers (such as rapid combining of desirable traits) that limit plant breeding. For example, there are now cases where the *Bacillus thuringiensis* crystal protein has been introduced into commercially viable material and provided highly effective control of a selected number of target insect species, several of which are associated with promoting mycotoxin problems in (for example) mostly cotton and corn. Reports have indicated much more effective control of target insects compared to conventional insecticide treatments, such as for European corn borer control in corn [196]. Corn expressing the Bt gene in all parts of the plant showed very little ear mold compared to corresponding wild-type material, which was highly correlated with the lack of target insect damage [81].

Although commercially released material in 1996 did not yet express the Bt gene in kernels and silks, expression in green tissues such as the husk and ear shank may prevent entry of European corn borers through the lower parts of the ear. As indicated earlier, these areas, because of growth of mold in pollen lodging in the axil and ear/leaf gap, and their protected area, provide highly attractive areas for European corn borer initial feeding on the pollen, and subsequent entry. They are also attractive to sap beetles, the larvae of which can enter through both entry and exit holes made by caterpillars. These damaged sites, on a percentage basis, have the highest levels of colonization by *Fusarium* spp. (Dowd, unpublished data). However, where insecticide control for caterpillars has ceased because of the efficacy of pheromone mating disruption (such as peach orchards in Australia), sap beetles have become the major pest [197]. There is obviously a potential for this situation to be repeated when Bt material effective only against caterpillars occurs, especially should this result in the growth of more open-husked varieties in areas where endemic mycotoxin problems occur. The potential for resistance development is also slowing adoption of the Bt material.

Other resistance mechanisms are also of use, and would be of greater value if they are effective against several pests. There are several mechanisms of potential resistance against both fungi and insects, including chemical factors such as phenolics, flavonoids, and hydroxamic acids [194,198,199] and proteins [199]. Resistance to corn earworm and *Fusarium* stalk rot have two chromosomal bin locations in approximately the same place [200]. Resistance to second-generation European corn borer and *Fusarium* stalk rot have five chromosomal bin locations in approximately the same place [200]. Thus, there are data at the genetic levels that also suggests common resistance mechanisms to insects and fungi. Some of these factors, when combined, may be additive, synergistic, or antagonistic. Introduction of resistance mechanisms may be by conventional breeding, genetic engineering, or a combination of the two. Multigenic (in spite of a preexisting background of resistant genes already present in plants) resistance mechanisms introduced by genetic engineering appear more promising than a monogenic approach based on theoretical modeling and laboratory selection studies.

Ultimately, integrated management will be involved. Although partial systems are in operation and/or under development, comprehensive systems have not yet been adopted. Aflatoxin management systems that include insect control appear important for peanuts [201] and corn [202–205]. Other integrated programs for general mycotoxigenic ear mold control in corn have been proposed [6]. Lack of comprehensive management strategies appears to stem not only from a lack of development of key components, but also from limited motivation. This appears to be due to a lack of understanding of the complexity of interactions involved by those empowered to promote adoption of such strategies. Conflicting information provided by different special-interest groups (despite relevant examples of other economically successful pest management strategies such as those for the boll weevil [206]) and other political concerns contribute to this lack of understanding. Hopefully increased understanding of the complexities of the problem will further help in the development of suitable strategies for reducing mycotoxins in food and feed.

ACKNOWLEDGMENTS

I thank M. A. Doster, T. Isakeit, R. E. Lynch, T. J. Michailides, G. P. Munkvold, N. W. Widstrom, and G. L. Windham for providing reprints and preprints of their work, and J. L. Richard and F. E. Vega for comments on this manuscript.

Disclaimer: Names are necessary to report factually on available data: however, the USDA neither guarantees nor warrants the standard of the product, and the use of the name by USDA implies no approval of the product to the exclusion of others that may also be suitable.

REFERENCES

1. UL Diener, RE Pettit, RJ Cole. Aflatoxins and other mycotoxins in peanuts. In: HE Pattee, CT Young, eds. Peanut Science and Technology. Yoakum, TX: American Peanut Research and Education Society, 1982, p 486.
2. UL Diener, RJ Cole, TH Sanders, GA Payne, LS Lee, MA Klich. Epidemiology of aflatoxin formation by *Aspergillus flavus*. Annu Rev Phytopathol 25:249, 1987.
3. PF Dowd. Nitidulids as vectors of mycotoxin-producing fungi. In: OL Shotwell, CR Hurburgh Jr, eds. Aflatoxin in Corn: New Perspectives. Ames: Iowa State Univ., 1991, p 335.
4. PF Dowd. Insect interactions with mycotoxin-producing fungi and their hosts. In: D Bhatnagar, EB Lillehoj, DK Arora, eds. Handbook of Mycology. Vol. 5. Mycotoxins in Ecological Systems. 1992, p 137.
5. PF Dowd. Detoxification of mycotoxins by insects. In: CA Mullin and JG Scott, eds. Molecular Mechanisms of Insecticide Resistance. Washington: American Chemical Society, 1992, p 264.
6. PF Dowd. Sap beetles and mycotoxins in maize. Food Add Contam 12:497, 1995.
7. EB Lillehoj, MS Zuber. Aflatoxin problem in corn and possible solutions. Proceedings of the Thirteenth Annual Corn and Sorghum Research Conference. Chicago, 1975, p 230.
8. EB Lillehoj, CW Hesseltine. Aflatoxin control during plant growth and harvest of corn. In: JV Rodricks, CW Hesseltine, MA Mehlman, eds. Mycotoxins in Human and Animal Health. Park Forest South, IL: Pathotox Publishers, 1977, p 107.
9. WW McMillian. Role of arthropods in field contamination. In: UL Diener, RL Asquith, JW Dickens, eds. Aflatoxin and *Aspergillus flavus* in Corn. Auburn, AL: Auburn University, 1983, pp 20–22.
10. WW McMillian. Relation of insects to aflatoxin contamination of maize grown in the southeastern USA. Proceedings of the Aflatoxin in Maize Workshop, El Batan, Mexico, 1987, pp 194–200.

11. WW McMillian, NW Widstrom, RW Beaver, DM Wilson. Aflatoxin in Georgia: factors associated with its formation in corn. In: OL Shotwell, CR Hurburgh Jr, eds. Aflatoxin in Corn: New Perspectives. Ames: Iowa State Univ., 1991, p 329.
12. GA Payne. Aflatoxin in maize. Crit Rev Plant Sci 10:423, 1992.
13. OL Shotwell. Aflatoxin in corn. J Am Oil Chem Soc 54:216A, 1977.
14. RN Sinha. The fungal community in the stored grain ecosystem. In: GC Carroll, DT Wicklow, eds. The Fungal Community. New York: Marcel Dekker, 1992, p 797.
15. DT Wicklow. Epidemiology of *Aspergillus flavus* in corn. In: OL Shotwell, CR Hurburgh Jr, eds. Aflatoxin in Corn: New Perspectives. Ames: Iowa State Univ., 1991, p 315.
16. DT Wicklow. The mycology of stored grain: an ecological perspective. In: DS Jayas, DDG White, WE Muir, eds. Stored Grain Ecosystems. New York: Marcel Dekker, 1995, p 197.
17. NW Widstrom. The role of insects and other plant pests in aflatoxin contamination of corn, cotton, and peanuts—a review. J Environ Qual 8:5, 1979.
18. NW Widstrom. Aflatoxin in developing maize: interactions among involved biota and pertinent econiche factors. In: D Bhatnager, EB Lillehoj, DK Arora, eds. Handbook of Applied Mycology. Vol. 5. Mycotoxins in Ecological Systems. New York: Marcel Dekker, 1992, p 23.
19. NW Widstrom. The aflatoxin problem with corn grain. Adv Agron 56:219, 1996.
20. NW Widstrom, MS Zuber. Prevention and control of aflatoxin in corn. In: UL Diener, RL Asquith, JW Dickens, eds. Aflatoxin and *Aspergillus flavus* in Corn. Auburn, AL: Auburn University, 1983, p 72.
21. MS Zuber, EB Lillehoj. Status of the aflatoxin problem in corn. J Environ Qual 8:1, 1979.
22. JG Leach. Insect Transmission of Plant Diseases. New York: McGraw-Hill, 1940.
23. AA Berryman. Adaptive pathways in scolytid-fungus associations. In: N. Wilding, NM Collins, PM Hammond, JF Webber, eds. Insect-Fungus Interactions. New York: Academic Press, 1989, p 145.
24. V Grant, KA Grant. Pollination of *Opuntia basilaris* and *O. littorallis*. Plant Syst Evol 132:321, 1979.
25. H Podoler, I. Galon, S. Gazit. The role of nitidulid beetles in natural pollination of *Annona* in Israel. Acta Oecol 5:369, 1984.
26. NH Nadel, JE Peña. Identity, behavior and efficacy of nitidulid beetles (Coleoptera: Nitidulidae) pollinating commercial *Annona* species in Florida. Environ Entomol 23:878, 1994.
27. CK Dorsey, JG Leach. The bionomics of certain insects associated with oak wilt, with particular reference to the Nitidulidae. J Econ Entomol 49:219, 1956.
28. FF Jewell. Insect transmission of oak wilt. Phytopathology 46:244, 1956.
29. WA Connell. Nitidulidae of Delaware. Del Agric Exp Sta Bull 318, 1956.
30. VF Wright, RF Vesonder, A Ciegler. Mycotoxins and other fungal metabolites as insecticides. In: E. Kurstak, ed. Microbial and Viral Pesticides. New York: Marcel Dekker, 1982, p 559.
31. TM Gradziel, D Wang. Susceptibility of California almond cultivars to aflatoxigenic *Aspergillus flavus*. HortScience 29:33, 1994.

32. MA Doster, TJ Michailides. The development of early split pistachio nuts and their contamination by molds, aflatoxins, and insects. Acta Horticulturae 419:359, 1995.
33. JA Beti, TW Phillips, EB Smalley. Effects of maize weevils (Coleoptera: Curculionidae) on production of aflatoxin B_1 by *Aspergillus flavus* in stored corn. J Econ Entomol 88:1776, 1995.
34. JJ Christensen, CL Schneider. European corn borer *Pyrausta nubilalis* Hbn.) in relation to shank, stalk and ear rots of corn. Phytopathology 40:284, 1950.
35. EB Lillehoj, WF Kwolek, MS Zuber, et al. Aflatoxin contamination of field corn: evaluation of regional test plots for early detection. Cereal Chem 55:1007, 1978.
36. RK Jones, HE Duncan, GA Payne, KJ Leonard. Factors influencing infection by *Aspergillus flavus* in silk-inoculated corn. Plant Dis 64:859, 1980.
37. SF Marsh, GA Payne. Preharvest infection of corn silks and kernels by *Aspergillus flavus*. Phytopathology 74:1284, 1984.
38. MC Shurtleff. Compendium of Corn Diseases. St. Paul, MN: APS Press, 1980.
39. GN Odvody, JC Remmers, NM Spencer. Association of kernel splitting with kernel and ear rots of corn in a commercial hybrid grown in the coastal bend of Texas. Phytopathology 80:1045, 1990.
40. GN Odvody. Drought stress-induced loss of kernel integrity and its contribution to preharvest aflatoxin in south Texas corn. Proceedings of the 1995 Aflatoxin Workshop, Atlanta, GA, 1995, p 74.
41. DG White, TR Rocheford, B Kaufman, AM Hamblin. Chromosome regions associated with resistance to *Aspergillus flavus* and inhibition of aflatoxin production in maize. Proceedings of the ARS Aflatoxin Workshop, Atlanta, GA, 1995, p 8.
42. CW Bacon, DM Hinton, RT Riley, IE Yates, F Meredith. *Fusarium moniliforme* is a symptomless endophyte of corn. Proceedings of the ARS *Fusarium*/fumonisin Workshop, Beltsville, MD, 1995, p 21.
43. RD Plattner, AE Desjardins, L Claflin. Measurement of fumonisin levels in individual corn ears from a two year field study. Proceedings of the ARS *Fusarium*/fumonisin Workshop, Beltsville, MD, 1995, p 26.
44. SM Kelly, JR Wallin. Systemic infection of maize plants by *Aspergillus flavus*. Proceedings of the Aflatoxin in Maize Workshop, El Batan, Mexico, 1987, pp 187–193.
45. D Barry. Insects of maize and their association with aflatoxin contamination. Proceedings of the Aflatoxin in Maize Workshop, El Batan, Mexico, 1987, pp 201–211.
46. H Garman, HH Jewett. The life history and habits of the corn-ear worm (*Chloridea obsoleta*). Ky Agric Exp Sta Bull 187:513, 1914.
47. FC Bishop. The boll worm or corn-earworm. USDA Farmers Bull 872:3, 1917.
48. JJ Taubenhaus. A study of the black and yellow molds of ear corn. Tex Agric Exp Sta Bull 290:3, 1920.
49. B Koehler. Natural mode of entrance of fungi into corn ears and some symptoms that indicate infection. J Agric Res 64:421, 1942.
50. DI Fennell, EB Lillehoj, WF Kwolek. *Aspergillus flavus* and other fungi associated with insect-damaged field corn. Cereal Chem 52:314, 1975.
51. GW Rambo, J Tuite, RW Caldwell. *Aspergillus flavus* and aflatoxin in preharvest corn from Indiana in 1971 and 1972. Cereal Chem 51:848, 1974.

52. NW Widstrom, BR Wiseman, WW McMillian. Toxicity of *Aspergillus* spp. on *Heliothis zea* and aflatoxin-contamination of corn in the South. Georgia Agron Abstr 18:9, 1975.
53. GC Kingsland. Ecology of *Aspergillus flavus* associated with *Zea mays* in the field in North Carolina. Proc Am Phytopathol Soc 3:217, 1976.
54. GC Kingsland. Epidemiological relationships between *Aspergillus flavus*, *Heliothis zea*, and *Zea mays*. Proc Am Phytopathol Soc 3:218, 1976.
55. EB Lillehoj, DI Fennell, WF Kwolek, et al. Aflatoxin contamination of corn before harvest: *Aspergillus flavus* association with insects collected from developing ears. Crop Sci 18:921, 1978.
56. DI Fennell, EB Lillehoj, WF Kwolek, et al. Insect larval activity on developing corn ears and subsequent aflatoxin contamination of seed. J Econ Entomol 71:624, 1978.
57. OH Calvert, EB Lillehoj, WF Kwolek, MS Zuber. Aflatoxin B_1 and G_1 production in developing *Zea mays* kernels from mixed inocula of *Aspergillus flavus* and *A. parasiticus*. Phytopathology 68:501, 1978.
58. EB Lillehoj, WW McMillian, WD Guthrie, D. Barry. Aflatoxin-producing fungi in preharvest corn: inoculum source in insects and soil. J Environ Qual 9:691, 1980.
59. BA Fortnum, A Manwiller. Effects of irrigation and kernel injury on aflatoxin B_1 production in selected maize hybrids. Plant Dis 69:262, 1985.
60. WW McMillian, DM Wilson, NW Widstrom. Aflatoxin contamination of preharvest corn in Georgia: a six-year study of insect damage and visible *A. flavus*. J Environ Qual 14:200, 1985.
61. GA Payne, DK Cassel, CL Adkins. Reduction of aflatoxin contamination in corn by irrigation and tillage. Phytopathology 76:679, 1986.
62. WW McMillian, NW Widstrom, DM Wilson, BD Evans. Annual contamination of *Heliothis zea* (Lepidoptera: Noctuidae) moths with *Aspergillus flavus* and incidence of aflatoxin contamination in preharvest corn in the Georgia coastal plain. J Entomol Sci 25:123, 1990.
63. J Lussenhop, DT Wicklow, Nitidulid beetles (Nitidulidae: Coleoptera) as vectors of *Aspergillus flavus* in pre-harvest maize. Trans Mycol Soc Jpn 31:63, 1990.
64. J Dunlap, R Meagher, S Maas, S Lyda, A Bockholt. The relationship of interacting biological and environmental stresses on the susceptibility of corn to aflatoxin contamination. Proceedings of the ARS Aflatoxin Workshop, Fresno, CA, 1992, p 19.
65. EB Lillehoj, WF Kwolek, MS Zuber, et al. Aflatoxin in corn before harvest: interactions of hybrids and locations. Crop Sci 20:731, 1980.
66. EB Lillehoj, WF Kwolek, ES Horner, et al. Aflatoxin contamination of preharvest corn: role of *Aspergillus flavus* inoculum and insect damage. Cereal Chem 57:255, 1980.
67. DM Wilson, NW Widstrom, LR Marti, BD Evans. *Aspergillus flavus* group, aflatoxin, and bright greenish yellow fluorescence in insect-damaged corn in Georgia. Cereal Chem 58:40, 1981.
68. EB Lillehoj, WW McMillian, NW Widstrom, et al. Aflatoxin contamination of maize kernels before harvest. Mycopathologia 86:77, 1984.
69. D. Barry, NW Widstrom, LL Darrah, et al. Maize ear damage by insects in relation to genotype and aflatoxin contamination in preharvest maize grain. J Econ Entomol 85:2492, 1992.

70. DT Wicklow. Kernel "splitting" and aflatoxin contamination for corn varieties grown under "mild" climatic conditions (Lincoln, IL—1994). Proceedings of the 1995 Aflatoxin Workshop, Atlanta, GA, 1995, p 2.
71. MT Jenkins. Corn hybrids for the South. Yearbook of Agriculture 1943–1947: Science in Farming. Washington: U.S. Department of Agriculture, 1947, p 389.
72. WW McMillian, NW Widstrom, DM Wilson. Insect damage and aflatoxin contamination in preharvest corn: influence of genotype and ear wetting. J Entomol Sci 20: 66, 1985.
73. WW McMillian, NW Widstrom, DM Wilson. Impact of husk type and species of infesting insects on aflatoxin contamination in preharvest corn at Tifton, Georgia. J Entomol Sci 22:307, 1987.
74. D Barry, EB Lillehoj, NW Widstrom, et al. Effect of husk tightness and insect (Lepidoptera) infestation on aflatoxin contamination of preharvest maize. Environ Entomol 15:1116, 1986.
75. HL Trenholm, DB Prelusky, JC Young, JD Miller. A practical guide to the prevention of *Fusarium* mycotoxins in grain and animal feedstuffs. Arch Environ Contam Toxicol 18:443, 1989.
76. DG Smeltzer. Relationship between *Fusarium* ear rot and corn earworm infestation. Agron J 51:53, 1959.
77. HW Anderson, EW Nehring, WR Wichser. Aflatoxin contamination of corn in the field. J Agric Food Chem 23:775, 1975.
78. EB Lillehoj, WF Kwolek, A Manwiller, et al. Aflatoxin production in several corn hybrids grown in South Carolina and Florida. Crop Sci 16:483, 1976.
79. NW Widstrom, EB Lillehoj, AN Sparks, WF Kwolek. Corn earworm damage and aflatoxin B_1 on corn ears protected with insecticide. J Econ Entomol 69:677, 1976.
80. MS Smith, TJ Riley. Direct and interactive effects of planting data, irrigation, and corn earworm (Lepidoptera: Noctuidae) damage on aflatoxin production in preharvest field corn. J Econ Entomol 85:998, 1992.
81. PF Dowd, GA Bennett, JL Richard, et al. IPM of aflatoxin in the corn belt—FY 1995 results. Proceedings of the ARS Aflatoxin Workshop, Atlanta, GA, 1995, p 76.
82. PF Dowd, RW Behle, GA Bennett, et al. Adherent malathion flour granules as an environmentally selective control for chewing insect pests of dent corn ears and associated problems (submitted). 1997.
83. DL Brooks, ES Raun. Entomogenous fungi from corn insects in Iowa. J Invert Pathol 7:79, 1965.
84. EB Lillehoj, DI Fennell, WF Kwolek. *Aspergillus flavus* and aflatoxin in Iowa corn before harvest. Science 193:495, 1976.
85. WD Guthrie, EB Lillehoj, D. Barry, et al. Aflatoxin contamination of preharvest corn: interaction of European corn borer larvae and *Aspergillus flavus*-group isolates. J Econ Entomol 75:265, 1982.
86. WW McMillian, NW Widstrom, D Barry, EB Lillehoj. Aflatoxin contamination in selected corn germplasm classified for resistance to European corn borer (Lepidoptera: Noctuidae). J Entomol Sci 23:240, 1988.
87. EA Sobek, GP Munkvold. European corn borer as a vector of *Fusarium moniliforme* in symptomatic and asymptomatic infection of corn kernels. Phytopathology 85: 1180, 1995.

88. GL Windham, WP Williams, FM Davis. Effects of southwestern corn borer on aflatoxin contamination and *Aspergillus flavus* kernel infection in corn. Proceedings of the 1996 Aflatoxin Workshop, Fresno, CA, 1996.
89. NW Widstrom, AN Sparks, EB Lillehoj, WF Kwolek. Aflatoxin production and lepidopteran insect injury on corn in Georgia. J Econ Entomol 68:855, 1976.
90. ML Schroeder, LJ Ashworth Jr. Aflatoxins in Spanish peanuts in relation to pod and kernel condition. Phytopathology 55:464, 1965.
91. LB Ashworth Jr, BC Langley. The relationship of pod damage to kernel damage by molds in Spanish peanut. Plant Dis Rept 48:875, 1964.
92. JW Dickens, JB Satterwhite. Aflatoxin-contaminated peanuts produced on North Carolina farms in 1968. J Am Peanut Res Educ Assoc 5:48, 1973.
93. RE Lynch, DM Wilson. Enhanced infection of peanut, *Arachis hypogaea* L., seeds with *Aspergillus flavus* group fungi due to external scarification of peanut pods by the lesser cornstalk borer, *Elasmopalpus lignosellus* (Zeller). Peanut Sci 18:110, 1991.
94. KL Bowen, TP Mack. Relationship of damage from the lesser cornstalk borer to *Aspergillus flavus* contamination in peanuts. J Entomol Sci 28:29, 1993.
95. PD Blankenship, RJ Cole, TH Sanders, RA Hill. Effect of geocarposphere temperature on pre-harvest colonization of drought-stressed peanuts by *Aspergillus flavus* and subsequent aflatoxin contamination. Mycopathologia 85:69, 1984.
96. JW Dorner, RJ Cole, B Yagen, B Christiansen. Bioregulation of preharvest aflatoxin contamination of peanuts; role of stilbene phytoalexins. In: PE Hedin, ed. Naturally Occurring Pest Bioregulators. Washington: American Chemical Society, 1991, p 352.
97. HS Bagga, ML Laster. Relation of insects to the initiation and development of boll rot of cotton. J Econ Entomol 61:1141, 1968.
98. MJ Lukefahr, DF Martin. Evaluation of damage to lint and seed of cotton caused by the pink bollworm. J Econ Entomol 56:710, 1963.
99. RK Kiyomoto, LJ Ashworth J. Status of cotton boll rot in the San Joaquin valley of California following simulated pink bollworm injury. Phytopathology 64:259, 1974.
100. JL McMeans, CM Brown, LL Parker, RL McDonald. Aflatoxins in cottonseed: effects of pink bollworm control. Crop Sci 16:259, 1976.
101. TE Russell, TF Watson, GF Ryan. Field accumulation of aflatoxin in cottonseed as influenced by irrigation termination dates and pink bollworm infestation. Appl Environ Microbiol 31:711, 1976.
102. TJ Henneberry, LA Bariola, T Russell. Pink bollworm: chemical control in Arizona and relationship to infestations, lint yield, seed damage, and aflatoxin in cottonseed. J Econ Entomol 71:440, 1978.
103. ME Simpson, LR Batra. Ecological relations in respect to a boll rot of cotton caused by *Aspergillus flavus*. In: H Kurata, Y Ueno, eds. Toxigenic Fungi—Their Toxins and Health Hazard. New York: Elsevier, 1984, p 24.
104. LS Lee, PE Lacey, WR Goynes. Aflatoxin in Arizona cottonseed: a model study of insect-vectored entry of cotton bolls by *Aspergillus flavus*. Plant Dis 71:997, 1987.
105. PJ Cotty, LS Lee. Aflatoxin contamination of cottonseed: comparison of pink bollworm damaged and undamaged bolls. Trop Sci 29:273, 1989.
106. SV Thompson, MC Mehdy. Occurrence of *Aspergillus flavus* in pistachio nuts prior to harvest. Phytopathology 68:1112, 1978.

107. NF Sommer, JR Buchanan, RJ Fortlage. Relation of early splitting and tattering of pistachio nuts in orchards. Phytopathology 76:692, 1986.
108. MA Doster, J Michailides. *Aspergillus* molds and aflatoxins in pistachio nuts in California. Phytopathology 84:583, 1994.
109. MA Doster, TJ Michailides. The development of early split pistachio nuts and their contamination by molds, aflatoxins and insects. California Pistachio Industry Annual Report Crop Year 1993–1994:68, 1994.
110. MA Doster, TJ Michailides. The relationship between date of hull splitting and decay of pistachio nuts by *Aspergillus* species. Plant Dis 79:766, 1995.
111. MA Doster, TJ Michailides, DA Goldhamer. Influence of cultural practices on occurrence of early split pistachio nuts. California Pistachio Industry Annual Report Crop Year 1992–1993:82, 1993.
112. MA Doster, TJ Michailides. Ecology of *Aspergillus* molds in pistachio orchards. California Pistachio Industry Annual Report Crop Year 1991–1992:101, 1992.
113. MA Doster, TJ Michailides. Development of *Aspergillus* molds in litter from pistachio trees. Plant Dis 78:393, 1994.
114. MA Doster, TJ Michailides. Characteristics of pistachios with high incidence of mold and naval orangeworm contamination for identification at processing plants. California Pistachio Industry Annual Report Crop Year 1994–1995:67, 1995.
115. DJ Phillips, SL Purcell, GI Stanley. Aflatoxins in almond. Fresno, CA: USDA SEA, ARM-W-20, 1980.
116. TM Gradziel. Breeding for resistance to aflatoxin contamination in almond. Proceedings of the 1993 ARS Aflatoxin Workshop, Little Rock, AR, p 55.
117. TM Gradziel, MA Thorpe, N Hirsch, J McDonnel. Breeding for resistance to aflatoxin contamination in almond. Proceedings of the 1994 ARS Aflatoxin Workshop, St. Louis, MO, p 10.
118. MA Doster, TJ Michailides. Factors involved in the development of kernel mold in walnut. Walnut Res Rept 1994:342, 1994.
119. MA Doster, TJ Michailides, DP Morgan. *Aspergillus* species and mycotoxins in figs from California orchards. Plant Dis 80:484, 1996.
120. TJ Michailides, DP Morgan. Positive correlation of fig smut in Calimyrna fruit with amounts of dust and propagules of *Aspergillus niger* accumulated on the trees. Phytopathology 81:1234, 1991.
121. AM Julian, PW Wareing, SI Phillips, et al. Fungal contamination and selected mycotoxins in pre- and post-harvest maize in Honduras. Mycopathologia 129:5, 1995.
122. PJ Cotty. Effects of cultivar and boll age on aflatoxin in cottonseed after inoculation with *Aspergillus flavus* at simulated exit holes of the pink bollworm. Plant Dis 73: 489, 1989.
123. RA Norton. *Aspergillus flavus* grown on glass fiber filters is a sensitive method for evaluating inhibitors of growth and aflatoxin production. Proceedings of the 1995 Aflatoxin Workshop, Atlanta, GA, 1995, p 24.
124. JW Dorner, RJ Cole, PD Blankenship. Use of a biocompetitive agent to control preharvest aflatoxin in drought stressed peanuts. J Food Prot 55:88, 1992.
125. PJ Cotty. Influence of field application of an atoxigenic strain of *Aspergillus flavus* on the population of *A. flavus* infecting cotton bolls and on the aflatoxin content of cottonseed. Phytopathology 84:1270, 1994.

126. PF Dowd, MR McGuire, RW Behle, et al. Insect IPM for mycotoxin control in midwest corn: 1994 studies. Proceedings of the ARS Aflatoxin Workshop, St. Louis, MO, 1994, p 68.
127. PF Dowd, FE Vega, TC Nelsen, JL Richard. Sap beetle transmission of *Bacillus subtilis* for reduction of *Aspergillus flavus* and aflatoxin in maize. (Submitted.) 1997.
128. DJ Phillips, B Mackey, WR Ellis, TN Hansen. Occurrence and interaction of *Aspergillus flavus* with almond. Phytopathology 69:829, 1979.
129. EH Phillips, EH Smith, RE Smith. Fig smut. Cal Agric Exp Sta Bull 387, 1925.
130. HN Hansen, AE Davey. Transmission of smut and molds in figs. Phytopathology 22: 247, 1932.
131. JR Buchanan, NF Sommer, RJ Fortlage. *Aspergillus flavus* infection and aflatoxin production in fig fruits. Appl Microbiol 30:238, 1975.
132. MA Doster, TJ Michailides, DP Morgan, DA Goldhamer. Cultural practices that reduce aflatoxin contamination and characteristics for removal of contaminated figs and pistachio nuts. Proceedings of the 1996 Aflatoxin Workshop, Fresno, CA, 1996.
133. RL Warner, MM Barnes, EF Laird. Reduction of insect infestation and fungal infection by cultural practice in date gardens. Environ Entomol 19:1618, 1990.
134. KG Tate, JM Ogawa. Nitidulid beetles as vectors of *Monilia fructicola* in California stone fruits. Phytopathology 65:977, 1975.
135. CF Monroe, AE Perkins, CE Knoop. Silage from drought-damaged corn. Res Bull Ohio Agric Exp Sta 673:48, 1947.
136. WH Luckmann. Observations on the biology and control of *Glischrochilus quadrisignatus*. J Econ Entomol 56:681, 1963.
137. WW McMillian, DM Wilson, NW Widstrom, WD Perkins. Effects of aflatoxin B_1 and G_1 on three insect pests of maize. J Econ Entomol 73:26, 1980.
138. PF Dowd, TC Nelsen. Seasonal variation of sap beetle (Coleoptera: Nitidulidae) populations in central Illinois cornfield-oak woodland habitat and potential influence of weather patterns. Environ Entomol 23:1215, 1994.
139. MW Miller, EM Mrak. Yeasts associated with dried-fruit beetles in figs. Appl Microbiol 1:174, 1954.
140. WJ Moller, JE deVay. Insect transmission of *Ceratocystis fimbriata* in deciduous fruit orchards. Phytopathology 58:1499, 1968.
141. CL Morris, HE Thompson, BL Hadley Jr, JM Davis. Use of radioactive tracer for investigation of the activity pattern of suspected insect vectors of the oak-wilt fungus. Plant Dis Rep 39:61, 1955.
142. FS Stickney, DF Barnes, P Simmons. Date palm insects in the United States. USDA Circ 846, 1950.
143. JW Sanford, WH Luckmann. Observations on the biology and control of the dusky sap beetle in Illinois. Proc North Central Br Entomol Soc Am 18:39, 1963.
144. LW Stephenson, TE Russell. The association of *Aspergillus flavus* with hemipterus and other insects infesting cotton bracts and foliage. Phytopathology 64:1502, 1974.
145. DI Fennell, WF Kwolek, EB Lillehoj, et al. *Aspergillus flavus* presence in silks and insects from developing and mature corn ears. Cereal Chem 54:770, 1977.
146. CE Windels, MB Windels, T Kommendahl. Association of *Fusarium* species with picnic beetles on corn ears. Phytopathology 66:328, 1976.

147. WA Attwater. The role of *Glischrochilus quadrisignatus* (Say) in the epidemiology of gibberella corn ear rot. M.S. Thesis, Univ. Guelph, Ontario, 1982, p 67.
148. WA Attwater, LV Busch. Role of the sap beetle *Glischrochilus quadrisignatus* in the epidemiology of gibberella corn ear rot. Can J Plant Pathol 5:158, 1983.
149. JG Rodriguez, CG Patterson, MF Potts, CG Poneleit, RL Beine. Role of selected arthropods in the contamination of corn by *Aspergillus flavus* as measured by aflatoxin production. In: UL Diener, RL Asquit, JW Dickens, eds. Aflatoxin and *Aspergillus flavus* in Corn. Auburn, AL: Auburn University, 1983, pp 23–26.
150. DT Wicklow. Unpublished data. Cited in PF Dowd, RJ Bartelt, DT Wicklow. Novel insect trap useful in capturing sap beetles (Coleoptera: Nitidulidae) and other flying insects. J Econ Entomol 85:772, 1992.
151. G Hoyos, DC McGee, LH Tiffany. Waste corn as an inoculum source for *Aspergillus flavus*. Phytopathology 82:1134, 1992.
152. DC McGee, M. Olanya, LH Tiffany. Waste corn as an inoculum source for *Aspergillus flavus*. Proceedings of the ARS Aflatoxin Workshop, Fresno, CA, 1992, p 33.
153. DC McGee, OM Olanya, LH Tiffany. Sources of *Aspergillus*, the cause of aflatoxin, in the corn ecosystem. Seed Trade News 114:6, 1993.
154. FE Vega, PF Dowd, RJ Bartelt. Dissemination of microbial agents using an auto-inoculating device and several insects species as vectors. Biol Control 5:545, 1995.
155. LJ Ashworth, RE Rice, JL McMeans, CM Brown. The relationship of insects to infection of cotton bolls by *Aspergillus flavus*. Phytopathology 61:488, 1971.
156. EB Lillehoj, WF Kwolek, RE Peterson, OL Shotwell, CW Hesseltine. Aflatoxin contamination, fluorescence, and insect damage in corn infected with *Aspergillus flavus* before harvest. Cereal Chem 53:505, 1976.
157. OL Shotwell, ML Goulden, EB Lillehoj, WF Kwolek, CW Hesseltine. Aflatoxin occurrence in 1973 corn at harvest. III. Aflatoxin distribution in contaminated, insect-damaged corn. Cereal Chem 54:620, 1977.
158. CW Hesseltine, OL Shotwell, WF Kwolek, EB Lillehoj, WK Jackson, RJ Bothast. Aflatoxin occurrence in 1973 corn at harvest. II. Mycological studies. Mycologia 68:341, 1976.
159. JC LaPrade, A Manwiller. Relation of insect damage, vector, and hybrid reaction to aflatoxin B_1 recovery from field corn. Phytopathology 67:544, 1977.
160. D Barry, MS Zuber, EB Lillehoj, et al. Evaluation of two arthropod vectors as inoculators of developing maize ears with *Aspergillus flavus*. Environ Entomol 14: 634, 1985.
161. WW McMillian, NW Widstrom, DM Wilson, RA Hill. Transmission by maize weevils of *Aspergillus flavus* and its survival on selected corn hybrids. J Econ Entomol 73:793, 1980.
162. TAP Hamsa, JC Ayres. Factors affecting aflatoxin contamination of cottonseed. I. Contamination of cottonseed with *Aspergillus flavus*, at harvest and during storage. J Am Oil Chem Soc 54:219, 1977.
163. ML Schroeder, JP Snow. Passage of boll rot fungi through alimentary canal of cotton boll weevil. Plant Dis 66:1049, 1982.
164. ML Schroeder, JP Snow. Cotton boll rot fungi associated with *Anthonomus grandis*. Phytopathology 66:570, 1982.

165. T Isakeit. The relationship of insect injury to aflatoxin contamination of cottonseed in south Texas. Proceedings of the 1996 Aflatoxin Workshop, Fresno, CA, 1996.
166. FE Escher, PE Koehler, JC Ayres. A study on aflatoxin and mold contaminations in improved variety pecans. J Food Sci 39:1127, 1974.
167. PE Koehler, RT Hanlin, L. Beraha. Production of aflatoxin B_1 and G_1 by *Aspergillus flavus* and *Aspergillus parasiticus* isolated from market pecans. Appl Microbiol 30: 581, 1975.
168. JM Wells, JA Payne. Toxigenic species of *Penicillium*, *Fusarium*, and *Aspergillus* from weevil-damaged pecans. Can J Microbiol 22:281, 1976.
169. JM Wells, RJ Cole. Production of penitrem A and of an unidentified toxin by *Penicillium lanoso-coeruleum* isolated from weevil-damaged pecans. Phytopathology 67:779, 1977.
170. LB Reed. Insects and diseases of vegetables in the home garden. USDA Home Garden Bull 46, 1968.
171. DM Porter, JC Smith. Fungal colonization of peanut fruit as related to southern corn rootworm injury. Phytopathology 64:249, 1974.
172. LT Palmer, T Kommendahl. Root-infecting *Fusarium* species in relation to rootworm infestations in corn. Phytopathology 59:1613, 1969.
173. JE Funderburk, RL Brandenburg. Management of insects and other arthropods in peanut. In: HA Melouk, FM Shakes, eds. Peanut Health and Management. St. Paul, MN: American Phytopathological Society, 1995, p 51.
174. D Atanasoff. Ergot of grains and grasses. US Dept Agric Bur Plant Indus, 1920, p 120.
175. FL Stevens, JG Hall. Three interesting species of Claviceps. Bot Gaz 50:460, 1910.
176. FF Dicke. The role of insects in some diseases of maize. Iowa State J Res 49:553, 1975.
177. JPF Sellschop, NPJ Kerid, JCG Dupreez. Distribution and degree of occurrence of alfatoxin in groundnut products. Symposium of Mycotoxins in Foodstuffs: Agricultural Aspects, Pretoria, South Africa, 1965, p 9.
178. RE Lynch, DM Wilson, IO Dicko, AP Ouedraogo. Aflatoxin contamination of peanut: interaction of environmental variables, soil moisture, and soil insects. Proceedings of the ARS Aflatoxin Workshop, Fresno, CA, 1992, p 5.
179. AE Davey, RE Smith. The epidemiology of fig spoilage. Hilgardia 7:523, 1933.
180. JL Aucamp. The role of mite vectors in the development of aflatoxin in groundnuts. J Stored Prod Res 5:245–249, 1969.
181. PD Caldis. Etiology and transmission of endosepsis (internal rot) of the fruit of the fig. Hilgardia 2:287, 1927.
182. RE Smith, HN Hansen. The improvement of quality of figs. Calif Agric Exp Sta Circ 331, 1927.
183. KV Subbarao, TJ Michailides. Virulence of *Fusarium* species causing fig endosepsis in cultivated and wild caprifigs. Phytopathology 83:527, 1993.
184. DW Ragsdale, AD Larson, LD Newsom. Microorganisms associated with feeding and from various organs of *Nezara viridula*. J Econ Entomol 72:725, 1979.
185. JA Payne, JM Wells. Toxic penicillia isolated from lesions of kernel-spotted pecans. Environ Entomol 13:1609, 1984.
186. TJ Michailides, MA Doster. Possible involvement of hemipterans in contamination

of pistachio nuts with *Aspergillus flavus*. California Pistachio Industry Annual Report Crop Year 1992–1993:98, 1993.
187. JJ Farrar, RM Davis. Relationships among ear morphology, western flower thrips, and *Fusarium* ear rot of corn. Phytopathology 81:661, 1991.
188. RM Davis, FR Kegel, WM Sills, JJ Farrar. Fusarium ear rot of corn. Calif Agric 43:4, 1989.
189. CY Warfield, RM Davis. Importance of the husk covering on the susceptibility of corn hybrids to *Fusarium* ear rot. Plant Dis 80:208, 1996.
190. L Mercier. Sur le rôle des insectes comme agents de propagation de l'"ergot" des graminés. C R Soc Biol (Paris) 70:300, 1911.
191. MC Shurtleff, BJ Jacobson, DG White. Corn ear and kernel rots. Rep Plant Dis 205: 1, 1979.
192. JR Raulston, SD Pair, FAP Martines, J Westbrook, AN Sparks, VMS Valdez. Ecological studies indicating the migration of *Heliothis zea*, *Spodoptera frugiperda*, and *Heliothis virescens* from northeastern Mexico and Texas. In: W Danthanaryana, ed. Insect Flight: Dispersal and Migration. Berlin: Springer Verlag, 1986, p 204.
193. PF Dowd. Sap beetles and their control. Proceedings of the ARS Aflatoxin Workshop, St. Louis, MO, 1990, p 5.
194. PF Dowd. An integrated approach for controlling sap beetles and mycotoxins in corn. Proceedings of the ARS Aflatoxin Workshop, Fresno, CA, 1992, p 12.
195. PF Dowd, FE Vega. Sap beetle IPM for mycotoxin control in corn: 1993 results. Proceedings of the ARS Aflatoxin Workshop, Little Rock, AR, 1993, p 70.
196. Mycogen Plant Sciences. NatureGuard insect resistant plants. Mycogen Plant Sciences, 1996, p 12.
197. DG James, RJ Bartelt, RL Faulder, A Taylor. Attraction of Australian *Carpophilus* spp. (Coleoptera: Nitidulidae) to synthetic pheromones and fermenting bread dough. J Aust Entomol Soc 32:339, 1993.
198. PF Dowd. Responses of *Carpophilus hemipterus* (L.) larvae and adults to selected secondary metabolites of maize. Entomol Exp Appl 54:29, 1990.
199. PF Dowd, RA Norton. Mechanisms of resistance to corn insects. Proceedings of the 35th Annual Corn Dry Milling Conference, Peoria, IL, 1994, p 8.
200. MD McMullen, KD Simcox. Genomic organization of disease and insect resistance genes in maize. Mol Plant Micro Int 8:811, 1995.
201. DM Wilson. Management of mycotoxins in peanut. In: HA Melouk, FM Shokes, eds. Peanut Health Management. St. Paul, MN: American Phytopathological Society, 1995, p 87.
202. LA Rodríguez-del-Bosque, CA Reyes Méndez, S Acosta Núñez. Aflatoxin in corn in northeastern Mexico: effect on environment and cultural practices. Proceedings of the ARS Aflatoxin Workshop, Fresno, CA, 1992, p 23.
203. Comite Estatal de Modernizacion del Campo del Agua (CEMCA). Paquete technologico integral para la siembra del maiz in en riego zona norte y centro de Tamaulipas, Secretaría Agricultura y Recursos Hidráulicos, 1993.
204. LA Rodríguez-del-Bosque, CA Reyes Méndez, S. Acosta Núñez, JR Girón Calderón, I Garza Cano, R García Villanueva. Control de aflatoxinas en maiz en Tampaulipas, Foll Téc Núm 17, 1995.

205. LA Rodríguez-del-Bosque. Impact of agronomic factors on aflatoxin contamination in preharvest field corn in northeastern Mexico. Plant Dis 80:988, 1996.
206. NBC Ahouissoussi, ME Wetzstein, PA Duffy. Economic returns of the boll weevil eradication program. J Agric Appl Econ 25:46, 1993.
207. Rand McNally. Classroom Atlas. New York: Rand McNally, 1995.
208. FE Blair. The Weather Almanac. Detroit: Gale Research, Inc., 1992.
209. USDA. Agricultural Statistics—1992, 1992.
210. USDA. Agricultural Statistics—1994, 1994.
211. Corn Refiners Association. 1995 Corn Annual. Washington: Corn Refiners Association, 1995.

11

Recent Advances in Preharvest Prevention of Mycotoxin Contamination

Robert L. Brown, Deepak Bhatnagar, Thomas E. Cleveland, and Jeffrey W. Cary
Southern Regional Research Center, United States Department of Agriculture, Agricultural Research Service, New Orleans, Louisiana

I. INTRODUCTION

Mycotoxins, fungal metabolites exhibiting toxic effects in higher organisms, are produced by over 100 fungal species [1]. Approximately 25% of the world's food crops are affected each year by mycotoxins [2]. Because there is a relatively high intake of cereal and oilseed crops and the products derived from them in the diet of farmed animals, such as poultry, pigs, and cattle, mycotoxin-contaminated feed poses serious safety implications for crop, livestock, and poultry producers; grain handlers; food and feed processors; consumers worldwide; and eventually national economies [2]. A list of mycotoxins significantly impacting agricultural commodities would include aflatoxins produced by *Aspergillus flavus* and *A. parasiticus*; zearalenone and the trichothecenes (in particular deoxynivalenol) produced by *Fusarium* spp.; ochratoxins produced by *A. ochraceus* and *Penicillium viridicatum*; and fumonisins produced by *Fusarium moniliforme*. Also of significance is cyclopiazonic acid (CPA), produced by *A. flavus* and possibly responsible for many cases of unexplained mycotoxicoses. More than 50 countries have established or proposed regulations for controlling aflatoxins in foods and feeds, and at least 15 have regulations for levels of other mycotoxins [3]. The U.S. Food and Drug Administration (FDA) has limits of 20 ppb total aflatoxins on interstate commerce of food and feed, and 0.5 ppb of aflatoxin B_1 on the sale of milk. Presently, FDA has an advisory level for deoxynivalenol in wheat and is considering whether regulations for fumonisins in corn are required.

The economic effects of mycotoxin contamination of food and feed can be staggering. The cost of aflatoxin contamination in corn during 1980 to all of the southeastern states was estimated at > $237 million [4]. New York Corn Growers Association representatives have conservatively estimated a $12 million loss by corn producers in New York in 1990, due to discount pricing caused by low to moderate contamination with deoxynivalenol and zearalenone [4]. The economic implications of the mycotoxin problem and its potential health threat to humans have clearly created a need to eliminate or at least control mycotoxin contamination of food and feed.

II. PREHARVEST CONTROL OF MYCOTOXINS OTHER THAN AFLATOXINS

While an association between mycotoxin contamination and inadequate storage conditions has long been recognized, studies have revealed that seeds are contaminated with mycotoxins at the preharvest stage [5]. Studies of preharvest control of mycotoxins other than aflatoxin are, however, relatively in their infancy. For example, there is a need to gain a clearer understanding of the *Fusarium*-corn interaction in asymptomatic, field-grown corn kernels, to devise strategies for controlling fusaria toxins [6]. This need is certainly underscored by previous studies on fescue toxicity where the grass was shown to be infected by an endophytic, toxic fungus which did, however, provide benefits to plant growth and development [6]. Recently, a bacterium, *Enterobacter cloacae*, was discovered as an endophytic symbiont of corn [7], and corn plants with roots endophytically infected by *E. cloacae* were observed to be fungus-free [7]. In vitro control of *Fusarium moniliforme* and other fungi with this bacterium was also demonstrated [7].

Control of fumonisins in corn, however, might be complicated by the existence of different mating populations belonging to *Fusarium* section *Liseola* [8]. Of the six designated mating types, probably representing different biological species, four—A, D, E, and F—are commonly found in asymptomatic corn plants [8]. Under laboratory conditions, members of the A mating population produce an average of 1786 ppm of fumonisin B_1; members of population D averaged 636 ppm; the E population averaged 33 ppm; and the F population 7.5 ppm [8]. The ability of other related *Fusarium* species to produce fumonisin B_1 has also been demonstrated [9]. In the United States, strains belonging to the A mating population are ubiquitous in agricultural fields, and it was discovered that 80% to 90% of *Fusarium* isolates from corn seed were members of this population [10]. This creates the potential for high fumonisin B_1 production and an outbreak of fumonisin B_1-related diseases [11]. Fumonisins appear to play a role in fungal virulence on corn seedlings; however, they are not necessary or sufficient for virulence [12]. Another study has shown that hyphae of *F moniliforme* are confined to the pedicel of a corn kernel and thus, research objectives have been outlined for the introduction, through genetic engineering, of antifungal proteins into the pedicel region of the kernel [13].

Investigations of fumonisin biosynthesis in *F. moniliforme* are currently underway, and a gene cluster has been discovered on chromosome 1 in *F. moniliforme* [14]. The pathway for the biosynthesis of trichothecenes, another *Fusarium* toxin which can contaminate corn and wheat, is also being characterized using biochemical and molecular genetic analysis. It was shown that *Tri6* is a gene that encodes a unique zinc finger protein involved in trichothecene biosynthesis regulation in *F. sporotrichioides* [15]. Also, *Tri3*, a gene encoding 15-*O*-acetyltransferase from *F. sporotrichioides* has been isolated and characterized [16]. Biochemical and molecular analysis has demonstrated that trichothecene biosynthetic genes are also clustered [17,18]. This may facilitate efforts to isolate pathway genes for molecular gene isolation strategies. A significant discovery is that trichothecene production by *F. graminearum*, which causes head blight (scab) of wheat, contributes to fungal ability to cause head blight [19]. This suggests that trichothecene resistance in wheat may provide a way to control contamination of this mycotoxin and thus the resulting head blight problem [19].

III. PREHARVEST CONTROL OF AFLATOXIN CONTAMINATION

Aflatoxins are the most widely investigated of all the mycotoxins, due to their key role in establishing the significance of mycotoxins in inducing animal diseases [20]; these compounds are potent toxins and carcinogens [21]. As a result, the most significant research progress in controlling mycotoxins has been made with aflatoxins. Recognition of the need to control contamination of food and feed grains with aflatoxins, their potential health hazards, and their strict regulation [22], has elicited various approaches from researchers working with aflatoxin-susceptible crops such as corn, cottonseed, peanuts, and tree nuts. Postharvest elimination of aflatoxin contamination has been achieved through decontamination of toxin-containing products by various physical, chemical, or biological means, including physical or chemical inactivation of aflatoxins [3]. Of these possibilities, the most promising technique has been ammoniation [3]. However, in the United States, the FDA does not permit ammoniation as a method for reducing aflatoxin levels in feedstuffs, although it is used domestically in a few states [3]. Preharvest prevention of aflatoxin contamination is probably the best and most widely explored strategy, since *A. flavus* infects all the affected crops prior to harvest [23]. The preeminence of this strategy is also due to several recent advances in aflatoxin elimination research, such as the identification of resistant genotypes in corn, peanut, and tree nuts through plant breeding. Advances in preharvest control are occurring rapidly in aflatoxins, relative to other mycotoxins.

A. Developing Host Resistance Through Plant Breeding Strategies

1. Corn

Host resistance may present a promising strategy for the preharvest elimination of aflatoxins in corn. Until recently, most efforts aimed at the identification of resistant corn genotypes were successful only to a limited extent [24]. However, a corn breeding population, GT-MAS:gk, with resistance to aflatoxin production by *A. flavus* was identified during extensive field testing [25]. This population was derived from visibly classified, segregating kernels obtained from a single fungus-infected hybrid ear [25]. A laboratory kernel screening assay (KSA) was developed and used to study resistance to aflatoxin production in GT-MAS:gk kernels [26]. Results from these experiments confirmed the resistance demonstrated in the field and showed that pin-wounded (through the pericarp to the endosperm) GT-MAS:gk kernels maintained resistance as well (Table 1) [26]. The kernel screening assay (KSA) has several advantages to complement traditional breeding techniques [27]: (1) it can be performed and repeated several times throughout the year outside of the growing season; (2) it requires few kernels; (3) it can detect/identify different kernel resistance mechanisms expressed; (4) it can dispute or confirm field evaluations (e.g., identify escapes); and (5) relationships between laboratory findings and inoculations in the field have been demonstrated.

TABLE 1 Aflatoxin Production in MAS Kernels Pinwounded Through the Pericarp to the Endosperm, Inoculated with *A. flavus* and Tested Using the KSA

Entry	Treatment	Aflatoxin B_1 (ng/g)[a]
MAS	Pinwounded	311*
	Nonwounded	400*
Control	Pinwounded	5253†
	Nonwounded	5438†

Source: From Ref. 26.

[a]Values followed by the same symbol are not significantly different by the least significant difference test.

Further examination of this resistant breeding population of corn using the KSA has demonstrated the role of pericarp waxes in kernel resistance [28,29], and has supported the existence of two types of resistance in this population—at the pericarp level as well as at the subpericarp level. Quantitative differences between GT-MAS:gk kernel wax and that of susceptible genotypes have been demonstrated; qualitative differences appear to exist as well [30].

Other natural sources of corn resistance have also been discovered. In Illinois field trials, 27 corn inbreds were identified as potentially resistant to aflatoxin production by *A. flavus* [31]. When tested with the KSA, these genotypes had the same ranking on the basis of aflatoxin accumulation as in the field trials [27]. Further examination of resistant inbreds CI2, T115, and MI82, using both the KSA and an *A. flavus* transformant expressing *Escherichia coli* β-glucuronidase (a GUS reporter gene linked to an *A. flavus* β-tubulin gene promoter), demonstrated visually, both nonwounded and wounded-kernel resistance to fungal infection, and demonstrated a positive relationship between the degree of fungal infection and aflatoxin levels [27]. Lower aflatoxin levels and lower GUS in resistant genotypes, relative to susceptible controls, were also shown when GUS was quantified using a fluorogenic technique [32]. The "GUS tester strain" is another tool that may assist plant breeders in the development of resistant germplasm by facilitating marker-assisted breeding.

In studies with the major objective of improving elite midwestern corn lines such as B73 and Mo17, several resistant inbreds, previously tested in Illinois field trials and with the KSA, were studied using a generation mean analysis mating design to determine the inheritance of resistance in crosses with B73 and/or Mo17 [25]. In the cases of several highly resistant inbreds—Tex6, LB31, CI2, and Oh513—genetic dominance was indicated. This study also indicated that selection for resistance to Aspergillus ear rot and aflatoxin production should be effective. This conclusion was based on the following results: (1) significant genetic additive and dominance effects in the generation mean analysis; (2) moderate to high heritabilities; and (3) low estimates of effective factors. Also the frequency distribution of ear rot ratings and aflatoxin production of F3 families and of backcrosses to the susceptible self families of Mo17 × Tex 6 and B73 × LB31 point to the possibility of success in the development of resistant inbreds for use in commercial hybrids [33].

Chromosome regions associated with resistance to *A. flavus* and inhibition of aflatoxin production in corn have been identified through RFLP analysis in three "Illinois-resistant" lines (R001, Lb31, and Tex6), after mapping populations were developed using B73 and/or Mo17 elite inbreds [34]. In some cases, chromosomal regions were associated with Aspergillus ear rot and not aflatoxin levels, and vice versa, whereas other chromosomal regions were found to be associated with both traits. These observations suggest that these two traits may be at least partially under separate genetic control. It was also observed that variation can exist in the chromosomal regions associated with Aspergillus ear rot and aflatoxin inhibition in different mapping populations. This suggests that there may be different genes for resistance in the different identified resistance germplasm. RFLP technology may provide the basis for employing a successful strategy of pyramiding different types of resistances into commercially viable germplasm, while avoiding the introduction of undesirable traits.

Other sources of corn germplasm with greater tolerance for high temperatures and humidity have been successfully tested for resistance to aflatoxin production in south Texas [35]. These tropical hybrids all developed tight husks. Also, kernels were shown to resist aflatoxin contamination unless kernel pericarps were breached. These heat-tolerant tropical hybrids may offer new choices for producers operating in the southern portion of the United States, where aflatoxin contamination is an ongoing problem.

Studies employing the KSA as well as other studies [36] have demonstrated that kernel embryos are preferred as a substrate over endosperm tissue by aflatoxin-producing fungi (Table 2). Embryo viability has also been shown to be necessary for the expression of kernel resistance [26]. This suggests that resistance, especially in wounded kernels, is a function of kernel ability to limit fungal colonization to a small area after wounding, thus providing kernels with a two-pronged defense. This defense may include: (1) preventing interruption of whole-kernel expression of an embryo-based resistance mechanism; and (2) denying the fungus easy access to the oil-rich embryonic substrate [32]. Whether or not the large levels of aflatoxins found in susceptible kernels are the primary result of fungal metabolic activity on the embryonic substrate, or of later activity in the endosperm, is under investigation in our laboratory.

2. Peanut

Several sources of resistant peanut germplasm have also been identified from a core collection representing the entire peanut germplasm collection [37]. Over

TABLE 2 Aflatoxin Production in MAS Kernels Pinwounded in the Embryo Region, Inoculated with *A. flavus* and Tested Using the KSA

Entry	Treatment	Aflatoxin B_1 (ng/g)[a]
MAS	Pinwounded	33,923*
	Nonwounded	392†
Control	Pinwounded	8924‡
	Nonwounded	3320§

Source: From Ref. 26.

[a]Values followed by the same symbol are not significantly different by the least significant difference test.

95% of this core have been preliminarily screened in a single environment; 16 genotypes tested over 3 years in two environments still display low levels of aflatoxins. A possible link between low linoleic acid content in peanut and low preharvest aflatoxin production has been indicated [37]. Significant correlations have also been observed between leaf temperature and aflatoxin levels and/or visual stress ratings and aflatoxin levels. The preliminary screening of peanut genotypes using either or both of these traits could greatly reduce expenses involved in developing resistant cultivars. Promising germplasm, however, has less than acceptable agronomic characteristics, and is thus being hybridized with those with commercially acceptable features. Resistant lines are also being crossed to pool resistances to aflatoxin production. Thus, some success has been achieved in identifying resistant peanut germplasm, and field studies are being conducted by various researchers to verify this trait.

Methods to improve screening of peanuts for resistance to *A. flavus* have been developed. A system of evaluating peanuts in the field through the manipulation of drought stress was successfully tested [38]. Also, an in vitro seed culture system, demonstrating water stress responses in peanuts similar to field responses, and variations in peanut phytoalexins and aflatoxin levels appear potentially useful [39].

3. Tree Nuts

Among tree nuts, strategies for controlling preharvest aflatoxin formation by breeding for host resistance have been mostly studied in almonds [40]. The approach employed in this effort is to integrate multiple genetic mechanisms for control of *Aspergillus* spp. as well as Navel orangeworm (*Paramyelois transitella* Walk), which appears important for initial fungal infection. Resistance to fungal colonization is being pursued by incorporating seed coat resistance to infection. Genotypes demonstrating inhibition to fungal infection in seed tissues have been inconsistent over different environments [40]. Studies have also been conducted with figs and pistachios to identify the mode of infection of these crops by *A. flavus*. Once this parameter is clearly understood, strategies could be developed to identify germplasm with agronomically desirable characteristics and resistance to fungal infection [41].

B. Developing Host Resistance Through Addition or Enhancement of Antifungal Genes by Genetic Engineering

There are several requirements that must be met in order to employ genetic engineering as a means of developing host resistance against aflatoxin contamination in crops. Resistance genes, native or foreign, must be identified that express inhibitory activity against *A. flavus*. Gene promoters also must be selected which will regulate the desired type of expression of antifungal genes at a desired time in the candidate crop. Lastly, the appropriate technology for genetic transformation must be developed or adapted to each specific situation [42].

Resistance to *A. flavus* infection in plants could consist of an interaction of multiple components and biochemical changes that are either preformed or induced upon invasion by pests. Several reports indicate that stimulation by elicitors results in changes of gene expression and induction of pathogenicity-related (PR) proteins in plants [43–45]. Both constitutive and inducible metabolic factors/mechanisms can coexist at different stages of seed/kernel maturation and be induced to even higher levels during periods following fungal invasion.

1. Candidate Antifungal Compounds

Identifying resistance (e.g., in corn) makes it possible to correlate resistance with many endogenous low-molecular-weight compounds and biomacromolecules in kernel tissues already implicated as antifungal at various stages of kernel development in grain crops [46–52]. However, several of these antifungal compounds which demonstrated activity against other fungal species have proved entirely ineffective against *A. flavus*, and thus it is important to select the best candidate genes for these inhibitory compounds before plant genetic engineering procedures are initiated. A list of candidate antifungal compounds include ribosome inactivating proteins, lectins, relatively small MW polypeptides, cell-surface glycoproteins, hydrolases, and certain basic proteins [53].

A factor, possibly a protein, that inhibits aflatoxin biosynthesis without affecting fungal growth was identified in the coats of young developing cottonseed [54]. In another study, a xylan partially purified from developing cottonseed coats was also shown to inhibit aflatoxin biosynthesis [55]. An investigation of maize kernels demonstrated that resistance to aflatoxin formation can be induced in susceptible kernels by preincubating them for 3 days at 100% relative humidity, prior to inoculation with *A. flavus* [56].

A putative peptide also has been partially purified from aqueous kernel extracts of resistant inbred, Tex6, which demonstrated antifungal activity against *A. flavus* [57]. These latter findings are interesting in light of the recent identification of corn kernel pathogenesis-related (PR) proteins [44]. PR proteins appear to have a function during the normal process of seed germination [49,50]. However, they are induced to accumulate in response to fungal infection, and their expression is tissue-specific [44,58]. An investigation of kernel PR proteins using resistant and susceptible genotypes to examine specific tissue expression of these proteins under varying kernel physiology may facilitate the isolation of factors responsible for the subpericarp resistance demonstrated by some corn genotypes.

Other low-molecular-weight antimicrobial plant substances, either preformed or produced in response to pathogen invasion [59], include phytoalexins, lignins, and hydroxyproline-rich glycoproteins. A number of potentially antifungal enzymes/proteins are produced either constructively or in response to fungal attack in plants. These include chitinases and β-1,3-glucanases [52,60–62], osmotins [63], protease inhibitors [64], and polygalacturonase inhibitor proteins (PGIPs) [65]. In addition, low-MW peptides have been isolated from organisms other than plants that also show promise as antifungal agents—for example, the cecropins [66] and magainins [67] of insect and amphibian origin, respectively, and their synthetic analogs.

Several recent studies have suggested the potential of manipulating/inducing the lipoxygenase (LOX) pathway in plants to ward off fungal attack. *A. flavus* exhibits strong lipolytic activity during infection on oilseeds [68] and at times causes substantial deterioration of the crop seeds and oils they contain. Lipase activity originating from fungal degradation of host plant membrane tissues releases fatty acids from bound triglycerides and triggers the LOX-hydroperoxide lyase (HPLS) enzyme pathway converting linoleic and linolenic acids into hexanal and cis-hexenal, respectively [69]. Cis-3-hexenal is usually then isomerized to trans-2-hexenal both enzymatically and nonenzymatically [69]. Recently, it has been reported that specific LOX decay products, such as jasmonic acid and C_6–C_{10} alkenals and alkanals, may function as important signal molecules in host-pathogen interactions [39,70–72]. Jasmonic acid has also been shown to inhibit aflatoxin production and delay spore germination of *A. flavus* [73]. The antifungal properties of small chain alkanals and alkenals (derived from the LOX pathway) produced by cotton leaves have been demonstrated in solid and liquid cultures of aflatoxigenic *Aspergillus* spp. [74–76]. Because of the mode of activation of these volatile aldehydes and because of the significant antifungal activity they exhibit, it has been suggested that these compounds may function as "gaseous phytoalexins" in the cotton plant [75]. It was also demonstrated that the individual volatile C_6–C_{10} alkenals could elicit the cotton plant sesquiterpenoid naphthol phytoalexins, 2,7-dihydroxy-cadalene and 2-hydroxy-7-methoxy cadalene, their oxidation products lacinilene C and lacinilene C 7-methyl ether, and the coumarin phytoalexin scopoletin, in artificially wounded cotton bolls. Because of this eliciting activity, it was suggested that C_6–C_{10} alkenals may function also as "volatile elicitors" in developing cotton bolls [71].

There are only a few candidate genes whose expression products demonstrate convincing inhibitory activity against *A. flavus* and which show promise for transformation of plants to reduce infection of seed by this particular fungal species. Included among these antifungal products are certain small lytic peptides. It is relatively easy to chemically synthesize genes (using an oligonucleotide synthesizer) for transformation of plants that encode small peptides, since only relatively small coding regions are required for their complete synthesis. Cecropins, for example, are lytic peptides of 22–23 amino acids in linear arrays that comprise antimicrobial systems found in insects and pig intestine [66,77]. The broad antibacterial activities of the cecropins are due to the formation of large pores in the cell membrane [78,79]. They apparently do not lyse erythrocytes or other higher eucaryotic cells [80], but have been shown to inhibit growth of *A. flavus* mycelia [81].

In addition to lytic peptide genes, a variety of other candidate antifungal genes from bacterial, plant, and mammalian sources have a good potential to be active against *A. flavus* upon transformation of plants with these genes. Genes are available encoding LOX from plant sources [82,83] that are involved in production of antifungal and/or aflatoxin pathway-interfering products such as 13-hydroperoxylinoleic acid and its breakdown products/volatiles such as hexenal and hexanal. Genes are also available for possible genetic engineering of cotton that encode animal myeloperoxidase [84] and yeast chloroperoxidase [85]. In bioassays using *A. flavus* as the test organism, addition of myeloperoxidase greatly enhanced (90-fold) the lethality of H_2O_2 by catalyzing its conversion to sodium hypochlorite [86]. H_2O_2 is induced in plants by wounding or injuring of plant tissues, processes often associated with pest attack; thus, the substrate for these unique peroxidases should be available in the specific host plant tissues under attack [87].

2. Gene Promoters

Promoter elements have been identified which should allow for either constitutive, wound-inducible, or tissue-specific expression of antifungal genes in plants. Characterized promoter elements that should be useful in obtaining optimum expression of antifungal genes in plants include the CaMV 35S [88] and ubiquitin-3 promoter elements (constitutive) [89], the protease inhibitor II promoter (wound-inducible) [90], and storage protein gene promoters (seed-specific expression) [91]. Peanut has successfully been transformed with a wound-inducible promoter from soybean vegetative storage protein (*vsp*) [92]. When *vsp*-promoter-GUS gene fusion is inserted into peanut, expression of the gene follows temporal and spatial patterns as would be predicted from soybean [92]. The *E. coli* β-glucuronidase (GUS) reporter gene has been used to assess the level of gene expression obtained under the control of some of the above promoters in transformed cotton [42].

3. Transformation Methods

Antifungal genes housed in suitable gene expression vectors have been used in the transformation of plants by a variety of methods. The two most common means include *Agrobacterium*-mediated gene transfer [93,94] and biolistic particle delivery or "gene gun" technology [42,95]. After gene transfer, transformed tissues are identified by growth on selective medium, and whole plants are regenerated from the selected, transformed cells. Cotton transformation has been accomplished using the *Agrobacterium*-based system and cotton hypocotyl sections, although the subsequent regeneration procedure in this system is not necessarily straightforward and can be lengthy.

Cotton regeneration from transformed hypocotyl tissue involves the development of transformed embryogenic cell lines, embryoid formation, dissection, desiccation of embryos, and subsequent germination of the embryos [96,97]. Coker cultivars 201 and 312 of cotton have been transformed with *Agrobacterium*-mediated systems and regenerated [98,99]. Transformation of other commercially important cotton varieties has proven difficult due to the inability to generate embryogenic cell lines. However, in a recent report, *Agrobacterium*-mediated transformation was successfully employed on elite Acala and Coker cultivars [100]. Nevertheless, problems remain regarding the efficiency of this method of transformation and its adaptability to a wide range of germplasm. To circumvent the problem of cultivar-dependent regeneration, investigators have used the biolistic approach to transform cotton [42,101,102]. Peanut is also currently being transformed with antifungal genes by microprojectile bombardment of embryogenic tissues [103]. Walnut has been successfully transformed with barley lectin and nettle lectin antifungal genes [104]. The somatic embryo is the targeted tissue in walnut, and a new technique for cryopreservation of walnut somatic embryos is being used for long-term storage of embryo lines and to prevent somaclonal variation and loss of lines to contamination [104].

C. Use of Biocompetitive Agents

Micro-organisms have often been suggested as agents of control for aflatoxin contamination [105,106]. The best biocompetitive agent to control *A. flavus* in the field would be atoxigenic strains of *A. flavus*, because these strains, as compared to other potential microbial biocompetitive agents, would be adaptable to environmental conditions identical to the toxigenic strains and would be biologically active at the same time as well [107]. It is also important to note that *A. flavus* does not require aflatoxins for infecting the crops, and there is no relationship between the production of high levels of aflatoxins and strain virulence [108].

1. Greenhouse and Field Trial Results

In greenhouse and field experiments where developing cotton bolls or developing corn ears were wounded and then inoculated with different combinations of toxigenic and atoxigenic strains, the atoxigenic strains reduced preharvest aflatoxin contamination by 80% to 90% [109,110]. Very large and significant reductions were also obtained when atoxigenic *A. flavus* strains colonizing wheat seed were applied to cotton rows in a cotton-growing region with a high incidence of aflatoxin contamination [111]. This reduction in aflatoxin contamination by atoxigenic strains was by spatial exclusion of toxigenic strains and also by competition for resources required for aflatoxin production [112]. The ability of atoxigenic strains to interfere with contamination in coinfecting bolls may have practical value, since individual plant parts are often coinfected by several *A. flavus* strains, and cotton bolls damaged by pink bollworms are infected by *A. flavus* strains in multiple vegetative compatibility groups (VCG), at least 50% to 80% of the time [113]. It is also important to note that pink bollworm-damaged bolls contain the majority of aflatoxins in commercial fields [114].

Peanut, as well as cotton, may be the beneficiary of reduced aflatoxin contamination, through the use of atoxigenic strains as biocompetitive agents [115]. When rice infected with atoxigenic color mutants of *A. parasiticus* and *A. flavus* was applied to peanut plots 3 weeks after planting, significant reductions in aflatoxin levels were observed for 1994 and 1995 crop-years [116]. There was good control of aflatoxin production in peanuts at all inoculum levels, but the highest inoculum level demonstrated the best control.

2. Strain Application and Population Dynamics

The application of atoxigenic strains to agricultural fields early in crop development may be an important element in the successful implementation of this biocompetitive strategy. In theory, it may permit these strains to compete with other resident strains for crop-associated resources [117], and may facilitate an increase in population size among atoxigenic and toxigenic strains when environmental conditions favor aflatoxin production [117–119]. At the same time, atoxigenic strains may compete for infection sites. When an atoxigenic strain was seeded on colonized wheat seed [118] into a field of developing cotton, prior to crop flowering [120], there were large and significant reductions in the aflatoxin content of the crop at maturity, and the aflatoxin content was inversely proportional to the incidence of the seeded (atoxigenic) VCG [120]. Similar tests performed in later years also demonstrate that atoxigenic strains can exclude toxigenic strains and subsequent aflatoxin contamination when applied early in crop development [119].

It was worth mentioning that early strain application is not associated with increased crop infection or increased *A. flavus* populations on the crop at maturity [120]. Using atoxigenic strains as biocompetitive agents also will not limit the amount of crop infection by the *A. flavus* group or the quantity of these fungi associated with the crop. Infection appears to depend more on the predisposition of the host and the environment than on the number of *A. flavus* propagules. The magnitude of the *A. flavus* fungi in the field is primarily dependent on the available resources and the environment. Thus, *A. flavus* populations can rapidly develop under drought conditions, even in areas with normally low aflatoxin contamination [121]. Properly timed applications of atoxigenic strains, however, might partially control the composition of these rapidly increasing populations.

A. flavus usually becomes associated with crops in the field during crop development and remains such during harvest, storage, and processing. Thus, applying atoxigenic strains into agricultural fields prior to harvest may provide postharvest protection from contamination as well. Atoxigenic strains applied both prior to harvest and after harvest have been shown to provide protection from aflatoxin contamination of corn [110], even when toxigenic strains are associated with the crop prior to application.

3. Genetic Engineering of Biocompetitive Strains

Biocompetitive strains used in the studies described above were native atoxigenic strains. However, the potential exists for the development of more effective biocontrol strains through genetic engineering. There are several fungal virulence factors that determine the ability of an *A. flavus* strain to infect and spread through host tissues. Engineered strains could also be constructed with (1) supplemented aggressiveness traits and the ability to optimize infection site occupation and competitiveness, while minimizing host tissue disruption, and (2) a deleted ability to produce aflatoxins by genetic engineering of the identified gene cluster responsible for aflatoxin production. Of importance to the property of enhanced aggressiveness is that *A. flavus* cell wall degrading enzymes, such as pectinases, have been identified and shown to be associated with aggressiveness and infection of cotton bolls [122,123]. These discoveries could be important in efforts to genetically regulate aggressiveness traits in the fungus to produce more effective biocompetitive agents.

With regard to the property of deleted aflatoxin-producing ability, several pathway genes have been disrupted to create nonproducing strains [124,125], and more recently, deletion mutants have been developed with more than one gene deleted [126,127]. These deletions can be performed in highly aggressive strains as well to ensure a well-characterized, specific, atoxigenic strain.

D. Targeting Gene Cluster Governing Aflatoxin Biosynthesis

Elimination of preharvest contamination through the development of novel biotechnological approaches [128,129] could be significantly benefited by additional knowledge about the fundamental molecular and biological mechanisms that regulate the biosynthesis of aflatoxin by the fungus.

Previous studies have determined that aflatoxins are synthesized by the polyketide metabolic pathway and are considered secondary metabolites because these compounds have no known function in the fungi that produce them [130]. The generally accepted scheme for aflatoxin biosynthesis is: polyketide precursor $\rightarrow$ norsolorinic acid, NOR $\rightarrow$ averantin, AVN $\rightarrow$ hydroxyaverantin, HAVN $\rightarrow$

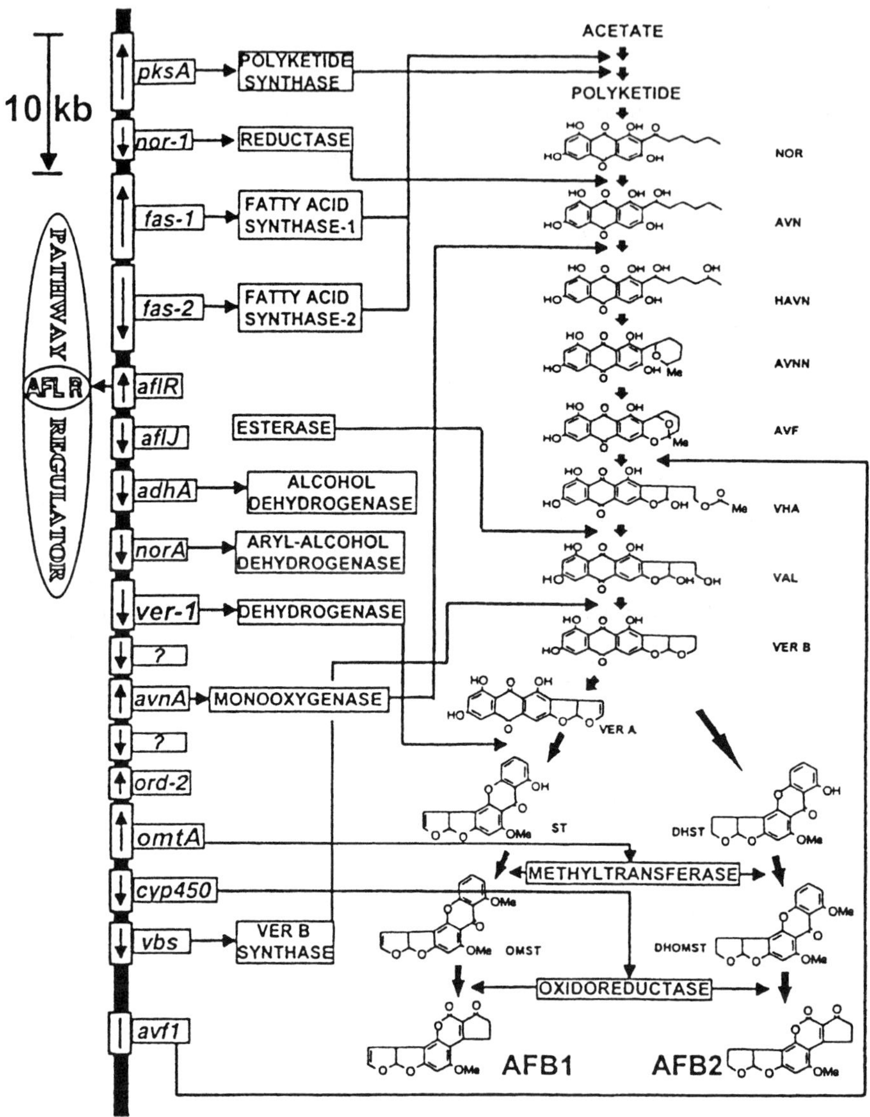
10 kb
pksA
POLYKETIDE SYNTHASE
nor-1
REDUCTASE
fas-1
FATTY ACID SYNTHASE-1
fas-2
FATTY ACID SYNTHASE-2
PATHWAY REGULATOR
AFL R
aflR
aflJ
ESTERASE
adhA
ALCOHOL DEHYDROGENASE
norA
ARYL-ALCOHOL DEHYDROGENASE
ver-1
DEHYDROGENASE
?
avnA
MONOOXYGENASE
?
ord-2
omtA
cyp450
vbs
VER B SYNTHASE
avf1
ACETATE
POLYKETIDE
NOR
AVN
HAVN
AVNN
AVF
VHA
VAL
VER B
VER A
ST
DHST
METHYLTRANSFERASE
OMST
DHOMST
OXIDOREDUCTASE
AFB1
AFB2

averufanin, AVNN → averufin, AVF → hydroxyversicolorone, HVN → versiconal hemiacetal acetate, VHA → versicolorin B, VER B → versicolorin A, VER A → demethylsterigmatocystin, DMST → sterigmatocystin,ST → O-methylsterigmatocystin, OMST → aflatoxin B_1, AFB_1 (Fig. 1). A branch point in the pathway has been established from VER B leading to different aflatoxin structural forms B_1 and B_2 [131–136].

In spite of early difficulties in identifying enzyme activities in cell-free extracts [137–140], several specific enzyme activities have been associated with precursor conversions in the aflatoxin pathway [130,132,133,136–144]; some of

---

FIGURE 1 Summary of the cluster of aflatoxin pathway genes, corresponding biosynthetic enzymes, and precursor intermediates involved in the aflatoxin B_1 and B_2 synthesis. The generally accepted aflatoxin B_1 and B_2 biosynthetic pathway in *A. parasiticus* and *A. flavus*, the identified enzymes for some specific conversion steps and cloned genes, are schematically presented. The regulatory gene, *aflR* coding for the pathway regulatory factor (AFLR protein) controls the expression of the structural genes at the transcriptional level. The *fas-1, fas-2* and *pksA* gene products, fatty-acid synthase and polyketide synthase, respectively, are involved in the conversion steps between the initial acetate unit to the synthesis of the decaketide backbone in aflatoxin synthesis. The *nor-1* gene encodes a reductase for the conversion of NOR to AVN. The *avnA* gene encodes a P450 monooxygenase for the conversion of AVN to HAVN. The *adhA* (homology to an alcohol dehydrogenase), *norA* (homology to an aryl-alcohol dehydrogenase), *ver-1* (encoding a dehydrogenase), *ord-2, cyp450*, and *avf1* gene products have been demonstrated to be functioning at various stages of the pathway, but their exact enzymatic role has not been fully characterized and is under investigation. The *omtA* gene encodes and *O*-methyltransferase for the conversion of ST to OMST and DHST to DHOMST. The *vbs* gene encodes a Ver B synthase (cyclase), which has been reported to be involved in the conversion of VHA to Ver B. The oxidoreductase and esterase have been characterized to be involved in the aflatoxin biosynthetic pathway; however, their corresponding genes have not been confirmed. The vertical bar on the left represents at least a 75-kb aflatoxin pathway gene cluster with identified genes shown in the open boxes. The names of the individual genes are labeled next to the open boxes, and the unnamed transcripts are labeled by question marks. Arrows inside the open boxes indicate the direction of transcription. Arrows indicate the relationships from the genes to the enzymes they encode; from the enzymes to the bioconversion steps they are involved in; and from the intermediates to products in the aflatoxin bioconversion steps. Abbreviations: NOR, norsolorinic acid; AVN, averantin; HAVN, 5′-hydroxyaverantin; AVNN, averufanin; AVF, averufin; VHA, versiconal hemiacetal acetate; VAL, versiconal; Ver B, versicolorin B; Ver A, versicolorin A; ST, sterigmatocystin; DHST, dihydrosterigmatocystin; OMST, *O*-methylsterigmatocystin; DHOMST, dihydro-*O*-methylsterigmatocystin; AFB_1, aflatoxin B_1; AFB_2, aflatoxin B_2. In addition, 1-hydroxyversicolorone is an intermediate between AVF, and VHA and demethylsterigmatocystin is an intermediate between Ver A and ST.

these activities have been partially purified [133,143,144]. Bhatnagar et al. [145] and Keller et al. [146] have purified two distinct methyltransferases (168 kDa and 40 kDa); both catalyze ST → OMST conversion. Yabe et al. [144] identified two distinct methyltransferase activities in cell-free extracts of *A. parasiticus* strain NRRL 2999 which migrated with the 180-kDa and 210-kDa fractions on a gel filtration column; the former activity (180 kDa) could correspond to the protein (168 kDa) purified by Bhatnagar et al. [145] because it catalyzed ST to OMST conversion, whereas the 210-kDa fraction methylated only demethylsterigmatocystin (DMST) to yield ST. In addition, a 43-kDa reductase that catalyzes the reduction of NOR to AVN has been purified [147]. It has been postulated [130] that alternative pathways may exist at several steps in the aflatoxin pathway, hence, different enzymes with similar catalytic functions may be isolated from pertinent fungal cells. A cyclase [148,149] and an esterase [150] have also been purified from *A. parasiticus* which are involved in aflatoxin biosynthesis; and Matsushima et al. [151] have purified and characterized two versiconal hemiacetal acetate reductases involved in toxin synthesis.

It has been demonstrated that independent reactions and different chemical precursors involved in AFB_1 and AFB_2 syntheses are catalyzed by common enzyme systems [133,135,142,146,152,153]; AFB_1 precursors are the preferred substrates for the relevant enzymes [133]. A desaturase activity has been demonstrated in cell-free fungal extracts at the branch (Fig. 1) in the AFB_1/B_2 biosynthetic pathways [131,132].

Progress in genetic studies of the imperfect-fungi, *A. flavus* and *A. parasiticus*, is adversely affected by (1) a lack of a sexual stage, and (2) a lack of hyphal anastomosis and nuclear exchange between all but the most closely related strains [154]. To overcome this, mutants of *A. parasiticus* and *A. flavus* strains have been analyzed through the parasexual cycle [155,156 for review; 157]. The genetics of *A. flavus* is better understood than that of *A. parasiticus*, and over 30 genes have been mapped to eight linkage groups [155–157]. Karyotype analysis for several *A. flavus* and *A. parasiticus* strains has been used to develop genetic maps of these fungi [158,159].

Genetic transformation systems have been developed for *A. flavus* [157] and *A. parasiticus* [160,161] to identify genes in the aflatoxin biosynthetic pathway by genetic complementation of blocked mutant strains of toxigenic *A. flavus* and *A. parasiticus*. Using this methodology, individual conserved structural and regulatory genes have been cloned from both species [162–168]. Molecular cloning of genes related to aflatoxin biosynthesis has also been attempted by using differential screening with different cDNA probes and colony hybridization procedures [169]. Identification and biochemical characterization of enzymes involved in the aflatoxin biosynthetic pathway have also led to the subsequent cloning and isolating of genes coding for these enzymes; the purified proteins are used to obtain antibodies as immunoscreening probes [170,171] and amino acid sequences for oligonucleotide probes [146]. Using these tools, cDNA libraries [139,152,172] were screened to identify and clone the gene for the 40-kDa methyltransferase protein required for the ST to OMST conversion activity [173]. Other genes identified using this approach include *norA* [174] and *vbs* [148].

All the identified genes related to aflatoxin biosynthesis have been found to be located within a 75-kb DNA region in both *A. parasiticus* and *A. flavus* [148,168,175–177]. Several additional transcripts within the gene cluster have also been identified [168,176,177]. Several of these genes located on the aflatoxin pathway gene cluster have recently been characterized by gene disruption techniques [176,178–180].

A. flavus and *A. parasiticus* genes shown to be involved in aflatoxin biosynthesis are conserved in the sterigmatocystin-producing strains of *A. nidulans* [181,182].

As described by Cleveland et al. [53], the successful cloning of genes involved in aflatoxin biosynthesis will allow the use of these genes as expression "probes" to identify phytological agents that may naturally inhibit the aflatoxin biosynthetic pathway and genes. Additional knowledge is needed about the identity, tissue sites, and mechanism of synthesis of plant components that influence regulatory mechanisms of the genes governing aflatoxin synthesis during the plant-fungus interaction. Plant chemicals from crops vulnerable to aflatoxin having modulatory effects on aflatoxin biosynthesis have been reported [75,183], however, little is known about the molecular mechanisms involved in the observed modulatory effects by host plant components on aflatoxin pathway expression during the plant-fungus interaction. Plant chemicals influencing pathway expression, identified using reporter gene technology, could be used as selectable markers for improvement of resistance to aflatoxin contamination through plant breeding or genetic engineering procedures.

IV. CONCLUSION

While aflatoxin contamination may be reduced somewhat by careful cultural practices, significant control of this problem will likely be linked to the introduction of resistant germplasm, that is, antifungal and/or acting against toxin biosynthesis, into the commercial market. This is especially true for corn, which has been shown to possess a large amount of natural genetic diversity with respect to fungal infection response. Naturally resistant corn germplasm provides us with not only a source of resistance, but also nature's lesson as to the specific requirements of resistance (e.g., antifungal compounds, regulation of these compounds, and physiological conditions for bioactivity). The identification of specific biochemical factors linked to resistance against *A. flavus* may not only contribute heavily toward eliminating aflatoxin contamination, but may also contribute to the control of other fungal diseases; *A. flavus* usually responds less to antifungal compounds than do many other fungi. Breakthroughs in understanding the aflatoxin biosynthetic pathway and biocompetition, however, have made possible a multiple approach to the control of aflatoxin contamination. This information allows us to understand and meet very specific conditions to eliminate aflatoxins. Biocompetition certainly may be more feasible in crops where the lack of genetic diversity limits the possibility of developing resistant germplasm (e.g., cotton).

The aflatoxin contamination process is, however, so complex that in most cases a combination of approaches will be required to eliminate or even control the toxin problem. The recent research advances made toward understanding the preharvest aflatoxin contamination process in crops not only has certainly provided a foundation for controlling aflatoxin contamination, but may provide insights into controlling other mycotoxin problems as well.

REFERENCES

1. RP Sharma, DK Salunkhe. Introduction to mycotoxins. In: RP Sharma, DK Salunkhe, eds. Mycotoxins and Phytoalexins. Boca Raton: CRC Press, 1991, pp 3–11.
2. J Mannon, E Johnson. Fungi down on the farm. New Sci 105:12–16, 1985.
3. F Haumann. Eradicating mycotoxins in food and feeds. Inform 6:248–256, 1995.
4. LL Charmley, HL Trenholm, DB Preluskey, A Rosenberg. Economic losses and decontamination. Natural Toxins 3:199–203, 1995.
5. N Lisker, EB Lillehoj. Prevention of mycotoxin contamination (principally aflatoxins and *Fusarium* toxins) at the preharvest stage. In: JE Smith, RS Henderson, eds. Mycotoxins and Animal Foods. Boca Raton: CRC Press, 1991, pp 689–719.
6. IE Yates, CW Bacon, DM Hinton. Targeting strategies for controlling fusaria toxins in corn. Proceedings of the USDA-ARS *Fusarium*/Fumonisin Workshop, Beltsville, MD, 1995, p 23.
7. DM Hinton, CW Bacon. *Enterobacter cloacae* is an endophytic symbiont of corn. Mycopathologia 129:117–125, 1995.
8. JF Leslie, RD Plattner, AE Desjardins, CA Klittich. Fumonisin B_1 production by strains from different mating populations of *Gibberella fujikuroi* (*Fusarium* section *Liseola*). Phytopathology 82:341–345, 1992.
9. PE Nelson, RD Plattner, DD Shackelford, AE Desjardins. Fumonisin B_1 production by *Fusarium* species other than *F. moniliforme* in section *Liseola* and by some related species. Appl Environ Microbiol 58:984–989, 1992.
10. JF Leslie, FJ Doe, RD Plattner, DD Shackelford, J Jonz. Fumonisin B_1 production and vegetative compatibility of strains from *Gibberella fujikuroi* mating population 'A'. Mycopathologia 117:37–45, 1992.
11. NPJ Kriek, TS Kellerman, WFO Marasas. Comparative study of the toxicity of *Fusarium verticillioides* (=*F. moniliforme*) to horses, primates, pigs, sheep, and rats. Onderstepoort J Vet Res 48:129–131, 1981.

12. AE Desjardins, RD Plattner, TC Nelson, JF Leslie. Genetic analysis of fumonisin production and virulence of *Gibberella fujikuroi* mating population A (*Fusarium moniliforme*) on maize (*Zea mays*) seedlings. Appl Environ Microbiol 61:79–86, 1995.
13. MJ Muhitch. A genetic engineering approach to lowering fumonisin levels in maize kernels. Proceedings of the USDA-ARS *Fusarium*/Fumonisin Workshop, Beltsville, MD, 1995 p 25.
14. AE Desjardins, RD Plattner, RH Proctor. Biochemistry and genetics of fumonisin biosynthesis. Proceedings of the USDA-ARS *Fusarium*/Fumonisin Workshop, Beltsville, MD, 1995 p 27.
15. RH Proctor, TM Hohn, SP McCormick, AE Desjardins. *Tri6* encodes an unusual, zinc finger protein involved in regulation of trichothecene biosynthesis in *Fusarium sporotrichioides*. Appl Environ Microbiol 61:1923–1930, 1995.
16. SP McCormick, TM Hohn, AE Desjardins. Isolation and characterization of *Tri3*, a gene encoding 15-*O*-acetyltransferase from *Fusarium sporotrichioides*. Appl Environ Microbiol 62:353–359, 1996.
17. TM Hohn, AE Desjardins, SP McCormick, RH Proctor. Biosynthesis of trichothecenes, genetic and molecular aspects. In: M. Eklund, JL Richard, K Mise, eds. Molecular Approaches to Food Safety: Issues Involving Toxic Microorganisms. Ft. Collins, CO: Alaken, Inc., 1995, pp 239–248.
18. TM Hohn, SP McCormick, AE Desjardins. Evidence for a gene cluster involving trichothecene-pathway biosynthetic genes in *Fusarium sporotrichioides*. Curr Genet 24:291–295, 1993.
19. RH Proctor, AE Desjardins, SP McCormick, TM Hohn. Trichothecene toxins and wheat head scab. Proceedings of the USDA-ARS *Fusarium*/Fumonisin Workshop, Beltsville, MD, 1995 p 30.
20. TD Wyllie, LG Morehouse. Introduction. In: TD Wyllie, LG Morehouse, eds. Mycotoxic Fungi, Mycotoxins, Mycotoxicoses, An Encyclopedic Handbook. Vol. 3. New York: Marcel Dekker, 1978, pp vii–xiii.
21. UL Diener, RJ Cole, TH Sanders, GA Payne, LS Lee, MA Klich. Epidemiology of aflatoxin formation by *Aspergillus flavus*. Annu Rev Phytopathol 25:249–270, 1987.
22. WF Busby, GN Wogan. Food-borne mycotoxins and alimentary mycotoxicosis. In: HP Reiman, FL Bryan, eds. Food Borne Infections and Intoxications. New York: Academic Press, 1979, pp. 510–610.
23. EB Lillehoj. The aflatoxin in maize problem: the historical perspective. Proceedings of the CIMMYT Aflatoxin in Maize Workshop, El Batan, Mexico City, 1987, pp 13–32.
24. ND Davis, CG Currier, UL Diener. Response of corn hybrids to aflatoxin formation by *Aspergillus flavus*. Ala Agric Exp Stn Bull 575:1–23, 1985.
25. NW Widstrom, WW McMillan, D Wilson. Segregation for resistance to aflatoxin contamination among seeds on an ear of hybrid maize. Crop Sci 27:961–963, 1987.
26. RL Brown, PJ Cotty, TE Cleveland, NW Widstrom. Living maize embryo influences accumulation of aflatoxin in maize kernels. J Food Prot 56:967–971, 1993.
27. RL Brown, TE Cleveland, GA Payne, CP Woloshuk, KW Campbell, DG White. Determination of resistance to aflatoxin production in maize kernels and detection of

fungal colonization using *Aspergillus flavus* transformant expressing *Escherichia coli* β-glucuronidase. Phytopathology 85:983–989, 1995.
28. BZ Guo, JS Russin, TE Cleveland, RL Brown, NW Widstrom. Wax and cutin layers in maize kernels associated with resistance to aflatoxin production by *Aspergillus flavus*. J Food Prot 58:296–300, 1995.
29. BZ Guo, JS Russin, TE Cleveland, RL Brown, KE Damann. Evidence for cutinase production by *Aspergillus flavus* and its possible role in infection of corn kernels. Phytopathology 86:824–829, 1996.
30. KM Tubajika, BZ Guo, JS Russin, RL Brown, TE Cleveland, NW Widstrom. Factors associated with resistance to aflatoxin production in maize. Phytopathology 85:512, 1995.
31. KW Campbell, DG White. Evaluation of corn genotypes for resistance to Aspergillus ear rot, kernel infection, and aflatoxin production. Plant Dis 79:1039–1045, 1995.
32. RL Brown, TE Cleveland, GA Payne, CP Woloshuk, DG White. Growth of *Aspergillus flavus* transformant expressing *Escherichia coli* β-glucuronidase in maize kernels resistant to aflatoxin production. J Food Prot 60:84–87, 1997.
33. DG White, TR Rocheford, B Kaufman, AM Hamblin. Further genetic studies and progress on resistance to aflatoxin production in corn. Proceedings of the USDA-ARS Aflatoxin Elimination Workshop, Atlanta, GA, 1995, p 7.
34. DG White, TR Rocheford, B Kaufman, AM Hamblin. Chromosome regions associated with resistance to *Aspergillus flavus* and inhibition of aflatoxin production in maize. Proceedings of the USDA-ARS Aflatoxin Elimination Workshop, Atlanta, GA, 1995, p 8.
35. JR Dunlap. Identifying heat tolerant sources of corn germplasm with reduced susceptibility to aflatoxin contamination. Proceedings of the USDA-ARS Aflatoxin Elimination Workshop, Atlanta, GA, 1995, p 12.
36. NP Keller, RAE Butchko, B Sarr, TD Phillips. A visual pattern of mycotoxin production in maize kernels by *Aspergillus* spp. Phytopathology 84:483–488, 1994.
37. CC Holbrook, DM Wilson, ME Matheron. An update on breeding peanut for resistance to preharvest aflatoxin contamination. Proceedings of the USDA-ARS Aflatoxin Elimination Workshop, Atlanta, GA, 1995, p 3.
38. VK Mehan, RC Nageswara Rao, D McDonald, JH Williams. Management of drought stress to improve field screening of peanuts for resistance to *Aspergillus flavus*. Phytopathology 78:659–663, 1988.
39. SM Basha, RJ Cole, SK Pancholy. A phytoalexin and aflatoxin producing peanut seed culture system. Peanut Sci 21:130–134, 1994.
40. T Gradziel, A Dandekar, M Alhumada, N Hirsh, J Driver, A Tang. Integrating fungal pathogen and insect vector resistance for comprehensive preharvest aflatoxin control in almond. Proceedings of the USDA-ARS Aflatoxin Elimination Workshop, Atlanta, GA, 1995, p 5.
41. MA Doster, TJ Michailides, DP Morgan. Aflatoxin control in pistachio, walnut, and figs: identification and separation of contaminated nuts and figs, ecological relationships, and agronomic practices. Proceedings of the USDA-ARS Aflatoxin Elimination Workshop, Atlanta, GA, 1995, pp 63–64.
42. CA Chlan, J Lin, JW Cary, TE Cleveland. A procedure for biolistic transformation

and regeneration of transgenic cotton from meristematic tissue. Plant Mol Biol Rep 13:31–37, 1995.
43. EE Farmer, CA Ryan. Octadecanoid precursors of jasmonic acid activate the synthesis of wound-inducible proteinase inhibitors. Plant Cell 4:129–134, 1992.
44. MJ Cordero, D Raventos, B San Segundo. Induction of PR proteins in germinating maize seeds infected with the fungus *Fusarium moniliforme*. Physiol Mol Plant Pathol 41:189–200, 1992.
45. HJM Linthorst. Pathogenesis-related proteins of plants. Crit Rev Plant Sci 10:123–150, 1991.
46. AJ Vigers, WK Roberts, CP Selitrennikoff. A new family of plant antifungal proteins. Mol Plant-Microbe Interact 4:315–323, 1991.
47. QK Huynh, JR Borgmeyer, JF Zobel. Isolation and characterization of a 22 kDa protein with antifungal properties from maize seeds. Biochem Biophys Res Commun 182:1–5, 1992.
48. SR Kumari, A Chandrashekar. Proteins in developing sorghum endosperm that may be involved in resistance to grain molds. J Sci Food Agric 60:275–282, 1992.
49. LJF Darnetty, S. Muthukrishnan, M Swegle, AJ Vigers, CP Selitrennikoff. Variability in antifungal proteins in the grains of maize, sorghum and wheat. Physiol Plant 88:339–349, 1992.
50. QK Huynh, CM Hironaka, EB Levine, CE Smith, JR Borgmeyer, DM Shah. Antifungal proteins from plants. J Biol Chem 267:6635–6640, 1992.
51. JN Neucere, TE Cleveland, C Dischinger. Existence of chitinase activity in mature corn kernels (*Zea Mays* L.). J Agric Food Chem 39:1326–1328, 1991.
52. JN Neucere, RL Brown, TE Cleveland. Correlation of antifungal properties and β-1,3-glucanases in aqueous extracts of kernels from several varieties of corn. J Agric Food Chem 43:275–276, 1995.
53. TE Cleveland, JW Cary, RL Brown, et al. Use of biotechnology to eliminate aflatoxin in preharvest crops. Bull Inst Compr Agric Sci, Kinki Univ 5:75–90, 1997.
54. SP McCormick, SP Bhatnagar, WR Goynes, LS Lee. An inhibitor of aflatoxin synthesis in developing cottonseed. Can J Bot 66:998–1002, 1988.
55. JE Mellon, PJ Cotty, MA Godshall, E Roberts. Demonstration of aflatoxin inhibitory activity in a cotton seed coat xylan. Appl Environ Microbiol 61:4409–4412, 1995.
56. BZ Guo, JS Russin, RL Brown, TE Cleveland, NW Widstrom. Effects of relative humidity and preincubation on growth of *Aspergillus flavus* and aflatoxin production in maize kernels. Phytopathology 84:1064, 1994.
57. Z-Y Huang, DG White, GA Payne. Characterization of inhibitory compounds to aflatoxin biosynthesis in the corn inbred line Tex6. Proceedings of the USDA-ARS Aflatoxin Elimination Workshop, Atlanta, GA, 1995, p 27.
58. JM Casacuberta, D Raventos, P Puigdomenech, B San Segundo. Expression of the gene encoding the PR-like protein PRms in germinating maize embryos. Mol Gen Genet 234:97–104, 1992.
59. NT Keen. The molecular biology of disease resistance. Plant Mol Biol 19:109–122, 1992.
60. BJC Cornelissen, LS Melchers. Strategies for the control of fungal diseases with transgenic plants. Plant Physiol 101:709–712, 1993.

61. K Broglie, I Chet, M Holliday, et al. Transgenic plants with enhanced resistance to the fungal pathogen *Rhizoctonia solani*. Science 254:1194–1197, 1991.
62. F Meins, J-M Neuhaus, C Sperisen, J Ryals. The primary structure of plant pathogenesis-related glucanohydrolases and their genes. In: T Boller, F Meins, eds. Genes Involved in Plant Defense. New York: Springer–Verlag, 1992, pp 245–282.
63. NK Singh, DE Nelson, D Kuhn, PM Hasegawa, RA Bressan. Molecular cloning of osmotin and regulation of its expression by ABA and adaptation to low water potential. Plant Physiol 90:1096–1101, 1989.
64. CA Ryan. Protease inhibitors in plants: genes for improving defenses against insects and pathogens. Annu Rev Phytopathol 28:425–449, 1990.
65. P Toubart, A Desiderio, G Salvi, et al. Cloning and characterization of the gene encoding the endopolygalacturonase-inhibiting protein (PGIP) of *Phaseolus vulgaris* L. Plant J 2:367–373, 1992.
66. HG Bowman, D Hultmark. Cell-free immunity in insects. Annu Rev Microbiol 31:103–126, 1987.
67. M Zasloff. Magainins, a class of antimicrobial peptides from *Xenopus* skin. Proc Natl Acad Sci USA 84:5449–5453, 1987.
68. KB Raper, DI Fennell. The *Aspergillus flavus* group. In: The Genus Aspergillus. Baltimore: Williams & Wilkins, 1965, pp 393–394.
69. BA Vick, DC Zimmerman. Oxidative systems for modification of fatty acids: The lipoxygenase pathway. In PK Stumpf, ed. The Biochemistry of Plants, a Comprehensive Treatise. Vol. 9. Lipids: Structure and Function. New York Academic Press, 1987, pp 53–90.
70. H Gundlach, MJ Miller, TM Kutchan, MH Zenk. Jasmonic acid is a signal transducer in elicitor-induced plant cell cultures. Proc Natl Acad Sci USA 89:2389–2393, 1992.
71. HJ Zeringue Jr. Effects of C_6-C_{10} alkenals and alkanals on eliciting a defense response in the developing cotton boll. Phytochemistry 31:2305–2308, 1992.
72. HJ Zeringue, RL Brown, JN Neucere, TE Cleveland. Relationships between C6-C12 alkanal and alkenal volatile contents and resistance of maize genotypes to *Aspergillus flavus* and aflatoxin production. J Agric Chem 44:403–404, 1996.
73. TM Goodrich, NE Mahoney, SB Rodriguez. The plant growth regulator methyl jasmonate inhibits aflatoxin production by *Aspergillus flavus*. Microbiology 141: 2831–2837, 1995.
74. RC Gueldner, DM Wilson, AR Heidt. Volatile compounds inhibiting *Aspergillus flavus*. J Agric Food Chem 33:411–413, 1985.
75. HJ Zeringue Jr, SR McCormick. Relationships between cotton leaf-derived volatiles and growth of *Aspergillus flavus*. JOACS 66:581–585, 1989.
76. HJ Zeringue Jr, SR McCormick. Aflatoxin production in cultures of *Aspergillus flavus* incubated in atmospheres containing cotton leaf-derived volatiles. Toxicon 28:445–448, 1990.
77. J-Y Lee, A Boman, S Chuanxin, et al. Antibacterial peptides from pig intestine: isolation of a mammalian cecropin. Proc Natl Acad Sci USA 86:9159–9162, 1989.
78. B Christensen, J Fink, RB Merrifield, D Mauzerall. Channel-forming properties of cecropins and related model compounds incorporated into planar lipids membranes. Proc Natl Acad Sci USA 85:5072–5076, 1988.

79. BL Kagan, ME Selsted, T Ganz, RI Lehrer. Antimicrobial defensin peptides form voltage-dependent ion-permeable channels in planar lipid membranes. Proc Natl Acad Sci USA 87:210–214, 1990.
80. H Steiner, D Hultmark, A Engstrom, H Bennich, HG Boman. Sequence and specificity of two antibacterial proteins involved in insect immunity. Nature 292:246–248, 1981.
81. AJ DeLucca, TJ Jacks, KA Brogden. Binding between lipopolysaccharide and Cecropin A. Mol Cell Biochem 151:141–148, 1995.
82. E Bell, JE Mullet. Lipoxygenase gene expression is modulated in plants by water deficit, wounding, and methyl jasmonate. Mol Gen Genet 230:456–462, 1991.
83. M Melan, X Dong, ME Endara, KR Davis, FM Ausubel, TK Peterman. An *Arabidopsis thaliana* lipoxygenase gene can be induced by pathogens, abscissic acid, and methyl jasmonate. Plant Physiol 101:441–450, 1993.
84. K Morishita, N Kubota, S Asano, Y Kaziro, S Nagata. Molecular cloning and characterization of cDNA for human myeloperoxidase. J Biol Chem 262:3844–3851, 1987.
85. C Wolffram, K-H van Pee, F Lingens. Cloning and high-level expression of a chloroperoxidase gene from *Pseudomonas pyrrocinia*. FEBS Lett 238:325–328, 1988.
86. TJ Jacks, PJ Cotty, O Hinojosa. Potential of animal myeloperoxidase to protect plants from pathogens. Biochem Biophys Res Commun 178:1202–1204, 1991.
87. TJ Jacks, O Hinojosa. Superoxide radicals in intact tissues and in dimethyl sulfoxide-based extracts. Phytochemistry 33:563–568, 1993.
88. CL Schardl, AD Byrd, G Benzion, MA Altschuler, DF Hildebrand, AG Hunt. Design and construction of a versatile system for the expression of foreign genes in plants. Gene 61:1–11, 1987.
89. JE Garbarino, WR Belknap. Isolation of a ubiquitin-ribosomal protein gene (*ubi3*) from potato and expression of its promoter in transgenic plants. Plant Mol Biol 24: 119–127, 1994.
90. RW Thornburg, G An, TE Cleveland, R Johnson, CA Ryan. Wound-inducible expression of potato inhibitor II—chloramphenicol acetyltransferase gene fusion in transgenic tobacco plants. Proc Natl Acad Sci USA 84:744–748, 1987.
91. M Burrow, P Sen, C Chlan, M Murai. Developmental control of the beta-phaseolin gene requires positive, negative, and temporal seed-specific transcriptional regulatory elements and a negative element for stem and root expression. Plant J 2:537–548, 1992.
92. P Ozias-Akins, H Fan, A Wang. Genetic engineering of peanut with Bt and function of a soybean promoter in peanut. Proceedings of the USDA-ARS Aflatoxin Elimination Workshop, Atlanta, GA, 1995, p 19.
93. RB Horsch, JB Fry, NL Hoffman, D Eichholtz, SG Rogers, RT Fraley. A simple and general method for transferring genes into plants. Science 227:1229–1231, 1985.
94. P Umbeck, G Johnson, K Barton, W Swain. Genetically transformed cotton (*Gossypium hirsutum* L.) plants. Bio/Technology 5: 263–266, 1987.
95. Bio-Rad Biolistic PDS-1000. He Particle Delivery System. Bulletin 1700. Hercules, CA: Bio-Rad Laboratories, 1992, pp 1–8.
96. G Davidonis, RH Hamilton. Plant regeneration from callus tissue of *Gossypium hirsutum* L. Plant Sci Lett 32:89–93, 1983.

97. N Trolinder, JR Goodin. Somatic embryogenesis and plant regeneration in cotton (*Gossypium hirsutum* L.). Plant Cell Rep 6:231–234, 1987.
98. E. Firoozabady, D DeBoer, D Merlo, et al. Transformation of cotton (*Gossypium hirsutum* L.) by *Agrobacterium tumefaciens* and regeneration of transgenic plants. Plant Mol Biol 10:105–116, 1987.
99. C Bayley, N Trolinder, C Ray, M Morgan, J Quisenberry, DW Ow. Engineering 2,4-D resistance into cotton. Theoret Appl Genet 83:645–649, 1992.
100. K Rajasekaran, JW Grula, RL Hudspeth, S Pofelis, DM Anderson. Herbicide-resistant Acala and Coker cottons transformed with a native gene encoding mutant forms of acetohydroxyacid synthase. Mol Breeding 2:307–319, 1996.
101. JJ Finer, MD McMullen. Transformation of cotton (*Gossypium hirsutum* L.) via particle bombardment. Plant Cell Rep 8:586–589, 1990.
102. DE McCabe, BJ Martinelli. Transformation of elite cotton cultivars via particle bombardment of meristems. Bio/Technology 11:596–598, 1993.
103. A Weissinger, R Cade, L Urban. Transformation of peanut cv. 'NC 7' with genes encoding defensive peptides. Proceedings of the USDA-ARS Aflatoxin Elimination Workshop, Atlanta, GA, 1995, p 20.
104. M Mendum, G. McGranahan, A Dandekar, S Uratsu. Progress in engineering walnuts for resistance to *Aspergillus flavus*. Proceedings of the USDA-ARS Aflatoxin Elimination Workshop, Atlanta, GA, 1995, p 21.
105. DT Wicklow, OL Shotwell. Intrafungal distribution of aflatoxins among conidia and sclerotia of *Aspergillus flavus* and *Aspergillus parasiticus*. Can J Microbiol 29:1–5, 1982.
106. N Kimura, S Hirano. Inhibitory strains of *Bacillus subtilis* for growth and aflatoxin-production of aflatoxigenic fungi. Agric Biol Chem 52:1173–1179, 1988.
107. PJ Cotty, P Bayman, DS Egel, KS Elias. Agriculture, aflatoxins, and *Aspergillus*. In: KA Powell, ed. The Genus *Aspergillus*. New York: Plenum Press, 1994, pp 1–27.
108. PJ Cotty. Virulence and cultural characteristics of two *Aspergillus flavus* strains pathogenic on cotton. Phytopathology 79:808–814, 1989.
109. PJ Cotty. Effect of atoxigenic strains of *Aspergillus flavus* on aflatoxin contamination of developing cottonseed. Plant Dis 74:233–235, 1990.
110. RL Brown, PJ Cotty, TE Cleveland. Reduction in aflatoxin content of maize by atoxigenic strains of *Aspergillus flavus*. J Food Prot 54:623–626, 1991.
111. PJ Cotty. Influence of field application of an atoxigenic strain of *Aspergillus flavus* on the population of *A flavus* infecting cotton bolls and on the aflatoxin content of cottonseed. Phytopathology 84:1270–1277, 1994.
112. PJ Cotty, P Bayman, D Bhatnagar. Two potential mechanisms by which atoxigenic strains of *Aspergillus flavus* prevent toxigenic strains from contaminating cottonseed. Phytopathology 80:944, 1990.
113. P Bayman, PJ Cotty. Vegetative compatibility and genetic diversity in the *Aspergillus flavus* population of a single field. Can J Bot 69:1707–1711, 1991.
114. PJ Cotty, LS Lee. Aflatoxin contamination of cottonseed. Comparison of pink bollworm damaged and undamaged bolls. Trop Sci 29:273–277, 1989.
115. JW Dorner, RJ Cole, PD Blankenship. Use of a biocompetitive agent to control preharvest aflatoxin in drought stressed peanuts. J Food Prot 55:888–892, 1992.

116. RJ Cole, JW Dorner, PD Blankenship. Update on biological control of preharvest aflatoxin contamination of peanuts. Proceedings of the USDA-ARS Aflatoxin Elimination Workshop, Atlanta, GA, 1995, p 53.
117. RJ Cole, PJ Cotty. Biocontrol of aflatoxin production by using biocompetitive agents. In: JR Robens, ed. A Perspective on Aflatoxin in Field Crops and Animal Food Products in the United States. Beltsville: Agricultural Research Service, 1990, pp 62–66.
118. PJ Cotty. Use of native *Aspergillus flavus* strains to prevent aflatoxin contamination. United States Patent 5,171,686, 1992.
119. PJ Cotty. *Aspergillus flavus*. Wild intruder or domesticated freeloader. In: JR Robens, ed. Aflatoxin Elimination Workshop. Beltsville: Agricultural Research Service, 1992, p. 28.
120. PJ Cotty. Prevention of aflatoxin contamination of cottonseed by qualitative modification of *Aspergillus flavus* populations. Phytopathology 81:1227, 1991.
121. JF Shearer, LE Sweets, NK Baker, LH Tiffany. A study of *Aspergillus flavus/parasiticus* in Iowa crop fields, 1988–1990. Plant Dis 76:19–22, 1992.
122. RL Brown, TE Cleveland, PJ Cotty, JE Mellon. Spread of *Aspergillus flavus* in cotton bolls, decay of intercarpellary membranes, and production of fungal pectinases. Phytopathology 82:462–467, 1992.
123. TE Cleveland, PJ Cotty. Invasiveness of *Aspergillus flavus* isolates in wounded cotton bolls is associated with production of a specific fungal polygalacturonase. Phytopathology 81:155–158, 1991.
124. F. Trail, P-K Chang, J Cary, JE Linz. Structural and functional analysis of the *nor-1* gene involved in the biosynthesis of aflatoxins by *Aspergillus parasiticus*. Appl Environ Microbiol 60:4078–4085, 1994.
125. P-K Chang, JW Cary, J Yu, D Bhatnagar, TE Cleveland. The *Aspergillus parasiticus* polyketide synthase gene *pksA*, a homolog of *Aspergillus nidulans wA*, is required for aflatoxin B_1 biosynthesis. Mol Gen Genet 248:270–277, 1995.
126. N Mahanti, D Bhatnagar, JW Cary, J Joubran, JE Linz. Structure and function of *fas-1A*, a gene encoding a putative fatty acid synthetase directly involved in aflatoxin biosynthesis in *Aspergillus parasiticus*. Appl Environ Microbiol 62:191–195, 1996.
127. CMH Watanabe, D Wilson, JE Linz, CA Townsend. Demonstration of the catalytic roles and evidence for the physical association of type 1 fatty acid synthases and a polyketide synthase in the biosynthesis of aflatoxin B_1. Chem Biol 3:463–469, 1996.
128. TE Cleveland, D Bhatnagar. Molecular strategies for reducing aflatoxin levels in crops before harvest. In: D Bhatnagar, TE Cleveland, eds. Molecular Approaches to Improving Food Quality and Safety. New York: Van Nostrand Reinhold, 1992, pp 205–228.
129. GA Payne. Aflatoxin in maize. Crit Rev Plant Sci 10:423–440, 1992.
130. D Bhatnagar, KC Ehrlich, TE Cleveland. Oxidation-reduction reactions in biosynthesis of secondary metabolites. In: D Bhatnagar, EB Lillehoj, DK Arora, eds. Mycotoxins in Ecological Systems. New York: Marcel Dekker, 1991, pp 255–286.
131. SM McGuire, SW Brobst, TL Graybill, K Pal, CA Townsend. Partitioning of tetrahydro- and dihydrobisfuran formation in aflatoxin biosynthesis defined by

cell-free and direct incorporation experiments. J Am Chem Soc 111:8308–8309, 1989.

132. K Yabe, Y Ando, T Hamasaki. Desaturase activity in the branching step between aflatoxins B_1 and G_1 aflatoxins and aflatoxins B_2 and G_2. Agric Biol Chem 55:1907–1911, 1991.
133. D Bhatnagar, TE Cleveland, DGI Kingston. Enzymological evidence for separate pathways for aflatoxin B_1 and B_2 biosynthesis. Biochemistry 30:4343–4350, 1991.
134. TE Cleveland, D Bhatnagar, CJ Foell, SP McCormick. Conversion of a new metabolite to aflatoxin B2 by *Aspergillus parasiticus.* Appl Environ Microbiol 53:2804–2807, 1987.
135. K Yabe, Y Ando, T Hamasaki. Biosynthetic relationship among aflatoxins B_1, B_2, G_1, G_2. Appl Environ Microbiol 54:101–2106, 1988.
136. K Yabe, Y Ando, T Hamasaki. A metabolic grid among versiconal hemiacetal acetate, versiconol acetate, versiconol and versiconal during aflatoxin biosynthesis. J Gen Microbiol 137:2469–2415, 1991.
137. MF Dutton. Enzymes and aflatoxin biosynthesis. Microbiol Rev 52:274–295, 1988.
138. D Bhatnagar, TE Cleveland, EB Lillehoj. Enzymes in aflatoxin B_1 biosynthesis—strategies for identifying pertinent genes. Mycopathologia 107:75–83, 1989.
139. TE Cleveland, AR Lax, LS Lee, D Bhatnagar. Appearance of enzyme activities catalyzing conversion of sterigmatocystin to aflatoxin B_1 in late growth-phase *Aspergillus parasiticus* cultures. Appl Environ Microbiol 53:1711–1713, 1987.
140. AA Chuturgoon, MF Dutton. The affinity purification and characterization of a dehydrogenase from *Aspergillus parasiticus* involved in aflatoxin B_1 biosynthesis. Mycopathologia 113:41–44, 1991.
141. DPH Hsieh, CC Wan, JA Billington. A versiconal hemiacetal acetate converting enzyme in aflatoxin biosynthesis. Mycopathologia 107:121–126, 1989.
142. JA Anderson, CH Chung, S-H Cho. Versicolorin A hemiacetal, hydroxydihydrosterigmatocystin and aflatoxin G2a reductase activity in extracts from *Aspergillus parasiticus.* Mycopathologia 111:39–45, 1990.
143. AA Chuturgoon, MF Dutton, RK Berry. The preparation of an enzyme associated with aflatoxin biosynthesis by affinity chromatography. Biochem Biophys Res Commun 166:38–42, 1990.
144. K Yabe, Y Ando, J Hashimoto, T Hamasaki. Two distinct O-methyltransferases in aflatoxin biosynthesis. Appl Environ Microbiol 55:2172–2177, 1989.
145. D Bhatnagar, AHJ Ullah, TE Cleveland. Purification and characterization of a methyltransferase from *Aspergillus parasiticus* SRRC 163 involved in aflatoxin biosynthetic pathway. Prep Biochem 18:321–369, 1988.
146. NP Keller, HC Dischinger, D Bhatnagar, TE Cleveland, AHJ Ullah. Purification of a 40-kilodalton methyltransferase active in the aflatoxin biosynthetic pathway. Appl Environ Microbiol 59:479–484, 1993.
147. D Bhatnagar, AR Lax, B Prima, JW Cary, TE Cleveland. Purification of a 43 kDa enzyme that catalyzes the reduction of norsolorinic acid to averantin in aflatoxin biosynthesis. FASEB J 10:A2164, 1996.
148. JC Silva, RE Minto, CE Barry III, KA Holland, CA Townsend. Isolation and characterization of the versicolorin B synthase gene from *Aspergillus parasiticus*: expansion of the gene cluster. J Biol Chem 271:13600–13608, 1996.

149. B-K Lin, JA Anderson. Purification and properties of versiconal cyclase from *Aspergillus parasiticus*. Arch Biochem Biophys 293:67–70, 1992.
150. K-I Kusumoto, DPH Hsieh. Purification and characterization of the esterases involved in aflatoxin biosynthesis in *Aspergillus parasiticus*. Can J Microbiol 42: 804–810, 1996.
151. K-I Matsushima, Y Ando, T Hamasaki, K Yabe. Purification and characterization of two versiconal hemiacetal acetate reductases involved in aflatoxin biosynthesis. Appl Environ Microbiol 60:2561–2567, 1994.
152. D Bhatnagar, TE Cleveland. Aflatoxin biosynthesis: developments in chemistry, biochemistry and genetics. In: OL Shotwell, CR Hurburgh Jr, eds. Aflatoxin in Corn: New Perspectives. Ames: Iowa State University Press, 1991, pp 391–405.
153. M Chatterjee, CA Townsend. Evidence for the probable final steps in aflatoxin biosynthesis. J Org Chem 59:4424–4429, 1996.
154. P Bayman, PJ Cotty. Vegetative compatibility and genetic diversity in the *Aspergillus flavus* population of a single field. Can J Bot 69:1707–1711, 1991.
155. JW Bennett, KE Papa. The aflatoxigenic *Aspergillus* spp. In GS Sidhu, ed. Advances in Plant Pathology: Genetics of Plant Pathogenic Fungi. New York Academic Press, 1988, Vol 6, pp 263–280.
156. NP Keller, TE Cleveland, D Bhatnagar. A molecular approach toward understanding aflatoxin production. In: D Bhatnagar, EB Lillehoj, DK Arora, eds. Mycotoxins in Ecological Systems. New York: Marcel Dekker, 1991, pp 287–310.
157. GA Payne, CP Woloshuk. The transformation of *Aspergillus flavus* to study aflatoxin biosynthesis. Mycopathologia 107:139–144, 1989.
158. NP Keller, TE Cleveland, D Bhatnagar. Variable electrophoretic karyotypes of members of *Aspergillus* section *flavi*. Current Genet 21:371–315, 1992.
159. KR Foutz, CP Woloshuk, GA Payne. Cloning and assignment of linkage group loci to a karyotypic map of the filamentous fungus *Aspergillus flavus*. Mycologia 87: 787–794, 1995.
160. CD Skory, JS Horng, JJ Pestka, JE Linz. Transformation of *Aspergillus parasiticus* with a homologous gene (*pyr G*) involved in pyrimidine biosynthesis. Appl Environ Microbiol 56:315–3320, 1990.
161. JS Horng, PK Chang, JJ Pestka, JE Linz. Development of a homologous transformation system for *Aspergillus parasiticus* with the gene encoding nitrate reductase. Mol Gen Genet 224:294–296, 1990.
162. JS Horng, JE Linz, JJ Pestka. Cloning and characterization of the *trpC* gene from an aflatoxigenic strain of *Aspergillus parasiticus*. Appl Environ Microbiol 55:2561–2568, 1989.
163. ER Seip, CP Woloshuk, GA Payne, SE Curtis. Isolation and sequence analysis of a β-tubulin gene from *Aspergillus flavus* and its use as a selectable marker. Appl Environ Microbiol 56:3686–3692, 1990.
164. GA Payne, GJ Nystrom, D Bhatnagar, TE Cleveland, CP Woloshuk. Cloning of the *afl-2* gene involved in aflatoxin biosynthesis from *Aspergillus flavus*. Appl Environ Microbiol 59:156–162, 1993.
165. P-K Chang, JW Cary, D Bhatnagar, et al. Cloning of the *Aspergillus parasiticus apa-2* gene associated with the regulation of aflatoxin biosynthesis. Appl Environ Microbiol 59:3273–3279, 1993.

166. P-K Chang, CD Skory, JE Linz. Cloning of a gene associated with aflatoxin biosynthesis in *Aspergillus parasiticus*. Curr Genet 21:231–233, 1992.
167. CD Skory, P-K Chang, J Cary, JE Linz. Isolation and characterization of a gene from *Aspergillus parasiticus* associated with the conversion of versicolorin A to sterigmatocystin in aflatoxin biosynthesis. Appl Environ Microbiol 58:3527–3537, 1992.
168. R Prieto, GL Yousibova, CP Woloshuk. Identification of aflatoxin biosynthesis genes by genetic complementation in an *Aspergillus flavus* mutant lacking the aflatoxin gene cluster. Appl Environ Microbiol 62:3567–3571, 1996.
169. GH Feng, FS Chu, TJ Leonard. Molecular cloning of genes related to aflatoxin biosynthesis by differential screening. Appl Environ Microbiol 58:455–460, 1992.
170. BH Liu, NP Keller, D Bhatnagar, TE Cleveland, FS Chu. Production and characterization of antibodies against sterigmatocystin *O*-methyltransferase. Food Agric Immunol 5:155–164, 1993.
171. RC Lee, JW Cary, D Bhatnagar, FS Chu. Production and characterization of polyclonal antibodies against norsolorinic acid reductase involved in aflatoxin biosynthesis. Food Agric Immunol 7:21–32, 1995.
172. JW Cary, TE Cleveland, D Bhatnagar. Regulation by thiamine of expression of a gene from *Aspergillus parasiticus* encoding norsolorinic acid reductase activity. FASEB J 6:A228, 1992.
173. J Yu, JW Cary, D Bhatnagar, TE Cleveland, N Keller, FS Chu. Cloning and characterization of a cDNA from *Aspergillus parasiticus* encoding an O-methyltransferase involved in aflatoxin biosynthesis. Appl Environ Microbiol 59:3564–3571, 1993.
174. JW Cary, M Wright, D Bhatnagar, R Lee, FS Chu. Molecular characterization of an *Aspergillus parasiticus* dehydrogenase gene, *norA*, located on the aflatoxin biosynthesis gene cluster. Appl Environ Microbiol 62:360–366, 1996.
175. J Yu, P-K Chang, JW Cary, et al. Comparative mapping of aflatoxin pathway gene clusters in *Aspergillus parasiticus* and *Aspergillus flavus*. Appl Environ Microbiol 61:2365–2371, 1995.
176. J Yu, P-K Chang, JW Cary, D Bhatnagar, TE Cleveland. *avnA*, A gene encoding a cytochrome P450 monooxygenase is involved in the conversion of averantin to averufin in aflatoxin biosynthesis in *Aspergillus parasiticus*. Appl Environ Microbiol 63:1349–1356, 1997.
177. F Trail, N Mahanti, M Rarick, S-H Liang, R Zhou, JE Linz. A physical and transcriptional map of an aflatoxin gene cluster in *Aspergillus parasiticus* and the functional disruption of a gene involved early in the aflatoxin pathway. Appl Environ Microbiol 61:2665–2673, 1995.
178. P-K Chang, JW Cary, D Bhatnagar, TE Cleveland. *Aspergillus parasiticus* polyketide synthase *pksA*, a homolog of *Aspergillus nidulans wA*, is required for aflatoxin B_1 biosynthesis. Mol Gen Genet 248:270–277, 1995.
179. GH Feng, TJ Leonard. Characterization of polyketide synthase gene (*pksl1*) required for aflatoxin biosynthesis in *Aspergillus parasiticus*. J Bacteriol 177:6246–6254, 1995.
180. N Mahanti, D Bhatnagar, JW Cary, J Joubran, JE Linz. Structure and function of *fas-1A*, a gene encoding a putative fatty acid synthetase directly involved in aflatoxin biosynthesis in *Aspergillus parasiticus*. Appl Environ Microbiol 62:191–195, 1996.

181. DW Brown, J-H Yu, HS Kelkar, et al. Twenty-five coregulated transcripts define a sterigmatocystin gene cluster in *Aspergillus nidulans*. Proc Natl Acad Sci USA 93: 1418–1422, 1996.
182. NP Keller, NJ Kantz, TH Adams. *Aspergillus nidulans verA* is required for the production of mycotoxin sterigmatocystin. Appl Environ Microbiol 60:1444–1451, 1994.
183. SP Kale, JW Cary, D Bhatnagar, JW Bennett. Characterization of experimentally induced, nonaflatoxigenic variant strains of *Aspergillus parasiticus*. Appl Environ Microbiol 62:3399–3404, 1996.

12
Detoxification of Mycotoxins and Food Safety

Kaushal K. Sinha
T. M. Bhagalpur University, Bhagalpur, India

I. INTRODUCTION

Mycotoxin contamination of agricultural commodities has become a natural phenomenon in most parts of the tropical countries. This may be because of the favorable climatic conditions prevalent in those regions coupled with the traditional methods of crop cultivation, harvesting, handling, and storage, all of which ultimately lead to severe mold growth and mycotoxin contamination in the agricultural commodities. Several guidelines have been suggested in the past to prevent mold growth and mycotoxin development in the crops, from field conditions to their consumption [1–3].

Even then the problem of mycotoxin contamination sometimes becomes unavoidable. Rapid detection methods as well as monitoring programs often recommend the diversion of contaminated grain lots to animal feed, but this is not a solution to the problem. Use of toxic or costly chemicals also cannot be recommended for this purpose. The only approach to alleviate mycotoxin problem is to use the resistant crop varieties. Extensive efforts are being made in recent years to identify the genes responsible for resistance to the production of mycotoxins by the toxigenic fungi and to evolve the varieties that can serve as "resistant crop varieties" [4,5]. This aspect has been discussed in detail by Dr. Brown and his colleagues in a separate chapter of this book. However, this approach is also not feasible, as none of the available crop varieties is found to be completely immune against mycotoxin production by all the toxigenic strains of a particular fungal species at different geographical regions. Production of mycotoxins can also be minimized if the grains are dried properly up to safe moisture level before storage, but this is also not possible as the grains are directly influenced by the fluctuating atmospheric conditions in most cases. Under these circumstances the preventive strategies have to be supplemented with detoxification methods in order to eliminate the toxins from the contaminated lots or at least reduce the toxin hazards by bringing down the level of mycotoxins up to permissible level.

Extensive investigations have been carried out on the detoxification of aflatoxins, particularly in maize and groundnut, and a massive literature is available on these aspects. Comparatively very few reports are available on the detoxification of other mycotoxins viz., zearalenone, trichothecenes, citrinin, and ochratoxin. In this communication emphasis has been given on the detoxification of aflatoxin-contaminated agricultural commodities and the acceptability of the processed commodities as food or feed. Some of these aspects are covered in Chapter 13.

Detoxification of mycotoxins can be achieved by removal or elimination of the contaminated commodities or by the inactivation of toxins present in the commodities through various physical, chemical, and biological means. For a successful detoxification procedure, Heathcote and Hibbert [6] have suggested the following basic criteria:

1. It must be economical, so that the detoxified commodities may still be sold at competitive prices.
2. The process must be relatively simple, not too time-consuming, and suitable for operation by unskilled labor.
3. It must be capable of removing all traces of active toxin and the chemical residues must not constitute a health hazard.
4. The nutritional quality must not be impaired.

The above criteria should also be kept in mind if one has to develop a new method for the detoxification of mycotoxins in agricultural commodities.

II. REMOVAL OR ELIMINATION OF MYCOTOXINS FROM CONTAMINATED COMMODITIES

The majority of the mycotoxins in the contaminated commodities usually reside in a relatively small number of seeds or kernels [7]. This affords an exceptional opportunity for effectively reducing mycotoxin content by removal of all those contaminated seeds/kernels [8], which can be done by any one of the following methods.

A. Physical Separation

Physical methods of separation are being used successfully in the peanut and other nutmeat industries throughout the world, including India. The principle is based on the identification of damaged kernels in the seed lots because of the variations in size, shape, color, and, more often, visible mold growth on the affected kernels. Aflatoxin-contaminated kernels are usually damaged, shriveled, or discolored [9]. Therefore, a combination of sieving and electronic sorting can be helpful in eliminating most of the undesired kernels and leave the remaining nuts virtually free from aflatoxin [10]. A final hand-picking is generally used to detect and remove the nuts that escaped the first two processing steps. Separation of toxic seeds can also be done directly by hand picking or by the use of electronic sorters [11]. A systematic electronic and hand-sorting of contaminated peanuts has been developed by the peanut industry in the U.S. to reduce the level of aflatoxins in peanut products used for human consumption. Park and Stoloff [15] succeeded in getting up to 99% removal of aflatoxins in peanut kernels by using different processes, viz., belt separation, shelling, color sorting, and blanching.

Mechanical sorting devices have also been suggested for the kernels of larger size such as almonds, Brazil nuts, and pistachio nuts. Separation of damaged almonds can be done on the basis of the energy reflected from the particles illuminated by UV light [13]. The nicked almond seeds are therefore separated from the sound ones, because fluorescent oil occurs in fissures in the seed coat. Schade et al. [13] also analyzed high levels of aflatoxins in the rejected lots of almonds. In another experiment, highly contaminated Brazil nut kernels were found to exhibit yellow fluorescences when illuminated under UV light at 360 nm [14]. Not all the fluorescent kernels, however, were contaminated with aflatoxins. While Brazil nut kernels showed different appearances, only brown or brown-spotted pistachio nut kernels were highly contaminated with aflatoxins. Separating "early split" from "normal" pistachio nuts for removal of nuts contaminated with aflatoxins has also been discussed by Pearson et al. [15].

Physical separation might be useful for cottonseed since there appears to be a good correlation between the presence of aflatoxins and the occurrence of a greenish-yellow fluorescence (GYF) in the contaminated seeds under UV light [16]. Electronic sorting devices can be used to eliminate contaminated seeds, but economics may not favor this approach. This type of separation was also not successful in pecans because of the inherent intense fluorescence in the kernels [17]. Bochelee-Morvan and Gilliver [18] concluded that aflatoxin contamination of unshelled peanuts can be significantly reduced by removing damaged pods by hand or by pneumatic sorting. Hand-sorting for unshelled edible products is now employed in Senegal. Lots of contaminated Brazil nuts have also been reconditioned by pneumatic sorting [19]. The reconditioning process is based on the assumption that moldy nuts are lighter than sound ones.

An effective method for reducing the levels of aflatoxin has also been reported through flotation and density segregation of toxic kernels in corn [20] and peanuts [21].

B. Removal by Filtration and Adsorption

Aflatoxin in crude peanut oil remains in finely suspended form and can easily be separated by filtration. Extensive studies on this aspect have been made at Central Food and Technological Research Institute (CFTRI), Mysore, India. Basappa and Sreenivasamurthy [22] developed a special filter pad system which can easily be adopted in oil mills to remove aflatoxin from crude oil. This filter pad can be prepared by impregnating Fuller's earth-salt slurry in between two filter cloth layers and dried completely at 100°C for 8 hours. The pads can be stored under desiccated conditions up to 2 months without loss in activity. Aflatoxin was removed up to 90% through single filtration, but on recirculation of the oil, they could achieve even up to 100% removal. Therefore, this appears to be a simple approach to the problem of aflatoxin in unrefined peanut oil.

Although aflatoxin in liquid materials could be adsorbed on certain adsorbents such as clay, kaolin, activated Fuller's earth, celite, activated charcoal, magnesium silicate, aluminum silicate, and asbestos, its practical application is limited to only certain foodstuffs [23,24].

C. Removal by Milling

The concentration of aflatoxin may be quite high in some of the raw materials, but in the processed products its level may vary with the nature of processing and food materials as well as the affinity and/or solubility of the toxin in the products.

In systemic studies conducted at CFTRI, Mysore, India, it has been found that 85% of the aflatoxins present in groundnut seeds goes into the cake after crushing in expeller oil mills or in Carver Laboratory Hydraulic Press, and only 15% remained in the oil [25]. This indicates that aflatoxin level in oil is far less because of the low solubility of aflatoxin in it, whereas the level in cake gets concentrated probably because of this partition effect.

Milling has also been found successful in reducing the levels of several fusarial toxins, viz., DON, nivalenol, and zearalenone in wheat [26,27].

D. Removal by Solvents/Extraction

A variety of solvents have been found capable of extracting aflatoxins from different commodities. However, application of these methods may have several advantages as well as limitations [28]. The obvious advantages are:

1. Aflatoxins can be removed completely under suitable conditions.
2. There is little chance of forming from the aflatoxins other products having adverse physiological activity.
3. Extraction can be carried out and the solvent recovered without nutritional loss in many cases.

Some of the limitations include:

1. Special solvent extraction equipment may be required.
2. There is a possibility of extraction of some of the soluble components with aflatoxins.
3. Added cost of additional processing.
4. Possible introduction of off-flavors.

It is not an easy task to find a solvent that fits all the above criteria. Some other factors, like cost of the solvent and its percentage recovery, toxicity of solvent and its residues, may also be taken into consideration.

Aflatoxins are insoluble in water and petroleum hydrocarbons and soluble in polar solvents. Aqueous isopropanol has been found to be an effective solvent for the removal of aflatoxins from both contaminated cottonseed and groundnut [29]. Six extractions with 80% aqueous isopropanol at 60°C resulted in complete removal of aflatoxins in both meals. Mixtures of hexane-methanol, hexane-ethanol, hexane-ethanol-water, and hexane-acetone-water were also evaluated for the extraction of aflatoxins by Vorster [30]. Maximum reduction was obtained with hexane-acetone-water and hexane-methanol. Using the same tertiary system (i.e., acetone-hexane-water), Goldblatt and Robertson [31] were able to remove aflatoxins from contaminated peanut meal without extracting large quantities of solids. Gardner et al. [32] also evaluated a tertiary solvent and binary aqueous acetone for aflatoxin removal from contaminated oilseed meal. Both the procedures offered an economical technique for removing aflatoxin to the 30-ppb level in the extracted materials.

Some polar solvents like methanol and ethanol effectively reduce aflatoxin levels in contaminated materials. But these are also capable of extracting significant quantities of solids from the materials [33,34]. Rayner et al. [35] reported that 80% aqueous isopropanol completely removed aflatoxins in cottonseed and peanut meal, but it also eliminated 8.7% and 9.5% meal solids, respectively. Glodblatt [34], however, reported complete removal of aflatoxins from peanut meal with only negligible removal of oils and proteins through aqueous (25% to 30%) acetone.

The most successful application is the removal of aflatoxin from oils during normal commercial processing [8]. Current processing of oilseeds either by mechanical expression or by extraction with hydrocarbon solvents leaves in the oil a portion of any aflatoxin that may have been present in the seed. The exact proportion, however, depends not only on the mode and conditions of processing, but also on the quality of raw materials. Through conventional commercial refining processes, Parker and Melnick [36] were able to remove all aflatoxins from the crude peanut oil. The unrefined peanut oil is preferred in many areas of the world, including India. Shantha and Sreenivasamurthy [37] reported that single extraction of crude oil with 10% aqueous sodium chloride (1:4) at 80°C for 30 min removed 80% of the toxin from the oil with minimum loss and maximum oil recovery. In another experiment, aflatoxin could not be detected in sodium chloride-boiled groundnut samples which were all contaminated [38]. The efficiency of this salt in removing aflatoxins was confirmed by cooking laboratory-inoculated batches of raw, unshelled groundnuts in 5% aqueous sodium chloride solution at 116°C and 0.7 bar for 30 min. Results of five batches of groundnuts containing aflatoxins at 19,992 to 66,320 μg/kg showed a removal of aflatoxins from 80% to nearly 100%.

Sreenivasamurthy et al. [39] also reported the removal of all aflatoxins from peanut meal by an aqueous solution of calcium chloride. Aibara and Yano [40] described a procedure for methoxy-methane extraction of aflatoxin from contaminated peanut meals that effectively removed oil and aflatoxin in a single step with no solvent residues in dried meals.

Decontamination of some fusarial toxins has been tried through simple washing procedures. Reduction in DON (65% to 69%) and zearalenone (2% to 61%) levels has been observed in contaminated barley and corn by single washing in distilled water [41]. Details of other physical methods for decontamination of *Fusarium* mycotoxins have been given by Charmley and Prelusky [42].

As discussed earlier, use of solvent for removal of aflatoxins will have its own impact on the processed commodities. Most of the organic solvents, which eliminate aflatoxins from the food materials, are also capable of extracting some of the vital nutritional components of the processed materials. In some cases these may also introduce flavor problems. All these conditions are bound to reduce the nutritive quality of the food commodities, ultimately leading to economic losses.

III. INACTIVATION OF MYCOTOXINS IN CONTAMINATED COMMODITIES

Sometimes removal or elimination of mycotoxins in the contaminated commodities is difficult to carry out effectively or is successful only on a small scale. In that case there is the possibility of inactivating or destroying the mycotoxins present in the agricultural commodities through various physical, chemical, and biological processes.

A. Physical Methods

1. Thermal Inactivation

Although initial reports indicated aflatoxins as heat-stable [43,44], later studies demonstrated that aflatoxins can be degraded partially at higher temperatures. Even different aflatoxins react in different ways to it. While aflatoxin B_1 is more stable, aflatoxin G_1 can be inactivated more easily by heat [45].

Dry heating was ineffective in destroying aflatoxins, but heating or steaming under pressure has been found to reduce the level of aflatoxins in many food and feed items.

Conventional processing of food like cooking, roasting, frying, spray drying, baking, etc. has also been shown to destroy aflatoxins present in the food [46]. Rehana et al. [47] compared different methods used by Indians for cooking of rice and reported that pressure cooking at 15 psi for 5 min gave maximum destruction (72%) of aflatoxins as compared to the method of ordinary cooking (50%) and cooking with excess water (50%). They also recommended pressure cooking of rice not only because of destroying maximum amount of aflatoxins but also for preserving the nutrients of rice.

Roasting resulted in a reduction in the levels of aflatoxins in many nuts and oilseed meals [48] and corn [49], but in no case was total destruction achieved [50].

Frying of foods in vegetable oils such as peanut oil, mustard oil, sesame oil, coconut oil, palm oil, etc. is very common in India. Dwarkanath et al. [51] reported that heating a crude oil having high level of aflatoxin (260 ppb) at 150°C destroyed 50% of the aflatoxin after 10 min and 61%, the maximum destruction, after 20 min. They also reported that the pooris (thin fried foods prepared from wheat flour) fried in oil picked up a significant level of aflatoxin which may pose a health hazard after consumption.

Zearalenone was not found to be affected by high heat. Even heating of contaminated corn at 150°C for 44 hours did not make any destruction of zearalenone in that substrate [52]. Microwave oven heating at 100 to 230°C or higher has, however, resulted in 50% to 100% reduction in DON concentration in corn [53,54]. Losses in the levels of citrinin have also been recorded in heated cereal grains [55].

The need to use elevated temperatures and pressures for effective detoxification of contaminated foods hinders the exploration of heat treatments as a practical means of mycotoxin detoxification [56]. The impairment of nutritional and organoleptic qualities and doubts concerning the generation of toxic pyrolysate at elevated temperatures discourage conventional heat treatment as an effective method for decontamination. However, as mentioned earlier, industrial treatments providing high temperatures and pressures such as steam flaking, extrusion cooking, microwave cooking, popping, and spray drying have given very promising results [57,58] and suggest redesigning them for effective decontamination of aflatoxin in foods.

2. Inactivation Through Light/Irradiation

Under suitable conditions aflatoxin is degraded by exposure to light, particularly ultraviolet radiations. Arthur and Robertson [59] were able to destroy aflatoxin in chloroform solution to the extent of 99% when exposed to UV light. As indicated by mass spectrometry, photodiamerization of the coumarin moiety might have occurred before the degradation of the aflatoxin B_1 molecule. Encouraging results have also been recorded with the exposure to sunlight of aflatoxin-contaminated vegetables oils. Unrefined groundnut oil containing aflatoxin (> 100 μg/kg) was exposed to bright sunlight, gas-filled tungsten lamp, or longwave UV light [60]. Sunlight destroyed 99% of the aflatoxin present in 15 min, whereas tungsten lamp light and ultraviolet light destroyed 82% to 85% of aflatoxin in 18 hours and 30% to 40% of aflatoxin in 2 hours' exposure, respectively.

The photolysed oil was not toxic to albino rats. They suggested the reason for this photodestruction to be a shift of absorption peak from 700 nm to 280 nm after exposure to sunlight. Samarajeewa et al. [61] also obtained similar results in cases of different edible oils, viz., coconut, peanut, soybean, and sesame oils after exposure to solar radiation. Complete destruction of aflatoxins was also obtained when aflatoxin-contaminated peanut oil kept in glass containers was exposed to direct sunlight (approximately 50,000 lux) for 1 hour [62]. They also reported that photodestructed aflatoxin was not regenerated in the oil during storage for 6 months in dark [63]. The quality and thickness of the container used could be the critical factor, as the sunlight has to penetrate to and within the oil. Sunlight has not, however, proved to be effective for destruction of aflatoxin in contaminated peanut seeds and cake, as the toxin is bound to the protein molecules in these substrates and there are problems of light penetration [64].

Removal of aflatoxin from crude peanut oil has also been reported in Senegal by using a local clay and sunlight [65]. They recorded complete decontamination of aflatoxins when a crude peanut oil having 160 ppb of aflatoxin was mixed with 5% of attapulgite, a local clay, for 3 min. This treatment did not significantly affect any of the chemical parameters of the oil. Complete detoxification was also reported when peanut oil having a contamination of 600 ppb aflatoxin was exposed to sunlight for 18 and 24 hours in transparent glass and transparent plastic bottles, respectively. The chemical parameters were also not affected by this treatment.

Recently Santamarina et al. [66] surveyed the important physical methods including visible and ultraviolet light, gamma rays, and chemical methods to destroy mycotoxins in food. The authors concluded that no single method can be effective for this purpose but that gamma irradiation in combination with other methods could be employed to achieve removal of mycotoxins.

The effect of gamma irradiation on the disappearance of mycotoxins was also observed in maize, wheat, and soybean [67]. Radiation doses of 5, 7.5, 10, and 20 kGy were applied to spiked grain samples, and the residual toxins were estimated using the ELISA technique. Radiation doses up to 20 kGy did not affect the level of aflatoxin B_1 in all the above three substrates. A significant reduction was, however, recorded in the levels of T-2, DON, and zearalenone at doses above 7.5 kGy. Significant losses in the levels of some of the essential amino acids were also observed due to this irradiation. Gamma irradiation (up to 7.5 Mrad) did not cause any decomposition of ochratoxin A in methanol solution [68], whereas destruction of sterigmatocystin was observed in dry state at a dose of 50 Mrad [69]. Destruction of aflatoxins by microwaves has also been reported in peanuts and yellow corn deliberately infected by *Aspergillus flavus* [70]. Risk evaluation and management in radiation-preserved foods have, however, been discussed in detail by Dr. Arun Sharma in a separate chapter of this book.

Detoxification (70% to 90%) of some trichothecenes has also been observed in the contaminated corn by applying ultrasonication in Austria [71]. The final product retained its original taste and appearance. The laboratory studies have further been implemented to a pilot plant study to record its feasibility in detoxifying different mycotoxins in the cereals. Besides DON, a significant reduction in the levels of ochratoxins and zearalenone has been recorded due to ultrasonication.

B. Chemical Methods

Numerous chemicals have been found to destroy or inactivate mycotoxins in naturally contaminated agricultural commodities. Chemical inactivation actually appears to offer the most promising and feasible approach to alleviate this problem in certain specific cases [72]. This detoxification procedure would, however, be technically and economically viable only when it satisfies the following criteria laid down by FAO [73]:

1. It must destroy or inactivate the mycotoxins.
2. It must not produce or leave any toxic or carcinogenic residue in the final product.
3. It should destroy fungal spores and mycelium, which under suitable conditions could grow and recontaminate the product.
4. It should preserve, as far as possible, the nutritive value and palatability of the starting material.
5. It should not significantly alter the important technological properties of the starting material.

In this case also, extensive studies have been carried out on aflatoxins; very limited information is available on other mycotoxins. Types of chemicals tested for the detoxification of aflatoxins include acids, alkalis, aldehydes, oxidizing agents, and gases like chlorine, sulfur dioxide, ozone, and ammonia [8]. Unfortunately, most of these chemicals do not satisfy all the above criteria [28]. Although they destroy aflatoxins, they may decrease significantly the nutritive value of the processed material or produce toxic products or products having undesirable side effects.

Some of the chemicals that have shown to be effective in destroying aflatoxins in the contaminated commodities besides satisfying the above criteria include: hydrogen peroxide or similar oxidizing agents [74,75]; calcium hydroxide/formaldehyde [76]; sodium hydroxide [34,77]; sodium hypochlorite [75,78]; sodium or potassium bisulfite [79,80], methylamine [81]; and ammonia [82–86]. Of the above chemicals, hydrogen peroxide, sodium hydroxide, and sodium hypochlorite would seem to have the ability for the production of aflatoxin-free protein isolates or products for edible use, while treatments with dimethylamine, methylamine, or ammonia would appear to have application for the detoxification of oilseed meals and maize [28].

Hydrogen peroxide has been utilized to develop practical methods for oxidative decontamination of aflatoxin in peanut meals. A promising laboratory method is the treatment of peanut meal at 80°C for 0.5 hour with hydrogen peroxide at pH 9.5 [74]. This procedure was patented by Sreenivasamurthy et al. [39] for the removal of aflatoxin from peanut seed meal. Chakravarti [87] also developed a 3% H_2O_2 technique which reduced aflatoxin level from 397 ppb to below 20 ppb in maize with less than 0.6% loss in protein and lipid levels. Paulsen et al. [88] also recorded reduction in aflatoxin content after spraying with hydrogen peroxide in skin-removed peanut kernels. An innovative method has been developed by Applebaum and Marth [89] to inactivate aflatoxin M_1 in milk by adding H_2O_2 and 0.5 mM riboflavin to the contaminated milk at 30°C for 30 min and then heating at 63°C for another 30 min. Hydrogen peroxide and benzoyl peroxide have also been found to be effective in destroying aflatoxins in peanuts [75,90].

Detoxification of aflatoxins from peanut meal has been reported with the treatment of formaldehyde alone or in combination with calcium hydroxide (lime) in a bench-scale reactor, operated both sealed and at atmospheric pressure [76]. In general, addition of calcium hydroxide to formaldehyde caused greater inactivation of the aflatoxins than did formaldehyde alone. Espoy [91] patented a process that uses finely divided calcium hydroxide for detoxification of aflatoxin in contaminated oilseed meals. Complete destruction of aflatoxin has also been reported in the spiked foods due to treatment with chlorine gas [92]. Reduction of mutagenicity and toxicity of aflatoxin has also been reported by this treatment.

Treatment of peanut meal with aldehydes to inactivate aflatoxins has been patented by Frayssinet and Lafarge [93]. Mann et al. [94] also screened more than 60 chemicals for the inactivation of aflatoxins in peanut and cottonseed meals and found calcium hydroxide, sodium hydroxide, formaldehyde, methylamine, and ammonia to be effective chemicals for this purpose. Use of sodium hydroxide to destroy aflatoxins in oilseed meals has also been reported by other workers [34,77]. A cooking procedure with aflatoxin-contaminated peanut meal was developed that utilized 2% sodium hydroxide and 30% moisture for 2 hours at 100°C. This procedure reduced total aflatoxin to < 5 ppb; however, this technique also reduced lysine content so that amendments were needed before treated material could be used as complete feed rations [95].

Provocative data have been obtained from the treatment of aflatoxin-contaminated peanut protein isolate and defatted peanut meal by sodium hypochlorite [78]. Results revealed that the sodium hypochlorite concentration and pH were the two important factors involved in reducing high toxin levels to non-detectable amounts; e.g., at pH 8.0, 0.4% sodium hypochlorite reduced aflatoxin B_1 from 725 ppb to trace amounts in ground raw peanuts, whereas at pH 9.0 only 0.3% sodium hypochlorite was required. Similar results were obtained with defatted peanut meal.

Sodium hypochlorite has also been utilized as a decontamination agent in the laboratories to degrade glass-bound toxin [33]. Excessive aflatoxin-sodium hypochlorite ratios can, however, produce aflatoxin B_1-2,3-dichloride [96]. Since this dihalide toxin derivative has been identified as an effective carcinogen [97], the use of hypochlorite was questioned. Even then, this technique has been used to eliminate aflatoxin with no residual toxicity in the treated materials [98,99].

Oxidative destruction of aflatoxin by ozone has been considered as a practical method for decontamination of oilseed meals [100]. Total aflatoxin B_1 was found to be inactivated after incubation with ozone for 2 hours at 100°C in cottonseed and peanut meals having 22% to 30% moisture. Aflatoxin B_2 and G_1 were, however, inactivated in the range of 78% to 90%. Because ozone reacts only with the olefinic bonds of B_1 and G_1 (B_2 and G_2 are less affected), its usefulness as detoxifying agent is reduced in the later stages [95].

Alkaline degradation of aflatoxins has also been obtained through methylamine, which brings about the hydrolysis of lactone ring in the aflatoxins more easily. This compound was effective in reducing the aflatoxin content of contaminated peanut and cottonseed meals. A 2% concentration of methylamine applied to peanut meal having 8.9% moisture content and subsequently heated for 90 min at 100°C, reduced the toxin concentration to < 5 ppb [82].

The treated meal has, however, been found to damage the normal functioning of rat liver. Mann et al. [101] increased the efficiency of aflatoxin degradation in cottonseed meal by simultaneous incorporation of 2% methylamine and 1% sodium hydroxide. Heating the oilseed meal with 2% monomethylamine at normal pressure or under mechanical pressure in an extruder brings about rapid degradation of the aflatoxins [81]. Unfortunately, under these conditions the detoxified meal has a sharp smell and an unpleasant taste which cannot be removed even after prolonged vacuum treatment. However, by adding 0% to 5% monomethylamine and 2% calcium hydroxide (w/w meal), it has been possible to produce an "acceptable" detoxified product. A detailed account of the residual aflatoxin content of groundnut meal after monomethylamine treatment and the nutritional evaluation of the detoxified meal has been given by Coker et al. [28].

Sodium bisulfite has been shown to form water-soluble products after reaction with major aflatoxins, viz., B_1, G_1, M_1, and aflatoxicol under various experimental conditions [102]. More than 45% reduction in the level of aflatoxin

M_1 has been recorded due to addition of 0.04 g potassium bisulfite per 10 ml milk [80]. The structure of sodium bisulfite adduct of aflatoxin has also been established.

Aqueous or gaseous ammonia with or without increased temperature and pressure seems to be the most efficient approach to detoxify aflatoxins from the contaminated agricultural commodities [95]. Masri et al. [83] patented a process of detoxification in which ammoniation of contaminated peanut meal containing 709 ppb aflatoxin B_1 moistened to 9.6% and 14.6% at 94°C for 60 min at 20 psig anhydrous ammonia pressure, reduced the level of aflatoxin B_1 by 96.4% and 97.6%, respectively.

Similar procedures of Dollear et al. [82] were also effective at somewhat milder conditions of temperature and time. Complete destruction of aflatoxin B_1 was obtained at 64°C for 30 min in peanut meal under ammonia pressure of 30 psig. This process, however, increased the overall nitrogen levels of the meal, but some reductions in lysine levels were also apparent.

In large-scale pilot-plant runs, Gardner et al. [84] specified the optimum processing conditions to detoxify aflatoxins in one or more ton lots of oilseed meals by ammoniation. The most effective conditions included adjustment of moisture levels to 12.5%, temperature from 114 to 122 °C, ammonia pressure of 50 psig, and incubation time of 60 min. Since a significant amount of ammonia was released at the end of the detoxification processes, monocalcium phosphate was added to the treated meal to absorb the residual ammonia and to make the detoxified material odor-free. The ammoniated cottonseed/peanut meal was eventually fed to the lactating cows, but none of the milk samples contained detectable amounts of aflatoxin M_1 [103,104].

A similar method has been suggested for the detoxification of aflatoxin-contaminated corn [105]. The rate and extent of detoxification were, however, dependent on the nature (i.e., gaseous/aqueous form) and concentration of ammonia used, moisture content, and temperature and retention time of the process. Residual aflatoxin B_1 has been found to be reduced from 1000 μg/kg to < 10 μg/kg if the corn having 17% moisture is added with 15% ammonia at 38°C for about 2 weeks. This process was subsequently expanded to larger lots (1000 bushels, or about 40 metric tons), using a standard grain bin with a perforated floor and a fan for recirculating the atmosphere above the grain [106]. Grain was also adjusted to 15% to 22% moisture level, with a target range of 18% to 19% after equilibrium [107,108]. When gaseous ammonia (0.5% to 15%) was introduced in the closed atmosphere for 24 hours and the treated grain was incubated for 2 weeks, aflatoxin level was found to be reduced from 100 to < 20 μg/kg.

Several distinct advantages and disadvantages have, however, been discussed for the utilization of anhydrous gaseous ammonia as decontaminating agent in maize [95]. A procedure for enzymatic (urease) release of ammonia from urea has also been developed successfully which detoxified aflatoxins in the contaminated peanut meal [109]. However, owing to a number of problems, this procedure could not be found suitable for industrial use.

Destruction of zearalenone and other fusarial toxins in the contaminated corn has been tried with the help of several chemicals [42,52,110,111]. No effect on this process was achieved when the contaminated corn was treated with propionic acid, acetic acid, hydrochloric acid, sodium bicarbonate, hydrogen peroxide, or ammonia. Complete destruction of zearalenone was recorded when the contaminated corn was treated with 3.7% formaldehyde solution for 16 hours at 50°C. Partial destruction was also reported by ammonium hydroxide and formaldehyde vapor.

Efficacy of different concentrations (3%, 5%, and 10%) of hydrogen peroxide has also been recorded for the destruction of zearalenone in contaminated corn [112]. Although the percentage of destruction was dependent on the concentration of hydrogen peroxide, temperature, and period of exposure, maximum destruction was recorded when the contaminated corn was treated with 10% hydrogen peroxide at 80°C for 16 hours.

Calcium hydroxide monomethylamine effectively decontaminated feeds containing T-2 toxin, diacetoxyscirpenol (DAS), or zearalenone at 10 to 20 mg/kg [113]. The success of the procedure was, however, dependent on the moisture content of the feed at the processing temperature.

Ammonia treatment, combined with heat and pressure, was used to decontaminate corn contaminated with 86 mg/kg of fumonisin [114]. A reduction of 79% in fumonisin level was recorded in this study.

C. Biological Methods

Among biological methods, microbial detoxification is the foremost approach for reducing the level of aflatoxins [48,72]. This aspect has been reviewed critically by Bhatnagar et al. [115]. Initial studies on microbial degradation were made by Ciegler et al. [116], who screened several micro-organisms including yeasts, molds, bacteria, actinomycetes, algae, and fungal spores for their activity to degrade and/or modify aflatoxins. Only one of the bacteria, *Flavobacterium aurantiacum* (NRRL B-184), was reported to significantly remove aflatoxin from the liquid medium without producing toxic byproducts. Toxin-contaminated milk, oil, peanut butter, groundnuts, and corn were completely detoxified, and contaminated soybean was partially detoxified by the addition of this micro-organism. Duckling assay revealed that detoxification of aflatoxin solution by the bacterium was complete with no new toxic product formed.

Removal of aflatoxin B_1 from growing cells of *F. aurantiacum* (NRRL B-184) has also been reported by Lillehoj et al. [117]. This bacterial strain also removed aflatoxin M_1 from aqueous solutions [118]. Toxin was completely removed from the liquid medium by incubating 5×10^{10} resting cells per milliliter with 8 μg/ml of aflatoxin M_1 from toxin-contaminated milk. The ability of *F. aurantiacum* to reduce aflatoxin B_1 concentration in peanut milk was also demonstrated by Hao and Bracket [119] and Hao et al. [120]. They also indicated that some of the toxins may be bound to the peanut protein and may not be available for the removal by *F. aurantiacum*.

Degradation of pure aflatoxins has also been reported by *Tetrahymena pyriformis* [121]. It decreased the concentration of aflatoxin B_1 by 58% in 24 hours, and by 67% in 48 hours. An unknown, bright-blue fluorescent compound was produced, with intensity about one half that of the unchanged B_1 and 0.55 for B_2 on a TLC plate, and with an ultraviolet spectrum showing maxima of 253, 261, and 328 mμ. No noticeable effect on aflatoxin G_1 was recorded in this treatment.

Aflatoxins have also been found to be degraded by several fungi. Certain acid-producing molds catalyzed the hydration of aflatoxin B_1 to a less toxic product, B_{2o} [122]. Two pathways (i.e., one directly and another through aflatoxin B_{2a}) have been suggested for the degradation of aflatoxin B_1 by *Aspergillus niger* [123]. Degradation of aflatoxin B_1 by *A. niger* and *Corynebacterium rubrum* was also analyzed by adding ^{14}C-labeled aflatoxin B_1 to cultures of these microorganisms [124]. Two blue fluorescent compounds, formed by *A. niger* from aflatoxin B_1 with Rf values 0.42 and 0.48 (Rf value of aflatoxin B_1 being 0.54), were accumulated and characterized by UV fluorescence and mass spectrometry as aflatoxin R_o. Under the same conditions, *Mucor ambiguus* and *Trichoderma viride* also produced aflatoxin R_o.

Doyle et al. [80] conducted detailed investigations on the biological degradation of mycotoxins in foods and other agricultural commodities. Mycelia of *Aspergillus parasiticus* was found to degrade aflatoxin, possibly via fungal peroxidase. Such degradation was, however, affected by strain of *A. parasiticus*, amount of mycelium, temperature, pH, and concentration of aflatoxin. Some other mycotoxins like patulin and rubratoxin were also shown to be degraded by actively fermenting yeasts and mycelium of *Penicillium rubrum*, respectively [125]. About 50% loss of aflatoxin B_1 and G_1 has been reported during the early stage of the miso fermentation, which was attributed to the degradation of toxin by the microbes [125]. Direct evidence of this degradation is, however, lacking. An atoxigenic strain of *A. flavus* has also been found to degrade the level of aflatoxin produced by a toxigenic strain on the seed [126].

An investigation on aflatoxin B_1 degradation by an isolate of *Rhizopus arrhizus* obtained from Georgia groundnuts and three known *Rhizopus* species revealed the accumulation of two fluorescent metabolites of aflatoxin B_1 during degradation [127]. These were identified as hydroxylated stereo isomers derived from reduction of the ketone function on the cyclopentane ring of aflatoxin B_1. The same species of *Rhizopus* also degraded aflatoxin G_1 [128]. An intermediate in this biological degradation was isolated and identified as previously reported metabolite of *A. flavus* (aflatoxin B_3) and *A. parasiticus* (parasitol), which was also proved conclusively using radioactive isotopes. Strains of *Rhizopus* and *Neurospora* also degraded aflatoxin B_1 in groundnut [129].

Besides aflatoxin B_1, the fate of some other mycotoxins, viz., ochratoxin A and citrinin in the contaminated grains used as substrates for the fermentative production of ethanol, has also been investigated by several workers [130–133]. Common results revealed (1) little degradation of the toxin during fermentation, (2) absence of toxin in the alcohol, and (3) accumulation of toxin in the spent grains. These results are of serious concern, as most of the spent grains are diverted toward animal feed. Similar results were also obtained for fumonisin B_1-contaminated corn [134]. They, however, suggested ethanol fermentation of fumonisin-contaminated corn coupled with effective detoxification of distiller's grains and aqueous stillage as a practical process strategy for salvaging contaminated corn.

Total degradation of DON by some micro-organisms has also been recorded in 6 days, although the toxicity test of the end product has not been carried out [135]. The ability of some mesophilic, thermophilic, and thermotolerant fungi to biodegrade ochratoxin A has been demonstrated by using species isolated from naturally self-heat Zambian maize [136]. Maximum reduction in toxin levels was recorded through *Thermoascus* isolate F51 (90%), followed by *Paecilomyces* isolate A23 (80%).

The potential for biological detoxification of aflatoxins also exists by the use of some domestic animals as a "filter" [8]. Several observations indicated that some farm animals can consume rations containing aflatoxin B_1 up to 200 ppb with no adverse effects on feed efficiency. Although some aflatoxin or its metabolites may be transmitted from the feed to animal products such as meat and eggs, the amount transmitted is only a very small percentage of that fed [137].

The application of biological methods of detoxification on commercial scale may, however, be doubtful because of the toxicity of the detoxifying microorganisms, nutritional losses during treatment, and the sophistication of the treatment.

IV. SAFETY ASSESSMENT OF DETOXIFIED COMMODITIES

Danger of mycotoxin contamination in food exists from the time the crop is grown in the field until the final food product is consumed. No practical and economical methods are so far available that can be recommended for the prevention or complete removal of mycotoxin contamination. However, in terms of human health, one of the most effective methods for reducing the levels of mycotoxins in the food supply is to encourage the diversion of moldy and contaminated grains and oilseeds to nonfood uses or to processing industries which recover one or more mycotoxin-free products.

If we consider the diversion of contaminated food as feed, we should also have to know the relationship between the concentration of mycotoxin in the rations fed to meat and dairy animals and poultry, and the concentration of mycotoxin or its animal metabolites that appear as residues in muscle, adipose and organ tissues, or in milk and eggs [73].

In the U.S., the responsibility of controlling the levels of contaminants such as mycotoxins is jointly shared by the USDA and the FDA, the former being responsible for the raw agricultural products as well as meat, milk, and eggs, and the latter being responsible for the consumer products [138]. Appropriate sampling plans have also been designed to minimize, to the extent possible, human exposure to the contaminant without affecting the marketability of the product; i.e., in the process, the processor's risk and the consumer's risk are taken into account.

As discussed earlier, the severity of mycotoxin exposure can also be minimized with the help of several physical, chemical, and biological methods of detoxification. In all cases, safety of the processed commodity is to be assessed before it is finally recommended for consumption as food or feed. The principles for the safety assessment of food additives and contaminants have been dealt with in a report by WHO [139]. Detailed information has also been given on the safety of mycotoxin-contaminated processed food, maybe through various postharvest procedures, and the role of individual and governmental agencies in implementing safety regulations as separate chapters in this book by Drs. Park, Wood, Trucksess, and Moy. These will not be repeated here. The main emphasis here will be on the acceptability of the processed commodity as food and feed.

Contaminated grains and feeds may contain a wide variety of mycotoxins of differing chemical characteristics including heat stability, solubility, and adsorbent affinity [140,141]. Consequently, it is difficult to develop a detoxification method that will be equally effective against each of the various mycotoxins present in the naturally contaminated grains. However, in a broader sense the physical method may be regarded as better than the chemical or biological method for the following reasons: (1) it is simple to operate; (2) it does not involve highly sophisticated equipment; (3) it can remove considerable amount of mycotoxins with no hazardous chemical residues; and (4) it does not impair the nutritive quality of the substrate [23]. Therefore, these methods should be preferred in order to retain the food value of the detoxified commodities.

As mentioned earlier, the use of hydrogen peroxide and ammonia has gained prominence among all the known chemical treatments reported for the detoxification of mycotoxins in agricultural commodities which may be due to their nontoxic effects. Hydrogen peroxide treatment of peanut meal has not been found to affect the protein efficiency factor (PER) in rat trials, and no toxicity was exhibited to chick embryo [74]. Toxicity of ammonia-treated commodities has been evaluated by several workers. Brekke et al. [105] could not record any toxic effect of ammonia on ducklings, broiler chicks, or trout when these were fed with the detoxified corn. In another study, no deleterious effects on egg production, egg quality, feed consumption, or mortality in poultries was recorded due to feeding of ammoniated commodities [142]. Pathological and histopathological examinations conducted on experimental and farm animals fed on ammoniated meals did not show any sign of aflatoxicosis [58]. Egg production and immunological responses were also found to be unaffected in the poultries in the same experiment.

On the basis of other toxicological studies also, it has been concluded that ammoniation of agricultural commodities contaminated with aflatoxins provides a practical and economical method for detoxification [143–146]. A small but significant incidence of liver cancer has, however, been reported in rainbow trout that has been fed milk from cows that had received ammoniated cottonseed meal in the rations [187]. Contrary to this report, ammonia-treated feed has been found to reduce the carcinogenicity of aflatoxin in trout [116]. Since ammonia converts parent aflatoxin B_1 to an almost nontoxic byproduct in the processed commodities, it is preferred as a successful detoxifying agent in different parts of the world [148]. Park et al. [149] have also given a comprehensive review on the ammonia detoxification process and outlined the current applications and regulatory status of the ammonia process for reducing the levels of aflatoxin in animal feeds.

Ammoniation of seeds is also authorized by the Food and Drug Administration in the U.S. Even in the U.S., Arizona, California, and Texas permit the ammoniation of cottonseed products. Mexico and South Africa have approved it for the detoxification of corn, while several European countries use ammonia-treated peanut meal as animal feed. The process is routinely used in France, Senegal, Sudan, and Brazil. Ammonia-treated peanut meal is also being imported on a regular basis by several member countries of the European Community [56,150]. However, further research is needed in order to declare the ammonia detoxification method as absolutely safe for human and animal health.

REFERENCES

1. FAO. Prevention of mycotoxins. FAO Food and Nutrition Paper 10, 1979, p 71.
2. TM Phillips, BA Clement, DL Park. Approaches to reduction of aflatoxins in foods and feeds. In: DL Eaton, JD Groopman, eds. The Toxicology of Aflatoxins: Human Health, Veterinary and Agriculture Significance. San Diego: Academic Press, 1994, p 383.
3. F Haumann. Eradicating mycotoxins in food and feeds. Inform 6:248, 1995.
4. VK Mehan. Screening groundnuts for resistance to invasion by *Aspergillus flavus* and to aflatoxin production. Aflatoxin Contamination of Groundnut: Proceedings of the International Workshop, ICRISAT, Patancheru, India, 1989, pp 323–334.
5. KW Campbell, DG White. Evaluation of corn genotypes for resistance to *Aspergillus* ear rot, kernel infection and aflatoxin production. Plant Dis 79:1039, 1995.

6. JG Heathcote, JR Hibbert. Biochemical effects, structure activity relationships. In: LA Goldblatt, ed. Aflatoxin: Chemical and Biological Aspects. Amsterdam: Elsevier, 1978, p 112.
7. JW Dickens. Aflatoxin control programme for peanuts. J Am Oil Chem Soc 54: 225A, 1977.
8. LA Goldblatt, FG Dollear. Review of prevention, elimination and detoxification of aflatoxins. Pure Appl Chem 49:1959, 1977.
9. KR Natrajan, KC Rhee, CM Cater, KF Mattil. Distribution of aflatoxins in various fractions separated from raw peanuts and defatted peanuts meal. J Am Oil Chem Soc 52:44, 1975.
10. A Ciegler. Detoxification of aflatoxin-contaminated agricultural commodities. In: R Rosenberg, ed. Toxin: Animal, Plant and Microbial. Oxford: Pergamon Press, 1976, p 729.
11. PJ Tiemstra. Aflatoxin control during food processing of peanuts. In: JV Rodricks, CW Hesseltine, MA Mehlman, eds. Mycotoxins in Human and Animal Health. Park Forest South, IL: Pathotox Pub., Inc., 1977, p 121.
12. DL Park, L Stoloff. Aflatoxin control—how a regulatory agency managed risk from an unavoidable natural toxicant in food and feed. Regul Toxicol Pharmacol 9:109, 1989.
13. JE Schade, K McGreevy, AD King Jr, G Fuller. Incidence of aflatoxin in California almonds. Appl Microbiol 29:48, 1975.
14. WE Steiner, K Brunschweiler, E Leimbacher, R Schneider. Aflatoxins and fluorescence in Brazil nuts and pistachio nuts. J Agric Food Chem 44:2453, 1992.
15. TC Pearson, DC Slaughter, HE Studer. Separating "early split" from "normal" pistachio nuts for removal of nuts contaminated on the tree with aflatoxin. Pap Am Soc Agric Eng 1993, p 11.
16. LJ Ashworth Jr, JL McMeans, JL Pyle, CM Brown, JW Osgood, RE Ponton. Aflatoxin in cotton seed: influence of weathering on toxin content of seeds and on a method for mechanically sorting seed lots. Phytopathology 58:102, 1968.
17. F Escher. Mycotoxin problems in the production and processing of peanuts and pecans in the USA. Lebensmitt Wiss Technol 7:255, 1974.
18. A Bockellee-Morvan, P Gillier. Trial of the elimination of aflatoxin in groundnuts by physical methods. Proc Conf on Animal Feeds of Tropical and Subtropical Origin. London: Trop Prod Inst, 1975, pp 291–296.
19. L. Stoloff. Occurrence of mycotoxins in foods and feed. In: JV Rodricks, ed. Mycotoxins and Other Fungal Related Food Problems. Adv Chem Series No 149. Washington: Amer Chem Soc, 1976, p 23.
20. WE Huff, WM Hagler Jr. Density segregation of corn naturally contaminated with aflatoxin, deoxyninalenol and zearalenone. J Food Prot 48:416, 1985.
21. RJ Cole. Technology of aflatoxin detoxification. In: S Natori, K Hashimato, Y Ueno, eds. Mycotoxins and Phycotoxins. Amsterdam: Elsevier, 1989, p 174.
22. SC Basappa, V Sreenivasamurthy. Decontamination of groundnut oil from aflatoxin. Indian J Technol 17:440, 1979.
23. SC Basappa. Physical methods of detoxification of aflatoxin contaminated materials. In: KS Bilgrami, T Prasad, KK Sinha, eds. Mycotoxins in Food and Feed. Bhagalpur, India: Allied Press, 1983, p 251.

24. DL Park, B Liang. Perspectives on aflatoxin control for human food and animal feed. Review. Trends Food Sci Technol 4:334, 1993.
25. SC Basappa, V Sreenivasamurthy. Partition of aflatoxin during separation of different constituents of groundnut kernel. J Food Sci Technol 11:137, 1974.
26. T Tanaka, A Hasegawa, YM Yamamoto, Y Ueno. Residues of *Fusarium* mycotoxins nivalenol, deoxynivalenol and zearalenone in wheat and processed food after milling and baking. J Food Hyg Soc Jpn 27:653, 1986.
27. LS Lee, HS Jang, T Tanaka, YJ Oh, CM Cho, Y Ueno. Effect of milling on decontamination of *Fusarium* mycotoxins nivalenol, deoxynivalenol and zearalenone in Korean wheat. J Agric Food Chem 35:126, 1987.
28. RD Coker, BD Jones, MJ Nagler. Aflatoxin control and detoxification. In: Mycotoxin Training Manual. London: Trop Dev Res Inst, 1984, Section 8, p 1.
29. ET Rayner, FG Dollear. Removal of aflatoxins from oilseed meals by extraction with aqueous isopropanol. J Am Oil Chem Soc 45:622, 1968.
30. LJ Vorster. Studies on the detoxification of peanuts contaminated by aflatoxin (in French). Rev Fr Cps Gras 13:7, 1966.
31. LA Goldblatt, JA Robertson Jr. Process for removing aflatoxin from peanuts. U.S. Patent 3,515,736, 1970.
32. HK Gardner Jr, SS Koltun, HLE Vix. Solvent extraction of aflatoxins from oilseed meals. J Agric Food Chem 16:990, 1968.
33. RW Detroy, EB Lillehoj, A Ciegler. Aflatoxin and related compounds. In: A Ciegler, S Kadis, SJ Ajl, eds. Microbial Toxins. Vol. VI. New York: Academic Press, 1971, p 53.
34. LA Goldblatt. Control and removal of aflatoxin. J Am Oil Chem Soc 48:605, 1971.
35. ET Rayner, SP Koltun, FG Dollear. Solvent extractions of aflatoxins from contaminated agricultural products. J Am Oil Chem Soc 54:242A, 1977.
36. WA Parker, D Melnick. Absence of aflatoxin from refined vegetable oils. J Am Oil Chem Soc 43:635, 1966.
37. T Shantha, V. Sreenivasamurthy. Detoxification of groundnut oil. J Food Sci Technol 12:20, 1975.
38. Z Farah, JJR Martins, MR Bachmann. Removal of aflatoxin in raw unshelled peanuts by a traditional salt boiling process practiced in the North East of Brazil. Lab-Wissen Technol 16:122, 1983.
39. V. Sreenivasamurthy, S. Srikantia, HAB Parpia. Removing aflatoxin from peanut seed meals. Indian Patent 120,257, 1971.
40. K Aibara, N Yano. New approach to aflatoxin removal. In: JV Rodricks, CW Hesseltine, MA Mehlman, eds. Mycotoxins in Human and Animal Health. Park Forest South, IL: Pathotox Pub. Inc., 1977, p 151.
41. HL Trenholm, LL Charmley, DB Perlusky, RM Warner. Washing procedures using water or sodium carbonate solutions for the decontamination of three cereals contaminated with deoxynivalenol and zearalenone. J Agric Food Chem 40:2147, 1992.
42. LL Charmley, DB Perlusky. Decontamination of *Fusarium* mycotoxins. In: JD Miller, HL Trenholm, eds. Mycotoxins in Grain—Compounds Other Than Aflatoxin. St. Paul: Eagan Press, 1994, p 421.
43. RD Hartley, BF Nesbitt, J O'Kelly. Toxic metabolites of *Aspergillus flavus*. Nature (Lond) 198:1056, 1963.

44. AJ Feuell. Aflatoxin in groundnuts. IX. Problems of detoxification. Trop Sci 8:61, 1966.
45. T. Asao, G. Buchi, MM Abdel-Kader, SB Chang, EL Wick, GN Wogan. The structure of aflatoxins B_1 and G_1. J Am Chem Soc 787:882, 1965.
46. CM Christensen, CJ Mirocha, RA Meronuck. Mold, Mycotoxins and Mycotoxicoses. Agric Exp Stn Misc Report 143. St. Paul: Univ of Minnesota, 1977.
47. R Rehana, SC Basappa, V Sreenivasamurthy. Destruction of aflatoxin in rice by different cooking methods. J Food Sci Technol 16:111, 1979.
48. EH Marth, MP Doyle. Update on molds: degradation of aflatoxin. Food Technol 33:81, 1979.
49. HF Conway, RA Anderson, EB Bagley. Detoxification of aflatoxin contaminated corn by roasting. Cereal Chem 55:115, 1978.
50. RD Coker. Control of aflatoxin in groundnut products with emphasis on sampling, analysis and detoxification. Aflatoxin Contamination of Groundnut: Proceedings of the International Workshop, ICRISAT, Patancheru, India, 1989, pp 123–132.
51. CT Dwarkanath, V Sreenivasamurthy, HAB Parpia. Aflatoxin in Indian peanut oil. J Food Sci Technol 6:107, 1968.
52. GA Bennett, OL Shotwell, CW Hesseltine. Destruction of zearalenone in contaminated corn. J Am Oil Chem Soc 57:245, 1980.
53. JC Young, LM Subryan, D Potts, ME McLaren, FH Gobran. Reduction in levels of deoxynivalenol in contaminated wheat by chemical and physical treatments. J Agric Food Chem 34:461, 1986.
54. HM Stahr, GD Osweiler, P Martin, M Domoto, B Debey. Thermal detoxification of trichothecene contaminated commodities. In: GC Llewellyn, CE O'Rear, eds. Biodeterioration Research 1. New York: Plenum Press, 1987, p 231.
55. PM Scott. Possibilities of reduction or elimination of mycotoxins present in cereal grains. In: J Chelkowski, ed. Cereal Grain—Mycotoxins, Fungi and Quality in Drying and Storage. Amsterdam: Elsevier, 1991, p 529.
56. SC Basappa, T Shantha. Methods for detoxification of aflatoxins in foods and feeds—a critical appraisal. J Food Sci Technol 33:95, 1996.
57. RS Applebaum, AE Brackett, EH Marth. Aflatoxin: toxicity to dairy cattle and occurrence in milk and milk products—a review. J Food Prot 45:752, 1982.
58. U Samarajeewa, AC Sen, MD Cohen, CL Wei. Detoxification of aflatoxins in food and feed. J Food Prot 53:489, 1990.
59. JC Arthur Jr, JA Robertson Jr. Detoxification of aflatoxin. U.S. Patent 3,506,452 (1970).
60. T. Shantha, V Sreenivasamurthy. Photodestruction of aflatoxin in groundnut oil. Indian J Technol 15:453, 1977.
61. U Samarajeewa, TV Gamage, SN Arseculeratne. Nontoxicity of solar-irradiated edible oils contaminated with aflatoxin B_1. Toxicon 26:38, 1988.
62. T Shantha, V Sreenivasamurthy. Storage of groundnut oil detoxified by exposure to sunlight. Indian J Technol 18:346, 1980.
63. T Shantha. Detoxification of groundnut seed and products in India. In: Aflatoxin Contamination in Groundnut. Proceedings of the International Workshop, ICRISAT, Patancheru, India, 1989, pp 153–160.
64. T Shantha, V Sreenivasamurthy. Use of sunlight to partially detoxify groundnut

(peanut) cakes flour and casein contaminated with aflatoxin B_1. J Assoc Off Anal Chem 62:291, 1981.
65. A Kane, NBA Diop, TS Diack. Removal of aflatoxin from crude peanut oil (Abst). IX Int IUPAC Symp on Mycotoxins and Phycotoxins, Rome, Italy, 1996, p 179.
66. FJ Santamarina, SJ Gimenez, C Sabater, V Sanchis. Measure to reduce and eliminate mycotoxins in food and feeds. Rev Iber Micol 12:52, 1995.
67. H Hooshand, CF Klopfenstein. Effects of gamma irradiation on mycotoxin disappearance and amino acid contents of corn, wheat and soybean, with different moisture contents. Plant Food Human Nutr 47:337, 1995.
68. N Paster, R Barkai-Golan, R Padova. Effect of gamma radiation on ochratoxin production by the fungus *Aspergillus ochraceus*. J Sci Food Agric 36:445, 1985.
69. T Kume, H Ito, H Iizuka, M Takehisa. Radiosensitivity of *Aspergillus versicolor* isolated from animal feeds and destruction of sterigmatocystin by gamma-irradiation. Agric Biol Chem 47:1065, 1983.
70. RS Farag, MM Rashed, AA Abo Hgger. Aflatoxin destruction by microwaves (Abstr.). IX Int IUPAC Symp on Mycotoxins and Phycotoxins, Rome, Italy, 1996, p 339.
71. W Lindner, K Hasenhuti. Decontamination and detoxification of corn which was contaminated with trichothecenes applying ultrasonication (Abstr.). IX Int IUPAC Symp on Mycotoxins and Phycotoxins, Rome, Italy, 1996, p 182.
72. A Ciegler. Detoxification of aflatoxin contaminated agricultural commodities. In: P Rosenberg, ed. Toxins: Animal, Plant and Microbial. London: Pergamon Press, 1978, p 729.
73. FAO. Mycotoxins. FAO Food and Nutrition Paper 2, 1977.
74. V Sreenivasamurthy, HAB Parpia, S Srikantia, A Shankar Murti. Detoxification of aflatoxin in peanut meal by hydrogen peroxide. J Assoc Off Anal Chem 50:350, 1967.
75. KC Rhee, KR Natrajan, CM Cater, KR Mattil. Processing edible peanut protein concentrates and isolates to inactivate aflatoxins. J Am Oil Chem Soc 54:245A, 1977.
76. LP Codifer, GF Mann, FG Dollear. Aflatoxin inactivation. Treatment of peanut meal with formaldehyde and calcium hydroxide. J Am Oil Chem Soc 53:204, 1976.
77. FG Dollear, HK Gardner. Inactivation and removal of aflatoxin. Proc Fourth Nat Peanut Res Conf, Tifton, GA, 1966, pp 72–81.
78. KR Natrajan, KC Rhee, CM Cater, KF Mattil. Destruction of aflatoxins in peanut protein isolates by sodium hypochlorite. J Am Oil Chem Soc 52:160, 1975.
79. WM Hagler Jr, JE Hutchins, PB Hamilton. Destruction of aflatoxin in corn with sodium bisulfite. J Food Prot 45:1287, 1982.
80. MP Doyle, RS Applebaum, RE Bracket, EH Marth. Physical, chemical and biological degradation of mycotoxins in foods and agricultural commodities. J Food Prot 45:964, 1982.
81. C Giddey, J Brandt, G Bunter. The detoxification of oil-seed cakes polluted by aflatoxins. Research and development of an industrial process. Ann Technol Agric 27:331, 1977.
82. FG Dollear, ME Mann, LP Codifer Jr, HK Gardner Jr, SP Koltun, HLE Vix. Elimination of aflatoxins from peanut meal. J Am Oil Chem Soc 45:862, 1968.

83. MS Masri, HLEE Vix, LA Goldblatt. Process for detoxifying substances contaminated with aflatoxin. U.S. Patent 3,429,709, 1969.
84. HK Gardner Jr, SP Koltun, FG Dollear, ET Rayner. Inactivation of aflatoxins in peanut and cotton seed meals by ammoniation. J Am Oil Chem Soc 48:70, 1971.
85. OL Brekke, RO Sinnhuber, AJ Peplinski, et al. Aflatoxin in corn: ammonia inactivation and bioassay with rainbow trout. Appl Environ Microbiol 34:34, 1977.
86. RD Coker, K Jewers, NR Jones, J Nabney, DH Watson. Process for the destruction of aflatoxin in agricultural products. U.K. Patent GB 2108365 B (1985).
87. AG Chakravarti. Detoxification of corn. J Food Prot 44:591, 1981.
88. MR Paulsen, GH Brusewitz, BL Chary, GV Odell, J Pominski. Aflatoxin content and skin removal by treatment with chemicals, water spray, heated air and nitrogen. J Food Sci 41:667, 1976.
89. R Applebaum, EH Marth. Inactivation of aflatoxin M_1 using hydrogen peroxide and hydrogen peroxide plus riboflavin or lactoperoxidase. J Food Prot 45:557, 1982.
90. W Trager, L Stoloff. Possible reactions for aflatoxin detoxification. J Agric Food Chem 15:679, 1967.
91. HM Espoy. Detoxification of oilseed meal with calcium hydroxide. U.S. Patent 3,689,275 (1972).
92. AC Sen, CI Wei, SY Fernando, J Toth, EM Ahmed, GE Dunaif. Reduction in mutagenicity and toxicity of aflatoxin B_1 by chlorine gas treatment. Food Chem Toxicol 26:745, 1988.
93. C Frayssinet, C Lafarge. Procede pour eliminer les aflatoxines dans les toruteaux d'arachide. French Patent 2,098,711 (1972).
94. GE Mann, LP Codifer, HK Gardner, SP Koltun, FG Dollear. Chemical inactivation of aflatoxins in peanut and cottonseed meals. J Am Oil Chem Soc 47:173, 1970.
95. EB Lillehoj, JH Hall. In: MS Zuber, EB Lillehoj, BL Renfro, eds. Aflatoxin in Maize—A Proceeding of the Workshop. CIMMYT, Mexico City, 1987, p 260.
96. M Castegnaro, M Friesen, J Michelon, EA Walker. Problems related to the use of sodium hypochlorite in the detoxification of aflatoxin B_1. J Am Indust Hyg Assoc 42: 389, 1981.
97. DH Swenson, JS Miller, EC Miller. The reactivity and carcinogenicity of aflatoxin B_1-2,3 dichloride, a model for the putative 2,3 oxide metabolite of aflatoxin B_1. Cancer Res 35:3811, 1975.
98. CY Yang. Comparative studies on the detoxification of aflatoxins by sodium hypochlorite and commercial bleaches. Appl Microbiol 24:885, 1972.
99. FA Draughon, EA Childs. Chemical and biological evaluation of aflatoxin after treatment with sodium hypochlorite, sodium hydroxide and ammonium hydroxide. J Food Prot 45:703, 1982.
100. CT Dwarkanath, ET Rayner, GE Mann, FG Dollear. Reduction of aflatoxin levels in cottonseed and peanut meals by ozonization. J Am Oil Chem Soc 45:93, 1968.
101. GE Mann, HK Gardner Jr, AN Booth, MR Gungmann. Aflatoxin inactivation: chemical and biological properties of ammonia and methylamine treated cottonseed meal. J Agric Food Chem 19:1155, 1971.
102. MP Doyle, EH Marth. Bisulfite degrades aflatoxin: effect of temperature and concentration of bisulfite. J Food Prot 41:774, 1978.

103. JD McKinney, GC Kavanagh, JT Bell, et al. Effects of ammoniation of aflatoxin in rations fed lactating cows. J Am Oil Chem Soc 50:79, 1973.
104. JM Fremy, JP Gautier, MP Herry, C Terrier, C Calet. Effects of ammoniation on the "carry-over" of aflatoxins into bovine milk. Food Addit Contam 5:39, 1988.
105. OL Brekke, AJ Peplinski, EL Griffin Jr. Cleaning trials for corn containing aflatoxin. Cereal Chem 52:198, 1975.
106. OL Brekke, AC Stringfellow, AJ Peplinski. Aflatoxin inactivation in corn by ammonia gas: laboratory trials. J Agric Food Chem 26:1383, 1978.
107. EB Bagley. Decontamination of corn containing aflatoxin by treatment with ammonia. J Am Oil Chem Soc 56:808, 1979.
108. WP Norred. Effect of ammoniation on the toxicity of corn artificially contaminated with aflatoxin B_1. Appl Pharmacol Toxicol 51:411, 1981.
109. E Santi, P Amedeo, G Piva. Detoxification of aflatoxin B_1 in peanut meal by urea treatment. Ann Chim 22:299, 1983.
110. JC Young. Reduction in levels of deoxynivalenol in contaminated corn by chemical and physical treatments. J Agric Food Chem 34:465, 1986.
111. JC Young, HL Trenholm, DW Friend, D Prelusky. Detoxification of deoxynivalenol with sodium bisulfite and evaluation of the effects when pure mycotoxin or contaminated corn was treated and given to rats. J Agric Food Chem 35:259, 1986.
112. El-Sayed AM Abdalla. Zearalenone: incidence, toxigenic fungi and chemical destruction in Egyptian cereals (Abstr.). IX Int. IUPAC Symp on Mycotoxins and Phycotoxins, Rome, Italy, 1996, p 180.
113. J Bauer, M Gareis, W Datzler, B Gedek, K Heinritzi, G Kabilka. Detoxification of mycotoxins in animal feeds. Tierarrtl Umsch 42:70, 1987.
114. DL Park, SM Rua, CJ Mirocha, ESAM Abdalla, CY Weng. Mutagenic potentials of fumonisin contaminated corn following ammonia decontamination procedure. Mycopathology 117:105, 1992.
115. D Bhatnagar, EB Lillehoj, JW Bennett. Biological detoxification of mycotoxins. In: JE Smith, RS Henderson, eds. Mycotoxins and Animal Foods. Boca Raton: CRC Press, 1991, p 815.
116. A Ciegler, EB Lillehoj, RE Peterson, HH Hall. Microbial detoxification of aflatoxin. Appl Microbiol 14:934, 1966.
117. EB Lillehoj, A Ciegler, HH Hall. Aflatoxin B_1 uptake by *Flavobacterium aurantiacum* and resulting toxic effects. J Bacteriol 93:464, 1967.
118. EB Lillehoj, RD Stubblefield, GM Shanon, OL Shotwell. Aflatoxin M_1 removal from aqueous solutions by *Flavobacterium aurantiacum*. Mycopathol Mycol Appl 45:259, 1971.
119. Y-Y Hao, RE Brackett. Removal of aflatoxin B_1 from peanut milk inoculated with *Flavobacterium aurantiacum*. J Food Sci 53:1384, 1988.
120. Y-Y Hao, RE Brackett, TOM Nakayama. Removal of aflatoxin B_1 from peanut milk by *Flavobacterium aurantiacum*. In: Aflatoxin Contamination of Groundnuts: Proceedings of the International Workshop, ICRISAT, Patancheru, India, 1989, pp 141–152.
121. DJ Teunisson, JA Robertson. Degradation of pure aflatoxin by *Tetrahymena pyriformis*. Appl Microbiol 15:1099, 1967.

122. A Ciegler, RE Peterson. Aflatoxin detoxification: hydroxydihydro aflatoxin B_1. Appl Microbiol 16:665, 1968.
123. H Tsubouchi, K Yamamoto, K Hisada, Y Sakabe, K Tsuchihira. Degradation of aflatoxin B_1 by *Aspergillus niger*. Proc Jpn Assoc Mycotoxicol 12:33, 1980.
124. R Mann, HJ Rehm. Degradation products from aflatoxin B_1 by *Corynebacterium rubrum*, *Aspergillus niger*, *Trichoderma viride* and *Mucor ambiguus*. Eur J Appl Microbiol 2:297, 1976.
125. M Manabe, S Matsurra. Studies on the fluorescent compounds in fermented foods. Part IV. Degradation of added aflatoxin during miso fermentation. J Food Sci Technol (Jpn) 16:275, 1972.
126. PJ Cotty. Influence of field application of an atoxigenic strain of *Aspergillus flavus* on the populations of *A. flavus* infecting cotton bolls and on the aflatoxin content of cottonseed. Phytopathology 84:1270, 1994.
127. RJ Cole, JW Kirksey, BR Blankenship. Conversion of aflatoxin B_1 to isomeric hydroxy compounds by *Rhizopus* spp. J Agric Food Chem 20:1100, 1972.
128. RJ Cole, JW Kirksey. Aflatoxin G_1 metabolism by *Rhizopus* species. J Agric Food Chem 19:222, 1971.
129. MJR Nout. Effect of *Rhizopus* and *Neurospora* spp. on growth of *Aspergillus flavus* and *A. parasiticus* and accumulation of aflatoxin in groundnut. Mycol Res 93:518, 1989.
130. EB Lillehoj, A Lagoda, WF Maisch. The fate of aflatoxin in naturally contaminated corn during the ethanol fermentation. Can J Microbiol 25:911, 1979.
131. P Krogh, B Hald, P Gijertsen, F Myken. Fate of ochratoxin A and citrinin during malting and brewing experiments. Appl Microbiol 28:31, 1974.
132. WK Nip, FC Chang, FS Chu, N Prentice. Fate of ochratoxin A in brewing. Appl Microbiol 30:1048, 1975.
133. RJ Bothast, GW Nofsinger, AA Lagoda, LT Black. Integrated process for ammonia inactivation of aflatoxin contaminated corn and ethanol fermentation. Appl Environ Microbiol 43:961, 1982.
134. RJ Bothast, GA Bennett, JE Vancau-Wenberge, JL Richard. Fate of fumonisin B_1 in naturally contaminated corn during ethanol fermentation. Appl Environ Microbiol 58:233, 1992.
135. J Binder, EM Horvath, J Heidegger, et al. Determination of the deoxynivalenol-detoxification capacity of micro-organisms (Abstr.). Int IUPAC Symp on Mycotoxins and Phycotoxins, Rome, Italy, 1996, p 217.
136. PF Mooney, JE Smith, JG Anderson, RD Coker. Thermophilic microbial biodegradation of ochratoxin A (Abstr.). IX Int IUPAC Symp on Mycotoxins and Phycotoxins, Rome, Italy, 1996, p 232.
137. IFH Purchase. Aflatoxin residues in food of animal origin. Food Cosmet Toxicol 10: 531, 1972.
138. AE Pohland. Overview of international mycotoxin and phycotoxin programs (Abstr.). IX Int IUPAC Symp on Mycotoxins and Phycotoxins, Rome, Italy, 1996, p 109.
139. WHO. Principles for the Safety Assessment of Food Additives and Contaminants in Food. Environ Health Criteria 70. Geneva: WHO, 1977, p 174.
140. BC Foster, GA Neish, DR Lauren, HL Trenholm, DB Prelusky, RMT Hamilton. Fungal and mycotoxin content of slashed corn. Microbiol Aliment Nutr 4:199, 1986.

141. BC Foster, HL Trenholm, DW Friend, BK Thompson, KE Hartin. Evaluation of different sources of deoxynivalenol (vomitoxin) fed to swine. Can J Anim Sci 6: 1149, 1986.
142. BL Hughes, BD Barnett, JE Jones, JW Dick. Safety of feeding aflatoxin-inactivated corn with the Leghorn layer-breeders. Poult Sci 58:1202, 1979.
143. LL Southern, AJ Clawson. Ammoniation of corn contamination with aflatoxin and its effects on growing rats. J Anim Sci 50:459, 1980.
144. WP Norred. Effect of ammoniation on the toxicity of corn artificially contaminated with aflatoxin B_1. Toxicol Appl Pharmacol 50:411, 1979.
145. WP Norred. Excretion and distribution of ammoniated ^{14}C-labeled aflatoxin B_1 contaminated corn (Abstr.). Fed Proc 40:280, 1981.
146. C Frayssinet, L Frayssinet. Effect of ammoniation on the carcinogenicity of aflatoxin contaminated groundnut oil cakes: long term feeding study in rat. Food Addit Contam 7:63, 1990.
147. RO Sinnhuber, DJ Lee, JH Wales, MK Landers, AC Keyl. Hepatic carcinogenesis of aflatoxin M_1 in rainbow trout (*Salmo gairdneri*) and its enhancement by cyclopropene fatty acids. J Natl Cancer Inst 53:1285, 1974.
148. DL Park. Perspectives on mycotoxin decontamination procedures. Food Addit Contam 10:49, 1993.
149. DL Park, LS Lee, RL Price, AE Pohland. Review of the decontamination of aflatoxin by ammoniation and current status. J Assoc Off Anal Chem 71:685, 1988.
150. RJ Cole, JW Dorner, PD Blankenship. Management strategies for prevention and control (Abstr.). IX Int IUPAC Symp on Mycotoxins and Phycotoxins, Rome, Italy, 1996, p 33.

13

Effectiveness of Postharvest Procedures in Management of Mycotoxin Hazards

Rebeca López-García and Douglas L. Park
Louisiana State University, Baton Rouge, Louisiana

I. INTRODUCTION

Naturally occurring toxicants are unavoidable and unpredictable and pose a unique challenge to food safety. The Food and Agriculture Organization of the United Nations has reported that the production of agricultural commodities is barely sustaining the world's increasing population; and every year, at least 25% of the world's food crops are contaminated with mycotoxins [1,2]. Thus, the destruction of contaminated products or diversion to nonhuman uses is not always practical, and would seriously compromise the world's food supply. Risk management through processing and decontamination/detoxification to avoid potential risks associated with mycotoxins and maintain the world's food supply became a major issue in food safety. Futhermore, an integrated food safety program in a given country will have to consider factors such as the public health significance of the contaminant, the effects on the availability of the countries' food supply, economy, legal infrastructure, analytical resources, producer/consumer education, the effectiveness of communication systems, and the availability of preharvest, harvest, and postharvest technologies.

Prevention through preharvest management is the best method for controlling mycotoxin contamination; however, should the contamination occur, the hazards associated with the toxin must be managed through postharvest procedures if the product is to be used for food or feed purposes. Ideally, there should be an integrated management system where mycotoxin hazards would be minimized in every phase of production. This chapter will concentrate on effectiveness of postharvest procedures for reducing risks associated with exposure to mycotoxin- contaminated food/feed.

Once a food commodity has been harvested, there should be routine steps to ensure the safety of the product. Figure 1 shows a decision-making process to develop a food safety program that will ultimately prevent potential hazards associated with exposure to mycotoxin-contaminated food and feed. This process involves several steps including the identification of the toxicant, hazard evaluation, exposure determination, risk determination, and risk management. A hazard evaluation provides a description of the expected lesion and of the dose/response relationship. The determination is made based on available data on comparative toxicological properties of the compound of interest, usually using laboratory animal and in vitro toxicological tests. Following the development of suitable analytical and sampling procedures, a determination must be made of the geographic and temporal distribution of the toxic agent in each major commodity susceptible to contamination. Combined with the commodity consumption data, this information can lead to a determination of exposure to the toxic agent by various human and animal populations. A risk determination is made using available information on hazard evaluation and exposure assessment [3]. Finally, after the risk has been determined, risk management decisions can be made. A good risk management program will involve the establishment of regulatory limits, a monitoring program, a sampling plan, and development of alternative processing/decontamination procedures or diversion to less-risk uses. This program should also include the development of strategies to destroy highly contaminated products.

Processing can play an important role in diminishing the potential risks of mycotoxin-contaminated food commodities. Thus, it is important to evaluate the effects of processing on individual toxins to determine the actual risk posed to humans/animals. A large amount of data have been generated through surveys on incidence and levels of mycotoxin contamination in agricultural products, such as nuts, oilseeds, and grains [4,5]. However, published information on mycotoxins in foods processed for human consumption is limited. More research is needed to provide food manufacturers and regulatory authorities with information on mycotoxin persistence and transformation during processing. This information would be helpful in the development of effective food safety programs.

In the past, the establishment of a regulatory limit for the most potent mycotoxin, i.e., aflatoxin, was enough to ensure a safe food supply. Any product that was found to be contaminated over the regulatory limit had to be diverted to less-risk uses or destroyed. At present, not only is more information available, but also the world's food supply is more limited. Furthermore, there have been several reports on the cocontamination of various toxins, i.e., aflatoxin B_1/fumonisin B_1

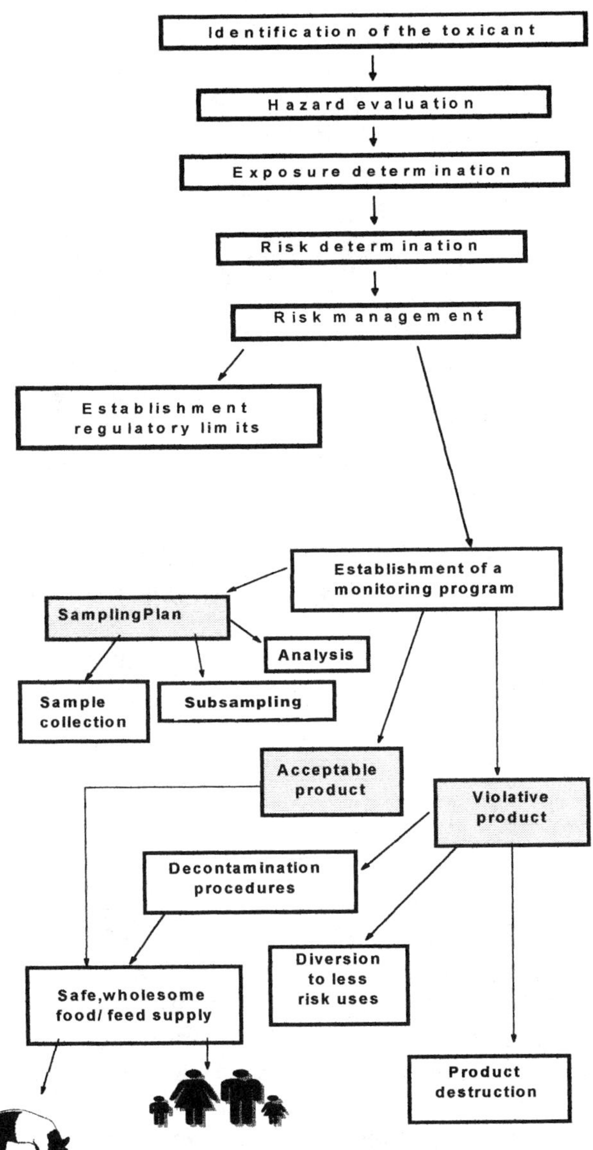

FIGURE 1 Risk assessment analysis and food safety program for naturally occurring toxicants.

[6], ochratoxin A/aflatoxin B_1 [7,8], ochratoxin A/citrinin [9,10], ochratoxin A/deoxynivalenol [11], ochratoxin A/penicillic acid [12], ochratoxin A/T-2 toxin [13], aflatoxin/cyclopiazonic acid [14], aflatoxin/kojic acid [15], aflatoxin B_1/deoxynivalenol [16], and aflatoxin B_1/T-2 toxin [17]. More documentation on the levels and types of cocontaminating mycotoxins occurring in different commodities is needed to more clearly define the mycotoxin combinations and concentrations to which humans and animals are likely to be exposed. A multitoxin model should be considered when evaluating a particular decontamination/detoxification process. This complex environment poses new regulatory concerns. When the effect of processing and modified processing to destroy/inactivate one or several toxins is considered, many decision-making paradoxes may rise. Should innovative processing designed to detoxify a product be allowed, or should traditionally processed products contaminated with low/medium doses of several mycotoxins stay in the market? More research is needed in the area of alternative processing that will decontaminate/detoxify mycotoxin-contaminated commodities and render products that will be approved by regulatory agencies and stay commercially competitive. Ultimately, the best decontamination/detoxification process would be the one that is approved by regulatory agencies, is cost-effective, and reduces the risks of exposure to multiple-toxin-contaminated commodities to acceptable levels.

When evaluating the effectiveness of a specific processing procedure on a given mycotoxin or multitoxin system, many factors need to be examined. These factors include:

1. Chemical stability of the mycotoxin(s)
2. Nature of the process
3. Type and interaction with the food matrix
4. Interaction between different mycotoxins present

Some processes involve the separation of the commodity into different fractions, i.e., milling or oil pressing. In these cases, it is important to determine which fraction remains contaminated. If most of the toxin(s) remain with process byproducts, and these are not used for human consumption, then the process may be deemed effective in detoxifying the main product. However, caution should be used when determining the end fate of the byproducts, i.e., animal feed. When traditional processing methods are not sufficient to decontaminate/detoxify the commodity, then decontamination procedures and/or modified processing must be considered (Fig. 2). In these cases, careful evaluation of the procedure must be performed before considering it as an alternative. It is well known that most mycotoxins would not remain chemically stable in extreme conditions, i.e., extreme pH, heat, etc. However, these procedures would not result in a usable food/

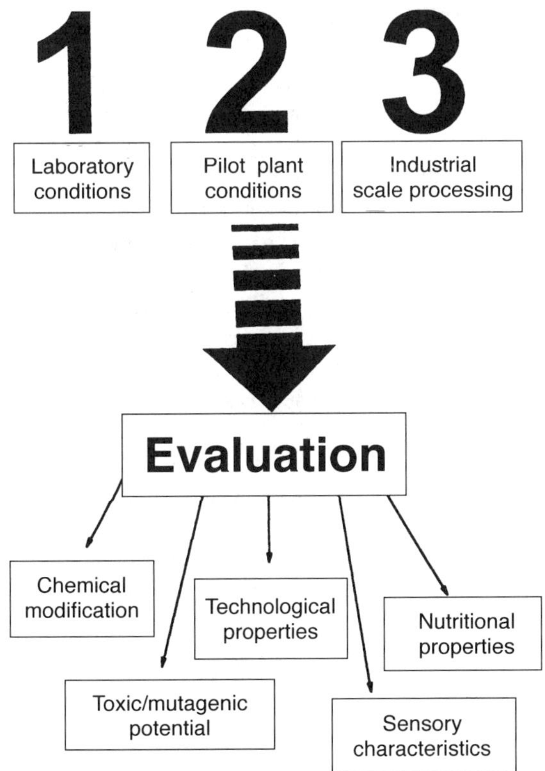

FIGURE 2 Development of decontamination/detoxification procedures.

feed product. Thus, special considerations must be made when developing decontamination/detoxification procedures. Park and co-workers [18] have proposed specific criteria for evaluating the acceptance of a given mycotoxin reduction or decontamination procedure. In order to be effective, the process must:

1. Inactivate, destroy, or remove the toxin
2. Not produce or leave toxic residues in the food/feed
3. Retain nutritive value and food/feed acceptability of the product
4. Not alter significantly the technological properties of the product
5. If possible, destroy fungal spores

II. RISKS ASSOCIATED WITH MYCOTOXIN CONTAMINATION

Managing the risks associated with mycotoxin contamination should involve an integrated system. Processed foods cannot be safe if prevention, control, good manufacturing practices, and quality control are not used at all stages of production. Figure 3 shows selected aspects of an integrated aflatoxin control program. This can be used as a model system when developing food safety programs for other mycotoxins or naturally occurring toxicants. Each phase would have to be adapted according to the toxin and the commodity in question. A Hazard Analysis Critical Control Point (HACCP) approach to processing mycotoxin-contaminated commodities should be considered. Control parameters would include time of harvesting, temperature, and moisture during storage and transportation, selection of agricultural products prior to processing, processing/decontamination conditions, temperature, addition of chemicals, and final product storage and transportation. This approach should be developed not only for specific single toxins but also for multitoxin models, i.e., aflatoxin B_1/fumonisin B_1.

As mentioned before, a food safety management program for naturally occurring toxicants should involve different phases such as the ones outlined below:

1. Setting of regulatory limits
2. Establishment of a monitoring program
    A. Establishment of a sampling plan
        1. Sample collection
        2. Sample preparation
        3. Analysis
    B. Permitted uses of mycotoxin-contaminated products
    C. Designation of use for violative products
3. Control through processing
    A. Good manufacturing practices
    B. Quality control
4. Decontamination through specific treatments
    A. Evaluation of the final product
    B. Designation of use of treated product

The establishment of regulatory limits may vary in each country depending on level of exposure and sociological, political, and economic factors. Due to the increased international trade, it has been suggested that an international standard be set. However, at present, different regulations are in place. Table 1 shows current regulatory status for aflatoxins in the United States. Currently, mycotoxin regulatory data are available for 90 countries. Seventy- seven countries are known to have some regulations on mycotoxins, and 13 countries are known to have no

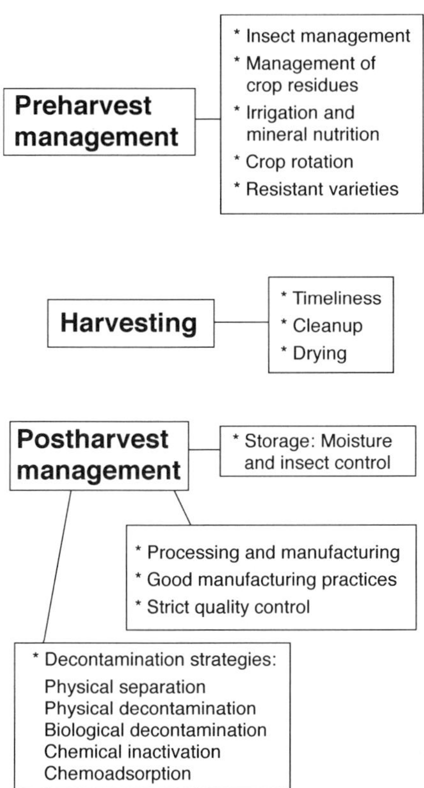

FIGURE 3 Integrated aflatoxin control management.

regulations. No data are available for about 40 countries with more than 5 million inhabitants, most of them in Africa [19]. Most of the existing mycotoxin regulations are for aflatoxins (Table 2).

Once a regulatory limit has been set, sampling plays an important role in determining the fate of a particular product. Adequate, random sampling techniques should be used at each point of analysis. It is important to consider the existence of "hot spots," or highly contaminated portions of the product. If these highly contaminated portions are not detected, the toxin(s) would then be distributed throughout the lot, rendering a highly contaminated product. Once the levels of contamination have been determined through a monitoring program, the final use of the commodity can be determined. The product may be deemed either safe for human consumption, safe for feed use through treatment, or totally unsuitable for use at any point in the food chain.

Table 1 Current Aflatoxin Action Levels (Total) Established by the Food and Drug Administration

Food or feed	Action level (μg/kg, ppb)
Human foods (except milk)	20
Milk	0.5
Animal feeds (except as listed below)	20
Cottonseed meal (used for mature beef, swine, and poultry rations)	300
Corn and peanut meal for breeding beef cattle, breeding swine, or mature poultry	100
Corn and peanut meal for finishing swine	200
Corn and peanut meal for feedlot beef cattle	300

Source: Ref. 20.

A good example of management through a monitoring program is the aflatoxin control program used by the state of Arizona [20]. During the year 1978, almost 910,000 pounds of milk were dumped with contamination levels as high as 10 ppb aflatoxin M_1. As a result of this huge commercial loss and the need to establish an effective aflatoxin control program, the state instituted a program to monitor aflatoxin levels in whole cottonseed and cottonseed products, the major source of aflatoxin contamination in feed. All cottonseed produced in the state is tested for aflatoxin content. The maximum size of the lots tested is 100 tons, and the testing is conducted in state-certified laboratories. The end use of the product is dictated by aflatoxin levels found. Cottonseed lots testing over 20 μg aflatoxin/kg are usually treated with ammonia to reduce aflatoxin levels. Cottonseed products containing < 20 ppb aflatoxin, with or without ammonia decontamination

Table 2 Medians and Ranges in 1995 of Maximum Tolerated Levels (μg/kg) for Some Aflatoxins and Number of Countries That Have Regulations for These

	Median	Range	No. of countries
B_1 in foodstuffs	4	0–30	33
B_1, B_2, G_1, G_2 in foodstuffs	8	0–50	48
B_1 in foodstuffs for children	0.3	0–5	5
M_1 in milk	0.05	0–1	17
B_1 in feedstuffs	20	5–1000	19
B_1, B_2, G_1, G_2 in feedstuffs	50	0–1000	21

Source: Ref. 19.

TABLE 3 Aflatoxin Residues in Milk Produced in Arizona

Year	No. of samples	No. of samples with aflatoxin M_1 concentration of: 0.2–0.5 μg/L	>0.5 μg/L	% Samples containing detectable levels of aflatoxin M_1
1979[a]	535	140	0	26
1980	972	169	0	17
1981	940	109	0	11
1982	802	227	0	28
1989	900	121	0	13

Source: Ref. 20.
[a]In 1978 909,442 lb of milk was destroyed with aflatoxin M_1 levels as high as 10 μg/L.

treatment, can be used for dairy rations. The key aspect of the aflatoxin decontamination program is that the product is tested for aflatoxin levels before and after the ammonia treatment, as appropriate. The use of this program has been a significant benefit in keeping Arizona's milk supply safe from aflatoxin contamination (Table 3).

In integrated mycotoxin management, each phase will help prevent the risk of exposure to the toxins. After preharvest control, harvesting, and monitoring, most of the contaminated product has been separated and destined to different uses or decontamination treatments. However, industrial processing is still an important phase of control. The effects of some selected processes are presented in the next section. The final goal of processing will not only be the decontamination of the commodity, but also the prevention of new mold colonization with its subsequent mycotoxin production. Thus, when the product reaches the consumer, the hazards associated with mycotoxin contamination of food will be minimal.

III. EFFECTS OF PROCESSING ON MYCOTOXIN-CONTAMINATED FOODS

Mycotoxins comprise a structurally diverse and chemically complex group that can contaminate a wide variety of food products. Table 4 presents a summary of some of the most commonly affected commodities, the potential toxic mold genera/species, and the mycotoxins that contaminate them.

Most of the research on the effects of processing has been related to the aflatoxins, but some processing and stability experiments have been done on other *Aspergillus* and *Penicillium* metabolites such as ochratoxin A, citrinin, penicillic acid, and patulin. More recently, research has been conducted on the

TABLE 4 Summary of Selected Reports of Potentially Toxic Molds from Various Food or Agricultural Commodities

Commodity	Potentially toxic genera/species found		Potential mycotoxins
Wheat, flour, bread, cornmeal, popcorn, tortillas, chips	*Aspergillus* *flavus* *ochraceus* *versicolor* *Fusarium* *moniliforme*	*Penicillium* *citrinum* *citreo-viride* *cyclopium* *martensii* *patulum* *pubertum*	Aflatoxins Ochratoxin A Sterigmatocystin Patulin Penicillic acid Deoxynivalenol Zearalenone Fumonisin
Peanut, in-shell pecans, nuts	*Aspergillus* *flavus* *ochraceus* *versicolor* *Fusarium* sp.	*Penicillium* *cyclopium* *expansum* *citrinum*	Aflatoxins Ochratoxin A Patulin Sterigmatocystin Trichothecenes Cytochalasins Oosporein
Meat pies, cooked meats, cocoa powder, hops, cheese	*Aspergillus* *flavus*	*Penicillium* *viridicatum* *roqueforti* *patulum* *commune*	Aflatoxins Ochratoxin A Patulin Penicillic acid
Aged meat products, cured ham, moldy meats, cheese	*Aspergillus* *flavus* *ochraceus* *versicolor*	*Penicillium* *viridicatum* *cyclopium*	Aflatoxins Ochratoxin A Patulin Penicillic acid Sterigmatocystin Penitrem
Black and red pepper, pasta noodles	*Aspergillus* *flavus* *ochraceus*	*Penicillium* sp.	Aflatoxins Ochratoxin A
Dry beans, soybeans, corn, sorghum, barley	*Aspergillus* *flavus* *ochraceus* *versicolor* *Alternaria* sp. *Fusarium* *moniliforme*	*Penicillium* *cyclopium* *viridicatum* *citrinum* *expansum* *islandicum* *urticae*	Aflatoxins Ochratoxin A Sterigmatocystin Penicillic acid Patulin Citrinin Griseofulvin Alternariol Altenuene

TABLE 4 Continued

Commodity	Potentially toxic genera/species found		Potential mycotoxins
Refrigerated and frozen pastries	*Aspergillus* *flavus* *versicolor*	*Penicillium* *cyclopium* *citrinum* *martensii* *citreo-viride* *palitans* *puberulum* *roqueforti* *urticae* *viridicatum*	Aflatoxins Sterigmatocystin Ochratoxin A Patulin Penicillic acid Citrinin Penitrem
Moldy supermarket foods	*Penicillium* *cyclopium* *Fusarium* *oxysporum* *solani*	*Aspergillus* sp.	Penicillic acid Trichothecenes Aflatoxins Possibly other *Aspergillus* and *Penicillium* toxins
Foods stored in homes, both refrigerated and nonrefrigerated	*Penicillium* sp.	*Aspergillus* sp.	Aflatoxins Kojic acid Ochratoxin A Penitrem Patulin Penicillic acid
Apples and apple products	*Penicillium* *expansum*		Patulin

Source: Ref. 21.

effect of processing on the *Fusarium* toxins, i.e., trichothecenes, fumonisins, deoxynivalenol (DON), T-2 toxin, and zearalenone [22]. A detailed description of the effects of processing will be provided in another chapter. This chapter will only discuss a few examples pertinent to the effectiveness of normal processing operations in reducing the risks associated with mycotoxin contamination and the industrial applicability of selected decontamination procedures.

A. Physical Methods of Mycotoxin Removal

1. Cleaning and Segregation

Once a contaminated product has reached a processing facility, cleaning and separating are the first alternatives of control. These procedures are usually noninvasive, and, except for milling, will not significantly alter the product. In some cases, these are the best methods to reduce mycotoxin presence in the finished products, i.e., a significant amount of aflatoxins in shelled peanuts can be removed by electronic sorting and hand-picking [23]. However, complete removal of all contaminated particles or aflatoxin cannot be expected with physical methods of separation. Since the toxin can diffuse into the interior of the kernel, residual contamination may be present at very low levels in the final product. If there is a high level of residual contamination, other procedures must be used to manage the residual contamination in the final product.

Flotation and density segregation of aflatoxin-contaminated corn and peanuts have been reported to significantly reduce aflatoxin concentrations [24]. Kirskey and co-workers [25] reported that 95% of the aflatoxin in 21 of 29 samples of peanuts was contained in kernels that floated in tap water. Furthermore, Phillips and co-workers [26] reported that the mean aflatoxin level was decreased from 301 ppb to 20 ppb using flotation as a separation mechanism. Although there is still an aflatoxin residue, separation has been shown to significantly reduce aflatoxin contamination and can be considered as an effective first line of defense for certain products.

Separation of fumonisin-contaminated corn, particularly screenings, based on size has been reported as a possible candidate for decontamination. Corn screenings are broken corn kernels that usually contain about 10 times the fumonisin content of intact corn [27]. However, screenings represent a significant proportion of the corn used for feed, so it is not possible to segregate fumonisin contents by screening size without compromising the feed supply. Further decontamination procedures need to be used to treat corn screenings used for feed.

Although significant quantities (ca. 25%) of deoxynivalenol (DON) can be removed by cleaning and polishing, the toxin remains in wheat flour at levels ranging from 60% to 80% of original toxin levels from the starting wheat [28,29]. Another study on wheat samples from several states showed a reduction in deoxynivalenol content in flour from spring wheat that had been cleaned by aspiration prior to milling into flour [30]. However, the toxin residues found in the flour indicate that physical separation should only be considered a step in a more elaborate risk management plan.

If apples rotted with *Penicillium expansum* are used for commercial juice production, the final product may contain patulin. It has been reported that trimming of apples to remove the rot reduces the patulin content by 93% to 99% [31]. In general, physical separation or trimming of apples before processing is the best method to reduce patulin contamination. In this case, physical separation should be considered the focal point of the risk management plan. The correct use of trimming and separation will render safe high-quality commercial juice.

Cleaning and segregation represent a primary source of the removal of hazard in mycotoxin-contaminated food. In some cases, these procedures will eliminate the problem. However, in most cases, further processing is needed to reduce the risks associated with mycotoxin contamination. Although some contamination may persist, physical separation is a good alternative for the industry. These processes require an initial investment to purchase adequate equipment, but their maintenance is rather inexpensive. Furthermore, physical removal will not affect the product's characteristics; so after the initial separation stage, there is no other significant alteration of the product. With well-calibrated separating equipment, this method is a good choice for products such as peanuts, other tree nuts (aflatoxin), corn (aflatoxin and fumonisin), and apples (patulin).

2. Wet Milling

Wet milling is a process widely used for corn. Thus, there have been several studies that evaluate the effects of this process on mycotoxins associated with this grain. In this case, an established processing method can be used to manage contaminated grains. It is important to identify the fraction(s) that remain toxic. These fractions may then be diverted to less-risk uses or subjected to decontamination procedures. Laboratory studies have reported that during wet milling of inoculated corn, aflatoxin B_1 was distributed in the milling fractions. It went primarily into the steep water (39% to 42%) and fiber (30% to 38%), with the remainder found in gluten (13% to 17%), germ (6% to 10%), and starch (only 1%) [32,33]. The distribution of mycotoxins in the different fractions poses new concerns. In a good risk management plan, an individual contaminated fraction needs to be considered as a new product because it will have its own unique characteristics. Some of the factors to be considered for a particular milling product include the end use, the chemical properties, and the level and type of contamination. Each product or byproduct should be approached differently, and risk management decisions should be made on an individual basis. Other mycotoxins are also distributed during wet milling of corn. Zearalenone concentrates in the gluten (49% to 56%) and milling solubles (17% to 26%), but is absent in the starch [32,34]. Recent wet milling laboratory-scale studies on naturally contaminated corn have shown that the toxin remains mostly in the gluten and the germ remains relatively free of the toxin. The starch fractions which account for 65% to 71% of milled products, were free of detectable zearalenone. Milling solubles contained one to four times the levels of zearalenone in the starting corn and gluten fractions. The most contaminated fraction contained two to seven times the levels in starting corn. Although these fractions account for only 14% to 19% of the corn, they account for 72% to 75% of the zearalenone. Fiber and germ fractions contained up to three times the levels of zearalenone in the starting corn, and these fractions account for 15% to 16% of the milled product. These results show that if a mycotoxin-contaminated lot of corn is processed by wet milling, toxin-free starch is produced; however, other products generally used in animal feed have much higher levels of toxin than the starting corn [34]. The concentration of toxin in a particular fraction simplifies the process of risk management. The other milling fractions can be considered safe, and decisions need to be taken on a single product. The management of only one toxic fraction can be simpler than the handling of several contaminated milling products.

An advantage of wet milling is that additional chemicals can be added to the steeping solution. So, a modified process could be used to ensure the safety of the final products. Traditional wet milling (using lactic acid-sulfurous acid steeping) of two samples of fumonisin B_1 naturally contaminated corn has been conducted by Bennet and co-workers [35]. In corn containing 13.9 μg/g of fumonisin B_1, a portion (22%) of recoverable toxin was found in the steep and process waters. Other fractions contained fumonisins with gluten having the highest concentration, followed by fiber and finally germ with the lowest concentration. The gluten and fiber from corn contaminated at 13.9 μg/g could pose a risk because they contain toxins at levels considered to be harmful to certain animals [36].

In wet milling of deoxynivalenol-contaminated corn, much of the DON went into the steep liquor, although measurable amounts remained in the starch [22]. Experimental wet-milling of corn contaminated with T-2 toxin showed that 67% of the toxin was removed by the steep and process water (T-2 toxin is more water-soluble than aflatoxin or zearalenone). Wet-milling of corn containing ochratoxins A and B showed that 43% of ochratoxin A remaining in the steeped corn went into the process water and solubles on subsequent processing, with 4% remaining in the germ and 51% in the grits [37].

Mycotoxins have diverse chemical characteristics; therefore, the distribution during wet milling will vary according to the commodity and the type and level of contamination. Each case should be considered separately when developing a risk management plan for wet milling operations.

3. Dry Milling

Dry milling will also fractionate the toxin(s). Thus, identification of the toxic fractions is needed to adequately use milling as a decontamination procedure. In dry milling of naturally contaminated corn, the highest levels of aflatoxin B_1 occurred in the germ and hull fractions; but distribution varied with the contamination levels. The grits, low-fat meal, and low-fat flour (the prime products) contained only 6% to 10% of the aflatoxin B_1. In artificially contaminated rice, aflatoxin was greatly reduced by milling, with more than 95% in the bran and polish fraction [38]. In durum wheat, aflatoxin B_1 was determined to be in peripheral parts of the kernel. Upon milling, the aflatoxin concentration in the flour varied according to the quality of the final product and was maximal in the bran; prior conditioning reduced the aflatoxin content by 47% [22].

For zearalenone-contaminated corn, all dried milled corn fractions contained zearalenone, and only 3% to 10% was removed [22]. Dry cleaning (screening) of the zearalenone-naturally contaminated corn does not significantly reduce toxin levels in corn lots. Bennet and co-workers [39] reported that all mill fractions from laboratory scale and commercial processes were contaminated with zearalenone. The highest levels of toxin were found in the hull and high-fat fractions. Prime product mix (grits, low-fat meal, and flour, which account for 57% to 63% of product yield) contained about 20% of the zearalenone from the starting corn samples [39].

No specific dry milling studies have been conducted on fumonisin-contaminated grains. However, assays on corn-based foods have shown that the toxin is present in human foodstuffs, i.e., corn grits and corn meal [35,40,41]. On the industrial perspective, milling can be used as an effective method of separation. If the distribution of the toxin is determined, and alternative management of the toxic fraction is developed, milling can be considered a good cost/benefit method. A combination of cleaning, sorting, and milling can significantly reduce the risks associated with mycotoxins.

B. Physical Methods of Detoxification

1. Thermal Inactivation

Thermal inactivation is a good alternative for products that are usually heat-processed. However, some of the mycotoxins are chemically stable at processing temperatures. Thermal inactivation should be evaluated for the conditions of a particular process. Additional processing steps, i.e., use of compounds that would promote the chemical degradation of a toxin, could be used to ensure the safety of the procedure. In some cases, a change in conditions such as pH or moisture content will contribute to the chemical modification of a particular toxin.

Aflatoxins are resistant to thermal inactivation and are not destroyed completely by boiling water, autoclaving, or a variety of food and feed processing procedures [42]. Aflatoxins may be destroyed partially by oil- and dry-roasting of peanuts. However, Stoloff reported that aflatoxins are generally stable in peanut materials at room temperature [5]. Baur [43] found no significant changes in levels of aflatoxins B_1, B_2, G_1, and G_2 in peanut meals, or in raw and roasted peanut butter stored at 23°C for 2 years. Lutter and co-workers [44] reported that alternative peanut roasting methods, i.e., microwave roasting, destroy aflatoxins almost completely. Unfortunately, this process would involve an increase in processing cost, making it commercially impractical. Thus, microwave roasting is not a good industrial solution, even if it yields a safe product.

Roasting has been a good method for certain commodities—i.e., peanuts. As mentioned before, if an operation that is regularly used in processing is an effective decontamination procedure, then that operation would be the first choice to manage that particular product. This would provide a safe, inexpensive, risk management method. An example of roasting as an effective means of control is its use in coffee processing. It has been reported that roasting green coffee beans containing added aflatoxin B_1 at 200°C produced a 79% concentration loss after 12 min and > 94% loss after 15 min [45]. Furthermore, experimental roasting of green coffee beans at 200°C for 20 min destroyed 68% of added sterigmatocystin under laboratory conditions [46]. It is important to mention that these studies were performed with spiked coffee beans. The use of spiked material for experimental processes gives useful information. However, the effects of roasting should be confirmed with naturally contaminated green coffee beans because naturally incurred toxins may have different interactions with the food matrix. It is important to confirm the effectiveness of a processing operation under industrial conditions before using it as part of a risk management plan.

Several studies reported that aflatoxin B_1 is moderately stable in heated peanut oil, corn oil and coconut oil. Therefore, frying in unrefined oils could add aflatoxin during processing [47,48]. Fortunately, oils are not of concern because only a small percentage of aflatoxin B_1 present in oilseeds passes into extracted or pressed oil, and refining and bleaching operations essentially eliminate it [49]. Gracian reported that the average loss of total aflatoxin concentration in pressed olive oil was 76% [50].

Conway and co-workers [51] reported that roasting corn (145 to 165°C) yielded a reduction of 40% to 80% of the original aflatoxin B_1. Studies on cooking of aflatoxin-contaminated rice found that normal cooking of rice destroyed about 49% of aflatoxin B_1; pressure cooking and cooking with excess water destroyed more of the aflatoxin B_1 in the rice (73% and 82%, respectively). These results provide evidence of the influence of water on stability of aflatoxin during heat processing. Furthermore, when pasta products contaminated with aflatoxins were cooked for 10 min in boiling water, 29% of the aflatoxin was found in the drained water and 66% in the cooked pasta [52].

Surveys have shown that thermally processed corn products, i.e., canned corn, tortillas, and ready-to-eat cereals, generally have lower concentrations of fumonisin B_1 (FB_1) than do unprocessed products, i.e., corn meal and grits [49,53,54]. Since most corn-based foods receive some sort of thermal processing, it has been suggested that this kind of procedure must have an effect on the fumonisin content in food. However, results reported by Jackson and co-workers [55], Alberts et al. [56], Dupuy et al. [57], and Scott and Lawrence [58] indicate that the fumonisins are fairly heat-stable compounds. In general, it has been reported that the loss of FB_1 and FB_2 is more rapid and extensive in alkaline or acidic environments than at neutral pH. Thermal processing operations such as boiling or retorting, which occur at temperatures < 125°C, have little effect on the fumonisin content of food; however, foods that reach temperatures > 150°C during processing may have significant losses of fumonisins. Although some of these studies show decomposition of fumonisins, the toxic potential is not necessarily eliminated (R. Lopez-Garcia and D.L. Park, unpublished data); therefore, detoxification of these compounds cannot be assumed. Further work is needed to determine the toxicity of thermally processed fumonisin-contaminated foods [59]. In this case, thermal processing is not a good choice for industry. Alternative methods for fumonisin decontamination need to be applied.

Studies on the effect of heat processing of ochratoxin A-contaminated flour have reported that considerable reduction of ochratoxin A (76%) occurred in samples of white flour heated to 250°C for 40 min [22]. Ochratoxin A appears to be more readily destroyed in dry cereal than in the presence of water (unlike aflatoxin B_1 and patulin). This was also evident in a study reported by Osborne, where ochratoxin A was not degraded during breadmaking, but 62% was lost after baking of biscuits, which have a lower water content [60].

Processing of other commodities naturally contaminated with ochratoxin A has shown that after blanching, salting, and heat-processing beans, about 53% of ochratoxin A remained [61]. In this case, it is important to determine if the heat processing will yield products with unchanged technological characteristics.

Several authors have studied the effect of making bread on aflatoxin present in wheat flour or in maize flour [63,64]. Variable destruction levels have been observed, but there are still conflicting results. Bennett and Richard [35] reported that baking does not destroy or significantly reduce levels of deoxynivalenol. DON has proven to be more heat-stable during food processing than any other mycotoxin tested. No DON was destroyed during the baking of different ethnical products such as Egyptian bread, Western-style bread, or cookies baked from hard wheat flour [22].

2. Irradiation

The effects of irradiation on most mycotoxins is relatively unknown. However, with the state of today's processing facilities, irradiation does not seem to be a good commercial method. The cost involved in setting up an irradiation facility and the reluctance of the consumer to accept irradiated products makes this method inaccessible to most industries.

Feuell reported that gamma irradiation (2.5 Mrad) did not degrade aflatoxin in contaminated peanut meal and that UV light produced no observable change in fluorescence or toxicity of the treated sample. However, exposure to contaminated peanut oil to short-wave and long-wave UV light resulted in reduction of aflatoxin levels [65,66].

3. Adsorption from Solution and Covalent Binding

Adsorption from solution and covalent binding of aflatoxins to nonbiological materials are good methods for aflatoxin decontamination. In industries such as the oil refineries, the use of adsorbent materials is readily available and is part of normal processing operations. A variety of adsorbent materials, i.e., activated carbon and clays, has been shown to bind aflatoxins in aqueous solutions, and certain aluminosilicates have been reported to bind aflatoxins in peanut oil and animal feeds. The long-term effects and safety of the aluminosilicates has not been determined [67–69]. However, a phyllosilicate clay (HSCAS® or NovaSil®) that is currently used as an anticaking agent in the feed industry has been studied for its ability to bind aflatoxin and reduce the toxicity/mutagenicity of the final product. Phillips and co-workers have reported that this clay (1) tightly binds aflatoxins in aqueous suspensions; (2) markedly diminishes aflatoxin uptake by the blood and distribution to target organs; (3) prevents acute aflatoxicosis in farm animals; and (4) decreases the levels of aflatoxin M_1 residues in milk [25]. In this case, phyllosilicate clay is an effective decontamination agent that could be used in the feed industry. However, its only approved use is as an anticaking agent, and its use for removal of aflatoxin from feed is not permitted. Thus, its beneficial effects are limited to a particular industrial process. The bentonite clays used in oil purification (to remove pigments) seem to adsorb aflatoxins present in the unrefined oil. Activated charcoal has also proven to be effective in reducing the patulin in naturally contaminated fruit juices [70,71]. More research is needed for the application of this method with other mycotoxins and commodities.

C. Biological Decontamination

Biological methods are a good option for the fermenting industry, i.e., breweries. In this case, however, it is important to determine the yeast's viability in the presence of mycotoxins. It is also important to determine the effect of mycotoxins on the yield and development of secondary characteristics associated with the final product. It is important to note that biological methods demonstrating effective decontaminating properties are usually the result of specific compounds produced by selected microorganisms. When this occurs, it is usually more efficient and economical to add the active agent directly.

Chu et al. [72] reported that aflatoxin B_1 is not completely removed during the beer brewing process; i.e., 18% and 27% remained from two contamination levels of starting material, 1 and 10 ppm, respectively. It was relatively stable in the cooker mash treatment, but significantly lost in the wort boiling and final fermentation steps. Dam and co-workers [73] found that aflatoxin B_1 is reduced by 47% after cooking and fermentation of corn, wheat, or corn:milo. Following distillation, slightly higher concentrations of aflatoxin B_1 were present in the wet solids (dry weight basis) than were in the starting corn [74]. Aflatoxin from contaminated corn does not go into the alcoholic distillate, but accumulation of aflatoxin in spent grains is a potential problem when using this material as animal feed. Further decontamination procedures must be used if these byproducts are to be used as animal feed [73,74].

Bennet and Richard [35] reported that although pure fumonisins exhibit considerable water solubility, residues of toxin remain in the distillers' dried grains. This fermentation product accounts for 31% to 35% of the total fumonisin B_1 in the starting corn. Facilitated by solubility in water, 51% to 54% of fumonisin B_1 in starting corn was extracted into whole stillage. No toxin could be detected in distilled ethanol. The fermentation process did not destroy fumonisin, and 85% of the toxin could be recovered in the products. Products from ethanol fermentation of fumonisin-contaminated corn generally used as animal feeds could be detrimental if consumed by pigs or horses, animals sensitive to relatively low levels of this toxin.

The use of zearalenone-contaminated corn for ethanol production was investigated by Bennet and co-workers [75]. The presence of toxin had no apparent effect on ethanol yields, and there was no carryover from zearalenone to distilled ethanol. However, the fermentation process did not destroy the toxin, and recovered solids contained 2 to 2.5 times the level of zearalenone in the starting corn. Traditional fermentation for brewing of corn beer has shown 51% carryover of zearalenone into the finished product [76]. Fermentation by *Saccharomyces cerevisiae* of wort-containing zearalenone resulted in conversion of 69% of the toxin to beta-zearalenone, a metabolite with less activity than the parent compound [77]. Chu and co-workers [72] reported that 91% to 96% of ochratoxin A remained after the cooker mash treatment of the beer brewing process, and none was destroyed during pasteurization and boiling of beer itself [72].

Almost complete destruction of patulin occurs during alcoholic fermentation of apple juice [78]. Fermentation of juice containing patulin at a level of 40 μg/ml using *S. cerevisiae* effectively eliminated patulin in 2 weeks [79]. Wine making has also been reported to destroy patulin [80].

In a survey performed by Scott and co-workers [81], deoxynivalenol was found in 29 of 50 samples of Canadian and imported beers. Nine samples contained > 5 ng/ml deoxynivalenol. There are conflicting results on DON stability during the beer making process. Data are dependent upon the extraction and the analytical method.

D. Chemical Inactivation

Ammoniation of corn, peanuts, cottonseed, and meals to alter the toxic and carcinogenic effects of aflatoxin contamination has been subject to intense research efforts. Results of these studies have been summarized by Park and co-workers [18]. Reports on the extensive evaluation of this procedure demonstrate support for the efficacy and safety of ammoniation as a practical solution to aflatoxin decontamination in animal feeds. This process has been successfully used for many years in the U.S., France, Senegal, Sudan, Brazil, Mexico, and South Africa. The practical applications along with research results strongly support the use of the ammonia treatment to reduce risks posed by aflatoxin contamination, particularly in the high-temperature/high-pressure treatment, where process parameters are clearly defined and followed. With the ambient temperature/atmospheric pressure method, caution must be exercised to assure even distribution of ammonia gas/water or aqueous ammonia solution in the product. This process is limited to whole-kernel seed/nut products and has not been successful for ground or meal products due to the inability of the ammonia to penetrate the commodity. Norred and co-workers [82] found that atmospheric pressure/ambient temperature ammoniation reduced fumonisin content of *F. moniliforme* culture material but did not reduce the toxicity of material when fed to rats. Park and co-workers [83] reported 79% reduction of fumonisin levels in corn after high pressure/ambient temperature ammoniation followed by a low-pressure/high-temperature treatment.

Nixtamalization, the traditional alkaline heat treatment of corn used in the manufacture of tortillas, reduces significantly the levels of aflatoxin [84,85]. However, subsequent studies have shown that much of the original aflatoxin is reformed on acidification of the products [86]. This treatment has also been proposed as a method to detoxify fumonisin-contaminated corn [40]. A feeding study reported by Hendrich and co-workers [87] showed that nixtamalization greatly reduces FB_1 content. However, when nixtamilized corn was fed to weaning rats, the toxicity of the inoculated corn was not reduced. The major toxic product of fumonisin hydrolysis during nixtamalization appears to be hydrolyzed fumonisin B_1 (HFB_1), but it is possible that other breakdown products, such as calcium-FB_1 complexes and other products, may also play a role in the toxicity of nixtamalized corn. The hydrolysis of FB_1 by treatment with $Ca(OH)_2$ and heat does not render a detoxified product. It produces a more toxic final product. Feeding weaning rats with culture material containing 8 to 11 ppm of HFB_1 produced some toxicity symptoms as severe as did 45–48 ppm FB_1. These data indicate that the production of hydrolyzed fumonisins in the processing of goods for human consumption may be of significant concern. Park and co-workers [6] have reported a "modified nixtamalization" procedure that involves the addition of hydrogen peroxide and sodium bicarbonate to the traditional process. Results from this study show 40% reduction of brine shrimp mortality when comparing the products of traditional nixtamalization and modified nixtamalization. Traditional nixtamalization is probably not a useful strategy for fumonisin detoxification [88]. However, the modified procedures where other agents such as hydrogen peroxide and/or sodium bicarbonate are used, pose a good alternative for Latin American countries where nixtamalized corn is a major component of the diet.

Sodium chloride has a marked influence on loss of aflatoxins in artificially contaminated unshelled peanuts following cooking in water in a pressure cooker for 0.5 h; loss in concentration was 80% to 100% with 5% sodium chloride, but only 35% without the salt [22]. Sodium bisulfite has been shown to react with aflatoxins (B_1, G_1, M_1, and aflatoxicol) under various conditions of temperature, concentration, and time to form water-soluble products [89].

Nonenzymatic browning has been proposed as a decontamination procedure for fumonisin B_1. When FB_1 was incubated with glucose or fructose to mimic nonenzymatic browning conditions, a first-order rate reaction was observed. This showed that the amine group of FB_1 was derivatized by the sugar. Preliminary cell tissue culture tests with these adducts suggest they are less toxic than FB_1 [88].

Both patulin and penicillic acid are moderately stable in apple juice, with a half-life of several weeks [90]. The rate of disappearance of these toxins increases markedly with the addition of ascorbate or a mixture of ascorbate and ascorbic acid to the juice [91].

Processed foods are very complex systems. Processing not only alters the food but also adds new ingredients and conditions. These new factors change the environment, so many new interactions can occur. Exploring the application of known food additives to control mycotoxins as an integral part of the process can open a lot of opportunities for risk management.

E. Other Industrial Processes

Since molds and their secondary metabolic products are ubiquitous in foods and feeds, there have been several reports on the effects of processing and storage on mycotoxins present in commodities other than grains and oilseeds—i.e., cured meats, dairy products, and fruit products. It has been reported that recovery of aflatoxins added to meats is only 16% from raw ham, and 19% from salami after storage for 6 to 8 weeks at 5 to 37°C [92]. Storage at 5°C for shorter periods of time (up to 1 week) did not destroy aflatoxins B_1 and G_1 in bologna or cheese [93], nor did curing and cooking bacons and hams greatly reduce aflatoxin levels [94]. Van Egmond [95] reported that sterigmatocystin was stable in cheese for at least 3 months at temperatures of 18 to 16°C. Aflatoxins B_1 and G_1 were completely unstable in berry jams stored for 6 months at 22°C, although 26% to 38% of aflatoxins B_2 and G_2 were recovered [96].

Aflatoxin M_1 is stable during pasteurization [97]. Aflatoxin M_1 has been reported to be more stable in naturally contaminated milk than in freeze-dried spiked milk. Evidence has been obtained that aflatoxin M_1 binds to the milk protein, casein [22].

In most cases, there is no single method that will ensure the complete removal of the contaminant. Integrated management with a combination of different methods can significantly reduce the risks associated with mycotoxin contamination and yield products that are safe and acceptable to the consumer.

IV. CONCLUSIONS

Concern about chemical and toxic residues in the food chain is not a new issue. Naturally occurring toxicants pose a unique challenge due to their diversity and random occurrence. A 100% elimination of mycotoxin contamination is not practical. However, risks associated with mycotoxin-contaminated commodities can be reduced by following an integrated mycotoxin prevention and control plan. The most effective and practical procedures include (1) physical removal of mold-damaged or incomplete kernels/seeds/nuts (cleaning and separation); (2) chemical inactivation such as the ammoniation procedure; (3) use of additional chemical agents normally used in industrial processes, i.e., nixtamalization; and (4) removing biological availability of the contaminant, i.e., chemisorbents.

There is an increased need for information on the effects of processing on mycotoxins and multitoxin systems. More research is needed not only to develop decontamination treatments, but also to determine the safety of the final processed products, as well as prevention of recontamination during storage.

REFERENCES

1. Food and Agriculture Organization of the United Nations. Basic facts of the world cereal situation. Food Outlook 5/6, 1996.
2. LG Rice, FP Ross. Methods for detection and quantitation of fumonisins in corn, cereal products and animal excreta. J Food Prot 57:536, 1994.
3. DL Park, L Stoloff. Aflatoxin control—how a regulatory agency managed risk from an unavoidable natural toxicant in food and feed. Regul Toxicol Pharmacol 9:109, 1989.
4. L Stoloff. Occurrence of mycotoxins in foods and feeds. In: JV Rodricks, ed. Mycotoxins and Other Fungal Related Food Problems. Adv Chem Ser 149. Washington: American Chemical Society, 1976, p 23.
5. L Stoloff. Mycotoxins as potential environmental carcinogens. In: HF Stich, ed. Carcinogens and Mutagens in the Environment. Boca Raton: CRC Press, 1982, p 97.
6. DL Park, R Lopez-Garcia, S Trujillo-Preciado, RL Price. Reduction of risks associated with fumonisin contamination in corn. In: LS Jackson, JW DeVries, LB Bullerman, eds. Fumonisins in Food. New York: Plenum Press, 1996, p 335.
7. RB Harvey, WE Huff, LF Kubena, TD Phillips. Evaluation of diets co-contaminated with aflatoxin and ochratoxin fed to growing pigs. Am J Vet Res 50:1400, 1989.
8. CF Brownie, C Brownie. Preliminary study on serum enzyme changes in Long Evans rats given parenteral ochratoxin A, aflatoxin B_1 and their combination. Vet Human Toxicol 30:211, 1988.
9. TP Brown, RO Manning, OJ Fletcher, RD Wyatt. The individual and combined effects of citrinin and ochratoxin A on renal ultrastructure in layer chicks. Avian Dis 30:191, 1986.
10. RP Glahn, RF Wideman, JW Evangelisti, WE Huff. Effects of ochratoxin A alone and

in combination with citrinin on kidney function of single comb white Leghorn pullets. Poult Sci 67:1034, 1988.
11. LF Kubena, WE Huff, RB Harvey, DE Corrier, TD Phillips, CR Creger. Influence of ochratoxin A and deoxynivalenol on growing broiler chicks. Poult Sci 67:253, 1984.
12. LF Kubena, TD Phillips, DA Witzel, ND Heidelbaugh. Toxicity of ochratoxin A and penicillic acid to chicks. Bull Environ Contam Toxicol 32:717, 1984.
13. LF Kubena, RB Harvey, WE Huff, DE Corrier, TD Phillips, GE Rottinghaus. Influence of ochratoxin A and T-2 toxin singly and in combination on broiler chickens. Poult Sci 68:867, 1989.
14. EE Smith, LF Kubena, CE Braithwaite, RB Harvey, TD Phillips, AH Reine. Toxicological evaluation of aflatoxin and cyclopiazonic acid in broiler chickens. Poult Sci 71:1136, 1992.
15. LE Giroir, WE Huff, LF Kubena, et al. The individual and combined toxicity of kojic acid and aflatoxin in broiler chickens. Poult Sci 70:1351, 1991.
16. WE Huff, LF Kubena, RB Harvey, et al. Individual and combined effects of aflatoxin and deoxynivalenol (DON, Vomitoxin) in broiler chickens. Poult Sci 66:2351, 1986.
17. RB Harvey, LF Kubena, WE Huff, DE Corrier, GE Rottinghaus. Effects of treatment of growing swine with aflatoxin and T-2 toxin. Am J Vet Res 51:1688, 1990.
18. DL Park, LS Lee, RL Price, AE Pohland. Review of decontamination of aflatoxin by ammoniation: current status and regulation. J AOAC 71:685, 1988.
19. Food and Agriculture Organization of the United Nations. Worldwide regulations for mycotoxins. A compendium. In: FAO Food and Nutrition Paper. 1997.
20. DL Park. Controlling aflatoxin in food and feed. Food Technol 47(10):92, 1993.
21. Council for Agricultural Science and Technology. Mycotoxins: economic and health risks. Task Force Rep 116:8–9, 1989.
22. PM Scott. Effects of food processing on mycotoxins. J Food Prot 47(6):489, 1984.
23. JW Dickens, TB Whitaker. Efficacy of electronic color sorting and hand picking to remove aflatoxin contaminated kernels from commercial lots of shelled peanuts. Peanut Sci 2:45, 1975.
24. RJ Cole. Technology of aflatoxin decontamination. In: S Natori, K Hashimoto, Y Ueno, eds. Mycotoxins and Phycotoxins '88. Amsterdam; Elsevier Scientific, 1989, p 177.
25. JW Kirskey, RJ Cole, JW Dorner. Relationship between aflatoxin content and buoyancy in Florunner peanuts. Peanut Sci 16:48, 1989.
26. TD Phillips, BA Clement, DL Park. Approaches to reduction of aflatoxins in foods and feeds. In: DL Eaton, JD Groopman, eds. The Toxicology of Aflatoxins, Human Health, Veterinary and Agricultural Significance. San Diego; Academic Press, 1994, p 383.
27. PA Murphy, LG Rice, PF Ross. Fumonisins B_1, B_2, and B_3 content of Iowa, Wisconsin, and Illinois corn and corn screenings. J Agric Food Chem 41:263, 1993.
28. LM Seitz, WD Eustace, HE Mohr, MD Shogren, WT Yamazaki. Cleaning, milling and baking tests with hard red winter wheat containing deoxynivalenol. Cereal Chem 63:146, 1986.
29. HL Trenholm, LL Charmley, PB Prelusky, RM Warner. Two physical methods for the decontamination of four cereals contaminated with deoxynivalenol and zearalenone. J Agric Food Chem 39:356, 1991.

30. GA Bennett, G Rottinghaus, TL Nelson. Deoxynivalenol in wheat and flour from wheat cleaned by aspiration. Proceedings of Am Bakers Assoc ARS Workshop, New Orleans, LA, 1992.
31. J Lovett, RG Thompson Jr, BK Boutin. Trimming as a means of removing patulin from fungus-rotted apples. J AOAC 58:909, 1975.
32. GA Bennett, EE Vandegraft, OL Shotwell, SD Watson, BJ Bocan. Zearalenone: distribution in wet-milling fractions from contaminated corn. Cereal Chem 55:455, 1978.
33. GM Wood, SJ Cooper, WB Chapman. Problems associated with laboratory simulation of effects of food processes on mycotoxins. Proceedings of V Int IUPAC Symp Mycotoxins, Vienna, Austria, 1982, p 142.
34. GA Bennett, RA Anderson. Distribution of aflatoxin and/or zearalenone in wet-milled corn products: a review. J Agric Food Chem 26:1055, 1978.
35. GA Bennett, JL Richard. Influence of processing on *Fusarium* mycotoxins in contaminated grains. Food Technol 50(5):235, 1996.
36. PF Ross, LG Rice, RD Plattner, et al. Concentrations of fumonisin B_1 in feeds associated with animal health problems. Mycopathology 114:129, 1991.
37. GM Wood. Effects of processing on mycotoxins in maize. Chem Ind 972, 1982.
38. HW Achroder, RA Boller, H Hein Jr. Reduction in aflatoxin contamination of rice by milling procedures. Cereal Chem 45:574, 1968.
39. GA Bennett, AJ Peplinski, OL Brekke, LK Jackson. Zearalenone: distribution in dry-milled fractions of contaminated corn. Cereal Chem 53:299, 1976.
40. EW Sydenham, GS Shephard, PG Thiel, WFO Marasas, S Stockenstrom. Fumonisin contamination of commercial corn-based human foodstuffs. J Agric Food Chem 39: 2014, 1991.
41. ME Stack, RM Eppley. Liquid chromatographic determination of fumonisins B_1 and B_2 in corn and corn-based products. J AOAC Int 75:834, 1992.
42. CM Christensen, CJ Mirocha, RA Meronuck. Mold, mycotoxin and mycotoxicoses. In: Agricultural Experiment Station, Report 142, St. Paul; University of Minnesota, 1977.
43. FJ Baur. Effect of storage upon aflatoxin levels in peanut materials. J Am Oil Chem Soc 52:263, 1975.
44. L Luter, W Wyslouzil, SC Kashyap. The destruction of aflatoxins in peanuts by microwave roasting. Can Inst Food Sci Technol J 15:236, 1982.
45. CP Levi. Mycotoxins in coffee. J AOAC 63:1282, 1980.
46. CP Levi, HL Ternk, JA Yeransian. Investigations of mycotoxins relative to coffee. Colloq Int Chim Cafes (CR) 7:287, 1975.
47. CT Dawarkanath, V Sreeniwasamurthy, HAB Parpia. Aflatoxin in Indian peanut oil. J Food Sci Technol (Mysore) 6:107, 1969.
48. FG Peers, CA Linsell. Aflatoxin contamination and its heat stability in Indian cooking oils. Trop Sci 17:229, 1975.
49. WA Parker, D Melnick. Absence of aflatoxin from refined vegetable oils. J Am Oil Chem Soc 43:635, 1966.
50. J Gracian, G Arevalo. Presencia de aflatoxinas en los productos del olivar. Grasas y Aceites 31:167, 1980.
51. HF Conway, RA Anderson, EB Bagley. Detoxification of aflatoxin-contaminated corn by roasting. Cereal Chem 55:115, 1978.

52. L Stoloff, MW Trucksess, PW Anderson, EF Glabe, JG Aldridge. Determination of the potential for mycotoxin contamination of pasta products. J Food Sci 43:228, 1978.
53. A Pittet, V Parisod, M Schellenberg. Occurrence of fumonisins B_1 and B_2 in corn-based products from the Swiss market. J Agric Food Chem 40:1352, 1992.
54. MB Doko, A Visconti. Occurrence of fumonisins B_1 and B_2 in corn and corn-based human foodstuffs in Italy. Food Addit Contam 11:433, 1994.
55. LS Jackson, JJ Hlywka, KR Senthil, LB Bullerman, SM Musser. The effects of time, temperature and pH in the stability of fumonisin B_1 in an aqueous model system. J Agric Food Chem 44:906, 1996.
56. JF Alberts, WCA Gelderblom, PG Thiel, WFO Marasas, DJ van Schalwyk, Y Behrend. Effects of temperature and incubation period on production of fumonisin B_1 by *Fusarium moniliforme*. Appl Environ Microbiol 56:1729, 1990.
57. J Dupuy, P LeBars, H Boudra, J LeBars. Thermostability of fumonisin B_1, a mycotoxin from *Fusarium moniliforme* in corn. Appl Environ Microbiol 59:2864, 1993.
58. PM Scott, GA Lawrence. Stability and problems in recovery of fumonisins added to corn-based foods. J AOAC Int 77:541, 1994.
59. LS Jackson, JJ Hlywka, KR Kannaki, LB Bullerman. Effects of thermal processing on the stability of fumonisins. In: LS Jackson, JW DeVries, LB Bullerman, eds. Fumonisins in Food. New York: Plenum Press, 1996, p 345.
60. BG Osborne. Reverse-phase high performance liquid chromatography determination of ochratoxin A in flour and bakery products. J Sci Food Agric 30:1065, 1979.
61. J Harwig, Y-K Chen, DL Collins-Thompson. Stability of ochratoxin A in beans during canning. Can Inst Food Sci Technol J 7:288, 1974.
62. BGE Jossefsson, TE Moeller. Heat stability of ochratoxin A in pig products. J Sci Food Agric 31:1313, 1980.
63. AA El-Banna, P-Y Lau, PM Scott. Fate of mycotoxins during processing of foodstuffs. II. Deoxynivalenol (vomitoxin) during making of Egyptian bread. J Food Prot 46:484, 1983.
64. J Reiss. Mycotoxins in foodstuffs. XI. Fate of aflatoxin B_1 during preparation and baking of whole-meal wheat bread. Cereal Chem 55:421, 1978.
65. AJ Feuell. Aflatoxin in groundnuts. IX. Problems of detoxification. Trop Sci 8:61, 1966.
66. T Shantha, M Sreenivasamurthy. Photo-destruction of aflatoxin in groundnut oil. Indian J Technol 15:453, 1977.
67. N Mansimanco, J Remancle, J Ramaut. Elimination of aflatoxin B_1 by absorbent clays in contaminated substrates. Ann Nutr Alim 23:137, 1973.
68. WJ Decker. Activated charcoal absorbs aflatoxin B_1. Vet Human Toxicol 22:388, 1980.
69. MD Machen, BA Clement, EC Shepherd, AB Sarr, RE Pettit, TD Phillips. Sorption of aflatoxins from peanut oil by aluminosilicates. Toxicologist 8:265, 1988.
70. DC Sands, JL McIntyre, GS Walton. Use of activated charcoal for the removal of patulin from cider. Appl Environ Microbiol 32:388, 1976.
71. GS Walton, DC Sands, JL McIntyre. Patulin removal from cider using activated charcoal. Proc Am Phytopathol Soc 3:255, 1976.
72. FS Chu, CC Chang, SH Ashoor, N Prentice. Stability of aflatoxin B_1 and ochratoxin A in brewing. Appl Microbiol 29:313, 1975.

73. RS Dam, SW Tam, LD Satterlee. Destruction of aflatoxins during fermentation and by-product isolation from artificially contaminated grains. Cereal Chem 54:705, 1977.
74. S Lindroth. Thermal destruction of patulin in berries and berry jams. J Food Safety 2:165, 1980.
75. GA Bennett, AA Lagoda, OL Shotwell, CW Hesseltine. Utilization of zearalenone-contaminated corn for ethanol production. J Am Oil Chem Soc 58:974, 1981.
76. ZSC Okoye. Stability of zearalenone in naturally contaminated corn during Nigerian traditional brewing. Food Addit Contam 4:59, 1987.
77. PM Scott, SR Kanhere, EF Dailey, JM Farber. Fermentation of wort containing deoxynivalenol and zearalenone. Mycotoxin Res 8:58, 1992.
78. LF Burroughs. Stability of patulin to sulfur dioxide and to yeast fermentation. J AOAC 60:100, 1977.
79. J Harwig, PM Scott, BPC Kennedy, Y-K Chen. Disappearance of patulin from apple juice fermented by *Saccharomyces* sp. Can Inst Food Sci Technol J 6:45, 1973.
80. PM Scott, T Fuleki, J Harwig. Patulin content of juice and wine produced by moldy grapes. J Agric Food Chem 25:434, 1977.
81. PM Scott, SR Kanhere, D Weber. Analysis of Canadian and imported beers for *Fusarium* mycotoxins by gas chromatography-mass spectroscopy. Food Addit Contam 10:381, 1993.
82. WP Norred, KA Voss, CW Bacon, RT Riley. Effectiveness of ammonia treatment in detoxification of fumonisin-contaminated corn. Food Che Toxicol 29(12):819, 1991.
83. DL Park, SM Rua, CJ Mirocha, EAM Abd-Alla, CY Weng. Mutagenic potentials of fumonisin contaminated corn following ammonia decontamination procedures. Mycopathology 117:105, 1992.
84. M Ulloa, T Herrera. Persistencia de las aflatoxinas durante la fermentacion del pozol. Rev Lat-Am Microbiol 12:19, 1970.
85. M Ulloa Sosa, HW Shroeder. Note on aflatoxin decomposition in the process of making tortillas from corn. Cereal Chem 46:397, 1969.
86. RL Price, KV Jorgensen. Effects of processing on aflatoxin levels and on mutagenic potential of tortillas made from naturally contaminated corn. J Food Sci 50:347, 1985.
87. S Hendrich, KA Miller, TA Wilson, PA Murphy. Toxicity of *Fusarium proliferatum*-fermented nixtamalized corn-based diets fed to rats: effects on nutritional status. J Agric Food Chem 41:1649, 1993.
88. PA Murphy, EC Hopmans, K Miller, S Hendrich. Can fumonisins be detoxified? In: WR Bidlack, ST Omaye, eds. Natural Protectants and Natural Toxicants in Foods. Lancaster, PA, Technomic Publishing Company, 1995, p 105.
89. WM Hagler JR, JE Hutchins, PB Hamilton. Destruction of aflatoxin in corn with sodium bisulfite. J Food Prot 45:1287, 1982.
90. AE Pohland, R Allen. Stability studies with patulin. J AOAC 53:688, 1970.
91. RE Brackett, EH Marth. Ascorbic acid and ascorbate cause disappearance of patulin from buffer solutions and apple juice. J Food Prot 42:864, 1979.
92. WT Stott, LB Bullerman. Instability of patulin in cheddar cheese. J Food Sci 41:201, 1976.
93. FY Lieu, LB Bullerman. Production and stability of aflatoxins, penicillic acid and patulin in several substrates. J Food Sci 42:1222, 1228, 1977.

94. RM Furtado, AM Pearson, JI Gray, MG Hogberg, ER Miller. Effects of cooking and/ or processing upon levels of aflatoxins in meat from pigs fed a contaminated diet. J Food Sci 46:1306, 1981.
95. HP van Egmond. Mycotoxins in dairy products. Food Chem 11:289, 1983.
96. O Pensala, A Niskanen, S Lindroth. Aflatoxin production in black currant, blueberry and strawberry jams. J Food Prot 41:344, 1978.
97. DW Wiseman, RS Applebaum, RE Brackett, EH Marth. Distribution and resistance to pasteurization of aflatoxin M_1 in naturally contaminated whole milk, cream and skim milk. J Food Prot 46:530, 1983.

14

Mycotoxins: Risk Evaluation and Management in Radiation-Processed Foods

Arun Sharma
Bhabha Atomic Research Centre, Mumbai, India

I. INTRODUCTION

Molds and mycotoxins have existed in nature since their evolutionary appearance on the face of Earth. Though the selective advantage mycotoxins offer to the producer organism is not very clear, their role in giving the organism an edge over its competitors is certain [1]. Mycotoxin-producing organisms have succeeded not only in competing for human and animal food but also in the process contaminating it with their toxins that jeopardize the health of their competitors [1–3]. The world had witnessed several outbreaks of mycotoxicoses before, but the seriousness of the problem was only realized in 1960 when in Great Britain aflatoxin demonstrated its devastating power on a living system, killing some 100,000 turkeys in 3 months, and affecting the related economy [4,5]. It is the economic impact and health implications of a food contaminant that best sensitize us to its importance.

It has now been realized that mycotoxins as food contaminants are important to us on both these counts. In view of the detrimental effects of mycotoxins on human and animal health, regulatory agencies have stipulated stringent controls on the presence of mycotoxins in food and feed. Often it has resulted in large-scale rejection of contaminated lots and caused economic losses. Mycotoxicoses could either be primary, through direct consumption of contaminated agricultural produce, or secondary, through consumption of products of animals raised on contaminated feed. Researchers around the world have studied the various aspects of the mycotoxin problem in food and agricultural commodities. They have studied conditions that would promote or suppress mold growth and mycotoxin formation in foods and laboratory-designed media. Many of these studies could help in identifying critical control points needed for hazard analysis and control of mycotoxin contamination during food processing and preservation.

For a long time fungi belonging to the genera *Aspergillus* and *Penicillium* were grouped as storage molds [6]; their toxins were essentially considered the products of saprophytic mode of fungal existence and a postharvest problem [7]. However, later reports of preharvest infection of standing crops, reflecting parasitic mode of fungal growth, added new dimensions to the problem [8]. Though the methods for detoxification of preformed toxin in harvested crops have evolved, they are too drastic to be used for human food [9,10]. Strategies for aflatoxin control based on the understanding of molecular mechanisms underlying interaction of mycotoxin-producing fungi with their host plants are only in their infancy [11,12].

A beginning has been made in the identification and cloning of genes involved in the pathway of aflatoxin production [13,14]. However, very little is known about the resistance of the host. Some recent efforts in the search for resistance to *Aspergillus flavus* in maize have been successful [15,16]. Therefore, it will be long before the genes conferring resistance to plants are identified, cloned, and used in the development of superior varieties using conventional breeding or transgenic procedures. Till that time an integrated approach has to be adopted for the control of mycotoxins in our food and animal feed. Any food-processing technology, existing or emerging, should as a part of good manufacturing practice not only monitor the presence of preformed mycotoxins but also eliminate postprocessing risk of mycotoxin formation in food.

II. FOOD IRRADIATION TECHNOLOGY

Food irradiation is the controlled application of the energy of ionizing radiations such as gamma rays, X-rays, and accelerated electrons to food and food ingredients, for improving hygiene and safety of food and increasing its storage and distribution life [17,18]. The technology holds promise because in many cases it has an edge over the conventional methods. It could be applied judiciously where conventional methods are inadequate, are uneconomical, or pose potential health risks. It can also be used as a complementary process with many existing procedures. The process helps in keeping chemical burden in food low and increases packaging possibilities. These benefits accrue from the cold nature of the process and high penetrating power of the ionizing radiations.

Irradiation has a multipurpose role in the conservation and processing of foods. Depending on the product and the applied dose, at least four functions can be achieved by the application of radiation. These include eradication of insect pests, inhibition of sprouting and delay in ripening, destruction of spoilage microbes, and elimination of pathogens and parasites. On the basis of radiation dose food irradiation applications could be classified into:

Low-dose applications (< 1 kGy*)
- Sprout inhibition in bulbs, tubers, and rhizomes
- Delay in ripening of fruits
- Disinfestation of grain, grain products, and pulses, and for quarantine purpose
- Elimination of parasitic protozoa and helminths in meat

Medium-dose applications (1–10 kGy)
- Reduction of spoilage microbes to improve refrigerated shelf-life of meat, poultry, and seafoods
- Elimination of pathogens in meat and seafoods
- Microbial decontamination of spices

High-dose applications (> 10 to < 70 kGy)
- Sterilization of packaged meat, poultry, and prepared foods to make them shelf-stable at ambient temperature
- Sterilization of hospital diets

Interest in food irradiation has emerged for many reasons. These include persistently high food losses from insect infestation, microbial contamination and spoilage, mounting concern over foodborne diseases, harmful residues of chemical fumigants and their impact on environmental pollution, and growing international trade in food products that must meet stiff standards of quality and quarantine. In all these areas food irradiation has demonstrated practical benefits and could be gainfully exploited to strengthen food security and food safety in many countries.

III. SAFETY OF IRRADIATED FOOD

No other food-processing technology has been studied so thoroughly and extensively as food irradiation for ensuring its toxicological, nutritional, and microbiological safety [19–25]. The data available from worldwide research collected over a period of more than four decades have been reviewed at a series of international meetings organized by the World Health Organization (WHO) jointly with the experts from Food and Agriculture Organization (FAO) and International Atomic Energy Agency (IAEA). In 1980, a Joint WHO/FAO/IAEA

*A Gray is a unit of absorbed dose of ionizing energy, and is equivalent to 1 joule/kg. The Gray replaces rad as the unit of absorbed dose. One Gray is equivalent to 100 rads.

Expert Committee on the Wholesomeness of Irradiated Food, convened in Geneva, concluded that any food irradiated up to an overall average dose of 10 kGy presents no toxicological hazard; hence, toxicological testing of foods so treated is no longer required, and irradiation up to 10 kGy introduces no special nutritional or microbiological problems [20].

IV. MICROBIOLOGICAL ASPECTS OF IRRADIATED FOODS

Microbiological aspects of any food-processing technology are of paramount importance from the standpoint of its efficacy, safety, and feasibility. With irradiation aimed at achieving commercial sterility in food, no public health problem of microbiological origin could be foreseen, provided the integrity of the container is not breached. For nonsterilizing doses, the various aspects of microbiological safety have been studied. These include the study of possible selective changes in microflora, changes in the virulence of food pathogens, changes in the toxigenicity of molds, development of radiation-resistant strains, and changes in diagnostic characters of the micro-organisms.

Though the question of microbiological safety of irradiated foods has been the subject of earlier reviews [19–25], the present article especially reviews aspects related to mycotoxins in detail. In irradiated foods the risk of mycotoxins could be best evaluated by taking into account interaction of radiation with mycotoxins, mycotoxin-producing fungi, competing microflora, and the antifungal constituents of food.

V. RADIATION SENSITIVITY OF MYCOTOXINS

Agriculture and food-processing industries are interested in finding successful means of destroying mycotoxins in food and feed. However, a completely satisfactory method is yet to emerge for destroying preformed mycotoxins. Though ultraviolet light and sunlight have been found to destroy aflatoxin [26,27], destruction of aflatoxin with gamma rays without the quality of food deteriorating is very difficult or impossible. To reduce the biological toxicity of aflatoxin B_1, as determined by chick embryo test, to half the original level, a dose of 0.4 kGy and 9 kGy in water and ethanol, respectively, was required. In a dry state a dose as high as 182 kGy could destroy only 10% of aflatoxin [28]. In another study [29], a dose range of 2.5 to 50 kGy, was reported to bring down the toxin level to between 50% and 100% of the original value in bread. In similar experiments irradiation of rice with aflatoxin showed no significant reduction of aflatoxin content even after irradiation with a dose of 30 kGy. Temcharoen and Thilly [30] found that after treatment with 50 to 100 kGy of ionizing energy from gamma rays, peanut meal had lost its toxic and mutagenic properties attributed to aflatoxin B_1 contamination. Treatment with 1.0 to 10 kGy removed 75% to 100% of the toxicity but not the mutagenicity of the toxin. The required doses for inactivation of aflatoxin are so high that the chemical and organoleptic changes brought about by these doses in food would be unacceptable. The detoxification of some basic food materials may be possible if irradiation is used in combination with chemical agents [28]. Recently, Santamarina et al. [31] surveyed the main physical methods including gamma rays, ultraviolet and visible light, and some chemical methods to destroy mycotoxins. The authors concluded that there was no single effective method that eliminated all mycotoxins from food and feed successfully, and probably gamma irradiation in combination with other methods could only be employed to achieve freedom from mycotoxins. In another recent study [32], the effect of gamma irradiation on mycotoxin disappearance in corn, wheat, and soybean has been investigated. Radiation doses of 5, 7.5, 10, and 20 kGy were applied to spiked grain samples and the residual toxins were measured using enzyme-linked immunosorbent assay (ELISA). It was found that radiation doses up to 20 kGy did not significantly affect aflatoxin B_1 in any of the three commodities. The authors, however, observed a significant reduction in the levels of trichothecene toxins such as T-2, deoxynevalenol, and zearalenone at doses above 7.5 kGy. These doses also caused small but significant losses in essential amino acids lysine, phenylalanine, methionine, and histidine. These observations are interesting since a food or feed low in nutrients would certainly be more acceptable than that laced with toxins, and the feasibility of radiation decontamination of specific toxins such as trichothecenes could be more closely examined.

In the absence of an efficient system to destroy preformed mycotoxins, it becomes important to control the growth of mycotoxin-producing molds in food and agricultural commodities. This control is required at both the preharvest and postharvest levels in the case of agricultural commodities and at the preprocessing and postprocessing stages in the case of foods meant for processing.

VI. RADIATION SENSITIVITY OF MYCOTOXIN-PRODUCING MOLDS

Fungi can exist in several morphological forms. Most commonly in foods meant for consumption, they exist in unicellular form as spores or conidia, but may grow to multicellular and multinuclear coenocytic hyphal or mycelial form, and occasionally make specialized structures such as sclerotia and cleistothecia. Any study on the assessment of the effect of radiation on molds should therefore take into account the mold morphology being dealt with, since the response of radiation to dormant and uninucleate conidia is different from that of germinating conidia and coenocytic mycelium. Many of the earlier studies on the effect of irradiation on mycotoxin-producing molds did not take this into account.

To study the response of toxigenic and nontoxigenic strains to radiation, Frank et al. [33] irradiated conidia bearing cultures of three aflatoxin-producing and three nontoxigenic strains of *A. flavus* 16 times in 6 months at a dose range of 1.6 to 2.4 kGy each time, and found no differences in radioresistance between toxigenic and nontoxigenic strains. While studying the survival of the spores of *Aspergillus* sp. using gamma rays and X-rays, the spores in dry state were found to be more resistant than those in aqueous system [34]. In similar studies, the dose required to bring down the surviving fraction of conidia by 1 log cycle or by 90%, D_{10} value, was found to increase when the conidia were irradiated in dry state [35]. Six-month-old conidia of *Penicillium expansum* were found to be more sensitive to gamma radiation than 1- to 2-week-old ones [36]. Studies using pure spore suspensions of toxigenic and nontoxigenic strains of *A. flavus* were carried out by Padwal-Desai et al. [37]; the spores of toxigenic strains of *A. flavus* were found to be more resistant to radiation than those of nontoxigenic strains. The composition, temperature, and pH of the medium acted as dose-modifying factors in both cases.

On the other hand, the initial number and age of the spores were not found to influence the inactivation dose. Germinating spores of both strains showed higher sensitivity to radiation than did the nongerminating ones. After attaining a multinucleate stage the mycelium showed increasing resistance to radiation. The D_{10} for *A. parasiticus* NRRL 3145 spores in water with 0.1% Tween 80 was found to be 0.4 kGy [38]. A study of the sensitivity of ochratoxin A-producing mold, *Aspergillus alutaceus* var. *alutaceus*, to gamma irradiation showed variation in D_{10} values from 0.24 to 0.27 kGy within the pH range 3.6 to 8.8. On the other hand, irradiation with 10 MeV electrons, D_{10} value for the same organism reduced to a range between 0.21 and 0.22 kGy [39]. A D_{10} value of 0.17 kGy was reported for the Korean strain of this organism, and a reduction in D_{10} value with dose rate was also observed [40]. Ito [40] studied the effects of repeated exposures of *A. flavus* var. *columnaris* 846 to gamma radiation and found that the original D_{10} value of 0.22 kGy increased a little after six times exposure to a dose of 0.8 kGy. Effectiveness of radiation in destruction of molds could be increased by combination with mild heat. Heat followed by irradiation caused greater destruction of spores than radiation followed by heat [42]. The striking synergism of low-dose irradiation (0.25 to 1.0 kGy) and mild heat (50°C/5 min) could be exploited for preservation of fruits and vegetables [43–47].

VII. QUESTIONS RELATED TO MYCOTOXIN FORMATION IN RADIATION-PROCESSED FOODS

As a consequence of exposure of agricultural commodities and foods to nonsterilizing doses of ionizing radiations, three prominent questions regarding mycotoxin biogenesis have emerged:

What will be the mycotoxin–producing potential of a fungus exposed to gamma radiation?

Gamma radiation is a known mutagenic agent. Hence, the concern that it would cause mutations in the fungus and mutants with enhanced toxin-producing ability would increase the risk of mycotoxin in irradiated food.

What will be the effect of elimination of competitive microflora on the mycotoxin-producing potential of the toxigenic fungi?

It is now well documented that competing microbes under the conditions of coculture or under natural competition reduce toxin accumulation by the toxigenic fungi. In fact, this competition is the basis for developing biological control strategies for prevention of mycotoxin formation in field crops. It is conjectured that elimination of competing microflora by radiation could, therefore, give better opportunity to the toxin producer and result in greater risk of toxin accumulation.

What will be the mycotoxin-producing potential of a nonirradiated fungus on an irradiated substrate?

Concern has been raised as to the possible increase in susceptibility of irradiated food to infection by fungi and decay following irradiation. Susceptibility could increase as a result of reduction in the physiological and biochemical resistance of the host tissue and radiolytic breakdown of the large molecules such as starch, protein, and lipid into smaller units, which is possible at relatively higher doses of radiation. The question basically relates to postirradiation contamination of a commodity by the toxigenic molds.

All these questions have been scrutinized by researchers and have been the subject of a number of scientific studies which are summarized in the following text.

VIII. MYCOTOXIN PRODUCTION BY FUNGUS EXPOSED TO IONIZING RADIATION

The earliest reports on mycotoxin production by fungus exposed to gamma irradiation were those of Jemmali and Guilbot [48]. They studied inactivation of "spores and mycelium" of *A. flavus* by gamma irradiation and reported that even single irradiation in the dose range of 0.1 to 5.0 kGy influenced the ability of *A. flavus* to produce aflatoxin B_1, both increased and decreased production being observed in different experiments. In studies with conidia-bearing cultures of three aflatoxin-producing and three nontoxigenic strains of *A. flavus* exposed 16 times to a dose range of 1.6 to 2.4 kGy over 6 months, it was observed that two of the three nontoxigenic strains remained aflatoxin-negative. Of the 14 subcultures of the third strain, two were positive after 2 years, and four more became positive after 3 years of subculturing [33]. These authors concluded that the observed variation in aflatoxin production was possibly due to recombinations in the heterokaryotic fungal system during the course of subculturing rather than due to mutations. With respect to the practical implications of their study to food irradiation, these authors remarked that they do not believe there is an increased health risk from the possible industrial application of food irradiation. In another study [49], spores and vegetative mycelia of *A. parasiticus* were irradiated to 1 to 2 kGy doses of gamma radiation. The growth and aflatoxin production were monitored after 7 days at 25°C. These authors reported a stimulation in toxin production following irradiation. A major drawback of many of these studies was that whole cultures were subjected to irradiation. Since the response to radiation of pure spores would be different from that of mycelium, these studies are probably best applicable to food with mold growth which under normal circumstances would be unfit for consumption.

Applegate and Chipley in a series of papers [50–52] reported an increased production of aflatoxin by a strain of *A. flavus* following gamma irradiation. The authors exposed the spores of the fungus to gamma rays in a dose range of 0.25 to 6.0 kGy. An increased aflatoxin formation was observed when the irradiated spore suspension was used as inoculum on cracked red wheat or in a synthetic liquid medium compared to the inoculum from the suspension of nonirradiated spores. The most notable increase was observed when the spores were exposed to gamma radiation doses in the range of 1.5 to 3.0 kGy. The spores exposed to a 4.0-kGy dose failed to germinate and produce any toxin, obviously because of the lack of survivors. The authors concluded their paper with a note: "From the practical standpoint, since cobalt irradiation at levels of 2 kGy is sanctioned by the USDA for irradiation of wheat as a means of controlling insects, the result of this investigation regarding the increased production of aflatoxin by irradiation of toxigenic strains of fungal spores over their nonirradiated counterparts could be of public health significance."

In a similar study, enhancement of ochratoxin formation by *A. ochraceus* NRRL 3174 after exposure to gamma radiation was reported by the same authors [53]. Enhanced aflatoxin production by *A. flavus* and *A. parasiticus* after irradiation of the spore inoculum was also reported and attributed to the formation of mutants with higher rates of toxin formation [54]. Recently, enhanced aflatoxin production by *A. parasiticus* and *A. flavus* after low-dose gamma irradiation has been reported by Ito [55]. According to the author, irradiation of spores at 0.05 kGy resulted in higher B_1 or G_1 production than the nonirradiated controls. However, spores of both strains irradiated at 0.2 or 0.4 kGy produced less aflatoxin than the nonirradiated controls. Apparent stimulation by low-dose irradiation was slight, and these enhanced effects were not observed after reinoculation in fresh medium [55]. Enhanced ochratoxin production as a result of gamma irradiation of

Aspergillus ochraceus were reported by Paster et al. [56]. In another study, irradiated (1 to 2 kGy) spores of patulin-producing mold *Penicillium patulum* were found to produce substantially less patulin than nonirradiated spores [57].

In our laboratory we have undertaken a systematic study to assess the mycotoxin-producing potential of the mold spores exposed to radiation [38]. In our experiments we stressed the need to employ pure spore suspensions free from any mycelial contamination. This was ensured by washing a lawn of a 10-day-old sporulating culture of the fungus in a Roux bottle with sterile water containing 0.01% Tween 80 and filtering it through 16 layers of cheesecloth. A suspension containing a known number of mycelium-free viable spores of the fungus was used in all experiments. Resistance of the pure spores of *A. parasiticus* to gamma radiation was initially studied. The D_{10} value was found to be 0.4 kGy [38]. When an aliquot from an irradiated suspension containing 10^6 spores/ml and exposed to a dose in the range of 1.6 to 2.0 kGy, i.e. four to five times the D_{10} value, was inoculated in a liquid synthetic medium, aflatoxin yield was found to almost double compared to an aliquot of nonirradiated suspension similarly used. This observation seemed to corroborate the findings of Applegate and Chipley [50]. The only difference between the two inocula appeared to be in the number of viable spores. The spore suspension exposed to radiation had a viable spore count lowered by an order of 4 to 5 log cycles. And yet it yielded almost double the quantity of the toxin. It meant that an inoculum with low viable spore count gave a higher yield of the toxin than an inoculum with a high viable spore count. A control experiment was undertaken to cross-check the results. Inocula containing different numbers of viable spores were prepared by using simple serial dilution in sterile water, and the same were used to inoculate the liquid synthetic medium. It was observed that aflatoxin production increased gradually with the increase in the dilution of inoculum.

Similar increase was observed when the suspensions exposed to increasing doses of gamma radiation were used as inocula. It could be concluded from these studies that the reported increase in aflatoxin production was brought about by radiation through mere reduction in spore numbers in the inoculum and not through any biochemical or genetic change in the fungus. A decrease in inoculum size, however, induced a small lag in both growth and aflatoxin production, though the final yield of the mycelium was about the same. The maximum accumulation of aflatoxin was observed on 14th day of incubation in a still culture. Also, a transition from a biphasic to monophasic pattern in aflatoxin production was observed in cultures with dilute inoculum. Such transitions were also observed earlier [50].

This dilute inoculum effect was later confirmed in other laboratories [58,59]. In a study on ochratoxin production in 0.5 to 4 kGy irradiated barley, an inoculum with 10^2 spores was reported to produce greater amount of mycotoxin than the one with 10^5 spores [60]. The effect of heat and gamma irradiation combination on the production of aflatoxin B_1 in a static liquid culture by *A. flavus* NRRL 5906 has also been studied. The combination treatment of moist heat (60°C at > 85% RH) and a radiation dose of 3.5 kGy was found to be effective in prevention of aflatoxin formation by a heavy inoculum of *A. flavus* spores [61].

IX. POSSIBILITY OF RADIATION-INDUCED HYPERTOXIGENIC MUTANTS

Though not the best of the known mutagenic agents, ionizing radiations have been used to increase the rate of mutations in living organisms for more than six decades. During irradiation of food a theoretical possibility exists that mutations might transform a nonpathogenic organism to a pathogenic one or a less virulent strain into a more virulent one. Many studies have been directed to address this question [62–68]. Physical treatments such as heating, drying, freezing, and treatment with chemicals are also known to enhance the rate of mutation in microorganisms [20]. In fact, mutants would also be produced spontaneously naturally. Mutants produced or selected by irradiation have been found to be less competitive than the wild types [66–68]. It has been shown that irradiation can more often lead to loss of virulence and infectivity of pathogens [69,70]. At the request of FAO and WHO, the subject was considered by the Committee on Food Microbiology and Hygiene of the International Union of Microbiological Societies, which concluded that the irradiation of food does not pose any special problem related to mutations leading to increased virulence or pathogenicity of food pathogen [20].

It is well known that mutants are less well adapted to the environment of the wild type and have to be selected and propagated in a more specialized medium. Mutated organisms generally need to be compensated for any loss in gene function. Mutations would usually lead to derailing of a multigene-multienzyme pathway rather than improving its efficiency. To test the possibility of gamma irradiation of *A. parasiticus* spores leading to mutants with enhanced capabilities to produce aflatoxin, Sharma et al. [71] studied the progeny of the survivors in a gamma-irradiated spore suspension for their ability to produce aflatoxin in synthetic glucose salt medium and rice. The greatest amount of aflatoxin was produced by the inoculum from the progeny of nonirradiated spore suspension, whereas the inoculum from the progeny of the survivors from irradiated spore suspensions produced a comparatively lower amount of the toxin in both glucose salt medium and rice. In a recent study with ochratoxin, Chelack et al. [60] obtained red, yellow, and white mutants of the parent ocher-colored strain of *A. ochraceus* using gamma irradiation and reported that yellow and white variants produced more toxin than the parent ocher-colored strain.

There is also a concern regarding increased radioresistance of the survivors which may pose problems related to food processing and environment. Despite the best efforts of researchers, very few radiation-resistant organisms have been isolated after treatment with gamma rays [73,78]. The effect of growth recycling and gamma irradiation on radiation resistance and aflatoxin production by *A. flavus* was also studied by Faizur-Rahman and Idziak [73]. These workers did not find any increase in the toxin-producing ability or radioresistance of the organism following repeated exposure of the inoculum to gamma radiation doses up to 0.8 kGy. On the contrary, a decrease in the toxin-producing ability was reported. Similarly, repeated cycles of sublethal irradiation of *A. flavus* were found to lead more frequently to complete loss or decrease of aflatoxin production than to increase [33,69]. Mutation ratio as measured by the observed morphological changes in the survivors at different doses and aflatoxin production were not remarkably changed even after six times exposure of *A. flavus* var. *curvularia* and *A. parasiticus* at 0.8 kGy [55].

Reduced ability to produce aflatoxin rather than the increased ability was found to be prevalent in the mutants. One reason for the observed variability in mycotoxin production by cultures could be the heterokaryotic nature of the fungal mycelium. Hyphal anastomosis during subculturing and recombinations between heterokaryons with producing and nonproducing ability could lead to hyper-, hypo-, or nonproducing ability reflected in conidia of the fungus [74–76]. Mycotoxins, like aflatoxins, are secondary metabolites requiring expression of several structural and regulatory genes for their elaboration [77]. In the absence of a conscious effort for selection, it is easier to have a negative effect than one with enhanced production. Though radiation has been used to improve strains of fungi for higher yield of secondary metabolites such as antibiotics, these strains were not developed overnight by a single exposure to radiation. It has been possible only through sustained selection in a phased manner.

X. STUDIES ON MYCOTOXIN FORMATION IN IRRADIATED FOODS AND AGRICULTURAL COMMODITIES

Though there are several studies on the assessment of mycotoxin, mainly the aflatoxin-producing potential of the fungus exposed to gamma radiation, the potential of the irradiated commodity to support mycotoxin formation by a non-irradiated fungus has been investigated by relatively few authors. Some of the studies are based on experimental design remotely connected with the technology of food irradiation. In an early study [78,79], authors reported that following irradiation, aflatoxin production was found to increase by 45.6% in wheat, 31.4% in maize, 80.8% in sorghum, 66% in pearl millet, 77.4% in potatoes, and 84% in onion. Thus in all the agricultural commodities, irradiated commodity consistently showed an appreciable increase in aflatoxin production. However, the experimental protocol said that irradiated (0.75 kGy) and nonirradiated wheat, maize, sorghum, and pearl millet, as well as irradiated (0.1 kGy) and nonirradiated unpeeled potato and onion sliced in two, in 50-g lots were sterilized at 121°C for 15 min. The flasks containing irradiated as well as nonirradiated commodities were subsequently inoculated with the spores of *A. parasiticus*. The authors concluded, "If irradiated foods become infected with *A. flavus* during storage as is possible, particularly if conditions of storage are not satisfactory and the moisture content increases, the risk of greater amount of toxin being formed must be considered as being very real."

As is obvious from the protocol, the experiment did not simulate the actual conditions of irradiation and storage of agricultural commodities. In a study on wheat, where the conditions of irradiation and storage were simulated as per the actual practice [80], gamma-irradiated (0.2 kGy) and nonirradiated wheat was stored for 6 months with and without inoculating with the spores of toxigenic *A. flavus*. It was found that aflatoxin production in wheat was a function of relative humidity during storage and the grain moisture. A similar type of relationship was also reported earlier [81] while studying the influence of relative humidity on production of aflatoxin in rice by *A. parasiticus*. Bullerman et al. [82] carried out studies on the effect of selected doses of gamma radiation on the ability of two toxigenic strains of *A. parasiticus* in fresh bread system. Bread slices were inoculated with 10^2 and 10^6 spores, packed in polyethylene bags, and exposed to 0, 1, and 2 kGy doses of gamma radiation. Aflatoxin content was analyzed after storage for 6 weeks.

A stimulation of aflatoxin production was observed only in the case of slices inoculated with 10^6 spores and irradiated at 1-kGy dose. However, the samples irradiated with 2-kGy dose did not show the toxin formation. Niles [83] studied growth rates of *A. flavus* and aflatoxin production in wheat sterilized by ethylene oxide fumigation, autoclaving, and irradiation. The fastest growth rates and in general the highest yields of aflatoxin B_1 were reported in wheat irradiated at 10, 25, and 40 kGy, but some very high aflatoxin levels were also recorded for one or two samples sterilized by other methods. No simple relationship was demonstrated between growth rates and aflatoxin levels. In these studies the author used disk inoculum of a growing fungus rather than the standard spore inoculum, and this may have been the cause of large variability observed in the study. Moreover, the doses used by the authors for sterilization of wheat were about 100-fold higher than those recommended for disinfestation of wheat. In many other studies that simulated conditions that actually exist in the practice of food irradiation, increased risk of mycotoxin formation has not been observed [84–86]. In a study with date fruits irradiation at a dose of 3.0 kGy was found to be more effective than other treatments for inhibition of growth of fungi and prevention of aflatoxin production. The author recommended irradiation over fumigation for the preservation of date fruits [84]. There have also been reports of mycotoxin formation in spices such as black and white pepper, capsicum, coriander, and fenugreek

[87,88]. However, irradiation of spices such as black pepper, white pepper, and red pepper did not seem to affect aflatoxin production on these spices [89,90].

XI. NEW APPROACHES TO MYCOTOXIN RISK EVALUATION

As far as the susceptibility of raw irradiated commodity to the attack of aflatoxin-producing molds is concerned, a different experimental approach was undertaken to evaluate the risk in our laboratory. Raw potato is not prone to attack of green mold [91,92]. Swaminathan and Koehler [93] attributed this to the presence of a phenolic compound identified as a derivative of *trans*-hydroxycinnamic acid, which has been shown to have inhibitory activity toward aflatoxin producing fungi. This compound is extractable in ethyl acetate. Ethyl acetate extracts from potato exposed to a sprout inhibition dose of 0.1 kGy of gamma radiation were tested for inhibitory activity toward *A. flavus* and *A. parasiticus* [92]. The radiation treatment was not found to adversely affect this inhibitor in potato. The inhibitory activity was tested both immediately after irradiation and after 4 weeks of storage. No differences were observed in the inhibitory potency of the extract from irradiated potato and that from nonirradiated potato. Potato samples both irradiated and nonirradiated and sterilized at 121°C/15 min, as described in the experiments of Priyadarshini and Tulpule [78,79], yielded an extract devoid of inhibitory activity. In fact, the inhibitor was reported to be labile at 60°C [93].

Onion has also been shown to possess antimicrobial properties [94–96]. Boiled water extracts of onion were reported to cause inhibition of *Alternaria tenuis*, *Helminthosporium* sp., and *Curvularia pernisata* [94]. In our laboratory various extractives of onion were tested for their inhibitory activity toward the growth of aflatoxin-producing fungi, *A. flavus*, and *A. parasiticus* [97,98]. Ether extracts and lachrymatory factor, identified as thiopropanol-s-oxide, were found to have a potent inhibitory activity against these fungi. Steam-distilled onion oil, which was devoid of lachrymatory factor, was not as potent. Its major component, dipropyl-disulfide, was found to be ineffective as the fungal inhibitor. Ethyl acetate fraction-containing phenolic fraction was also found to be ineffective. Exposure of onion to a sprout-inhibiting dose of gamma radiation (0.06 kGy) did not alter the inhibitory potency of the onion extractives, which again were found to be heat-labile. There are no reports of the green mold attack on onion as a saprophyte or parasite. These studies clearly demonstrated that irradiated potato and onion could not become prone to the attack of aflatoxin-producing fungi by the exposure to sprout-inhibiting doses of gamma radiation, and the doses of radiation employed for the preservation of these commodities are so small to have any significant impact on the composition of these commodities.

For similar reasons we have also studied the effect of gamma irradiation on the antifungal properties of certain spices. The essential oils of clove and cinnamon, eugenol and cinnamon aldehyde, respectively, exhibited potent inhibitor potency against aflatoxin-producing fungi. Though gamma irradiation did not affect the antifungal inhibitory potency of clove, reduction in antifungal activity of cinnamon was observed [99]. This could be attributed to the susceptibility of aldehyde group in cinnamon aldehyde to gamma radiation [100]. It was recently reported that levels of aflatoxin B_1 and G_1 increased from 15% to 90% by inoculation of 1-kGy-irradiated spores of *A. flavus* in autoclaved polished rice, black pepper, white pepper, and red pepper. However, irradiation of spices such as black pepper, white pepper, and red pepper at the same dose did not affect aflatoxin production on these spices [89,90].

XII. EFFECT OF IRRADIATION ON COMPETITIVE MICROFLORA AND FORMATION OF MYCOTOXINS

Inhibition of aflatoxin formation under competition from natural microflora in foods and coculture of the toxigenic strain with micro-organisms, both bacteria and molds including atoxigenic molds, has been observed in a number of studies [101–111]. Boller and Schroeder [101] made an early observation that species of the natural mycoflora influenced production and detection of aflatoxin in stored rice inoculated with *A. parasiticus* Speare, and the competition from *A. chevalieri* significantly reduced the quantities of aflatoxin. Subsequently, these authors reported that *A. candidus* which dominated the mycoflora in rice, also suppressed aflatoxin formation by *A. parasiticus* [102]. Similar findings were reported later [103–111]. *Rhizopus nigricans*, *Aspergillus*, *Penicillia*, and *Saccharomyces cerevisiae* were found to inhibit the growth and aflatoxin production by *A. parasiticus* [103–106]. Among bacteria, *Streptococcus lactis* inhibited and *Acetobacter aceti* stimulated aflatoxin formation in culture [107,108].

A possibility exists that the nonsterilizing doses of gamma irradiation eliminate competition for the toxin producers and provide them a better chance for proliferation and toxin production. Moreover, in many studies for carrying out aflatoxin risk assessment, commodities were irradiated and heat-sterilized and inoculated with irradiated or nonirradiated toxigenic cultures, making interpretation of such results in relation to normal grain or food handling practices rather difficult. Experiments have therefore been undertaken to assess the effect of competitive microflora on aflatoxin production in irradiated foods. Production of aflatoxin B_1 and G_1 and zearalenone, respectively, by *A. flavus* and *Fusarium graminearum* was measured when they were cultured alone and in pairs with other filamentous fungi in radiation-sterilized maize seeds at three water activities (0.98, 0.95, and 0.90) and two temperatures (25 and 16°C). *A. flavus* was paired with *A. niger*, *A. oryzae*, *P. viridicatum*, and *F. graminearum*. In another set of experiments *F. graminearum* was paired only with *A. flavus*.

Compared to pure culture, aflatoxin in mixed fungal cultures was decreased at high water activities but was enhanced when the water activity was low (0.90). More aflatoxin was found to be produced at 25 than 16°C by the presence of *A. flavus*. Zearalenone production was markedly decreased at 16°C by the presence of *A. flavus* but was little affected at 25°C [109]. In another study by the same authors, stimulation of aflatoxin production by *A. flavus* in coculture with *Hyphopichia burtoni* and *Bacillus amyloliquifaciens* was reported in irradiated maize [110]. Barley with 25% moisture inoculated with conidia of *A. alutaceus* var. *alutaceus* and then irradiated with either electron beam or gamma rays and incubated for prolonged periods at 28°C showed less ochratoxin accumulation with increasing doses of gamma irradiation.

Inoculation of barley following irradiation with ochratoxin-producing mold resulted in enhanced ochratoxin levels compared to that in nonirradiated controls [60]. The authors concluded that a reduction in the competing microflora by irradiation was responsible for the enhanced toxin production. The authors never encountered any enhancement of mycotoxin production in grain that was inoculated prior to irradiation. Similar findings were reported when barley was fumigated with phosphine or methyl bromide rather than irradiated [111]. Obviously, the effect of competitive microflora cannot be unique to radiation processing. It must, however, be noted that radiation resistance of vegetative cells of bacteria and spores of fungi does not vary greatly. So relative proportions of competing micro-organisms in the surviving fraction may not change so drastically as to have a profound effect on mycotoxin formation. Moreover, for formation of mycotoxin the fungus has to grow and proliferate, in the process making the commodity visibly spoiled and suspect, irrespective of any treatment.

XIII. PRACTICE OF FOOD IRRADIATION

Food irradiation is essentially a need-based technology. Some of its advantages are very specific, unlike conventional technologies. Treatment of food with ionizing radiations is different from other food-processing technologies in that it achieves different objectives depending on the nature of food and the dose employed. Moreover, all foods are not amenable to irradiation. As no processing technology would accept substandard or already spoiled raw material for treatment, in the same way commodity meant for treatment by radiation has to meet all the requisite standards of quality and comply with the prescribed conditions of postirradiation storage of a commodity. Good Manufacturing Practices (GMP) was well as Good Irradiation Practices (GIP) are required to be adopted by the processors to guarantee quality product to the consumer. The Codex General Standard for Irradiated Foods [19,20] and the Recommended International Code of Practice for the Operation of Irradiation Facilities Used for the Treatment of Food [19,20] provide authoritative guidelines that are recognized by regulatory authorities and industry around the world as a basis for safe and effective irradiation practice. Irradiation facilities that process food are also covered by the General Principles of Food Hygiene, advocated by the FAO/WHO Codex Alimentarius Commission as a basic recommendation to ensure hygienic food handling and processing.

In addition, all the Codex Codes of Hygiene and/or Technological Practice recommended for specific food commodities will apply as appropriate. Together with legislation and regulations adopted by countries that have approved the use of radiation in the processing of food, these recognized standards will help to ensure that the benefits of this technique are safely and productively realized throughout the world [112]. Most of the laboratory studies with heat-sterilized commodities inoculated with irradiated toxigenic cultures have little or no practical relevance to the practice of food irradiation. However, these studies have raised and also answered interesting questions related to safety of irradiated foods, and have provided a large database for evaluation of the safety of the process.

XIV. MANAGEMENT OF MYCOTOXINS IN IRRADIATED FOODS

Only an effective quality control program can ensure general food safety during any food-processing procedure including food irradiation. The Hazard Analysis and Critical Control Point (HACCP) system will make the quality control most effective in a food irradiation plant. The basis of HACCP is a regular online check made at preconceived control points in the process, right from the procurement of raw material stage through the processing stage, finished product stage to the market shelf. Though HACCP in a food irradiation plant would vary with the commodity to be treated, it must exercise control at least at four major points, the first being the incoming product quality. The identity, quantity, and quality of the incoming product is of foremost importance, beside other treatment parameters such as dimensions, bulk density, and configuration required for giving uniform dose to the product.

For example, grain lots coming for radiation disinfestation must be checked for mycotoxin contamination and molding. The grain moisture should be checked and should be within the prescribed limits. Second, during processing, the processing parameters must be specified. This should ensure delivery of the required dose, time of irradiation, and the conditions of irradiation. Obviously, for delivering higher doses more time would be required if the dose rate of the source is low. In that case the possibility of product quality deterioration during the treatment should be considered and appropriate steps taken. Many a time it may require controlling of conditions such as temperature and atmosphere of the commodity during irradiation. Third, the postirradiation quality inspection of the products and packages should be carried out to ensure the efficacy of the treatment. Postirradiation storage conditions should be specified for each treatment. For example, irradiation of onion and potato will ensure that the commodity does not sprout, but it is required that the commodity so treated be stored appropriately to discourage microbial rot. It is axiomatic that grains treated for disinfestation need to be stored in insect proof silos or packed in polyethylene lined bags to prevent reinfestation. Retail products treated by radiation for disinfestation should be packed in packaging material impervious to outside storage insects. Fortunately, formation of mycotoxins accompanies mold growth and any molded lots would be visibly spoiled and rejected before getting processed. This can happen if grain is stored under adverse conditions irrespective of the treatment. Radurized, or radiation pasteurized, meat and seafoods need to be stored under refrigerated conditions below 3°C to prevent the growth of *Clostridium botulinum* type E and toxin production. Breach of packaging of food treated for microbial decontamination would expose the contents to atmospheric contaminants in the same way as happens to a canned or sterilized product. Ready-to-eat food in prepacked form preserved by radiation will need all the necessary precautions undertaken for a product treated by other methods for a similar purpose. A radiation-sterilized product is essentially in no way different from the heat or chemically sterilized product as regards its postprocessing storage. Fourth, at the market shelf product quality should be assessed for the success of the treatment. The conditions for shelf storage should be identified and prescribed to the retailers. Finally, only feedback from the consumer can reassure faith in the technology.

XV. CONCLUSION

Advantages of radiation processing of foods accrue from the cold and highly penetrating nature of ionizing radiations which allow preservation and hygienization of food and agricultural commodities in bulk or packaged form without drastically affecting their fresh-like natural attributes. Experiments which simulate conditions akin to actual practice of food irradiation have not given any evidence of increased risk of mycotoxin formation in irradiated foods. Future food security and food safety scenarios demand rapid adoption of food irradiation technology by the food-processing industry, and irradiated foods pose no special problems or hazards in relation to mycotoxin production.

ACKNOWLEDGMENTS

I am grateful to Dr. P. C. Kesavan, Director, Biosciences Group, BARC, for his keen interest in the subject and consistent encouragement, and to Dr. Paul

Thomas, Head, Food Technology Division, Bhabha Atomic Research Centre, Mumbai, for critically reviewing the manuscript and providing helpful suggestions.

REFERENCES

1. JW Bennett, SB Christensen. New perspectives in aflatoxin biosynthesis. Adv Appl Microbiol 29:53, 1983.
2. JI Pitt. The current role of *Aspergillus* and *Penicillium* in human and animal health. J Med Vet Mycol 32:17, 1994.
3. PS Steyn. Mycotoxins, general view, chemistry and structure. Toxicol Lett 82:843, 1995.
4. K Sargeant, A Sheridan, J O'Kelly, RBA Carnaghan. Toxicity associated with certain samples of groundnuts. Nature 192:1096, 1961.
5. LA Goldblatt. Aflatoxins: Scientific Background, Control and Implications. New York: Academic Press, 1969.
6. CM Christensen. Deterioration of stored grains by fungi. Bot Rev 23:108, 1957.
7. PM Scott. Mycotoxins in stored grain, feed and other cereal products. In: RN Sinha, WE Muir, eds. Grain Storage: Part of a System. Westport, CT: Avi Publishing Co., 1973.
8. GA Payne. Aflatoxin in Maize. Crit Rev Plant Sci 10:423, 1992.
9. DL Park, LS Lee, RL Price, AE Pohland. A review of the decontamination of aflatoxin by ammoniation: current status and regulation. J Assoc Off Anal Chem 71: 685, 1988.
10. G Piva, F Galvano, A Pietri, A Piva. Detoxification methods of aflatoxin—a review. Nutr Res 15:767, 1995.
11. TE Cleveland, D Bhatnager, eds. Molecular strategies for reducing aflatoxin levels in crops before harvest. In: Molecular Approaches to Improving Food Quality and Safety. New York: Academic Press, 1994.
12. D Bhatnagar, T Cleveland, J Linz, G Payne. Molecular biology to eliminate aflatoxins. INFORM 6:262, 1995.
13. F Trail, N Mahanti, M Patrick, et al. Physical and transcriptional map of an aflatoxin gene cluster in *Aspergillus parasiticus* and functional disruption of gene involved early in the aflatoxin pathway. Appl Environ Microbiol 61:2665, 1995.
14. GA Payne, D Bhatnagar, TE Cleveland, JE Linz. Genomic organization and regulation of aflatoxin biosynthesis in *Aspergillus flavus* and *A. parasiticus*. In: M Eklund, JL Richard, K Mise, et al., eds. Molecular Approaches to Food Safety: Issues Involving Toxic Microorganisms. Alken, Inc., 1996.
15. RL Brown, TE Cleveland, GA Payne, CP Woloshuk, KW Campbell, DG White. Determination of resistance to aflatoxin production in maize kernels and detection of fungal colonization using and *Aspergillus flavus* transformant expressing *E. coli* B-glucuronidase. Phytopathology 85:983, 1995.
16. KW Campbell, DG White. Inheritance of resistance to *Aspergillus* ear rot and aflatoxin in corn genotypes. Phytopathology 85:887, 1995.

17. A Sharma, PM Nair. Food irradiation. Encycl Agric Sci 2:293, 1994.
18. WM Urbain. Food Irradiation. Orlando: Academic Press, 1986.
19. World Health Organization. Safety and Nutritional Adequacy of Irradiated Foods. Geneva: WHO, 1994.
20. World Health Organization. Wholesomeness of irradiated food: a report of the Joint FAO/IAEA/WHO Expert Committee. WHO Technical Report Series 659, 1981.
21. International Commission on Microbiological Specifications for Foods, Ionizing Radiations. In: Microbial Ecology of Foods. Vol I. New York: Academic Press, 1980.
22. A Brynjolfson. Wholesomeness of irradiated foods. J Food Safety 7:107, 1985.
23. Council for Agricultural Science and Technology. Ionizing energy in food processing and pest control. I. Wholesomeness of food treated with ionizing energy. Report 109, 1986.
24. T Radomyski, EA Murano, DG Olson, PS Murano. Elimination of pathogens of significance in food by low-dose irradiation. J Food Prot 57:73, 1994.
25. JF Diehl. Safety of Irradiated Foods. New York: Marcel Dekker, 1995.
26. T Shantha, V Sreenivasamurthy. Use of sunlight to partially detoxify groundnut cake flour and casein contaminated with aflatoxin B_1. J Assoc Anal Chem 64:29, 1981.
27. I Nkama, JH Nobbs, HG Muller, Destruction of aflatoxin B_1 in rice exposed to light. J Cereal Sci 5:167, 1987.
28. K Aibara, K Miyaki. Aflatoxin and its radiosensitivity. Proceedings of panel on Radiation Sensitivity of Toxins and Animal Poisons. IAEA, Bangkok, 1970.
29. HK Frank, T Grunewald. Radiation resistance of aflatoxins. Food Irradiat 11:15, 1970.
30. P Temcharoen, WG Thilly. Removal of aflatoxin B_1 toxicity but not mutagenicity by 1 megarad of gamma radiation of peanut meal. J Food Safety 4:199, 1982.
31. MP Santamarina, FJ Gimenez, C Sabater, V Sanchis. Measures to reduce and eliminate mycotoxins in food and feeds. Rev Iber Micol 12:52, 1995.
32. H Hooshmand, CF Klopfenstein. Effects of gamma irradiation on mycotoxin disappearance and amino acid contents of corn, wheat, and soybeans with different moisture contents. Plant Foods Human Nutr 47:337, 1995.
33. HK Frank, R Munzner, JF Diehl. Response of toxigenic and non-toxigenic strains of *Aspergillus flavus* to irradiation. Saboraudia 9:21, 1971.
34. RE Zirkle, DF Marchbank, KD Kuc. Exponential and sigmoidal survival curves resulting from alpha and X-irradiation of *Aspergillus* spores. J Cell Comp Physiol 39:75, 1952.
35. T Kume, H Ito, H Izuka, M Takehisa. Radiosensitivity of *Aspergillus versicolor* isolated from animal feeds and decomposition of sterigmatocystine by gamma radiation. Shokuhin Shosha 18:5, 1983.
36. TW Chou, B Singh, DK Salunkhe, WF Campbell, Effects of gamma radiation on *Penicillium expansum* L. I. Some factors influencing the sensitivity of the fungus. Rad Bot 10:511, 1970.
37. SR Padwal-Desai, AS Ghanekar, A Sreenivasan. Studies on *Aspergillus flavus*. I. Factors influencing radiation resistance of non-germinating conidia. Environ Exp Bot 16:45, 1976.

38. A Sharma, AG Behere, SR Padwal-Desai, GB Nadkarni. Influence of inoculum size of *Aspergillus parasiticus* spores on aflatoxin production. Appl Environ Microbiol 40:989, 1980.
39. WS Chelack, J Borsa, JG Szekley, RR Marquardt, AA Frohlich. Variants of *Aspergillus alutaceus* var. *alutaceus* with altered ochratoxin A production. Appl Environ Microbiol 57:2487, 1991.
40. EH Choi, HL Kim, SR Lea. Radiation sensitivity of some toxigenic molds isolated from deteriorated rice. Korean J Food Sci Technol 7:148, 1975.
41. H Ito, J Bunnak, ZM DeGuzman, I Ishigaki. Effect of irradiation on aflatoxin production by *Aspergillus flavus*. Food Irradiat Jpn 26:43, 1991.
42. SR Padwal-Desai, AS Ghanekar, A Sharma. Studies on *Aspergillus flavus*. III. Chemical sensitization of *Aspergillus flavus* spores to thermoradiation. Acta Alimentaria 11:343, 1982.
43. NF Sommer, RJ Fortlage. Ionizing radiation for control of post-harvest diseases of fruits and vegetables. Adv Food Res 15:428, 1967.
44. SR Padwal-Desai, AS Ghanekar, A Sreenivasan. Heat radiation combination for control of mold infection in harvested fruits and processed cereal foods. Acta Alimentaria 2:189, 1973.
45. P Thomas. Radiation preservation of foods of plant origin. III. Tropical fruits: banana, mangoes and papayas. CRC Crit Rev Food Sci Nutr 23:147, 1986.
46. P Thomas. Radiation preservation of foods of plant origin. IV. Temperate fruits: pome fruits, stone fruits and berries. CRC Crit Rev Food Sci Nutr 24:337, 1986.
47. P Thomas. Radiation preservation of foods of plant origin. VI. Mushrooms, tomatoes, minor fruits and vegetables, dried fruits and nuts. CRC Crit Rev Food Sci Nutr 26:313, 1986.
48. M Jemmali, A Guilbot. Influence of gamma irradiation on the tendency of *A. flavus* spores to produce toxins during culture. Food Irradiat 10:15, 1970.
49. LB Bullerman, TE Hartung. Effect of low dose gamma irradiation on growth and aflatoxin production by *Aspergillus parasiticus*. J Milk Food Technol 37:430, 1974.
50. KL Applegate, JR Chipley. Increased aflatoxin G_1 production by *Aspergillus flavus* via gamma irradiation. Mycologia 65:1266, 1973.
51. KL Applegate, JR Chipley. Effect of ^{60}Co gamma irradiation on aflatoxin B_1 and B_2 production by *Aspergillus flavus*. Mycologia 66:436, 1974.
52. KL Applegate, JR Chipley. Increased aflatoxin production by *Aspergillus flavus* via cobalt irradiation. Poult Sci 52:1492, 1973.
53. KL Applegate, JR Chipley. Production of ochratoxin A by *Aspergillus ochraceus* NRRL-3174 before and after exposure to ^{60}Co irradiation. Appl Environ Microbiol 31:349, 1976.
54. AF Schindler, AN Abadie, RE Simpson. Enhanced aflatoxin production by *Aspergillus flavus* and *Aspergillus parasiticus* after gamma irradiation of spore inoculum. J Food Prot 43:7, 1980.
55. H Ito. Enhanced aflatoxin production by *A. parasiticus* and *A. flavus* after low dose gamma irradiation. Food Irradiat Jpn 27:27, 1992.
56. N Paster, R Barkai-Golan, R Padova. Effect of gamma irradiation on ochratoxin production by the fungus *Aspergillus ochraceus*. J Sci Food Agric 36:445, 1985.

57. LB Bullerman, TE Hartung. Effect of low level gamma irradiation on growth and patulin production by *Penicillium patulum*. J Food Sci 40:195, 1975.
58. G Clevstrom, H Ljunggren. Aflatoxin formation and the dual phenomenon in *Aspergillus flavus* Link. Mycopathologia 92:129, 1985.
59. GT Odamtten, V Appiah, DI Langerak. Influence of inoculum size of *Aspergillus flavus* on production of aflatoxin in maize medium before and after exposure to combination treatment of heat and gamma radiation. Int J Food Microbiol 4:119, 1987.
60. WS Chelack, J Borsa, JG Szekley, RR Marquardt, AA Frohlich. Role of competitive microbial flora in the radiation-induced enhancement of ochratoxin production by *Aspergillus alutaceus* var. *alutaceus* NRRL 3174. Appl Environ Microbiol 57:2492, 1991.
61. GT Odamtten, V Appiah, DI Langerak. Preliminary studies on the effects of heat and gamma irradiation on the production of aflatoxin B_1 in static liquid culture, by *Aspergillus flavus* Link NRRL 5906. Int J Food Microbiol 3:339, 1986.
62. E Erdman, FS Thatcher, KF Mcqueen. Studies on irradiation of microorganisms in relation to food preservation. II. Radiation resistant mutants. Can J Microbiol 7:206, 1961.
63. FS Thatcher. Appendex VII. Some public health aspects of the microbiology of irradiated foods. Int J Appl Rad Isotope 14:51, 1963.
64. JJ Licciardello, JTR Nickerson, SA Goldblith, CA Shanon, WW Bishop. Development of radiation resistance in *Salmonella* cultures. Appl Microbiol 18:24, 1969.
65. NA Epps, ES Idziak. Radiation treatment of foods. II. Public health significance of irradiation recycled *Salmonella*. Appl Microbiol 19:338, 1970.
66. TJ Previte, Y Chang, W Scrutchfield, HM El-Bisi. Effect of radiation pasteurization on *Salmonella*. II. Influence of repeated radiation growth cycles on virulence and resistance to radiation and antibiotics. Can J Microbiol 17:505, 1979.
67. DO Cliver. Unlikelihood of mutagenic effects of radiation on viruses. In: Wholesomeness of Irradiated Food. Annex 2. Report of a joint FAO/IAEA/WHO expert committee. WHO Report Series 604. Geneva: WHO, 1977, pp 43–44.
68. RB Maxcy. Comparative viability of unirradiated and gamma irradiated bacterial cells. J Food Sci 42:1056, 1977.
69. M Ingram, J Farkas. Microbiology of foods pasteurized by ionizing radiation. Acta Alimentaria 6:123, 1977.
70. J Farkas. Microbiological safety of irradiated foods. Int J Food Microbiol 9:1, 1989.
71. A Sharma, SR Padwal-Desai, PM Nair. Aflatoxin producing ability of spores of *Aspergillus parasiticus* exposed to gamma irradiation. J Food Sci 55:275, 1990.
72. B Moseley. Radiation, microorganisms and radiation resistance. In: DE Johnston, MH Stevenson, eds. Food Irradiation and the Chemist. London: Royal Society of Chemistry, 1992.
73. ATM Faizur-Rahman, ES Idziak. Gamma irradiation recycling of *Aspergillus flavus* and its effect on radiation resistance and toxin production. Can J Food Sci Technol 10:5, 1977.
74. KE Papa. The parasexual cycle in *Aspergillus flavus*. Mycologia 65:1201, 1973.
75. KE Papa. The parasexual cycle in *Aspergillus parasiticus* Mycologia 70:766, 1978.

76. JW Bennett, KE Papa. The aflatoxin *Aspergillus* sp. In: Genetics of Plant Pathogenic Fungi. London: Academic Press, 1988.
77. SP Kale, D Bhatnagar, JW Bennett. Isolation and characterization of morphological variants of *A. parasiticus* deficient in secondary metabolite production. Mycol Res 98:642, 1994.
78. E Priyadarshini, PG Tulpule. Aflatoxin production on irradiated foods. Food Cosmet Toxicol 14:293, 1976.
79. E Priyadarshini, PG Tulpule. Effect of graded doses of gamma irradiation on aflatoxin production by *Aspergillus parasiticus* in wheat. Food Cosmet Toxicol 17:505, 1979.
80. AG Behere, A Sharma, SR Padwal-Desai, GB Nadkami. Production of aflatoxins during storage of gamma irradiated wheat. J Food Sci 43:1102, 1978.
81. RA Boller, HW Schroeder. Influence of *Aspergillus candidus* on production of aflatoxin in rice by *Aspergillus parasiticus*. Phytopathology 64:121, 1974.
82. LB Bullerman, HM Banhart, TE Hartung. Use of gamma irradiation to prevent aflatoxin in bread. J Food Sci 38:1238, 1973.
83. EV Niles. Growth rate and aflatoxin production of *Aspergillus flavus* in wheat sterilized by gamma irradiation, ethylene oxide fumigation and autoclaving. Trans Br Mycol Soc 70:239, 1978.
84. UA Emam, SEA Farag, AI Hammad. Comparative studies between fumigation and irradiation of semi dry date fruits. Nahrung 38:612, 1994.
85. G Ogbadu. Effect of gamma irradiation on aflatoxin production by *Aspergillus flavus* growing on some Nigerian food stuffs. Food Irradiat News Lett 2:39, 1978.
86. G Ogbadu. Influence of gamma irradiation of aflatoxin B_1 production by *Aspergillus flavus* growing on some Nigerian food stuffs. Microbios 27:19, 1980.
87. N Pal, AK Kundu. Studies on *Aspergillus* spp. from Indian spices in relation to aflatoxin production. Science Culture, 38:252, 1972.
88. M Seenappa, AG Kempton. Application of minicolumn detection method for screening spices for aflatoxin. J Environ Sci Health B15:219, 1980.
89. G Ogbadu. Effect of low dose gamma irradiation on the production of aflatoxin B_1 by *Aspergillus flavus* growing on *Capsicum annum*. Microbios Lett 10:139, 1979.
90. H Ito, H Chen, J Bunnak. Aflatoxin production by microorganism of the *Aspergillus flavus* group in spices and the effect of irradiation. J Sci Food Agric 65:141, 1994.
91. JD Wildman, L Stoloff, R Jacobs. Aflatoxin production by a potent *Aspergillus flavus* Link isolate. Biotechnol Bioeng 9:429, 1967.
92. A Sharma, AJ Shrikhande, SR Padwal-Desai, GB Nadkarni. Inhibition of aflatoxin producing fungi by ethyl acetate extracts from gamma irradiated potatoes. Potato Res 21:31, 1978.
93. B Swaminathan, PE Koehler. Isolation of an inhibitor of *Aspergillus parasiticus* from white potatoes (*Solanum tuberosum*). J Food Sci 41:313, 1976.
94. I Abdou, AS Abdou Zeids, MR El-Sherbeeny, ZM Aby El Gheats. Antimicrobial activity of *Allium sativum*, *Allium cepa*, *Raphanus sativus*, *Capsicum frutescens*, *Eruca sativa*, and *Allium kurrat* on bacteria. Qual Plant Mater Veg 22:29, 1972.
95. NF Lewis, BYK Rao, AR Shah, GM Tewari, C Bandyopadhyay. Antibacterial activity of volatile components of onion (*Allium cepa*). J Food Sci Technol 14:35, 1977.

96. PS Shekhawat, R Prasada. Antifungal properties of some plant extracts. 2. Growth inhibition studies. Sci Cult 37:40, 1971.
97. A Sharma, GM Tewari, AJ Shrikhande, SR Padwal-Desai, C Bandyopadhyay. Inhibition of aflatoxin producing fungi by onion extract. J Food Sci 44:1545, 1979.
98. A Sharma, SR Padwal-Desai, GM Tewari, C Bandyopadhyay. Factors affecting antifungal activity of onion extractives against aflatoxin producing fungi. J Food Sci 46:741, 1981.
99. A Sharma, AS Ghanekar, SR Padwal-Desai, GB Nadkarni. Microbiological status and antifungal properties of irradiated spices. J Agric Food Chem 32:1081, 1984.
100. JF Diehl, S Adams, H Delincee, V Jakubick. Radiolysis of carbohydrates and of carbohydrate containing food stuffs. J Agric Food Chem 26:15, 1978.
101. RA Boller, HW Schroeder. Influence of *Aspergillus chevalieri* on production of aflatoxin in rice by *Aspergillus parasiticus*. Phytopathology 64:17, 1973.
102. RA Boller, HW Schroeder. Influence of *Aspergillus candidus* on production of aflatoxin in rice by *Aspergillus parasiticus*. Phytopathology 64:121, 1974.
103. LS Weckbach, EH Marth. Aflatoxin production by *Aspergillus parasticus* in a competitive environment. Mycopathologia 62:1, 1977.
104. A Sharma. Studies on biogenesis of aflatoxin. Ph.D. thesis, Bombay University, 1984.
105. SM El-Gendy, EH Marth. Growth of toxigenic and non-toxigenic aspergilli and penicillia at different temperatures and in the presence of lactic acid bacteria. Arch Lebensmittelhyg 31:192, 1980.
106. PB Mislivec, MW Trucksess, L Stolloff. Effect of other mold species on aflatoxin production by *Aspergillus flavus* in sterile broth shake cultures. J Food Prot 51:449, 1988.
107. DW Wiseman, EH Marth. Growth and aflatoxin production by *Aspergillus parasiticus* when in the presence of *Streptococcus lactis*. Mycopathologia 73:49, 1980.
108. J Coallier-Asch, ES Idziak. Interaction between *Streptococcus lactis* and *Aspergillus flavus* on production of aflatoxin. Appl Environ Microbiol 49:163, 1985.
109. RG Cuero, JE Smith, J Lacey. Stimulation of *Hyphopichia burtoni* and *Bacillus amyloliquifaciens* of aflatoxin production by *Aspergillus flavus* in irradiated maize and rice grains. Appl Environ Microbiol 53:1142, 1987.
110. RG Cuero, JE Smith, J Lacey. Mycotoxin formation by *Aspergillus flavus* and *Fusarium graminearum* in irradiated maize grains in the presence of other fungi. J Food Prot 51:452, 1988.
111. J Borsa, WS Chelack, RR Marquardt, AA Frohlich. Comparison of irradiation and chemical fumigation used in grain disinfestation on production of ochratoxin A by *Aspergillus alutaceus* in treated barley. J Food Prot 55:990, 1992.
112. World Health Organization and Food and Agricultural Organization. Food Irradiation: A Technique for Preserving and Improving the Safety of Food. Geneva: WHO, 1988.

15

Regulatory Control Programs for Mycotoxin-Contaminated Food

Garnett E. Wood and Mary W. Trucksess
Food and Drug Administration, Washington, D.C.

I. INTRODUCTION

The presence of various chemicals in foods is a matter of utmost concern to many consumers. Modern food technology relies heavily on the addition of chemicals not only to preserve foods but also to produce appealing colors, flavors, and textures. Because these chemicals are intentionally added to food and declared on the label, they are not considered to be adulterants or contaminants and are referred to as food additives. In contrast, there are other chemicals, environmental contaminants, that inadvertently find their way into the human food supply. They can enter food directly or indirectly during the production, processing, storage, or transportation of food; their presence in food is never intended. Some of the naturally occurring environmental contaminants in foods include fungal toxins, called mycotoxins. Small amounts of these substances may be legally permitted if they are unavoidable under good manufacturing practices and if the amounts involved are not injurious to health. This chapter will review the application of food regulations and strategies to control exposure to mycotoxins. It is imperative that control programs be implemented to restrict the presence of environmental contaminants in foods to the lowest practical level.

II. FOOD REGULATIONS

In the United States, the Food and Drug Administration (FDA) is the federal government's primary consumer protection agency with regard to food [1]. The

FDA is responsible for enforcing a number of statutes passed by Congress; one of the most important is the Federal Food, Drug, and Cosmetic Act of 1938. This statute and its amendments serve as the legal basis for regulating poisonous or deleterious substances in foods, and prohibit the entry of adulterated food into interstate commerce. It is the most extensive law of its kind in the world. Many of the states in the U.S. have laws similar to the federal law, and some have provisions to add automatically any new federal requirements. The statute is intended to assure consumers that foods are pure and wholesome and produced under sanitary conditions. The strategies used by the FDA in enforcing the Act include:

1. Monitoring the marketplace to ensure compliance with the laws and regulatory limits
2. Initiating appropriate enforcement action against violators
3. Taking steps to prevent problems or situations which might expose the public to food hazards
4. Providing guidance to the food industry
5. Cooperating with state and international governments and other federal agencies in regard to the safety of foods.

Regulations issued by FDA are an important part of the food and drug law. Especially important are such regulations as:

1. Current good manufacturing practice (GMP) regulations, which set requirements for sanitation, inspection of materials and finished products, and other quality controls
2. FDA food standards, which set specifications for many food products
3. Food additive regulations that apply to substances which may, by their intended uses, become components of food, either directly or indirectly, or which may otherwise affect the characteristics of the food
4. Labeling regulations under the Nutrition Labeling and Education Act.

Such regulations help both consumers and industry by providing instruction on what must be done to ensure safe, acceptable products. All FDA regulations are published annually by the Government Printing Office in Title 21, Code of Federal Regulations. Mycotoxins are regulated under Section 402(a)(1) of the Food, Drug, and Cosmetic Act which considers a food to be adulterated if it contains any poisonous or deleterious substance which may render, or ordinarily render, it injurious to health.

III. MYCOTOXINS

Mycotoxins are toxic metabolites produced by certain fungi growing on agricultural commodities in the field and/or during storage. The occurrence of these toxins on grains, nuts, and other commodities susceptible to mold infestation is unavoidable and influenced by environmental factors such as temperature, humidity, and extent of rainfall during the preharvesting, harvesting, and postharvesting periods.

The level of contamination of a commodity with a particular mycotoxin varies with geographic location; agricultural and agronomic practices; and the susceptibility of the plants to fungal invasion during all phases of growth, storage, and/or processing. Some mycotoxins may exhibit toxicological manifestations in humans and susceptible animals. Some mycotoxins are also teratogenic, mutagenic, and/or carcinogenic in certain susceptible animal species and are associated with various diseases in domestic animals, livestock, and humans in many parts of the world.

Because environmental factors (weather conditions) play a major role in the occurrence of mycotoxins on grains and other commodities, it is not surprising that the incidence of mycotoxin contamination of a particular food crop can vary not only from region to region, but also from year to year.

Unfortunately, limited information is available on a large number of mycotoxins that have been identified, particularly in reference to the extent of their occurrence, stability in foods, and toxicity. It is known that mycotoxins can enter the human food chain by two major routes: direct contamination resulting from the growth of the fungi on the food, or indirect contamination resulting from the incorporation of contaminated ingredients into food. Human exposure to direct contamination can be a significant problem in tropical areas and some undeveloped countries where the consumption of moldy food may be unavoidable because of inadequate storage and processing facilities or actual shortages of good-quality foods. In developed countries, moldy food is usually discarded or fed to animals; therefore indirect exposure is more significant from the consumption of processed foods made from contaminated grains or edible animal products from animals that have consumed contaminated feed.

The mycotoxins that occur in naturally contaminated foods include aflatoxins, deoxynivalenol, fumonisins, ochratoxins, zearalenone, patulin, cyclopiazonic acid, penicillic acid, and sterigmatocystin. In order to establish the need for regulatory control programs for these and other toxins, background exposure data along with toxicological data must be obtained and carefully evaluated. Of all the mycotoxins that have been identified since 1960, more studies have been reported on aflatoxins than on any of the others. The major impetus for this trend was a series of findings which demonstrated that these toxins were potent liver toxins, were carcinogens in susceptible laboratory animals, and were present as contaminants in a variety of foods consumed by humans. The aflatoxins are the only mycotoxins for which a regulatory limit has been established in the U.S. Currently the FDA mycotoxin program includes efforts to control aflatoxins, deoxynivalenol, fumonisins, ochratoxins, and patulin levels in food.

IV. CONTROL PROGRAM FOR AFLATOXINS

A history of the developments that resulted in the establishment of an aflatoxin control program in the U.S. has been reviewed [2–4].

During the early 1960s, the FDA was alerted to an outbreak of a disease in England that resulted in the death of young turkeys, calves, and other animals. The disease was associated with the use of imported peanut meal from Brazil as a feed ingredient. Further studies resulted in the isolation of a fungal toxin that affected the liver of animals and was believed to be a hepatocarcinogen [5]. The major fungus involved was identified as *Aspergillus flavus*; hence the toxins in the meal were named aflatoxins. This information was given serious consideration by representatives of the peanut industry in the U.S. and the FDA. In 1963, some U.S. peanuts and peanut meal were found to be contaminated with aflatoxins. Liver cancer was observed in laboratory rats fed a diet of the contaminated meal [6]. The toxins were characterized as consisting of four structurally related compounds referred to as aflatoxins B_1, B_2, G_1, and G_2. These designations were based on the chromatographic fluorescent characterization of these compounds on thin-layer chromatography plates.

The finding of aflatoxins as contaminants in the peanut supply prompted peanut manufacturers in the U.S. to initiate a joint effort with the U.S. Department of Agriculture (USDA) and FDA to develop practical analytical methodology for these toxins. These methods were later used by various laboratories to conduct surveys to determine the extent of aflatoxin contamination in peanuts. Further studies revealed that aflatoxins were not readily degraded during routine processing techniques.

Although no direct evidence linked aflatoxins to liver cancer in humans in the U.S., the Commissioner of FDA concluded that the observations of severe carcinogenic effects in experimental animals and reports of positive correlations between dietary aflatoxins and primary human liver cancer in other parts of the world were sufficient justification to take action to control human exposure to aflatoxins to the lowest possible level.

A. Establishment of Regulatory Levels

FDA controls dietary contaminants by establishing and enforcing action levels. Action levels are intended for use during interim periods when there is the possibility that good agronomic practices or improved technological changes may be forthcoming to cause changes in the contamination patterns.

Action levels are also established as informal guidelines for FDA's field staff in enforcement actions. Action levels do not have the force and effect of law; therefore FDA must prove aspects of the statutory violation in each enforcement action. In 1965, an action level of 30 μg/kg total aflatoxins (B_1, B_2, G_1, and G_2)

was established for peanuts and peanut products. The selection of that limit was based on practical considerations relating to the capabilities of the current sampling procedures, analytical methodology, agronomic practices, and technological procedures to minimize human exposure to aflatoxins in contaminated products. In 1969, the action level was reduced to 20 μg/kg and was applied to all commodities (foods and feeds) susceptible to aflatoxin contamination.

Over the years, the action levels for aflatoxins have been periodically reviewed as newer scientific knowledge regarding analytical, toxicological, and technological procedures became available. In 1982, an action level of 300 μg/kg was established for aflatoxins in cottonseed meal as an animal feed ingredient for beef cattle, swine, and poultry [7]. This change was made as a result of an FDA evaluation of its overall policy for regulating aflatoxin contamination of feeds for food-producing animals. In 1988, FDA announced action levels for aflatoxins in corn shipped in interstate commerce and intended for certain food-producing animals [8]. The levels are 100 μg/kg aflatoxins for corn intended for breeding beef cattle, breeding swine, or mature poultry; 200 μg/kg aflatoxins for corn intended for finishing swine (i.e., 100 lb or greater); and 300 μg/kg aflatoxins for corn intended for finishing (feedlot) beef cattle.

The decision to take this action was made after very careful consideration of all the available data on the ratios of aflatoxins in feed to aflatoxins in various animal tissues used for human food (e.g., meat or eggs). The FDA concluded from the available scientific literature that the above levels for aflatoxins in corn used for animal feed were not expected to result in any significant increase in aflatoxin residues in human food and thus would have no adverse effect on animal health or the safety of the nation's food supply. In 1990, FDA established action levels for aflatoxins in peanut products used in animal feed, and set levels which followed the action levels announced for corn in 1988 for certain food-producing animals [9,10]. The available scientific information revealed that the inclusion of aflatoxin-contaminated peanut products in animal feed at those levels would have a negligible effect on the tissue residues.

Aflatoxin M_1 is a major metabolite of aflatoxin B_1 that is produced in the liver of animals that have ingested feed contaminated with aflatoxin B_1 [11]. It may be excreted in the urine and also in the milk of lactating mammals. FDA established an action level of 0.5 μg/kg for aflatoxin M_1 in fluid milk and milk products in 1977 [12]. This level was selected based upon the practical consideration of analytical capabilities, the need to minimize human exposure, and the finding in transmission studies that feed containing 20 μg/kg of B_1 would result in < 0.5 μg/kg of aflatoxin M_1 in the milk.

The current action levels for aflatoxins in foods and feeds are shown in Table 1. As a result of a decision by the U.S. Court of Appeals in a case involving aflatoxin contamination, FDA announced in 1988 [13] that its current action levels are not binding on the courts, the public (including food processors), or the

TABLE 1 FDA Action Levels for Total Aflatoxins[a]

Commodity	Level (ng/g)
All products, except milk, designated for humans	20
Corn for immature animals and dairy cattle	20
Corn and peanut products for breeding beef cattle, swine, and mature poultry	100
Corn and peanut products for finishing swine	200
Corn and peanut products for finishing beef cattle	300
Cottonseed meal (as a feed ingredient)	300
All other feedstuffs	20
Milk	0.5[b]

[a]Food and Drug Administration (FDA) Compliance Policy Guides 7120.26, 7106.10, 7126.33 (revised 1994).
[b]Aflatoxin M_1.

agency. The levels do represent the best guidance available on levels that FDA considers to be of regulatory interest and will thus enhance the safety of the food supply when implemented and adhered to by the food industry.

B. Analytical Methodology

Modern scientific methods are required to enforce laws such as the Food, Drug, and Cosmetic Act. Ensuring the safety and wholesomeness of foods would be impractical without reliable methods of laboratory analysis to determine whether products are up to standard.

Many methods are available for the analysis of foods for aflatoxins; however, it is important to choose the method which gives the most reliable results for the commodity being analyzed. Larger variability is expected in data obtained from methods designed for quantitation of extremely low levels of toxins. Of the many analytical methods published for aflatoxins, only a relatively small number have been subjected to a formal interlaboratory (collaborative) study. The traditional methods for analysis include thin-layer chromatographic (TLC) and high-performance liquid chromatographic (HPLC) procedures.

The basic steps in each include extraction, liquid partition, purification, separation, and quantitation. Recently, several relatively simple but sensitive, enzyme-linked immunosorbent assays (ELISAs) have been developed, evaluated, and approved for qualitative screening and quantitation of aflatoxins [14]. These latter methods offer the advantage of shorter analytical time compared with traditional methods. All commodities collected for official testing for aflatoxins by FDA are analyzed by collaboratively studied methods found in The Official

Methods of Analysis of the Association of Official Analytical Chemists International. This compendium of analytical methods is the leading internationally recognized guide to analytical procedures for law enforcement. The limit of detection for most methods within laboratories is 1 ng/g for aflatoxins in grains, nuts, and their products, and 0.05 ng/ml for aflatoxin M_1 in milk.

C. Monitoring Activity

The food supply is monitored routinely by FDA through compliance programs to ensure adherence to the action levels that have been established for aflatoxins in various commodities. The strategies used by FDA for implementing compliance programs have been published elsewhere [15]. Three basic steps are involved in the testing procedure associated with any monitoring program for aflatoxins:

1. Sampling a given lot of foodstuff to get a laboratory sample representative of that lot
2. Preparation of the sample for analysis
3. Analysis of the test sample for the toxin [16,17].

Although errors are associated with each step, the greater errors are associated with step 1. The first step is the most difficult to achieve due to the heterogeneous nature of aflatoxin contamination of agricultural commodities; hence variability encountered at this step is the most significant in the total monitoring scheme. For example, aflatoxin levels as high as 400,000 μg/kg have been found in individual corn kernels [18]. A high concentration of aflatoxin in individual kernels contributes to the variability observed in the reported results from sequential samples taken from a particular lot. An excellent review of accepted sampling and sample preparation procedures recommended for the identification and quantification of natural toxins in foods and feeds has been published [19].

In general, monitoring data obtained over the years show that the aflatoxins are frequent and major contaminants of corn, peanuts, and cottonseed and an occasional contaminant of almonds, pecans, pistachio nuts, and walnuts [15,20, 21]. Some monitoring data obtained for selected commodities for the years 1992 to 1995 are presented in Tables 2 to 5. The data reflect the unpredictable occurrence of aflatoxins in foods and the relatively frequent low levels present in certain commodities. Of the three peanut products examined in Table 2, roasted in-shell peanuts are always the least contaminated. The roasting process destroys at least 50% of the aflatoxins present in contaminated peanuts. The data in Table 3 reflect the use of current techniques: sorting tree nuts immediately after harvest and storing them in a cool, dry place. This is an effective method for maintaining a lower incidence of contamination in these commodities. The data in Tables 4 and 5 show that the incidence and levels of aflatoxins in corn and milled corn products were generally low; a few lots contained aflatoxins above the action level.

TABLE 2 Peanut Products Examined for Aflatoxins and Levels

Peanut product	Year	Total[a]	Determinable aflatoxins % of products >1 ng/g	% of products >20 ng/g	Maximum ng/g
Peanut butter	1992	82	22.0	0.0	9
	1993	64	3.1	0.0	5
	1994	77	30.0	0.0	15
	1995	70	41.0	0.0	18
Shelled, roasted	1992	89	9.0	2.2	31
	1993	69	2.9	0.0	2
	1994	79	3.8	1.3	44
	1995	82	7.3	0.0	5
In-shell roasted	1992	19	0.0	0.0	0
	1993	13	0.0	0.0	0
	1994	19	0.0	0.0	0
	1995	5	0.0	0.0	0

[a]Total number of products examined.

TABLE 3 Peanut Products Examined for Aflatoxins and Levels

Tree nut product	Year	Total[a]	Determinable aflatoxins % of products >1 ng/g	% of products >20 ng/g	Maximum ng/g
Almond	1992	55	1.8	0.0	11
	1993	48	4.2	4.2	64
	1994	64	1.6	0.0	8
	1995	36	5.6	0.0	3
Pecan	1992	66	1.5	0.0	2
	1993	47	4.2	0.0	8
	1994	35	2.8	0.0	2
	1995	55	7.3	0.0	15
Pistachio	1992	18	5.5	0.0	2
	1993	22	4.5	0.0	12
	1994	24	16.7	0.0	16
	1995	11	9.0	0.0	5
Walnut	1992	51	5.9	2.0	87
	1993	46	0.0	0.0	0
	1994	63	6.3	0.0	16
	1995	44	15.9	6.8	44

[a]Total number of products examined.

TABLE 4 Aflatoxins in Shelled Corn Designated for Human Consumption

Area of U.S.	Year	Total[a]	Determinable aflatoxins % of products >1 ng/g	% of products >20 ng/g	Maximum ng/g
Southeast[b]	1992	53	20.7	11.3	82
Corn belt[c]		114	12.3	0.0	17
AR-OK-TX		35	17.1	5.7	77
Rest of U.S.		37	2.7	2.7	34
Southeast[b]	1993	33	12.1	0.0	12
Corn belt[c]		119	10.1	7.6	139
AR-OK-TX		45	2.2	0.0	15
Rest of U.S.		51	1.9	0.0	5
Southeast[b]	1994	50	12.0	0.0	18
Corn belt[c]		97	5.1	0.0	10
AR-OK-TX		56	32.0	10.7	114
Rest of U.S.		33	3.0	0.0	4
Southeast[b]	1995	23	8.7	0.0	4
Corn belt[c]		113	0.9	0.0	8
AR-OK-TX		44	20.4	4.5	681
Rest of U.S.		18	0.0	0.0	0

[a]Total number of products examined.
[b]AL, FL, GA, LA, MS, NC, SC, TN.
[c]IA, IL, IN, KS, MI, MN, MO, NE, OH, SD, WI.

Further processing of milled corn products before human consumption will decrease any level of aflatoxin contamination present. These data are biased because the samples collected under compliance programs target those commodities and areas where one is most likely to find contamination. Although a relatively small number of samples are involved, the data enable one to detect trends in the aflatoxin contamination pattern in various parts of the country.

Aflatoxin exposure data, obtained from FDA monitoring activities and other sources, are periodically reviewed along with available toxicological data to determine if the current action levels for aflatoxins in human foods should be revised. Additionally, the program of continuous monitoring is a signal to the food industry that FDA is concerned about the presence of these mycotoxins in the food supply. The regulatory program that has been developed for aflatoxins in foods and feeds is relatively comprehensive. It includes consideration for the health of animals ingesting contaminated feed as well as the health of humans that may consume edible tissues and milk from such animals. A rationale for the control of aflatoxins in animal feed has been published elsewhere [22].

TABLE 5 Aflatoxins in Milled Corn Products

Area of U.S.	Year	Total[a]	Determinable aflatoxins		
			% of products >1 ng/g	% of products >20 ng/g	Maximum ng/g
Southeast[b]	1992	49	22.4	10.2	100
Corn belt[c]		71	7.0	0.0	8
AR-OK-TX		11	0.0	0.0	0
Rest of U.S.		27	3.7	0.0	38
Southeast[b]	1993	40	7.5	10.0	191
Corn belt[c]		59	0.0	0.0	0
AR-OK-TX		9	0.0	0.0	0
Rest of U.S.		25	0.0	0.0	0
Southeast[b]	1994	41	4.8	0.0	16
Corn belt[c]		68	1.5	0.0	3
AR-OK-TX		15	26.6	6.6	52
Rest of U.S.		44	2.4	0.0	2
Southeast[b]	1995	46	0.0	0.0	0
Corn belt[c]		52	0.0	0.0	0
AR-OK-TX		18	33.3	0.0	9
Rest of U.S.		24	8.3	0.0	7

[a]Total number of products examined.
[b]AL, FL, GA, LA, MS, NC, SC, TN.
[c]IA, IL, IN, KS, MI, MN, MO, NE, OH, SD, WI.

D. Factors Associated with Imposing a Regulatory Level

Many countries have attempted to limit exposure to aflatoxins and other selected mycotoxins by imposing legal restrictions or regulatory limits on susceptible commodities in the domestic market and on imports. Regulatory limits for aflatoxins in human foods may range from 0 to 50 μg/kg. Very few countries have presented a rationale for the levels imposed based on some type of risk assessment or available scientific information [23].

In the case of aflatoxins, the wide range of limits may reflect the difficulty in drawing definitive conclusions about the risk to human health resulting from chronic exposure to low levels of carcinogens in foods. Some factors that may influence the decisions made by countries to establish regulatory limits include: availability of analytical methods, toxicological and analytical survey data, the need to maintain an adequate supply of food items at reasonable cost to consumers, and legislation in other countries with which trade contacts exist [24].

Analytical methods for quantitation of aflatoxins at low levels (< 5 ng/g) have been developed, but this information cannot be used directly as a basis for changing regulatory limits. Regulatory levels should be practical and achievable by the regulated food and feed industries because they must be able to monitor their processing steps at even lower levels than the level set by the regulating agency. A particular laboratory may be able to achieve accurate and precise results with a particular method; however, interlaboratory studies using the same method may produce large variations in results.

A recent report [25] indicated that no improvement has been made among laboratories in the precision of aflatoxin assays over the past 20 years. All of the improvement in precision has occurred within laboratories. This suggests a need for laboratories to refer their measurements to reference materials and to operate under the aegis of a strong external quality assurance program. Various organizations have tried to improve the quality of analytical data through check sample programs in which a sample of known contamination level is sent to each participating laboratory for analysis by the method routinely used in that laboratory. The results obtained enable the laboratory director to judge the capability of the analysts in his laboratory and to identify problems that may be otherwise unnoticed. An example is the Smalley Check Sample Series sponsored by the American Oil Chemists' Society (AOCS). Various FDA, USDA, and many private laboratories in the U.S. have participated in this and other programs. A detailed review of the various elements of an Analytical Quality Assurance Program and its impact on the enforcement of regulatory limits was recently published [24].

V. SELECTED MYCOTOXINS

Reports on the occurrence of mycotoxins other than aflatoxins in foods and feeds are increasing. Some of these toxins are reported to be associated with adverse effects in humans as well as animals and are a public health concern. In view of the random, unpredictable contamination of food by mycotoxins, the control of these toxins is expected to be difficult. The experience and knowledge obtained through the development of an effective control program for aflatoxins in the U.S. has been utilized by FDA to address regulatory considerations involving other selected mycotoxins. In each case, the essential information needed includes the availability of analytical methodology, background exposure data, and toxicological manifestations in animal species.

A. Deoxynivalenol

Deoxynivalenol (DON), commonly called vomitoxin, is one of the trichothecene mycotoxins produced by several molds of the genus *Fusarium*, especially *F. graminearum*, which is a common contaminant of several grains, including wheat, corn, barley, and rye. The fungus strain varies and may result in simultaneous production of other trichothecene toxins. In the U.S. and Canada, DON, nivalenol, 3-acetyl-DON, and 15-acetyl-DON have been found in wheat, sorghum, and corn with an occasional report of T-2 toxin [26]. The fungus causes head blight or scab disease in wheat and other grains used for making bread. *Fusarium graminearum* thrives in cool, wet weather conditions in some midwestern grain-producing states. It is not possible to completely avoid the presence of DON even in wheat and other grains grown and harvested under normal weather conditions.

DON is probably one of the most widely distributed *Fusarium* mycotoxins. Exposure occurs through dietary consumption of contaminated grain products. During milling operations, DON is concentrated in the bran. Cooking contaminated flour-based products does not reduce the level of contamination appreciably. Attempts have been made to remove DON from naturally contaminated grains by physical and chemical treatments [27–29]; none of the treatments were effective.

In 1982, the spring wheat crops in four midwestern states were found to be heavily infected with pink scab; high levels of DON were reported with a labor-intensive gas-liquid chromatographic procedure. This finding resulted in the rapid development by several laboratories of simpler methods that could be used for analyzing large numbers of samples. A thin-layer chromatographic method was developed and used by FDA to conduct surveys of wheat and wheat products from many parts of the U.S. After reviewing the survey data and the available toxicological data, FDA issued an "advisory" to state and federal officials. The statement recommended a level of concern of 1 μg DON/g in finished wheat products for human consumption, 2 μg/g for wheat entering the milling process, and 4 μg/g for wheat and wheat milling byproducts used in animal feeds [30]. Advisory levels are issued to provide guidance to the industry concerning levels of DON present in food or feed that are believed by the agency to provide an adequate margin of safety to protect human and animal health. Although the levels involved are advisory in nature, FDA reserves the right to take regulatory enforcement action against interstate shipments of products that substantially exceed the various advisory levels. Limited surveys for DON in wheat, wheat products, corn, and corn products from many areas of the U.S. have been reported [31–34].

In 1993, adverse weather conditions prevailed in many areas of the Midwest thereby providing favorable conditions for *F. graminearum* to proliferate on the hard red spring wheat crop. A total of 630 wheat and barley samples were collected from 25 states; the analyses revealed that about 40% of the wheat samples contained DON at levels $> 2\mu g/g$ [35]. As a result, FDA provided an "update" of its original advisory [36]. This advisory eliminated the level for raw wheat entering the milling process for human food because of the variability in the reduction processes that could be achieved by millers in producing flour. Some milling procedures can result in an approximately two- to eightfold reduction of DON in the final milled product. The updated advisory level for DON in finished wheat products, e.g., flour, bran, and germ, for human food use is 1 μg/g.

In the case of grains and grain products for use as animal feed, the levels are: 10 μg/g for ruminating beef and feedlot cattle older than 4 months and for chickens, not to exceed 50% of the diets; 5 μg/g for swine, not to exceed 20% of the diet; and 5 μg/g for all other animals, not to exceed 40% of the diets. A survey designed to gain some insight into the effectiveness of currently employed industrial milling procedures in reducing the levels of DON in wheat products was recently completed [37]. A total of 525 samples of milled wheat products (white flour, whole-wheat flour, and wheat bran for human consumption) were collected and analyzed. Selected data from that survey are shown in Table 6.

Generally, 10.3% of white flour, 15.6% of whole-wheat flour, and 12.3% of the bran samples contained DON at levels > 1 μg/g. The industry was informed of these findings by letter and advised to increase efforts to reduce the levels of DON in wheat products.

FDA recently completed a survey of 250 corn samples from the 1993 crop harvested in the Midwest. The analytical results revealed that 6% of the samples contained DON at levels > 2 μg/g, 13% contained levels between 1 and 2 μg/g, and 70% contained levels between 0 and 1 μg/g. FDA is studying the presence of DON and hopes to make a determination as to whether regulatory limits are needed to control DON in food and feed products.

B. Fumonisins

The fumonisins are a structurally related group of secondary metabolites produced on corn by *Fusarium moniliforme* [38,39], *Fusarium proliferatum* [40], and several other fungi [41,42]. There are seven naturally occurring fumonisins (A_1, A_2, B_1, B_2, B_3, B_4, and C_1); of these, fumonisin B_1 (FB_1) and fumonisin B_2 (FB_2) are the most abundant [43]. FB_1 and FB_2 were initially isolated from *Fusarium moniliforme* cultures in 1988 and were found to exhibit cancer-promoting activity in rats [39]. In a survey of corn, collected in the midwestern states between 1988 and 1991, it was found that > 60% of the samples contained fumonisins; no

TABLE 6 Deoxynivalenol (DON) in Milled Wheat Products, 1994 Crop

		Determinable deoxynivalenol (μg/g)[a]		
Commodity	Total[b]	% of products >0.1	% of products >1.0	Range
White flour	272	51.8	10.3	0.10–2.63
Whole-wheat flour	90	40.0	15.6	0.15–3.80
Wheat bran	163	50.3	12.3	0.20–2.92

[a]Detection limit 0.02 μg/g.
[b]Total number of products examined.

correlation could be made between the climatic conditions and the fumonisin levels observed [44].

The FDA, along with other laboratories, has developed analytical methods for determining fumonisins in food products. Analytical methodology available for the identification and quantitation of fumonisins in corn and corn-based products include TLC, gas chromatography, HPLC, gas chromatography/mass spectrometry, and immunochemical assays [45–48]. These methods have been used in conducting limited surveys of processed corn-based food products, including commercially prepared canned and frozen sweet corn and tortillas [49]. The finding of a high incidence of fumonisins in corn-based products (cornmeals) purchased in the Washington, D.C., area, in addition to their classification as possible human carcinogens, enhanced the urgency by the FDA to initiate procedures to restrict the presence of these toxins to the lowest levels possible in human foods.

FDA has been participating, with USDA and several state veterinary toxicology offices, in a formal working group to coordinate information gathering. The agency has also been meeting with industry representatives to discuss monitoring needs and possible cooperative activities that can be instituted to reduce human and animal exposure to the fumonisins. The fumonisins are currently being monitored in corn-based products through FDA compliance programs. Fumonisin data obtained for 1994 and 1995 are shown in Table 7. The data confirm that shelled corn, cornmeal, and corn flour are frequently contaminated by fumonisins. The significance of the levels observed must await the acquisition of more data regarding the fate of fumonisins during various industrial processing operations.

No advisories, action levels, or regulations have been issued by FDA for fumonisins. The Mycotoxin Committee of the American Association of Veterinary Laboratory Diagnosticians [50] suggested in 1993 the following guidelines for fumonisins in livestock feed:

Horses: nonroughage portion of diet	$\leqslant$ 5 ppm
Swine: total ration	$\leqslant$ 10 ppm
Poultry: total ration	$\leqslant$ 50 ppm
Beef cattle: nonroughage portion of diet	$\leqslant$ 50 ppm
Dairy cattle: no recommendation	

These guidelines have not been sanctioned by FDA.

The FDA, in cooperation with USDA and the National Toxicology Program/National Center for Toxicological Research, in Arkansas, has initiated a chronic study of fumonisin B_1. The purified fumonisin B_1 needed for the chronic study, as well as other studies, was prepared in FDA laboratories. The data from various

TABLE 7 Corn and Corn Products Examined for Fumonisins B_1 and B_2

			Determinable fumonisins, B_1 and B_2 (ng/g)[a]				
Commodity	Year	Total[b]	% of products >25	% of products >250	% of products >500	% of products >1000	Max.
Corn meal	1994	39	66.6	15.3	5.1	7.6	1302
Corn flour		18	55.5	27.8	5.6	0.0	627
Corn grits		8	50.0	0.0	0.0	0.0	128
Corn, shelled		41	46.3	19.5	17.0	2.4	1132
Popcorn (unpopped)		17	58.8	5.9	0.0	0.0	282
Corn meal	1995	64	61.0	7.8	4.6	3.2	2526
Corn flour		15	53.0	0.0	0.0	0.0	185
Corn grits		13	38.0	0.0	0.0	0.0	215
Corn, shelled		78	46.1	11.5	14.1	10.2	4364
Popcorn (unpopped)		15	33.3	0.0	0.0	0.0	93

[a]Detection limit 25 ng/g.
[b]Total number of products examined.

surveys of corn and corn-based products, along with toxicological studies and studies of the transmission in human food, will be used to determine the need for regulatory programs and to assess human health risks to fumonisins.

C. Ochratoxins

The ochratoxins (ochratoxin A and ochratoxin B) are secondary metabolites produced by certain species of the genera *Aspergillus* and *Penicillium*. Ochratoxin A (OA) is the major metabolite found as a natural contaminant of cereal grains such as corn, barley, wheat, oats, and rye [51,52]. It is also found in beans (coffee, soya, cocoa), peanuts, and meat in some countries [53]. Surveys of various commodities for OA show a worldwide occurrence with large differences by country in incidence and levels. Large differences are also found within countries. The toxin is not completely destroyed during the normal processing and cooking of food.

The FDA and USDA conducted several surveys of cereal grains between 1967 and 1985 to determine the incidence of ochratoxin in this country; ochratoxin A was found in 46 of 2313 (2%) samples examined [54]. In those surveys, the highest incidence was in barley but the highest level was found in corn. In surveys involving 771 samples of green coffee beans, only five (0.6%) of the samples contained OA [55]. In a small survey conducted on pork-containing sausages in the U.S., no OA was found [56]. FDA initiated another survey of selected processed food products (e.g., dried peas/beans, whole barley, barley cereals/malt, green coffee beans, corn cereals/meal, oatmeal/crackers, whole rice, rice cereals, rye flour, wheat flour, and soya-based baby food products) to determine the incidence and levels of OA. None of the 351 samples analyzed contained identifiable levels of the toxin [21]. The results from the latter survey, when coupled with the results from earlier surveys, suggest that OA contamination is not a major problem for U.S. food products. Consequently, FDA has not established a regulatory limit for the toxin.

Although there is no regulatory limit for OA in the U.S., under the Food, Drug, and Cosmetic Act, any food containing any substance at a level that may be injurious to health is considered adulterated and may be subject to regulatory action. If OA is found in a food product, FDA would determine on a case-by-case basis whether the level found in that product may pose a health hazard such that it would be subject to regulatory action.

D. Patulin

Patulin is a toxic metabolite of several species of fungi including *Penicillium expansum*, a causal organism of soft rot in apples and other fruit [57,58]. Patulin can be produced by a variety of different fungi on apples, grains, and refrigerated products under laboratory conditions. Under natural conditions, it has been isolated almost exclusively from apples or apple products [58,59]. The main source of patulin in the human diet is probably apple juice prepared from apples contaminated with *P. expansum*.

Recent studies have demonstrated that patulin has immunosuppressive effects [60]. This, coupled with its potent antibiotic activity, may be of concern especially in certain compromised populations. At least 10 countries regulate patulin in apple juice at levels of 30 to 50 μg/kg with lower levels in infant foods [61]. In the absence of a rationale for these levels, one can speculate, to some degree, that these regulations are for quality control purposes. FDA initiated a limited survey in 1993 to obtain up-to-date information on the incidence and levels of occurrence of patulin in apple juice, to assess the current potential risk associated with its presence in apple juice in the U.S., and to determine if there was a need for regulatory action.

The results of that survey, which consisted of 101 samples of apple juice collected from retail outlets in various parts of the U.S., revealed that 26 (26%) of the samples examined contained patulin at levels > 50 μg/l (F. Thomas, FDA, personal communication). Twenty of the 101 samples collected were juice designated for infants; the patulin levels for these samples ranged from < 2 to 17.6 μg/L. These data, along with other information, are being reviewed to assess the need for regulatory action or conducting additional research regarding the incidence and levels of patulin in apple juice.

E. Other Mycotoxins of Concern

Other mycotoxins that have been of concern to the agency include zearalenone, penicillic acid, sterigmatocystin, cyclopiazonic acid, citrinin, and the alternaria toxins. No regulatory standards have been initiated for these toxins because it has been found (via surveys) that the extent of their natural occurrence in foods is very low and the possibility for human exposure is limited. FDA continuously follows the development of newer information and new data regarding these mycotoxins; therefore the need for the establishment of regulatory standards is consistently evaluated by the agency.

VI. DISCUSSION AND CONCLUSIONS

The maintenance of a wholesome food supply is a major responsibility that must be shared by the food industry (producers and processors) and the regulating agencies involved. It is incumbent upon the regulating agencies as well as the food industry to keep abreast of technological advances in processing techniques and agronomic practices that are available. Because mycotoxins are unavoidable contaminants that cannot be eliminated from the food supply by current good manufacturing practices, any regulatory level proposed or established for a mycotoxin should reflect the lowest level of exposure to the toxin that can be effectively monitored and controlled by the industry without causing a significant risk to the health of consumers. When the action level for aflatoxins in peanut and peanut products was originally established, the level was based on several important practical considerations. These included the capabilities of the available sampling procedures and analytical methodology to identify and quantitate aflatoxins, and the capability of the current agronomic practices to reduce the levels of contamination in crops without significant adverse effects on the economy of the peanut industry.

One way the food supply can be improved is by use of effective monitoring techniques and the establishment of quality control safeguards for food processing operations. The regulatory programs of the FDA for mycotoxins are designed to keep the amount in foods at the lowest levels that are attainable and consistent with maintaining an adequate commodity supply at a reasonable cost. The monitoring data obtained over the years, particularly in reference to aflatoxins, reveal that human exposure to these toxins is relatively low. The data in Tables 2 to 5 reflect that aflatoxins rarely occur on food, and when they do, they are present at relatively low levels. FDA's efforts are complemented by control programs carried out by USDA, state departments of agriculture, and various trade associations. These cooperative efforts are fairly effective at minimizing the levels of aflatoxins to which consumers are exposed.

As an example, the aflatoxin control program for raw peanuts produced in the U.S. is administered by the Peanut Administrative Committee (PAC) under the provisions of a USDA marketing agreement that require the analysis and certification of each lot of raw shelled peanuts for aflatoxin content. The objective of the agreement is to ensure the wholesomeness of peanuts moving into commercial channels for human consumption. The marketing agreement requirement for domestic and imported edible peanuts is 15 μg/kg or less of total aflatoxins. This agreement plays a very important role in the peanut industry's quality control effort to minimize aflatoxin contamination in peanuts and peanut products.

The random, unpredictable occurrence of mycotoxins in foods necessitates the exploration of different or novel approaches to minimize the levels to which consumers are exposed. One approach that might be considered is the use of the Hazard Analysis and Critical Control Point (HACCP) system [62]. The system involves a science-based analysis of potential hazards involved in the production of safe foods, determination of where the hazards can occur in processing techniques, institution of preventive measures, and corrective actions if they do occur. This system is designed to critically evaluate the effectiveness of controls at each major step involved in processing a food. This system was recently initiated by FDA to create a more effective and efficient system for ensuring the safety of seafood products [63], but the concepts involved could be adaptable for use with other potential contaminants, including mycotoxins. The details of a HACCP program that has been implemented by the grain industry for the control of aflatoxins has been reported [64].

Current technology cannot prevent mycotoxin contamination of field crops before harvest. Research is under way in many laboratories with the objective of controlling preharvest contamination of peanuts and corn specifically through genetic manipulations, use of irrigated growing plots, the application of various chemicals, and other innovative measures. Some progress has been made in these endeavors; however, no commercial, large-scale applications have been adopted.

Currently, there are no international regulatory guidelines for mycotoxins in food. The Codex Alimentarius Commission (a subsidiary body of the Food and Agriculture Organization [FAO] and the World Health Organization [WHO]) is concerned with setting standards for foods involved in international trade and providing guidance to countries that would like to create their own national food laws and regulations [65]. The FAO/WHO Codex Committee on Food Additives and Contaminants (CCFAC) and the Joint Expert Committee on Food Additives (JECFA) are currently interested in establishing international guidelines for the control of aflatoxins in susceptible commodities throughout the world.

An FAO Technical Consultation on Sampling Plans for Aflatoxin Analysis in Peanuts and Corn was convened in Rome, Italy, in 1993. The report from that meeting has been published [66]. If the recommendations from that consultation are adopted, they will allow for international uniformity in sampling various commodities based on the guideline level a country chooses for aflatoxins. Countries also have the option of using the sampling plans in the reverse manner, by first deciding which plan to use and then regulating at the appropriate aflatoxin level. It has been pointed out that harmonization of food standards on an international basis is one of the essential elements necessary to ensure food safety on a worldwide basis [65].

REFERENCES

1. YH Hui. United States Food Laws, Regulations, and Standards. New York: John Wiley & Sons, 1979, p 341.
2. L Stoloff. Aflatoxin control: past and present. J Assoc Off Anal Chem 63:1067, 1980.
3. L Stoloff. A rationale for the control of aflatoxin in human food. In: PS Steyn, R Vleggaar, eds. Mycotoxins and Phycotoxins. Amsterdam: Elsevier Science Publishers, 1986, p 457.
4. DL Park, L Stoloff. Aflatoxin control—how a regulatory agency managed risk from an unavoidable natural toxicant in food and feed. Regul Toxicol Pharmacol 9:109, 1989.

5. F Dickens, HEH Jones. The carcinogenic action of aflatoxins after its subcutaneous injection in the rat. Br J Cancer 19:691, 1963.
6. WD Salmon, PM Newberne. Occurrence of hepatomas in rats fed diets containing peanut meal as a major source of protein. Cancer Res 23:571, 1963.
7. FDA. Aflatoxins in cottonseed meal: revised action level. Fed Reg 47:33007, 1982.
8. FDA. Corn shipped in interstate commerce for use in animal feeds: action levels for aflatoxins in animal feeds—revised compliance policy guide. Fed Reg 54:22622, 1989.
9. Anonymous. FDA revises policy on aflatoxin from peanuts in feed. Food Chem News 32:3, 1990.
10. FDA. Action levels for aflatoxins in animal feeds: revised compliance policy guide. Fed Reg 59:17383, 1994.
11. TC Campbell, JR Hayes. The role of aflatoxin metabolism in its toxic lesion. Toxicol Appl Pharmacol 35:199, 1976.
12. FDA. Aflatoxin contamination of milk: establishment of action level. Fed Reg 42: 61630, 1977.
13. FDA. Action levels for added poisonous or deleterious substances in food. Fed Reg 53:5043, 1988.
14. MW Trucksess, GE Wood. Recent methods of analysis for aflatoxins in foods and feeds. In: DL Eaton, JD Groopman, eds. The Toxicology of Aflatoxins: Human, Veterinary, and Agricultural Significance. San Diego: Academic Press, 1994, p 409.
15. GE Wood. Aflatoxins in domestic and imported foods and feeds. J Assoc Off Anal Chem 72:543, 1989.
16. TB Whitaker, JW Dickens, RJ Monroe. Variability of aflatoxin test results. J Am Oil Chem Soc 51:214, 1974.
17. TB Whitaker, FE Dowell, WM Hagler, FG Giesbrecht, J Wu. Variability associated with sampling, sample preparation, and chemical testing for aflatoxin in farmer's stock peanuts. J AOAC Int 77:107, 1994.
18. OL Shotwell, ML Goulden, CW Hesseltine. Aflatoxin: distribution in contaminated corn. Cereal Chem 51:492, 1974.
19. DL Park, AE Pohland. Sampling and sample preparation for detection and quantitation of natural toxicants in food and feed. J Assoc Off Anal Chem 72:399, 1989.
20. GE Wood. Mycotoxins in foods and their safety ramifications. In: JW Finley, SF Robinson, DJ Armstrong, eds. Food Safety Assessment. ACS Symposium Series 484, American Chemical Society, Washington, D.C., 1992, p 261.
21. GE Wood, Mycotoxins in foods and feeds in the United States. J Anim Sci 70:3941, 1992.
22. DL Park, AE Pohland. A rationale for the control of aflatoxin in animal feeds. In: PS Steyn, R Vleggaar, eds. Mycotoxins and Phycotoxins. Amsterdam: Elsevier Science Publishers, 1986, p 473.
23. L Stoloff, HP van Egmond, DL Park. Rationales for the establishment of limits and regulations for mycotoxins. Food Addit Contam 8:213, 1991.
24. HP van Egmond. Mycotoxins: regulations, quality assurance and reference material. Food Addit Contam 12:321, 1995.
25. W Horwitz, R Albert, S Nesheim. Reliability of mycotoxin assays—an update. J Assoc Off Anal Chem 76:461, 1993.
26. PM Scott. Trichothecenes in grains. Cereal Foods World 35:661, 1990.

27. LP Hart, WE Braselton. Distribution of vomitoxin in dry milled fractions of wheat infected with *Gibberella zeae*. J Agric Food Chem 31:657, 1983.
28. PM Scott, SR Kanhere, P-Y Lau, JE Dexter, R Greenhalgh. Effects of experimental flour milling and bread baking on retention of deoxynivalenol (vomitoxin) in hard spring wheat. Cereal Chem 60:421, 1983.
29. LM Seitz, WT Yamazaki, RL Clements, HE Mohr, L Andrews. Distribution of deoxynivalenol in soft wheat mill streams. Cereal Chem 62:467, 1985.
30. Anonymous. FDA advises USDA on levels of concern set for vomitoxin. Food Chem News 24:23, 1982.
31. RM Eppley, MW Trucksess, S Nesheim, CW Thorpe, GE Wood, AE Pohland. Deoxynivalenol in winter wheat: thin layer chromatographic method and survey. J Assoc Off Anal Chem 67:43, 1984.
32. MW Trucksess, GE Wood, RM Eppley, et al. Occurrence of deoxynivalenol in grain and grain products. Proceedings of the 6th International Biodeterioration Symposium, Washington, D.C., 1986, p 243.
33. GE Wood, L Carter. Limited survey of deoxynivalenol in wheat and corn in the United States. J Assoc Off Anal Chem 72:38, 1989.
34. C Fernandez, ME Stack, SM Musser. Determination of deoxynivalenol in 1991 U.S. winter and spring wheat by high performance thin-layer chromatography. J AOAC Int 77:628, 1994.
35. MW Trucksess, F Thomas, K Young, ME Stack, WJ Fulgueras, SW Page. Survey of deoxynivalenol in U.S. 1993 wheat and barley crops by enzyme-linked immunosorbent assay. J AOAC Int 78:631, 1995.
36. Anonymous. FDA issues advisory on midwest hard red wheat. Food Chem News 35:55, 1993.
37. MW Trucksess, DE Ready, MK Pender, CA Ligmond, GE Wood, SW Page. Determination and survey of deoxynivalenol in flour, whole wheat flour and bran. J AOAC Int 79:883, 1996.
38. SC Bezuidenhout, WCA Gelderblom, CP Gorst-Allman, et al. Structure elucidation of the fumonisins, mycotoxins from *Fusarium moniliforme*. J Chem Soc Chem Commun 11:743, 1988.
39. WCA Gelderblom, K Jaskiewic, WFO Marasas, et al. Fumonisins—novel mycotoxins with cancer-promoting activity produced by *Fusarium moniliforme*. Appl Environ Microbiol 54:1806, 1988.
40. PF Ross, PE Nelson, JL Richard, et al. Production of fumonisins by *Fusarium moniliforme* and *Fusarium proliferatum* isolates associated with equine leukoencephalomalacia and a pulmonary edema syndrome in swine. Appl Environ Microbiol 56: 3225, 1990.
41. PE Nelson, RD Plattner, DD Shackelford, AE Desjardins. Fumonisin B_1 production by *Fusarium* species other than *F. moniliforme* in section *Liseola* and by some related species. Appl Environ Microbiol 58:984, 1992.
42. PG Thiel, WFO Marasas, EW Sydenham, GS Shepherd, WCA Gelderblom, J Nieuwenhuis. Survey of fumonisin production by *Fusarium* species. Appl Environ Microbiol 57:1089, 1991.
43. BE Branham, RD Plattner. Isolation and characterization of a new fumonisin from liquid cultures of *Fusarium moniliforme*. J Nat Prod 56:1630, 1993.

44. PA Murphy, LG Rice, PF Ross. Fumonisin B_1, B_2, and B_3 content of Iowa, Wisconsin, and Illinois corn and corn screenings. J Agric Food Chem 41:263, 1993.
45. LG Rice, PF Ross. Methods for detection and quantitation of fumonisins in corn cereal products and animal excreta. J Food Prot 57:536, 1994.
46. EW Sydenham, GS Shephard, PG Thiel. Liquid chromatographic determination of fumonisin B_1, B_2 and B_3 in foods and feeds. J Assoc Off Anal Chem 75:313, 1992.
47. JJ Pestka, JI Azcona-Olivera, RD Plattner, F Minervine, MB Doko, A Visconti. Comparative assessment of fumonisin in grain-based foods by ELISA, GC-MS, and HPLC. J Food Prot 57:169, 1994.
48. MW Trucksess. Immunochemical methods for fumonisins in corn. In: JM Van Emon, CL Gerlach, JC Johnson, eds. Environmental Immunochemical Methods. ACS Symposium Series 646, American Chemical Society, Washington, D.C., 1996, p 326.
49. MW Trucksess, ME Stack, S Allen, N Barrion. Immunoaffinity column coupled with liquid chromatography for determination of fumonisin B_1 in canned and frozen sweet corn. J AOAC Int 78:705, 1995.
50. Anonymous. High levels of fumonisin reported in three states. Food Chem News 37:57, 1995.
51. OL Shotwell, CW Hesseltine, EE Vandegraft, ML Goulden. Survey of corn from different regions for aflatoxins, ochratoxin, and zearalenone. Cereal Sci Today 16:266, 1971.
52. OL Shotwell, ML Goulden, CW Hesseltine. Survey of US wheat for ochratoxin and aflatoxin. J Assoc Off Anal Chem 59:122, 1976.
53. P Krogh. Ochratoxins in food. In: P Krogh, ed. Mycotoxins in Food. London: Academic Press, 1987, p 97.
54. AE Pohland, S Nesheim, L Friedman. Ochratoxin A: a review. Pure Appl Chem 64: 1029, 1992.
55. C Levi. Mycotoxins in coffee. J Assoc Off Anal Chem 63:1282, 1980.
56. AE Pohland, GE Wood. Occurrence of mycotoxins in food. In: P Krogh, ed. Mycotoxins in Food. London: Academic Press, 1987, p 35.
57. J Harwig, YK Chen, PC Kennedy, PM Scott. Occurrence of patulin and patulin producing strains of *Penicillium expansum* in natural rots of apples in Canada. Can Inst Food Sci Technol 6:22, 1973.
58. WT Stott, LB Bullerman. Patulin: a mycotoxin of potential concern in foods. J Milk Food Technol 38:695, 1975.
59. SJ Kubacki. The analysis and occurrence of patulin in apple juice. In: PS Steyn, R Vleggaar, eds. Mycotoxins and Phycotoxins. Amsterdam: Elsevier Science, 1986, p 293.
60. RP Sharma. Immunotoxicity of mycotoxins. J Dairy Sci 76:892, 1993.
61. HP van Egmond. Current situation on regulations for mycotoxins. Overview of tolerances and status of standard methods of sampling and analysis. Food Addit Contam 6:139, 1989.
62. DA Corlett. Importance of the hazard analysis and critical control point system in food safety evaluation and planning. In: JW Finley, SF Robinson, DJ Armstrong, eds. Food Safety Assessment. ACS Symposium Series 484, American Chemical Society, Washington, D.C., 1992, p 120.

63. FDA. Procedures for the safe and sanitary processing and importing of fish and fishery products. Fed Reg 60:65096, 1995.
64. KD Brenner. Grain handling and processing procedures to reduce and eliminate aflatoxins. In: A Perspective on Aflatoxin in Field Crops and Animal Food Products in the United States. A Symposium. USDA Agricultural Research Service, Washington, D.C., Jan. 23–24, 1990, p 82.
65. C Canet. Importance of international cooperation in food safety. Food Addit Contam 10:97, 1993.
66. FAO. Sampling plans for aflatoxin analysis in peanuts and corn. FAO Food and Nutrition Paper 55. Food and Agriculture Organization of the United Nations, Rome, Italy, 1993, p 1.

16
Roles of National Governments and International Agencies in the Risk Analysis of Mycotoxins

Gerald G. Moy
World Health Organization, Geneva, Switzerland

The health of a people is really the foundation upon which all their happiness and all their powers as a State depend.

—Benjamin Disraeli

I. INTRODUCTION

Since its inception in 1949, the World Health Organization has had a major interest in food safety as part of its overall health mandate. For example, in 1953 WHO published two monographs concerning important food safety issues at that time, one on milk pasteurization [1] and another on pesticides [2]. This longstanding interest has been strengthened by the growing recognition by WHO and its member states of the importance of food safety to both health and development [3]. The United Nations Conference on Environment and Development (UNCED), which met in Rio de Janeiro in June 1992, adopted a global strategy entitled Agenda 21, which, inter alia, emphasized that "particular attention should be directed towards food safety, with priority placed on the elimination of food contamination." Later that year, the International Conference on Nutrition convened in December 1992 by WHO in collaboration with the Food and Agriculture Organization of the United Nations (FAO) went further in recognizing that "access to nutritionally adequate and safe food is a right of each individual" [4].

In almost all societies, however, food safety has been long recognized as one of the most important prerequisites for health. Indeed, records of ancient civilizations from around the world demonstrate that the earliest governments were concerned about the quality and safety of food. While some of these early concerns were connected to religious precepts, a number of the provisions were based on sound hygienic observations still valid today. Unfortunately, the correlation between the consumption of moldy food and severe mycotoxicoses, as occurred at various times in history, was overlooked, often with profound health and social consequences [5]. Nonetheless, the historically recent recognition of the potential hazards posed by mycotoxins in food has brought into play well-established scientific and legal mechanisms which have evolved to assure the safety of the food supply. This paper will examine the use of risk analysis in addressing the problems of mycotoxins in food and the role of national governments and international organizations, in particular WHO, in promoting consistent and scientifically sound decision-making.

II. NATIONAL RESPONSIBILITIES FOR FOOD CONTROL

At the national level, food safety is viewed as a shared responsibility among government, industry, and consumers, but government has an important leadership role in this relationship. Almost all governments have enacted basic legislation to assure the safety of the food supply in order to promote and protect the health of their people. However, an effective enforcement of national food safety legislation requires an infrastructure which serves the following three basic functions: administration; inspection and sampling; and analytical services [6].

A. Administration

One of the most important functions of the central food control administration is to develop food policy and regulations. The statutory provision for the establishment of a central advisory and/or coordination body has been found to be very useful in many countries. Experience has shown that centralized national food administration at the top government level is more efficient and more economical than at state or local government levels. This will not only assure that internal barriers to trade are not inadvertently introduced, but will also facilitate the effective participation of the national government in the harmonization efforts being undertaken under the auspices of various international organizations. However, responsibilities for promoting food production and processing should be separated from responsibilities for food safety and public health.

With appropriate coordination and supervision from a national service, local food authorities are in a good position to provide cost-effective inspection services and efficient enforcement. This is also true for the control of mycotoxins, where policy development should be based on sound scientific risk assessment and be consistent across the country but inspection and enforcement should be implemented at the local level.

B. Inspection and Sampling

Effective food control systems must provide for the inspection of premises where foods are produced, prepared, packed, stored, or held for sale and for the sampling of foods. In this regard, local government bodies play an important role in enforcement due to their local knowledge and closer contact with local producers, manufacturers, traders, and consumers. In cases of violations, inspections provide the direct observation and documentation necessary for the imposition of regulatory sanctions, including criminal penalties. In the case of mycotoxins, such evidence often includes samples of suspected food, at times taken from storage facilities on the farm. Random food sampling is also an integral part of the surveillance of the food supply for mycotoxins, which can only be revealed by competent laboratory analysis.

All of the above relies on a well-trained, knowledgeable, and motivated food safety staff. Such officers play a key role not only in enforcing food safety legislation but also in encouraging voluntary compliance by the food industry and in promoting good agricultural and manufacturing practices which can minimize contamination.

C. Analytical Services

Any food control system should, as an essential part, have an appropriate analytical service suitable for the stage of development of the country. Many of the hazards, such as mycotoxins, associated with foods cannot be reliably detected with the senses. Acceptable methods of sampling and analysis are required for compliance monitoring to assure that established limits for mycotoxins are not exceeded. In cases of suspected violations, analytical results submitted by an accredited laboratory are considered prima facie evidence under most prosecution proceedings. Analytical services are also necessary for health-oriented, population-based monitoring for contaminants in individual foods and in the total diet as part of ongoing surveillance programs.

III. RISK ANALYSIS OF MYCOTOXINS BY NATIONAL GOVERNMENTS

While the enactment and enforcement of food legislation are often thought to be the most important functions of government, national governments must largely rely on the cooperation of the food industry to voluntarily comply with minimum food safety requirements prescribed by law or recommended by guidelines.

Voluntary compliance by the industry, in turn, is based on the credibility of the government's risk analysis of potential hazards and on the soundness of recommended interventions. Therefore, risk analysis is one of the most important activities of a good compliance program and is increasingly being seen as the preferred approach for addressing emerging food safety problems, especially for chemicals such as mycotoxins.

The evaluative framework offered by risk assessment not only permits an estimation of human risk, but also provides a means of organizing data and allocating responsibilities. Common structure permits a transparent and relatively uniform methodology to assess the risk posed by specific mycotoxins and provides useful information to risk managers in selecting options. Risk assessment can also identify areas where the available data are insufficient to reach reasonable decisions. However, while more than 70 countries have enacted regulations for mycotoxins in food and feed, in most countries these regulations have not been based on sound scientific risk assessment or science has not been fully utilized [7]. National limits for aflatoxins are often more influenced by a country's trading position (exporter/importer), climate (temperate/tropical), and other nonrisk factors.

With few exceptions, most countries have not fully applied risk analysis to the problems of mycotoxins in food. To some extent, the essential information for the risk analysis of many mycotoxins is often not available and considerable research needs to be undertaken. Improvement in the characterization of the toxic properties of mycotoxins is a major task which requires significant resources. Consequently, international organizations such as WHO have an important role to play in assuring coordination of the global research effort and in evaluating and disseminating results of that research.

On the other hand, the estimation of exposure to mycotoxins by the general population and certain sensitive groups is the function of national governments. Government risk management interventions tend to focus on the prevention and control of mycotoxins through improved agricultural practices, particularly those related to drying and storage. National governments also support research into various prevention measures, including the application of biotechnology methods to control the formation of fumonisins, trichothecenes, and aflatoxins. In addition, research is continuing on detoxification methods for heavily contaminated crops. However, in the absence of quantitative risk assessment, the cost/benefit evaluation of any government action would contain great uncertainties.

IV. RISK ANALYSIS BY INTERNATIONAL AGENCIES

The Codex Alimentarius Commission is an intergovernmental body established by WHO and FAO in 1963 with the purpose of, inter alia, "protecting the health of consumers and ensuring fair practices in the food trade" [8]. Codex, which now has more than 150 countries as members, has developed a considerable body of voluntary standards, guidelines, and other recommendations which serve to facilitate international trade in food. With the establishment of the World Trade Organization (WTO) in 1995, Codex-adopted texts are viewed as representing the international consensus for health and safety requirements for food. While Codex-adopted texts will technically remain voluntary until they are accepted or used by countries, the agreements being implemented by WTO provide a mechanism for the collective adoption of Codex standards, guidelines, and recommendations by all WTO member countries.

In order to assure the consistent and transparent application of risk analysis across all of the Codex committees, a Joint FAO/WHO Expert Consultation on the Application of Risk Analysis to Food Standards Issues was convened in Geneva in March 1995 [9]. The consultation made a number of recommendations for improving Codex risk assessment procedures, including proposed definitions for risk analysis of foodborne hazards, and these were adopted, on an interim basis, by the 22nd session of the Codex Alimentarius Commission in June 1997 [10]. The model for risk analysis recommended by the consultation was essentially that proposed by the National Research Council, U.S. National Academy of Sciences [11], but with some modification of the definitions. Note, however, that Codex-recommended definitions are still being discussed by Codex committees and member governments. In order to promote an international consensus on the definition of terms for food safety risk analysis, the following definitions have been proposed by WHO:

Hazard: A biological, chemical, or physical agent with the potential to cause an adverse health effect when present at an unacceptable level in food.

Risk: An estimate of the likelihood of occurrence of an adverse health effect, including injury, that may result from exposure to a biological, chemical, or physical agent in food.

Risk analysis: A process to scientifically evaluate the nature and likelihood of occurrence of known or potential adverse health effects resulting from human exposure to foodborne hazards (risk assessment), to weigh policy alternatives in light of the results of risk assessment, and, if required, to select and implement appropriate control options (risk management) and to exchange information and opinion interactively among risk assessors, risk managers, consumers, and other interested parties (risk communication).

Risk assessment: A scientific process for evaluating the likelihood of occurrence and severity of known or potential adverse health effects resulting from exposure to a biological, chemical, or physical agent in food. It consists of the following steps: (1) hazard identification; (2) hazard characterization; (3) exposure assessment; and (4) risk characterization. It includes qualitative and/or quantitative assessments of risk and their attendant uncertainties.

Hazard identification: A preliminary confirmation of a biological, chemical, or physical agent as a potential hazard in food, based on known or potential adverse health effects in humans, on known or potential levels in food and on any other available relevant information.

Hazard characterization: The qualitative and/or quantitative evaluation of the nature and severity of the adverse health effects associated with biological, chemical, and physical agents which may be present in food. For chemical agents, a dose-response assessment with attendant uncertainties should be performed. For biological or physical agents, a dose-response assessment should be performed if the data are obtainable.

Dose-response assessment: A determination of the relationship between the amount of exposure to a chemical, biological, or physical agent (dose) and the magnitude, frequency, and/or severity of associated adverse effects (response).

Exposure assessment: A scientific evaluation of the intake of a biological, chemical, or physical agent through food, taking into account exposure from other sources if relevant. The evaluation includes a quantitative and/or qualitative estimation of exposure and attendant uncertainties.

Risk characterization: An estimation of the likelihood of occurrence and severity of adverse health effects in a given population that may result from exposure to a biological, chemical, or physical agent in food, based on hazard identification, hazard characterization, and exposure assessment. It includes a qualitative and/or quantitative expression of risk and attendant uncertainties.

Risk management: A process of weighing policy alternatives in the light of the results of risk assessment and, if required, selecting and implementing appropriate control options. It includes overall responsibility for the risk analysis process and primary responsibility for risk communication.

Risk communication: An interactive exchange of information and opinions concerning risks among risk assessors, risk managers, consumers, and other interested parties.

V. RISK ANALYSIS OF MYCOTOXINS BY INTERNATIONAL AGENCIES

The scientifically sound risk assessment procedures which have been developed to support the Codex decision-making process are likely to become increasingly important as the basis for arbitration involving health-related trade disputes among countries. At present, Codex is in the process of developing recommendations for aflatoxins, including maximum limits in commodities, sampling and analytical methods, and codes of practice to reduce levels of aflatoxins in food.

However, progress has been hampered by differences in views concerning levels of aflatoxins that are reasonably achievable under good agricultural practices. Codex is periodically reviewing the need for recommendations for other mycotoxins as well, such as ochratoxin A and trichothecenes. It would therefore be instructive to examine the process by which Codex and other international organizations are applying the risk analysis model. The risk analysis process generally follows the chronology as discussed below, but, in reality, the process is iterative as new information relevant to the risk assessment becomes available.

A. Hazard Identification

The Codex process of controlling contaminants in the food supply often begins with the identification of potential health problems. Within Codex, "contaminant" is defined as any substance not intentionally added to food, which is present in such food as a result of the production (including operations carried out in crop husbandry, animal husbandry, and veterinary medicine), manufacture, processing, preparation, treatment, packing, packaging, transport, or holding of such food or as a result of environmental contamination. The term does not include insect fragments, rodent hairs, and other extraneous matter [12].

Ranking of problems by priority is based mainly on factors related to the toxicity of the substance and estimated dietary exposure as well as exposure through other sources. Hazard identification often includes input from both risk assessors and risk managers. With the isolation of aflatoxin as the cause of Turkey X disease in the 1950s, the identification of human health hazards posed by mycotoxins was initiated in many countries. However, human health problems caused by the consumption of most mycotoxins are complex and are still poorly understood. Nonetheless, available human epidemiology data suggest that mycotoxins may be responsible for a range of human diseases resulting from acute and chronic exposures [13]. Diseases thought to be associated with acute exposure to mycotoxins include ergotism (ergot alkaloids), alimentary toxic aleukia (trichothecenes), yellow rice disease (citrinin), sugarcane poisoning (3-nitroproprionic acid), acute hepatitis (aflatoxins), and kwashiorkor (aflatoxins). Diseases with possible chronic mycotoxin exposure etiologies include esophageal cancer (fumonisin), Indian childhood cirrhosis (aflatoxin), hepatocellular carcinoma (aflatoxin), and Balkan endemic nephropathy (ochratoxin A).

The International Program on Chemical Safety, which is jointly sponsored by WHO, the United Nations Environment Program, and the International Labor Organization, has established a steering group in collaboration with the International Life Sciences Institute, Europe, to develop a preliminary list of naturally occurring plant toxins considered constituting a hazard to the consumer. The steering group collects and evaluates available information using standard criteria and encourages and supports research on topics on which further information may be needed. A number of mycotoxins are under consideration by the steering committee, including many of those mentioned above [14]. Once sufficient information is available to document a potential hazard, the substance is referred to the Joint FAO/WHO Expert Committee on Food Additives (JECFA) for hazard characterization.

B. Hazard Characterization

Beginning in 1956, JECFA has been engaged in collecting and evaluating scientific data on food additives and making recommendations on safe levels of use. In 1972 the scope of the evaluations was extended to include contaminants in food. One of the major uses of JECFA recommendations is to serve as the scientific basis for developing Codex standards, guidelines, and other recommendations. In general terms, the purpose and functions of JECFA include: (1) reviewing the latest knowledge and expert information and making it available to FAO, WHO, and their member countries; (2) formulating technical recommendations; and (3) making recommendations designed to initiate, stimulate, and coordinate the research necessary to reach conclusions concerning the toxicological acceptance or otherwise of the presence of a substance in food [15].

For many mycotoxins, insufficient data are available to permit an evaluation by JECFA. However, in several cases, JECFA evaluations have provided useful guidance. For example, a Provisional Tolerable Weekly Intake (PTWI) has been established for ochratoxin A of 0.1 μg/kg body weight [16–19]. In addition, a Provisional Tolerable Daily Intake (PTDI) of 0.4 μg/kg body weight [20–22] has been established for patulin.

The International Agency for Research on Cancer (IARC) has undertaken extensive studies to assess the carcinogenicity of the various aflatoxins. IARC has concluded that there is sufficient evidence in humans for the carcinogenicity of naturally occurring mixtures of aflatoxins and that aflatoxin M_1 is possibly carcinogenic to humans [23]. However, no estimate of aflatoxin B_1 potency in humans was made. In recognizing the weaknesses inherent in extrapolating animal bioassays to humans, JECFA declined to establish a tolerable intake for aflatoxins, but instead recommended that their presence in food should be reduced to irreducible levels [24]. An irreducible level is defined as that concentration of a substance that cannot be eliminated from a food without involving the discarding of that food altogether, severely compromising the ultimate availability of major food supplies. JECFA is considering models based on epidemiology studies to assess the carcinogenic potency of aflatoxin B_1, but these efforts are confounded in some countries by the prevalence of hepatitis B, which is also a risk factor for primary hepatocellular carcinoma.

IARC has also reviewed a number of other mycotoxins; a summary of their evaluations is given in Table 1. Several mycotoxins have been classified by IARC as possibly carcinogenic to humans (category 2B).

TABLE 1 Summary of IARC Evaluations of Mycotoxins

Agent	Degree of evidence of carcinogenicity: Human	Degree of evidence of carcinogenicity: Animal	Overall evaluation of carcinogenicity to humans
Aflatoxins, naturally occurring mixture	S	S	1
Aflatoxin B_1	S	S	
Aflatoxin B_2		L	
Aflatoxin G_1		S	
Aflatoxin G_2		I	
Aflatoxin M_1	I	S	2B
Citrinin	I-NDA	L	3
Cyclochlorotin	I-NDA	I	3
Griseofulvin	I-NDA	S	2B
Luteoskyrin	I-NDA	L	3
Ochratoxin A	I	S	2B
Patulin	I-NDA	I	3
Penicillic acid	I-NDA	L	3
Rugulosin	I-NDA	I	3
Sterigmatocytin	I-NDA	S	2B
Toxins derived from *Fusarium graminearum*, *F. culmorum*, and *F. crookwellense*	I-NDA		3
Zearalenone		L	
Deoxynivalenol		I	
Nivalenol		I	
Fusarenone X		I	
Toxins derived from *Fusarium moniliforme*	I	S	2B
Fumonisin B_1		L	
Fumonisin B_2		I	
Fusarin C		L	
Toxins derived from *F. sporotrichioides*	I-NDA		3
T-2 Toxin		L	

Evidence:
S = sufficient
L = limited evidence
I = inadequate
I-NDA = inadequate, no data available

Overall evaluation:
1 = carcinogenic to humans
2B = possibly carcinogenic to humans
3 = not classifiable as to its carcinogenicity to humans

Source: Refs. 23, 27, 28, and 29.

C. Exposure Assessment

At the international level, the need for a global assessment of levels of chemicals in food and in the total diet led to the establishment in 1976 of the Global Environment Monitoring System/Food Contamination Monitoring and Assessment Program, which is commonly known as GEMS/Food. GEMS/Food, which now includes participating institutions in nearly 70 countries throughout the world, is intended to inform governments, the Codex Alimentarius Commission, and other relevant institutions, as well as the public, on levels and trends of contaminants in food, their contribution to total human exposure, and significance with regard to public health and trade. Nineteen contaminants are currently monitored by GEMS/Food, including aflatoxins. The program is an important part of national and international efforts to provide assurance regarding the safety of the food supply and provides the basis—where appropriate—for remedial actions, for food control, for industry and public education, and for resource management. Risk analysis of food contaminants is especially dependent on adequate and reliable monitoring programs to provide accurate assessments of exposure.

Periodically the GEMS/Food database is evaluated to assess levels and trends in food contamination. The most recent assessment of aflatoxin data from 30 countries conducted by GEMS/Food concluded that many countries, especially the developed nations, were taking effective steps to limit aflatoxins in food and feeds. However, the instances where relatively high levels were reported emphasized the necessity to intensify efforts in controlling aflatoxin levels in the global food supply, especially maize, groundnuts, tree nuts, pulses, and animal feeds. This was particularly true for many of the developing countries, where the climate is conducive to the growth of molds and resources are limited for preventing formation of mycotoxins in commodities and for controlling their levels in the food supply [25].

More recently, sampling plans for aflatoxin in large particle commodities, such as maize and groundnuts, have been revised to reflect the extreme heterogeneity of aflatoxin contamination [26]. Consequently, data which were based on a large number of relatively small samples with null results cannot be considered reliable. GEMS/Food has requested that new data be developed using the sampling plans which are being developed within Codex.

D. Risk Characterization

The risk characterization of ochratoxin A by JECFA concluded that, based on preliminary toxicological evaluation and exposure assessment, adverse health effects might be expected for certain segments of the population. Therefore, more detailed information on dietary exposure is needed to identify population groups at greater risk. Further studies are also encouraged to elucidate the role of ochratoxin A and other mycotoxins in nephropathy in pigs and humans, the mechanism of induction of tumors, and the role of phenylalanine in antagonizing the adverse effects of ochratoxin A. The committee noted that grain should be stored under suitable conditions to keep levels of ochratoxin A to a minimum.

In regard to patulin, JECFA noted that patulin levels in apple juice are generally below 50 μg/L which result in estimated intakes of 0.2 μg/kg of body weight per day for children and 0.1 μg/kg of body weight for adults. These are well below the PTDI of 0.4 μg/kg of body weight established by JECFA. However, because apple juice can occasionally be heavily contaminated, the committee recognized that continuing efforts are needed to minimize exposure to this mycotoxin by avoiding the use of rotten or moldy fruit in processing.

As stated earlier, because no quantitative risk characterization of aflatoxins could be developed, JECFA has recommended that aflatoxins be reduced to a level that cannot be further reduced in a food without involving the discarding of that food altogether, severely compromising the ultimate availability of major food supplies. However, this advice has proved difficult to implement, as discussed below.

E. Risk Management

As the risk becomes sufficiently characterized, various control options can be considered. However, risk management decisions must also include consideration of economic, social, and political factors (nonrisk analyses). If the contaminant is of relevance to the international trade in food, risk management options may be considered by the Codex Committee on Food Additives and Contaminants (CCFAC) for elaboration within the Codex system.

The CCFAC has been attempting to develop maximum limits (MLs) for aflatoxins in general, but in 1991 decided instead to develop MLs for individual commodities. However, this strategy has met little success because of differences in interpreting JECFA's recommendation regarding irreducible levels. In general, countries with climatic conditions that favor the growth of mycotoxins wanted higher levels in order to avoid disruption of their food trade. Countries in temperate regions that are largely free of aflatoxin problems wanted lower levels. Currently, CCFAC is only considering an ML for raw groundnuts and an ML for aflatoxin M_1 in milk. Development of MLs for processed groundnuts and maize for supplementary feeding stuffs for milk-producing animals has been discontinued. The CCFAC has requested further guidance from JECFA to improve the risk characterization in terms of both toxicology and dietary exposure. Codex is also developing a code of practice on the reduction of aflatoxins.

In the absence of reliable data and scientific consensus, there is often disagreement between importing and exporting countries which establish regulatory levels based on their perceptions of what levels are achievable by good agricultural and good manufacturing practices. Not surprisingly, national governments have established very different tolerances for aflatoxins. For example, tolerances for aflatoxin B_1 in groundnuts range from 0 to 50 μg/kg. Enforcement of these tolerances has considerable economic costs, such as condemnation or downgrading of commodities to animal feed, operation of compliance and quality control programs, and detoxification or disposal of commodities. The rejection of food shipments has resulted in significant economic costs, particularly for developing countries. In addition, national governments and international organizations have devoted considerable resources to research on methods to prevent, reduce, or eliminate the presence of mycotoxins in food.

VI. CONCLUSIONS

The hazards posed by mycotoxins in food present an enormous challenge to national governments and international agencies responsible for their control. Deficiencies in the risk assessment database and inconsistencies in risk management decisions that have confounded any international consensus on tolerable levels for aflatoxins may foretell further trade problems involving other mycotoxins. It is therefore essential that national governments adopt a consistent, science-based risk analysis approach to address the problem posed by mycotoxins in food. In this regard, the Codex Alimentarius Commission should be used to mediate competing risk management viewpoints to assure consumer protection while preventing the erection of unwarranted nontariff barriers to trade in food. This must be supported by a strengthened JECFA and improved information for hazard characterization and exposure assessment.

WHO remains committed to the elucidation of the possible role of mycotoxins in chronic diseases, such as atherosclerosis, for which clear etiologies have not been established. In addition, WHO continues to expand and improve its global food contamination monitoring program for mycotoxins under the auspices of GEMS/Food and in collaboration with its member states.

In light of the potential risks posed by mycotoxins, it would be prudent to continue to explore methods to prevent, reduce, or eliminate their presence in food, as is recommended by JECFA. Global research efforts should be strengthened, especially in developing countries. Dissemination of information and establishment of contact points, such as WHO Collaborating Centers for Food Contamination, can facilitate this process. Financial and human resources for research into the problem of mycotoxins should be pooled to accelerate the most promising new developments. In this regard, sophisticated technologies being developed in countries like the U.S. should be transferred to developing countries so that these technologies can be tested on a worldwide basis.

REFERENCES

1. Milk Pasteurization. WHO Monograph Series, No. 14. Geneva: World Health Organization, 1953.
2. Toxic Hazards of Certain Pesticides. WHO Monograph Series, No. 16. Geneva: World Health Organization, 1953.
3. The Role of Food Safety in Health and Development. Report of a Joint FAO/WHO Expert Committee on Food Safety, Technical Report Series No. 705. Geneva: World Health Organization, 1984.
4. World Declaration on Nutrition. FAO/WHO International Conference on Nutrition, Rome, December, 1992. Geneva: Food and Agriculture Organization of the United Nations/World Health Organization, 1993.
5. MK Matossian. Poisons of the Past: Molds, Epidemics and History. New Haven: Yale University Press, 1989.
6. Guidelines for Developing an Effective National Food Control System. WHO Food Control No. 1. Geneva: World Health Organization, 1976.
7. HP Van Egmond. Rationale for regulatory programs for mycotoxins in human food and animal feeds. Food Addit Contam 10:29, 1993.
8. Procedural Manual. Joint Food and Agriculture Organization of the United Nations/World Health Organization Codex Alimentarius Commission, Joint FAO/WHO Food Standards Program, FAO/WHO Secretariat. 10th ed. Rome: Food and Agriculture Organization of the United Nations, 1997, p 4.
9. Report of the Joint FAO/WHO Expert Consultation on the Application of Risk Analysis to Food Standards Issues, 13–17 March 1995, Geneva, Switzerland. WHO/FNU/FOS/95.3. Geneva: World Health Organization, 1995.
10. ALINORM 97/37. Report of the Twenty-second Session of the Joint FAO/WHO Codex Alimentarius Commission, 23–28 June 1997, Geneva. Rome: Food and Agriculture Organization of the United Nations, 1997.
11. Risk Assessment in the Federal Government: Managing the Process. Committee on Institutional Means for Assessment of Risks to Public Health. National Research Council, National Academy of Sciences. Washington: National Academy Press, 1983.
12. Procedural Manual. Joint Food and Agriculture Organization of the United Nations/World Health Organization Codex Alimentarius Commission, Joint FAO/WHO Food Standards Program. 10th ed. Rome: Food and Agriculture Organization of the United Nations, 1997, p 42.
13. CP Wild, AJ Hall. Epidemiology of mycotoxin-related disease. In: Howard, Miller, eds. The Mycota VI—Human and Animal Relationships. Berlin: Springer-Verlag, 1996.
14. Final Report. Steering Group on Naturally Occurring Toxins of Plant Origin, Carshalton, United Kingdom, 20–22 May 1992, International Program on Chemical Safety and International Life Sciences Institute, Europe, WHO/PCS/92.69. Geneva: World Health Organization, 1992.
15. Summary of Evaluations Performed by the Joint FAO/WHO Expert Committee on Food Additives (JECFA). Food and Agriculture Organization of the United Nations,

International Life Sciences Institute, International Program on Chemical Safety. Geneva: World Health Organization, 1996.
16. Evaluation of Certain Food Additives and Contaminants. Thirty-seventh report of the Joint FAO/WHO Expert Committee on Food Additives. WHO Technical Report Series No. 806. Geneva: World Health Organization, 1991.
17. Toxicological Monographs: Toxicological Evaluation of Certain Food Additives and Contaminants. WHO Food Additive Series, No. 28. Geneva: World Health Organization, 1991.
18. Evaluation of Certain Food Additives and Contaminants. Forty-fourth report of the Joint FAO/WHO Expert Committee on Food Additives. WHO Technical Report Series, No. 859, Geneva: World Health Organization, 1995, pp 35–36.
19. Toxicological Monograph: Toxicological Evaluation of Certain Food Additives and Contaminants. WHO Food Additive Series, No. 35. Geneva: World Health Organization, 1996.
20. Evaluation of Certain Food Additives and Contaminants. Thirty-fifth report of the Joint FAO/WHO Expert Committee on Food Additives, WHO Technical Report Series, No. 789. Geneva: World Health Organization, 1990.
21. Evaluation of Certain Food Additives and Contaminants. Forty-fourth report of the Joint FAO/WHO Expert Committee on Food Additives, WHO Technical Report Series, No. 859. Geneva: World Health Organization, 1995, pp 36–38.
22. Toxicological Monograph: Toxicological Evaluation of Certain Food Additives and Contaminants. WHO Food Additive Series, No. 35. Geneva: World Health Organization, 1996.
23. IARC Monographs on the Evaluation of Carcinogenic Risks to Humans. Vol. 56. Some Naturally Occurring Substances: Food Items and Constituents, Heterocyclic Aromatic Amines and Mycotoxins. Lyon, France: International Agency for Research on Cancer, 1993.
24. Evaluation of Certain Food Additives and Contaminants. Twenty-eighth report of the Joint FAO/WHO Expert Committee on Food Additives. WHO Technical Report Series, No. 759. Geneva: World Health Organization, 1987.
25. Assessment of Chemical Contaminants in Food. Report on the results of the UNEP/FAO/WHO program on health-related environmental monitoring. London: United Nations Environment Program Monitoring and Assessment Research Centre, 1988.
26. Sampling Plans for Aflatoxin Analysis in Peanuts and Corn. Report of an FAO technical consultation, 3–6 May 1993, Rome. FAO Food and Nutrition Paper, No. 55. Rome: Food and Agriculture Organization of the United Nations, 1993.
27. IARC Monographs on the Evaluation of Carcinogenic Risks to Humans. Vol. 10. Some Naturally Occurring Substances. Lyon, France: International Agency for Research on Cancer, 1976.
28. IARC Monographs on the Evaluation of Carcinogenic Risks to Humans. Vol. 31. Some Food Additives, Feed Additives and Naturally Occurring Substances. Lyon, France: International Agency for Research on Cancer, 1983.
29. IARC Monographs on the Evaluation of Carcinogenic Risks to Humans. Vol. 40. Some Naturally Occurring and Synthetic Food Components, Furocoumarins and Ultraviolet Radiation. Lyon, France: International Agency for Research on Cancer, 1986.

Index